RECENT ADVANCES IN FRACTURE MECHANICS

RECENT ADVANCES IN FRACTURE MECHANICS

Honoring Mel and Max Williams

Edited by

W.G. KNAUSS

Graduate Aeronautical Laboratories,
California Institute of Technology,
Pasadena, U.S.A.

and

R.A. SCHAPERY

Department of Aerospace Engineering and Engineering Mechanics,
The University of Texas at Austin, U.S.A.

Reprinted from *International Journal of Fracture*, Vol. 93, 1998

SPRINGER-SCIENCE+BUSINESS MEDIA, B.V.

A C.I.P. Catalogue record for this book is available from the Library of Congress.

ISBN 978-90-481-5266-7 ISBN 978-94-017-2854-6 (eBook)
DOI 10.1007/978-94-017-2854-6

Printed on acid-free paper

TABLE OF CONTENTS

International Journal of Fracture **93:** vii, 1998.

Publisher's Foreword

This special volume of the *International Journal of Fracture* marks the end of an era. In 1965 Max Williams initiated the journal, and for most of the time since then both he and Mel have served (and are still serving) the large community of scientists devoted to research into all aspects of fracture.

The first issue of the journal (at that time referred to as the *International Journal of Fracture Mechanics*) was published by Wolters-Noordhoff in Groningen, The Netherlands. Later, Martinus Nijhoff took over the publication of the Journal, and subsequently merged with Reidel in Dordrecht to form Kluwer Academic Publishers.

During 1996, Max transferred his editorial responsibilities to Dick Schapery, coincidentally one of the contributors to the first volume of the journal, and, a little later, Mel's involvement in the *Blue Pages* was taken over by Mark Kachanov.

At the Ninth International Conference on Fracture in Sydney, 1997, Kluwer Academic presented the first issue (Volume 83, issue 1) completely assembled under the editorial responsibility of Dick Schapery. This was celebrated with an editorial board meeting and dinner.

At the very same conference also a special session took place honouring Max and Mel, the fruit of which is assembled here in this special volume of *IJF*, edited by Dick Schapery and Wolfgang Knauss – both former students of Max's. We are very happy that Dick and Wolfgang took the trouble to edit this special volume, containing contributions from uniformly highly respected researchers in the field.

The undersigned have had the pleasure of meeting Max and Mel on various occasions over the years, and we always enjoyed their kind hospitality and friendship. When we prepared our visits we were always looking forward to the challenges that, especially, Max posed to us.

Our heartfelt thanks go to Max and Mel for making the journal what it is today, and we very much appreciate the help and advice they have given Dick and Mark to learn the tricks of the trade, and support them in successfully taking over the challenge of editorial responsibility.

We look forward to our continued cooperation with Dick, Mark and the Regional Editors to maintain *IJF* as the primary outlet for high-quality papers on all aspects of the physics and mechanics of fracture.

Kluwer Academic Publishers
Dordrecht, The Netherlands
March 1999

DAVID LARNER
Publishing Director

KAREL NEDERVEEN
Publishing Editor

International Journal of Fracture **93:** ix–x, 1998.

Dedication to Mel and Max Williams

It is with deep admiration and respect that we dedicate this volume of the International Journal of Fracture to Mel and Max Williams. Through their combined efforts, with Max as founder and Editor-in-Chief for thirty-one years, and Mel as Associate Editor for twenty-four years, this Journal has become the pre-eminent publication in the field of fracture. Although they retired from these positions in 1996, the Journal continues to benefit from their past wise and dedicated efforts and continuing editorial advice.

The Journal was founded in 1965, a time when the field of fracture mechanics was quite small and narrow, and many questioned the need for a publication covering such a limited field. Max, already having made seminal contributions to the mechanics of materials and structures, including fracture, recognized what broad impact a better understanding of fracture would have in science and engineering. Clearly, his vision has been realized. Indeed, the Journal has played a major part in fostering and communicating many of the basic and applied developments. This discipline has become so important that few significant structures are designed or materials selected without its input. The range of applications spans aircraft and space structures, pressure vessels, automotive components, ships, with forays into geomechanics on the large scale and mechanical strength performance of electronic chips at the other extreme.

Besides being one of the founders of the modern discipline of fracture mechanics, Max has repeatedly exercised seminal influence in the field. While working at the California Institute of Technology in the 1950's on stress concentrations at the junction of swept wings with the fuselage, he demonstrated the universal (elastic) singularity character of the stress field at the base of a stationary crack and called attention to the now also classical oscillatory character in the stress and small deformation fields at the tips of interfacial cracks. Besides his fundamental understanding and outlook, he has always been a firm believer in the realistic application of scientific principles to the engineering endeavor. Through his leadership role in the development of solid rocket structural integrity, he had a defining influence in the evolution of time dependent fracture mechanics in viscoelastic materials. This was followed by initiating studies in dynamic fracture through building on Schardin's contributions to that field, and the derivatives of that beginning are still internationally strong and alive today. Though his academic and engineering activities gradually broadened at the expense of the research oriented university environment, he has maintained a keen and guiding interest in the forefront of fracture topics and its applications to engineering problems, his latest concerns addressing fracture aspects at the submicron scale.

Having had the exceptionally good fortune of studying for our doctorates under Professor Williams during the late 50s and early 60s, we can speak personally about the exciting environment in Max's group at Caltech. He was (and still is) a master at identifying the core issues of a problem and stimulating those around him to participate in finding solutions. It was a time when the rocket industry was trying to deal with the cracking problem in solid propellant (particle-filled, viscoelastic rubber), but the practical means for its stress and fracture analysis were not yet available. A great national effort was underway at that time to make solid rocket motors more reliable, and Max was industry-wide clearly the dominant influence through engineering ideas for solutions to the major problems. For example, besides work on viscoelasticity he was the first to introduce fracture computations based on finite element calculations into rocket design, at a time when numerical fracture calculations in other engineering fields were not yet practiced. The methods developed during the early 60s are still largely in place today and form the backbone of past and current motor designs. Certainly, the longevity of many strategic and tactical rocket motor designs in use today are, in large part, due to the engineering input provided at the national level through Max.

We mention the specific application of fracture mechanics to solid propellant because it shows how Max's leadership extended from guiding the solution of fundamental problems to taking a dominant national role in solving the engineering problems – and we can speak of this from first hand experience during and for many years subsequent to our graduate student days.

It is tempting to go into more detail on this topic and the many others where Max has had a major impact. Instead, however, we refer the interested reader to the extended Foreword in the Special Triple Issue, Nos. 1–3, Volume 39, 1989 of the Journal. It was prepared in recognition of the many contributions of M.L. Williams on the occasion of his 65th birthday.

Mel and Max – what a fantastic team! As Associate Editor since 1973, Mel Williams has had a major role in processing papers, which included coordinating reviews and interacting with authors and reviewers. Her B.A. in English and M.A. in Information Science provided the perfect background for her work on the Journal. For some of us, at least, the only papers we have ever published with perfect English were those that appeared in the Journal, because Mel reviewed and corrected the English in all of its papers! In 1989, upon the untimely death of Professor Jerold L. Swedlow, Mel assumed the additional responsibility of Editor of the blue pages, then titled Reports of Current Research. Mel's dedicated work on the Journal has been a labor of love, and was a major factor in establishing it as a publication of the highest quality.

Dear Mel and Max,

We believe we can speak for the members of the Editorial Board, the authors of the papers in this volume, and your many other friends and acquaintances in thanking you for all of your efforts. They have had a profound influence on the field of fracture, and always have been accompanied by a most remarkable and greatly appreciated personal grace and warmth. We wish you many more enjoyable years, doing all those things you could not previously find time for.

W.G. Knauss

R.A. Schapery

International Journal of Fracture **93:** xi–xiii, 1998.

Foreword

The papers in this volume represent a considerable cross section of the field of fracture mechanics, a testimony to the breadth of interest that Mel and Max's friends share with them. They address different materials aspects, experimental vis-à-vis theoretical expositions, as well as a range of size scales. Several are expanded versions of papers that were given in special sessions honoring Mel and Max at the 1997 Ninth International Conference on Fracture in Sydney Australia.

The first group of papers focuses on dynamic fracture problems as viewed primarily from the classical continuum point of view. The contribution by K.B. Broberg speaks to the need of understanding crack propagation in anisotropic materials, either because such questions arise in micro-crack growth through crystals or in connection with the macroscopic response of anisotropic composites. Considering the problem in an antiplane strain deformation, he finds, in contrast to the isotropic case, that a singular stress transverse to the crack plane arises at the edge of the crack. The M.S. Breitenfeld and P.H. Geubelle paper attacks dynamic problems of stationary or propagating cracks along an interface by developing and applying a highly effective numerical method of spectral representation for cracks in infinite domains. Special attention is paid to the limiting speed in two- and three-dimensional problems, with the result that this interface crack speed is found to vary strongly, depending on the mode-mixity of loading. For mostly crack-normal loading the crack speed is subsonic with respect to the more compliant material, but moves at transonic speeds as the shear component increases, moderately or weakly related to the stresses developed in an accompanying cohesive zone. It is of interest to note that such transonic crack motion has been observed in bimaterial plates (Al-PMMA) at Caltech by Rosakis and co-workeres. Although steady motion of dynamically moving cracks is the easiest to treat analytically, crack motion under transient conditions poses a recurring problem. J.G. Williams examines periodic crack speed variations through perturbations superposed on steady crack motion. It is shown that such perturbation solutions exist, but their character depends on the configurations examined and on the initial conditions. While the previous papers used classical elastodynamics, the contribution by J. Lee, M. Kokaly and A.S. Kobayashi takes into account ductility for sheet structures relevant to airframe and other designs. Comparisons between computations and experiments (dynamic moiré interferometry) suggest that either a crack tip opening angle or a T-integral my be a suitable criterion for evaluating stable dynamic crack growth in thin aluminum sheets.

A group of three papers centers on analysis of relatively general crack geometries. D.D. Ang and D.D. Trong contribute the finding that electrostatic analysis may be used to identify irregular (star-shaped) cracks in the interior of an opaque body. S. Li and M. Mear reformulate three-dimensional, elastodynamic and elastostatic crack problems to alleviate the computational difficulties often associated with solving singular integral equations. Ultimately the problem of stress analysis of cracked media is reduced to a weakly singular set of integral equations, the solution of which may be obtained very efficiently by a numerical method that does not even require introduction of Cauchy principal values for evaluation. B. Yang and K. Ravi-Chandar pass beyond the purely elastic material behavior (as did the Breitenfeld/Geubelle paper) through incorporating a 'cohesive zone' possessing nonlinear, but otherwise general properties. This allows the authors to determine the stable/unstable phases of the combined feature of the cohesive zone and the crack as they change simultaneously under increasing loads.

The next group of five papers considers fracture problems of polymers and other relatively ductile materials. The cohesive zone is the focus in the first three papers by G. Pulos and W.G. Knauss that address the measured evolution of crack growth in PMMA under fatigue loading. With emphasis placed on the details of the evolution of the cohesive zone and its role in slow crack growth by way of accelerating and decelerating phases, it is found that this behavior is governed by cycle-induced changes of the cohesive material. Optical interferometric measurements yield deformation resolution that allows the determination of the cohesive stress-strain behavior (quasi-steady crack growth) and the cohesive force distribution during a cycle that exhibits a minimum in its middle. Of particular concern for engineering applications is the (observed) possibility that increased stress intensity does not necessarily translate into increased crack speed – and vice versa. Rate effects are clearly present and govern, together with the crack tip stress amplitude, whether the crack/craze produces secondary crazes at its tip or whether fracture occurs through the craze middle to render a very smooth surface. The fracture behavior of an epoxy above and below the glass transition is investigated by K. Ogawa, A. Misawa and M. Takashi. The lower limit of an extended J-integral for predicting time-dependent crack propagation is determined and the crack growth behavior is studied once that limit is exceeded. It is found that while only stable crack growth occurs above Tg, both stable and unstable phenomena occur below that temperature. The failure of an adhesive bond is simulated by C.H. Wang, who studies the role of rate- or time-independent 'plastic' deformations around the crack in a thin layer of material bounded by two elastic solids possessing relatively high moduli. This problem is similar to that considered numerically and earlier by D. Hunston and S.S. Wang; it also shares commonality with the contribution later in this Issue by Y. Wei and J.W. Hutchinson on peel separation of two solids under conditions when plastic deformations play a defining role in the energy balance of the failure process. The essence of the paper explores the role of plastic energy expenditure as a function of the layer thickness and plastic zone size.

Further important considerations in fracture mechanics are scaling rules that allow extension of results obtained at one size scale to be translated into behavior at different size scales. Two contributions are devoted to work at the macro-scale. A special aspect of this question is the (often convenient) substitution of homogeneous isotropic properties for those of the inhomogeneous and locally anisotropic properties at the grain size scale (for polycrystalline materials). In this study by K. Abdel-Tawab and G.J. Rodin, the effect of polycrystalline inhomogeneities on the stress distribution, and particularly on the fracture behavior, are investigated. Much to the advantage of practical engineering aspects, as well as in agreement with numerous experimental investigations, it is found that the size effect is weak. However, there is a significant statistical influence of grains adjacent to the crack tip. An additional fracture consideration at the micro-scale has been supplied by H.W. Liu, whos considers the role of dislocation motion in the fatigue process. Certain limiting behavior, such as the threshold for fatigue crack growth, is explained analytically in terms of barriers to dislocation motion, often associated with grain boundaries, which introduces the grain size as part of the analysis. Thus microstructure-sensitive crack growth behavior can result, depending on the size scale as dictated by the grains.

A large class of problems in fracture mechanics deals with interactions that produce complex stress fields (mode I, II and III interactions). Those problems arise naturally when a cracked structure is subjected to complex loading so as to evoke similarly compound stress fields at the crack front. Alternatively, the situation prevails typically in the fracture along interfaces, where compound mode interaction is the rule rather than the exception. Building on failure along interfaces, this problem is commonplace in the failure of composite materials, be they of the particulate or fiber type, chopped or continuous. The first two papers in this group of contributions which have mode interaction as a common thread, address the process of separation of thin members from a substrate by either a blister or a peel configuration. A. Shirani and K.M. Liechti measure blister failure response and use the results to derive an adhesive model incorporating a cohesive zone, which ultimately defines the cohesive force law in terms of the energy determined in blister experiments. A similar objective is pursued by T. Wei and J.W. Hutchinson, who consider the energy expenditure in a peel configuration, in which special attention is paid to the incorporation of energy loss to plastic deformations. While it is typical that the peeled member sustains plastic deformation, one example cited here allows the substrate to yield (elasto-plastic behavior), while the peeling member remains elastic.

Moving to fracture problems directly appropriate to composite materials, two contributions offer analyses and models for crack generation and propagation. E.S. Folias and M. Hohn consider the initiation of cracks in fiber reinforced materials as the result of thermally induced stresses in the matrix and fiber components of the composite. Theoretical results are used to identify locations of possible failure and to derive a local criterion of crack initiation that involves the residual stresses. Experimental evidence for matrix cracking in intermetallics confirms some of the predictions. The paper by A.S Argon, M.L. Seleznev, C.F. Shih and X.H. Liu considers unbonding along the fiber matrix interface in a metal matrix composite, with an objective of optimizing the longitudinal and transverse tensile strengths of the composite by means of a novel, specially tailored, mixed-mode traction/separation law.

The next two contributions deal with analysis of stress concentrations in anisotropic, elastic solids. Both are by C. Blanco, J.M. Martinez-Esnaola and C. Atkinson. The first addresses stress singularities at the root of sharp notches, and establishes the singularity exponent similar to the situations derived by M.L. Williams in 1952 for isotropic solids. The second paper considers the stress intensities at the tip of kinked-cracks in anisotropic materials, and determines the three modes of stress intensity factors.

It seems appropriate to have as the concluding paper in this volume one which revisits the problem of cracks in thin plates in bending, and has a title that begins with 'Williams meets von Karman'. This paper, by C.Y. Hui, A.T. Zehnder and Y.K. Potdar, allows for large deformations using von Karman plate theory and shows that the character of the near-tip field is identical to that of the linear plate theory developed by Williams in 1961. Explicit results are given which bring out the effect on stress intensity factors of nonlinear coupling associated with membrane and bending loads.

International Journal of Fracture **93:** 1–12, 1998.

Crack expanding with constant velocity in an anisotropic solid under anti-plane strain

K.B. BROBERG
Department of Mathematical Physics, University College Dublin, Belfield, Dublin 4, Ireland.
e-mail: bertram.broberg@ucd.ie

Received 15 July 1997; accepted in revised form 2 February 1998

Abstract. An analytic solution is given for a crack expanding with constant velocity from zero length in an anisotropic material under anti-plane strain. Not all anisotropic materials can support anti-plane strain, and the study is therefore by necessity limited to a certain class of materials, including monoclinic materials. A double Laplace transform is used and the inversion technique is based on the self-similarity of the problem. The result shows that the crack shape is elliptic, as in the corresponding isotropic case. The displacement on the crack plane outside the crack is found to be zero. Expressions are given for the stresses, the stress intensity factor and the energy flux into the crack edge. In contrast to the isotropic case a transverse normal stress may appear, singular at the crack edge.

Key words: Anisotropic crack expansion, anti-plane strain, Barenblatt region, energy flux.

1. Introduction

Crack propagation in engineering structures usually takes place in the opening mode. Thus, even if a pre-existing crack in an isotropic material is subjected to pure mode III loading, an extension will in general occur under a change of direction towards mode I conditions. However, in an anisotropic material a propagation under anti-plane strain may be more likely, because crack formation and propagation often occurs along weak planes. Crack growth under anti-plane strain is also one familiar mode at seismic slip propagation.

Here an elastically anisotropic material is considered. The presence of a strength anisotropy is not directly involved in the investigation. Some general relations are given here, but can be found in textbooks, for instance (Ting, 1996). Steady state crack propagation in anisotropic materials has been considered by Wu (1989, 1991). The problem of an expanding mode I crack in an orthotropic material was solved by Atkinson (1959).

The generalized Hooke's law reads

$$\sigma_{ij} = C_{ijkl}u_{k,l}, \quad i, j, k, l = 1, 2, 3, \tag{1}$$

where C_{ijkl} are the elastic stiffnesses, satisfying the symmetry conditions

$$C_{ijkl} = C_{jikl} = C_{klij} = C_{ijlk}. \tag{2}$$

The equations of motion are

$$\sigma_{ij,j} = C_{ijkl}u_{k,lj} = \varrho\frac{\partial^2 u_i}{\partial t^2}, \tag{3}$$

where ϱ is the density of the solid. Anti-plane strain is defined by

$$u_1 = u_2 = 0, \qquad u_{3,3} = 0, \tag{4}$$

where u_1, u_2 and u_3 are the displacements in the x_1, x_2 and x_3 directions, respectively. Insertion into the equations of motion gives

$$C_{i131}u_{3,11} + C_{i231}u_{3,21} + C_{i132}u_{3,21} + C_{i232}u_{3,22} = \varrho\frac{\partial^2 u_i}{\partial t^2}. \tag{5}$$

In contracted notation, (Nye, 1957), these three equations read

$$\begin{aligned} &c_{15}u_{3,11} + (c_{56} + c_{14})u_{3,12} + c_{46}u_{3,22} = 0,\\ &c_{56}u_{3,11} + (c_{25} + c_{46})u_{3,12} + c_{24}u_{3,22} = 0,\\ &c_{55}u_{3,11} + 2c_{45}u_{3,12} + c_{44}u_{3,22} = \varrho\frac{\partial^2 u_3}{\partial t^2}. \end{aligned} \tag{6}$$

Thus three equations are found for the single unknown displacement u_3. These cannot all be satisfied for general anisotropy, because they are linearly independent if all coefficients are nonzero and independent. However, the first two equations are satisfied for a material with

$$c_{14} = c_{15} = c_{24} = c_{25} = c_{46} = c_{56} = 0. \tag{7}$$

These relations are fulfilled for a monoclinic material oriented so that its symmetry plane is $x_3 = 0$, but express, in fact, a somewhat more general material. Specializations of a monoclinic material are, for instance, orthotropic and transversely isotropic materials.

It can be shown that the matrix $c_{\alpha\beta}$ is symmetric and positive definite as a consequence of the fact that the stress-strain energy is positive if not all strain or stress components vanish. Then the diagonal elements of the matrix and its principal minors (determinants made up of elements symmetrically situated with respect to the diagonal) are all nonzero and positive. Note in particular that the coefficients c_{44} and c_{55} in the third of (6) are nonzero and positive. After a change of notations to (x, y, z) for coordinates and w for the nonzero displacement, this equation can be written in the form

$$c_{55}\frac{\partial^2 w}{\partial x^2} + 2c_{45}\frac{\partial^2 w}{\partial x\,\partial y} + c_{44}\frac{\partial^2 w}{\partial y^2} = \varrho\frac{\partial^2 w}{\partial t^2} \tag{8}$$

to compare with the equation $\mu\Delta w = \varrho\partial^2 w/\partial t^2$ for isotropic materials, where μ is the modulus of rigidity. For orthotropic and transversely isotropic materials $c_{45} = 0$, so that the mixed derivative disappears from the equation, and for isotropic materials $c_{44} = c_{55} = \mu$.

For anti-plane strain the constitutive equations reduce to

$$\tau_{xz} = c_{55}\frac{\partial w}{\partial x} + c_{45}\frac{\partial w}{\partial y}, \qquad \tau_{yz} = c_{45}\frac{\partial w}{\partial x} + c_{44}\frac{\partial w}{\partial y}, \tag{9}$$

$$\sigma_z = c_{35}\frac{\partial w}{\partial x} + c_{34}\frac{\partial w}{\partial y} \tag{10}$$

Note that a full description of the elastic properties of the material involves 15 elastic stiffnesses, whereas only 5 elastic stiffnesses appear for anti-plane strain.

2. Statement of the problem

Consider a crack with the edges at $|x| = Vt = a$, where V is the velocity of each crack edge, t is the time and a is the current half-length of the crack. Thus the crack starts moving at constant velocity from zero length. The crack faces are traction free and the stress at infinity is $\tau_{yz} = \tau_{yz}^{\infty}$.

The equation of motion is now written in the form

$$\frac{\partial^2 w}{\partial y^2} + 2\frac{c_{45}}{c_{44}} \cdot \frac{\partial^2 w}{\partial x\, \partial y} + \frac{c_{55}}{c_{44}} \cdot \frac{\partial^2 w}{\partial x^2} = \frac{\partial^2 w}{\partial \tau^2}, \tag{11}$$

where $\tau = \sqrt{c_{44}/\varrho}\, t$ is a time variable with dimension length.

Consider the upper half $y \geqslant 0$. The boundary conditions are taken to be

$$\tau_{yz} = 0 \quad \text{for } y = 0, |x| < a = \gamma\tau, \qquad w = 0 \quad \text{for } y = 0, |x| > a = \gamma\tau, \tag{12}$$

$$\tau_{yz} \to \tau_{yz}^{\infty} \quad \text{as } \sqrt{x^2 + y^2} \to \infty. \tag{13}$$

Here $\gamma = V/\sqrt{c_{44}/\varrho}$ is a dimensionless crack velocity. Note that the condition $w = 0$ for $y = 0$ is applied tentatively. For the isotropic case it follows from the anti-symmetry $w(x, -y, t) = -w(x, y, t)$, but this anti-symmetry requires that $c_{45} = 0$. If, nevertheless, the condition $w = 0$ for $y = 0$ is assumed, a correct result is obtained if the stress $\tau_{yz}(x, 0, t)$ turns out to be even in x, i.e., equal to $\tau_{yz}(-x, 0, t)$, and independent of the sign of c_{45}. Then a solution for each half-plane gives matching stresses on $y = 0$. As will be evident later, this happens to be the case.

3. Solution of the problem

Introduce the double Laplace transforms

$$W_+ = \mathcal{L}_{qx}\mathcal{L}_{p\tau} w_+ = q \int_{-\infty}^{+\infty} \mathrm{e}^{-qx} p \int_0^{\infty} \mathrm{e}^{-p\tau} w_+ \,\mathrm{d}\tau\, \mathrm{d}x, \tag{14}$$

$$T_{yz} = \mathcal{L}_{qx}\mathcal{L}_{p\tau} \tau_{yz} = q \int_{-\infty}^{+\infty} \mathrm{e}^{-qx} p \int_0^{\infty} \mathrm{e}^{-p\tau} \tau_{yz} \,\mathrm{d}\tau\, \mathrm{d}x, \tag{15}$$

where index plus is used to indicate displacements in the upper half-plane, $y \geqslant 0$. Note that the Laplace transforms are introduced so that the dimension of the transform is the same as for the original. After transformation, (11) becomes

$$\frac{\mathrm{d}^2 W_+}{\mathrm{d}y^2} + 2\frac{c_{45}q}{c_{44}} \cdot \frac{\mathrm{d}W_+}{\mathrm{d}y} + \frac{c_{55}q^2}{c_{44}} W_+ = p^2 W_+ \tag{16}$$

with the solution

$$W_+ = (W_+)_{y=0}\, \mathrm{e}^{-[\kappa q + (p^2 - \lambda^2 q^2)^{1/2}]y}, \tag{17}$$

after removal of the term that violates the radiation condition. Here $\kappa = c_{45}/c_{44}$ and $\lambda = \sqrt{c_{44}c_{55} - c_{45}^2/c_{44}}$. Note that the diagonal elements c_{44} and c_{55} as well as the principal minor $c_{44}c_{55} - c_{45}^2$ are all positive. Thus, λ is real and positive.

Use of (9) gives, after double Laplace transformation

$$T_{yz} = c_{45}qW_+ + c_{44}\frac{\mathrm{d}W_+}{\mathrm{d}y} \tag{18}$$

and thus, by insertion of (17)

$$T_{yz} = -c_{44}(p^2 - \lambda^2 q^2)^{1/2} W_+. \tag{19}$$

The problem is characterized by symmetry and self-similarity, which implies that the stresses τ_{xz} and τ_{yz} as well as the derivative $\partial w/\partial \tau$ (a dimensionless velocity) can be written as functions of τ/x and τ/y. On $y = 0$ both τ_{yz} and $\partial w/\partial \tau$ are even functions. The double Laplace transform of such functions is of the form $F(q/p)$, where $F(\cdot)$ is an odd function. This is shown in the Appendix, where it is also shown how functions of the kind

$$qF(q/p)\,\mathrm{e}^{-[\kappa q + (p^2 - \lambda^2 q^2)^{1/2}]y} \tag{20}$$

are inverted.

For $y = 0$ the inversion formula is particularly simple (66)

$$f(x, 0, \tau) = -\frac{1}{2\pi i x}[F_+(\xi) - F_-(\xi)]_{\xi = -\tau/x}, \tag{21}$$

where $F_+(\xi)$ is the value of $F(q/p)$ when $F(\zeta)$ approaches the real axis from above, and $F_-(\xi)$ its value when $F(\zeta)$ approaches the real axis from below. Here the substitution $q/p = \zeta = \xi + i\eta$ has been made.

Symmetry and self-similarity allows the following representations of the double Laplace transforms of $(\tau_{yz})_{y=0}$ and $(\partial w_+/\partial \tau)_{y=0}$:

$$T_{yz} = \tau_{yz}^{\infty}\frac{q}{p}H(q/p) \quad \text{for } y = 0, \tag{22}$$

$$pW_+ = \frac{\tau_{yz}^{\infty}}{c_{44}} \cdot \frac{q}{p}G(q/p) \quad \text{for } y = 0, \tag{23}$$

where the proportionality constants τ_{yz}^{∞} and $\tau_{yz}^{\infty}/c_{44}$ have been introduced for later convenience. Insertion into (19) gives

$$H(\zeta) = -(1 - \lambda^2\zeta^2)^{1/2}G(\zeta), \tag{24}$$

where $\zeta = \xi + i\eta = q/p$. Branch cuts are made along the portions $|\xi| > 1/\lambda$ of the real axis, and the branch is chosen for which $(1 - \lambda^2\zeta^2)^{1/2} = \sqrt{1 - \lambda^2\xi^2}$ for $|\xi| < 1/\lambda$. Note that $\gamma < \lambda$ if $V < \sqrt{(c_{44}c_{55} - c_{45}^2)/(c_{44}\varrho)}$, which is assumed to be the case.

Use of the inversion formula (21) gives

$$\frac{\partial \tau_{yz}}{\partial \tau} = -\frac{\tau_{yz}^{\infty}}{2\pi i x}[H_+(\xi) - H_-(\xi)]_{\xi = -\tau/x}, \tag{25}$$

$$\frac{\partial^2 w}{\partial \tau^2} = -\frac{\tau_{yz}^{\infty}}{2\pi i c_{44} x}[G_+(\xi) - G_-(\xi)]_{\xi=-\tau/x}. \tag{26}$$

From the boundary conditions (12)–(13) it follows that

$$H_+(\xi) - H_-(\xi) = 0 \quad \text{for } |\xi| > 1/\gamma, \tag{27}$$

$$G_+(\xi) - G_-(\xi) = 0 \quad \text{for } |\xi| < 1/\gamma. \tag{28}$$

Considering (24) the first of these can be replaced by

$$G_+(\xi) + G_-(\xi) = 0 \quad \text{for } |\xi| > 1/\gamma. \tag{29}$$

Equations (28)–(29) form a Hilbert problem, the solution of which is

$$G(\zeta) = \frac{P(\zeta)}{(1/\gamma^2 - \zeta^2)^{3/2}}, \tag{30}$$

where $P(\zeta)$ is a polynomial, which, due to the boundedness of the mass velocity at $x = 0$, reduces to a constant, $P(0)$. The exponent $\frac{3}{2}$ gives the strongest singularity that does not violate the condition of bounded stress-strain energy. Branch cuts are made along the portions $|\xi| > 1/\gamma$ along the ξ axis, and the branch is chosen so that $(1/\gamma^2 - \zeta^2)^{3/2} = 1/\gamma^3$ for $\zeta = 0$. Then, from (26)

$$\left(\frac{\partial^2 w_+}{\partial \tau^2}\right)_{y=0} = -\frac{\tau_{yz}^{\infty} P(0)}{2\pi i c_{44} x}[(1/\gamma^2 - \xi^2)_+^{3/2} - (1/\gamma^2 - \xi^2)_-^{3/2}]_{\xi=-\tau/x}. \tag{31}$$

The right member is zero for $\tau/|x| < 1/\gamma$, i.e., for $|x| > Vt = a$. For $\xi = \tau/|x| > 1/\gamma$ the relation can be written in the form

$$\begin{aligned}\left(\frac{\partial^2 w_+}{\partial \tau^2}\right)_{y=0} &= -\frac{\tau_{yz}^{\infty} P(0)}{\pi i c_{44} x} \cdot \frac{1}{\sqrt{(1/\xi^2 - \gamma^2)^3}} \\ &= -\frac{\tau_{yz}^{\infty} P(0)}{\pi i c_{44}} \cdot \frac{\gamma^2 x}{\sqrt{(\gamma^2\tau^2 - x^2)^3}}.\end{aligned} \tag{32}$$

Two integrations result in

$$(w_+)_{y=0} = \frac{\tau_{yz}^{\infty} P(0)\gamma}{\pi c_{44}}\sqrt{a^2 - x^2} \quad \text{for } |x| < a, \tag{33}$$

where $a = \gamma\tau = Vt$. Thus the crack shape is elliptic as in the corresponding isotropic case.

From (24) and (30) it follows that

$$H(\zeta) = -\frac{P(0)\lambda(1/\lambda^2 - \zeta^2)^{1/2}}{(1/\gamma^2 - \zeta^2)^{3/2}}. \tag{34}$$

Note that $H_+(\xi) = H_-(\xi)$, and thus $\partial\tau_{yz}/\partial\tau = 0$, for $|\xi| < 1/\lambda$ and for $|\xi| > 1/\gamma$. Then, considering $x < 0$, the stress $(\tau_{yz})_{y=0}$ is found by integration of (25) from $\xi = 1/\lambda - 0$, where $(\tau_{yz})_{y=0} = \tau_{yz}^{\infty}$. Because $\partial/\partial\tau = -(\partial/\partial\xi)/x$, the result is

$$(\tau_{yz})_{y=0} = \tau_{yz}^{\infty} + \frac{\tau_{yz}^{\infty}}{2\pi i}\int_{1/\lambda-0}^{-\tau/x} [H_+(\xi) - H_-(\xi)]\,\mathrm{d}\xi = \tau_{yz}^{\infty} + \frac{\tau_{yz}^{\infty}}{2\pi i}\int_{\Gamma} H(\zeta)\,\mathrm{d}\zeta$$

$$= \tau_{yz}^{\infty} - \frac{\tau_{yz}^{\infty}P(0)\lambda}{2\pi i}\int_{\Gamma}\frac{(1/\lambda^2-\zeta^2)^{1/2}}{(1/\gamma^2-\zeta^2)^{3/2}}\,\mathrm{d}\zeta, \tag{35}$$

where Γ is the path along the lower side of the real axis from $\xi = -\tau/x = \tau/|x|$ to $\xi = 1/\lambda - 0$, corresponding to integration of the second term inside [] in the middle member of (35), then rounding $\zeta = 1/\lambda$ and returning to $\xi = \tau/|x|$ along the upper side of the real axis, which corresponds to integration of the first term.

The constant $P(0)$ is now determined by extending Γ to Γ_∞, beginning and ending at $\xi = \infty$, which corresponds to $x = 0$, where $(\tau_{yz})_{y=0} = 0$. Hence

$$(\tau_{yz})_{x=y=0} = \tau_{yz}^{\infty} - \frac{\tau_{yz}^{\infty}P(0)\lambda}{2\pi i}\int_{\Gamma_\infty}\frac{(1/\lambda^2-\zeta^2)^{1/2}}{(1/\gamma^2-\zeta^2)^{3/2}}\,\mathrm{d}\zeta = 0. \tag{36}$$

All singular points of the integrand are situated along the portion $|\xi| > 1/\lambda$ of the real axis and are avoided by infinitesimally small indentations. The path of integration can thus be deformed to the imaginary axis, i.e., from $\zeta = i\eta = -i\infty$ to $\zeta = i\eta = i\infty$, so that

$$\tau_{yz}^{\infty} = \frac{\tau_{yz}^{\infty}P(0)\lambda}{\pi}\int_0^{\infty}\frac{(1/\lambda^2+\eta^2)^{1/2}}{(1/\gamma^2+\eta^2)^{3/2}}\,\mathrm{d}\eta. \tag{37}$$

After integration this gives

$$P(0) = \frac{\pi}{\lambda\gamma\boldsymbol{E}(\sqrt{1-\gamma^2/\lambda^2})}, \tag{38}$$

where $\boldsymbol{E}(\cdot)$ is the complete elliptic integral of the second kind.

4. Displacement and stresses on the crack plane

Insertion of $P(0)$ into (33) gives

$$(w_+)_{y=0} = \frac{\tau_{yz}^{\infty}}{\mu_a\boldsymbol{E}(\sqrt{1-\gamma^2/\lambda^2})}\sqrt{a^2-x^2} \quad \text{for } |x| < a, \tag{39}$$

where $\mu_a = \sqrt{c_{44}c_{55}-c_{45}^2}$. In the isotropic case $\lambda = 1$ and $\mu_a = \mu$, the modulus of rigidity.

Return now to (35), which can be written in the form

$$(\tau_{yz})_{y=0} = \tau_{yz}^{\infty} + \frac{\tau_{yz}P(0)\lambda}{\pi}\int_{1/\lambda}^{\tau/|x|}\frac{\sqrt{\xi^2-1/\lambda^2}}{\sqrt{(1/\gamma^2-\xi^2)^3}}\,\mathrm{d}\xi. \tag{40}$$

Integration and insertion of $P(0)$ gives

$$(\tau_{yz})_{y=0} = \tau_{yz}^{\infty} + \frac{\tau_{yz}^{\infty}}{\boldsymbol{E}(\sqrt{1-\gamma^2/\lambda^2})} \times \left[\frac{|x|}{\lambda\tau}\sqrt{\frac{\lambda^2\tau^2 - x^2}{x^2 - a^2}} - E(\phi, \sqrt{1-\gamma^2/\lambda^2}) \right], \tag{41}$$

where $E(\cdot,\cdot)$ is the incomplete elliptic integral of the second kind and $\phi = \text{asin}\left(\sqrt{\lambda^2 - x^2/\tau^2}/\sqrt{\lambda^2-\gamma^2}\right)$.

Recall that $\tau_{yz}(x, 0, t)$ must be an even function of x and independent of the sign of c_{45} (which equals the sign of κ) for the boundary condition $(w_+)_{y=0} = 0$ for $|x| > a$ to be appropriate. This indeed, is the case.

It follows from (9)–(10) that

$$(\tau_{xz})_{y=+0} = \begin{cases} \dfrac{c_{44}c_{55} - c_{45}^2}{c_{44}} \left(\dfrac{\partial w_+}{\partial x}\right)_{y=0} & \text{for } |x| < a \\ \dfrac{c_{45}}{c_{44}}(\tau_{yz})_{y=0} & \text{for } |x| > a \end{cases} \tag{42}$$

$$(\sigma_z)_{y=+0} = \begin{cases} \dfrac{c_{35}c_{44} - c_{34}c_{45}}{c_{44}} \left(\dfrac{\partial w_+}{\partial x}\right)_{y=0} & \text{for } |x| < a \\ -\dfrac{c_{34}}{c_{44}}(\tau_{yz})_{y=0} & \text{for } |x| > a \end{cases} \tag{43}$$

because $\partial w/\partial x = 0$ on the crack plane outside the crack, and $\tau_{yz} = 0$ on the crack faces.

The stress intensity factor is

$$K_{\text{III}} = \lim_{x\to a}[\sqrt{x-a}(\tau_{yz})_{y=0}] = \frac{\tau_{yz}^{\infty}\sqrt{\pi a}\sqrt{1-\gamma^2/\lambda^2}}{\boldsymbol{E}(\sqrt{1-\gamma^2/\lambda^2})} \tag{44}$$

and the energy flux into each crack edge is found from the general expression for inverse square root stress singularity at a crack edge

$$\mathcal{G} = \pi AB/2 \tag{45}$$

if $\tau_{yz} \to A/\sqrt{r}$ and $w_+ \to B\sqrt{r}$ as $r \to 0$, where r is the distance to the crack edge. Hence

$$\mathcal{G} = \frac{K_{\text{III}}^2}{2\mu_a\sqrt{1-\gamma^2/\lambda^2}} = \frac{\pi(\tau_{yz}^{\infty})^2 a\sqrt{1-\gamma^2/\lambda^2}}{2\mu_a[\boldsymbol{E}(\sqrt{1-\gamma^2/\lambda^2})]^2}. \tag{46}$$

Figure 1 shows the stress intensity factor, normalized with respect to the static stress intensity factor, as function of the crack edge velocity in relation to the sound velocity in the crack propagation direction. Similarly, Figure 2 shows the energy flux into the crack edge, normalized with respect to the energy flux at infinitesimally small crack velocity, as a function

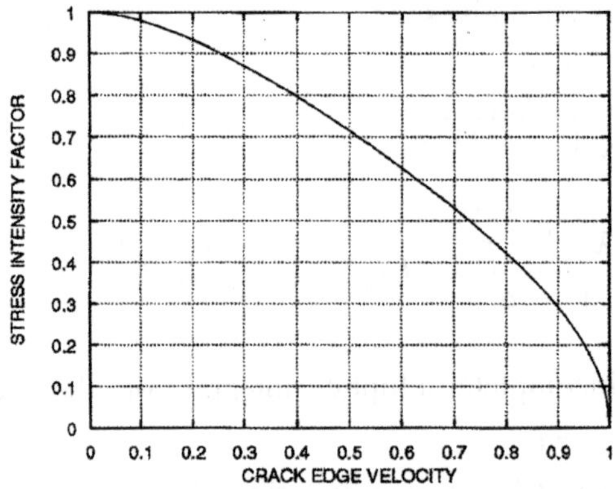

Figure 1. The normalized stress intensity factor $K_{\text{III}}/[\tau_{yz}^{\infty}(\pi a)^{1/2}]$ as function of the relative crack edge velocity γ/λ.

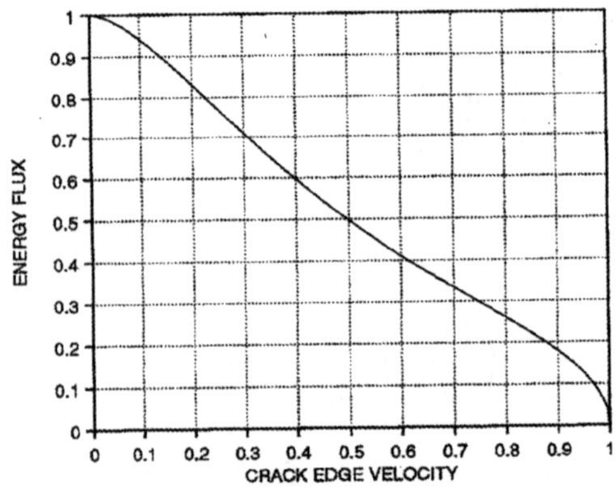

Figure 2. The normalized energy flux $2\mu_a\mathcal{G}/[\pi(\tau_{yz}^{\infty})^2 a]$ as function of the relative crack edge velocity γ/λ.

of the relative crack edge velocity. Note that, by using appropriate nondimensional quantities, both the stress intensity factor and the energy flux into the crack edge can be represented by one single curve for all possible value combinations of elastic stiffnesses, including those for isotropic materials.

5. Offside displacement and stresses

The displacement w_+ and the stress τ_{yz} offside the crack plane are found from the inversion formula (65) in the Appendix. Double Laplace transforms of $\partial^2 w_+/\partial\tau^2$ and $\partial\tau_{yz}/\partial\tau$ are found from (17) and (19) after use of (23) and (22)

$$p^2 W_+ = \frac{\tau_{yz}^{\infty}}{c_{44}} q G(q/p)\, \mathrm{e}^{-[\kappa q+(p^2-\lambda^2 q^2)^{1/2}]y}, \tag{47}$$

$$p T_{yz} = \tau_{yz}^{\infty} q H(q/p)\, \mathrm{e}^{-[\kappa q+(p^2-\lambda^2 q^2)^{1/2}]y}. \tag{48}$$

According to (65), the inversions are

$$\frac{\partial^2 w_+}{\partial\tau^2} = -\frac{\tau_{yz}^{\infty} U(\tau - r_0/\lambda)}{2\pi i c_{44}}$$

$$\times\{g_+(-\tau)[G_+(\zeta)]_{\zeta=\zeta_+(-\tau)} - g_-(-\tau)[G_-(\zeta)]_{\zeta=\zeta_-(-\tau)}\}, \tag{49}$$

$$\frac{\partial \tau_{yz}}{\partial \tau} = -\frac{\tau_{yz}^{\infty} U(\tau - r_0/\lambda)}{2\pi i}$$
$$\times\{g_+(-\tau)[H_+(\zeta)]_{\zeta=\zeta_+(-\tau)} - g_-(-\tau)[H_-(\zeta)]_{\zeta=\zeta_-(-\tau)}\}, \tag{50}$$

where $U(\cdot)$ is the unit step function, and r_0, $g_\pm(\cdot)$ and $\zeta_\pm(\cdot)$ are found in the Appendix.

Introduce a cylindrical coordinate system (r, φ, z), such that $x = a + r\cos\varphi$, $y = r\sin\varphi$, and consider the vicinity $r/a \ll 1$, $0 \leqslant \varphi \leqslant \pi$ of the right crack edge. Then, asymptotically as $r/a \to 0$,

$$\zeta_\pm(-\tau) \to -\frac{\tau}{x}\left[1 - \frac{r}{a}(\cos\varphi - \kappa\sin\varphi \pm i\sqrt{\lambda^2 - \gamma^2}\sin\varphi)\right], \tag{51}$$

$$g \pm (-\tau) \to \frac{1}{a}. \tag{52}$$

Insertion into (49) gives, after two integrations with respect to τ

$$w_+ \to \frac{\tau_{yz}^{\infty}\sqrt{ar}}{\mu_a \boldsymbol{E}(\sqrt{1-\gamma^2/\lambda^2})}$$
$$\times\sqrt{\sqrt{1 - 2\kappa\cos\varphi\sin\varphi - \theta(\gamma)\sin^2\varphi} - \cos\varphi + \kappa\sin\varphi}, \tag{53}$$

where $\theta(\gamma) = \gamma^2 + 1 - \lambda^2 - \kappa^2 = \gamma^2 + 1 - c_{55}/c_{44}$.

The same calculation may be made for the vicinity of the right crack edge for $-\pi \leqslant \varphi < 0$, but a simpler way is to make use of the antisymmetry relation $w_-(x, -y) = w_+(-x, y)$, noting that $w_+(-x, y)$ is found from $w_+(x, y)$ after changing κ to $-\kappa$. This implies that the validity of expression (53) is extended to the whole vicinity of the right crack edge.

The stresses in the vicinity of the right crack edge may be determined from (53) by using relations (9)–(10). The stress $\tau_{\varphi z}$ might be most interesting. After transformation to cylindrical coordinates, it is found that

$$\tau_{\varphi z} \to \frac{\tau_{yz}^{\infty}}{2\mu_a \boldsymbol{E}(\sqrt{1-\gamma^2/\lambda^2})}\sqrt{\frac{a}{r}}$$
$$\times\left\{F_1 D + \frac{F_2}{D}\left[\sin\varphi + \kappa\cos\varphi - \frac{\kappa\cos 2\varphi + \theta(\gamma)\cos\varphi\sin\varphi}{C}\right]\right\}, \tag{54}$$

as $r/a \to 0$. Here

$$F_1 = 2c_{45}\cos 2\varphi + (c_{44} - c_{55})\sin 2\varphi,$$

$$F_2 = c_{44}\cos^2\varphi + c_{55}\sin^2\varphi - c_{45}\sin 2\varphi,$$

$$C = \sqrt{1 - 2\kappa\cos\varphi\sin\varphi - \theta(\gamma)\sin^2\varphi},$$

$$D = \sqrt{C - \cos\varphi + \kappa\sin\varphi}.$$

6. Conclusion and discussion

In agreement with the corresponding isotropic case the crack shape is elliptic with the major axis in the direction of crack expansion, and the displacement is zero on the crack plane outside the crack. On the other hand, the stress τ_{xz} does not vanish on the crack plane outside the crack. Furthermore, the front of the radiated stress is not in general symmetric with respect to the crack plane, but consists of the slanting ellipse

$$\frac{(x/a-\kappa y/a)^2}{(\lambda/\gamma)^2}+\frac{(y/a)^2}{1/\gamma^2}=1. \tag{55}$$

This follows from the factor $U(\tau-r_0/\lambda)$ in the expressions for displacements and stresses. In the isotropic case this defaults to $x^2+y^2=a^2/\gamma^2$.

The stress σ_z may not vanish, as in the corresponding isotropic case, but exhibits an inverse square root singularity at the crack edge. This might be of some physical significance. If planes parallel to the z plane are weak planes, cracks may open up along such planes.

The energy flux into the crack edge approaches zero as $\gamma\to\lambda$, Thus the maximum theoretical velocity for an expanding crack is $V=\mu_a/\sqrt{c_{44}\varrho}$, where $\mu_a=\sqrt{c_{44}c_{55}-c_{45}^2}$. Note that this velocity also is obtained from (55) by putting $y=0$. Thus, the limiting crack velocity equals the propagation velocity of waves travelling in the crack direction. In the isotropic case this velocity defaults to $V=\sqrt{\mu/\varrho}$.

The self-similarity of the problem was taken advantage of in essentially the same way as used by Broberg (1959) for a half-plane problem. This procedure appears to be very simple. Other methods were devised by Smirnov and Sobolev (1932), using an analysis based on wave front structures, by Craggs (1963), who used the Chaplygin transformation for the problem of an expanding mode I crack, and by Willis (1973), who made use of wave arrivals for several different classes of self-similar problems, including three-dimensional and anisotropic problems. The method used by Atkinson (1959) is very similar to the present approach, except that he did not use Laplace transforms. Laplace transforms may be more convenient in contexts of self-similarity, because immediate inversion is possible. Although Atkinson solved a problem concerning an expanding crack in an orthotropic material, his basic approach is quite general and appears also to be applicable for cracks along a symmetry plane in materials that can support plane strain, which are the same materials as considered in the present work.

Appendix

A. Inversion procedure for double Laplace transforms at self-similarity

The double Laplace transform of a function $f(\tau/x)$ is

$$\begin{aligned}\mathcal{L}_{qx}\mathcal{L}_{p\tau}f(\tau/x) &= q\int_{-\infty}^{+\infty}\mathrm{e}^{-qx}p\int_0^{\infty}\mathrm{e}^{-p\tau}f(\tau/x)\,\mathrm{d}\tau\,\mathrm{d}x\\ &= q\int_{-\infty}^{0}\mathrm{e}^{-qx}h_1(px)\,\mathrm{d}x+q\int_0^{\infty}\mathrm{e}^{-qx}h_2(px)\,\mathrm{d}x,\end{aligned} \tag{56}$$

where $h_1(px) = \mathcal{L}_{p\tau} f(\tau/x)$, $x < 0$ and $h_2(px) = \mathcal{L}_{p\tau} f(\tau/x)$, $x > 0$. It is convenient to assume that p is real and positive. Then, by substituting integration variable x by px, it is obvious that the double Laplace transform can be written in the form

$$\mathcal{L}_{qx}\mathcal{L}_{p\tau} f(\tau/x) = F(q/p). \tag{57}$$

It can be shown that $F(\cdot)$ is odd if $f(\cdot)$ is even, and *vice versa*.

Consider now the inversion of the double Laplace transform

$$\mathcal{L}_{qx}\mathcal{L}_{p\tau} f(x, y, \tau) = qF(q/p)\, \mathrm{e}^{-[\kappa+(p^2-\lambda^2 q^2)^{1/2}]y}, \quad y \geqslant 0, \tag{58}$$

which is of a type that often appears in self-similar problems. A formal inversion with respect to q gives

$$\begin{aligned} \mathcal{L}_{p\tau} f(x, y, \tau) &= \frac{1}{2\pi i}\int_{-i\infty}^{i\infty} \mathrm{e}^{qx-[\kappa+(p^2/k^2-q^2)^{1/2}]y} F(q/p)\,\mathrm{d}q \\ &= \frac{1}{2\pi i}\int_{-i\infty}^{i\infty} p\,\mathrm{e}^{p\{\zeta x-[\kappa\zeta+(1-\lambda^2\zeta^2)^{1/2}]y\}} F(\zeta)\,\mathrm{d}\zeta, \end{aligned} \tag{59}$$

where the substitution $q = p\zeta$, with $\zeta = \xi + i\eta$, was made. It is assumed that the strip of convergence for the Laplace transform includes the imaginary axis. Branch cuts are made along $|\xi| > 1/\lambda$, $\eta = 0$, and the branch is chosen so that $(1 - \lambda^2\zeta^2)^{1/2} = \sqrt{1 - \lambda^2\xi^2}$ for $|\xi| < 1/\lambda$, $\eta = 0$. This implies that $(1/\lambda^2 - \zeta^2)^{1/2} \to \mp i\lambda\zeta$ as $|\zeta| \to \infty$, where minus sign refers to $\Im\zeta > 0$, plus sign to $\Im\zeta < 0$.

Put $s = \zeta x - [\kappa\zeta + (1 - \lambda^2\zeta^2)^{1/2}]y$ and find the path in the ζ plane for which s is real. By taking the real and imaginary parts of the relation

$$s - \zeta(x - \kappa y) = (1 - \lambda^2\zeta^2)^{1/2} y \tag{60}$$

and eliminating s, the following hyperbola in the ζ plane is obtained

$$\frac{\xi^2}{(x - \kappa y)^2} - \frac{\eta^2}{\lambda^2 y^2} = \frac{1}{\lambda^2 r_0^2}, \tag{61}$$

where $r_0 = \sqrt{(x - \kappa y)^2 + \lambda^2 y^2}$. Compare the isotropic case, $\kappa = 0$, $\lambda = 1$, for which the hyperbola $(\xi/x)^2 - (\eta/y)^2 = 1/\sqrt{x^2 + y^2}$ is obtained for self-similar problems (Broberg, 1959).

The left branch of the hyperbola (61) corresponds to $x - \kappa y \geqslant 0$ and the right branch corresponds to $x - \kappa y \leqslant 0$. The part of these branches in the upper half-plane will be denoted Γ_+, the part in the lower half-plane will be denoted Γ_-. From

$$\zeta = \frac{x - \kappa y}{r_0^2} s \pm iy\frac{(\lambda^2 s^2 - r_0^2)^{1/2}}{r_0^2} = \zeta_\pm(s) \quad \text{on } \Gamma_\pm, \tag{62}$$

it follows that s goes from $-r_0/\lambda$ to $-\infty$ as ζ goes from $-(x - \kappa y)/r_0^2$ to infinity along either branch Γ_+ or Γ_-. Index plus is used to denote quantities on Γ_+, index minus for quantities on Γ_-.

Thus

$$\mathcal{L}_{p\tau} f(x, y, \tau) = \frac{1}{2\pi i}\left[\int_{\Gamma_+} p\, e^{ps} F(\zeta)\, d\zeta + \int_{\Gamma_-} p\, e^{ps} F(\zeta)\, d\zeta\right]$$

$$= \frac{1}{2\pi i}\left[-\int_{-\infty}^{-r_0/\lambda} p\, e^{ps} F_+(\zeta)\frac{d\zeta_+}{ds}\, ds + \int_{-\infty}^{-r_0/\lambda} p\, e^{ps} F_-(\zeta)\frac{d\zeta_-}{ds}\, ds\right], \quad (63)$$

where ζ is chosen according to (62). $F_+(\zeta)$ is the value of $F(\zeta)$ in the upper half-plane, and $F_-(\zeta)$ is its value in the lower half-plane. The derivative

$$\frac{d\zeta_\pm}{ds} = \frac{x - \kappa y}{r_0^2} \pm \frac{i y \lambda^2 s}{r_0^2\sqrt{\lambda^2 s^2 - r_0^2}} = g_\pm(s), \quad (64)$$

where plus sign refers to Γ_+ and minus sign to Γ_-. Now inversion with respect to p involves only $p\exp(ps)$, the inversion of which is $\delta(\tau + s)$, where $\delta(\cdot)$ is the Dirac delta function. Hence, because the delta function sifts out the integrand for $s = -\tau$, the full inversion of the double Laplace transform (58) is

$$f(x, y, \tau) = -\frac{U(\tau - r_0/\lambda)}{2\pi i} \times\{g_+(-\tau)[F_+(\zeta)]_{\zeta=\zeta_+(-\tau)} - g_-(-\tau)[F_-(\zeta)]_{\zeta=\zeta_-(-\tau)}\}. \quad (65)$$

As $y \to 0$ the inversion formula takes the form

$$f(x, 0, \tau) = -\frac{U(\tau - |x|/\lambda)}{2\pi i x}[F_+(\xi) - F_-(\xi)]_{\xi=-\tau/x} \quad (66)$$

and the hyperbola has degenerated to the upper (+) and lower (−) sides of the part $\xi < -1/\lambda$ of the real axis for $x > 0$ and of the part $\xi > 1/\lambda$ for $x < 0$. The unit step function is optional for $y = 0$: omitting it is equivalent to extending the path Γ to the origin, $\xi = \eta = 0$.

References

Atkinson, C. (1965). The propagation of a brittle crack in anisotropic material. *International Journal of Engineering Science* **3**, 77–91.

Broberg, K.B. (1959). A problem on stress waves in an infinite elastic plate. *Transactions of the Royal Institute of Technology, Stockholm, Sweden* **139**, 1–27.

Craggs, J.W. (1963). Fracture criteria for use in continuum mechanics. *Fracture of Solids* (Edited by D.C. Drucker and J.J. Gilman), Interscience Publishers, N.Y., 51–63.

Nye, J.F. (1957). *Physical Properties of Crystals: Their Representation by Tensors and Matrices*, Oxford, Clarendon Press.

Smirnov, V.I. and Sobolev, S.L. (1932). Novyi metod resheniia ploskoi zadachi uprugikh kolebanii. (New method for solving the planar problem of elastic oscillations.) *Trudy seismich. in-ta Akad. Nauk SSSR*, No. 20, **5**.

Ting, T.C.T. (1996). *Anisotropic Elasticity: Theory and Applications*. Oxford University Press, New York.

Willis, J.R. (1973). Self-similar problems in elastodynamics. *Philosophical Transactions of the Royal Society (London)* **274A**, 435–491.

Wu, K.-C. (1989). On the crack-tip fields of a dynamically propagating crack in an anisotropic solid. *International Journal of Fracture* **41**, 253–266.

Wu, K.-C. (1991). Explicit crack-tip fields of an expanding interface crack in an anisotropic bimaterial. *International Journal of Solids and Structures* **27**, 455–466.

International Journal of Fracture **93:** 13–38, 1998.

Numerical analysis of dynamic debonding under 2D in-plane and 3D loading

M. SCOT BREITENFELD and PHILIPPE H. GEUBELLE*
Department of Aeronautical and Astronautical Engineering, University of Illinois at Urbana–Champaign, Urbana, IL 61801, U.S.A.

Received 23 September 1997; accepted in revised form 2 July 1998

Abstract. We present a numerical scheme specially developed for 2D and 3D dynamic debonding problems. The method, referred to as spectral scheme, allows for a precise modeling of stationary and/or spontaneously expanding interfacial cracks of arbitrary shapes and subjected to an arbitrary combination of time- and space-dependent loading conditions. It is based on a spectral representation of the elastodynamic relations existing between the displacement components along the interface plane and the corresponding dynamic stresses. A general stress-based cohesive failure model is introduced to model the spontaneous progressive failure of the interface. The numerical scheme also allows for the introduction of a wide range of contact relations to model the possible interactions between the fracture surfaces. Simple 2D problems are used to investigate the accuracy and stability of the proposed scheme. Then, the spectral method is used in various 2D and 3D interfacial fracture problems, with special emphasis on the issue of the limiting speed for a spontaneously propagating debonding crack in the presence of frictional contact.

Key words: Dynamic fracture, spectral method, interface, boundary integral method.

1. Introduction

In a recent paper, Geubelle and Breitenfeld (1997) have proposed a numerical scheme specially developed for dynamic debonding problems. The method, referred to as *spectral scheme*, allows for the simulation of a wide range of fundamental fracture problems involving stationary or *spontaneously* expanding cracks present at the interface between two dissimilar linearly elastic semi-infinite media. The numerical method consists of a spectral form of the boundary integral equations relating the traction stresses acting on the interface plane and the associated displacement components. It is inspired by an earlier version introduced by Geubelle and Rice (1995) for 3D dynamic fracture problems in homogeneous elastic media.

While similar in various ways to the homogeneous implementation, the bimaterial version of the spectral scheme involves a series of important differences which have been described in detail by Geubelle and Breitenfeld (1997) in the simpler framework of mode III (anti-plane shear) debonding. That paper also contains a limited list of references to existing experimental, theoretical and numerical papers dedicated to topics associated with the propagation of interfacial cracks and faults, starting from the pioneering work of Williams (1959) to the series of analytical papers spawn by the recent observations of transonic crack propagation along interfaces (Lambros and Rosakis, 1995; Singh and Shukla, 1996; Singh et al., 1997).

The maximum speed at which interfacial cracks propagate spontaneously is one of the many fundamental issues associated with the mechanics of dynamic interfacial fracture which

* Corresponding author.

have motivated the development of the present numerical scheme. Another topic of interest, which was not addressed in the scalar mode III case, is that of a contact zone present in the vicinity of an interface crack under in-plane loading, and of the relative importance of the associated frictional contact in the failure process. This issue might become especially relevant for problems involving fast interfacial cracks, as the size of the contact zone is expected to increase as the crack speed approaches that of the slower Rayleigh wave (Yang et al., 1991). While it can be argued that the existence of a contact zone predicted by the linearly elastic asymptotic solution described by Yang et al. (1991) is an artifact of the solution itself, the presence of actual contact behind the rapidly propagating crack tip has been observed experimentally by Lambros and Rosakis (1995) and by Singh and Shukla (1996), especially under shear-dominated conditions.

Also of interest is the intrinsic combination of tensile and shear failure modes present in the great majority of in-plane interface fracture problems: unlike in the homogeneous situation where the contributions of the modes I and II can be clearly distinguished, the near-tip stress fields present in the bimaterial case always involve some mode mixity, the amount of which varies with the distance to the crack tip. This mode mixity inherent in any bimaterial problem has repercussion even outside of the field of interfacial fracture, as illustrated by the frictional instability appearing in certain bimaterial problems involving two dissimilar elastic bodies sliding relative to each other (Adams, 1995).

In this paper, we develop and implement the bimaterial spectral scheme for 2D in-plane (modes I/II) and fully 3D fracture problems. As indicated in the preliminary mode III paper (Geubelle and Breitenfeld, 1997), two approaches can be used to investigate bimaterial situations: the first one, referred to as the *combined spectral formulation*, combines the elastodynamic responses of the two half spaces in a single integral equation involving the displacement discontinuities (or crack opening displacements) across the fracture plane. This was the approach used by Geubelle and Rice (1995) in the homogeneous situation. The second approach, referred to as the *independent spectral formulation*, consists in modeling the elastodynamic response of each half space separately before applying the interface conditions. While the two approaches have very similar stability and precision characteristics in the antiplane shear case, this similarity disappears in the in-plane situation investigated hereafter, and the independent formulation appears to be much more stable and precise than the combined formulation, especially in shear-dominated situations.

The paper is organized as follows: the derivation of the bimaterial independent spectral formulation is described in Section 2, followed by details on the implementation of the numerical scheme. Section 4 presents various simulations performed to validate the numerical method, with special emphasis on the comparative study of the stability and precision of the two aforementioned spectral formulations. Some of the aforementioned fundamental issues associated with dyamic interfacial fracture are investigated in the last three sections of the paper. Section 5 focuses on problems involving the dynamic loading of stationary cracks. Section 6 addresses the issue of frictional contact instability, and Section 7 summarizes the results of spontaneous crack propagation simulations.

2. Spectral formulation

The geometry of the interfacial fracture problem is described in Figure 1 and involves a planar crack of arbitrary shape located at the interface between two semi-infinite linearly elastic

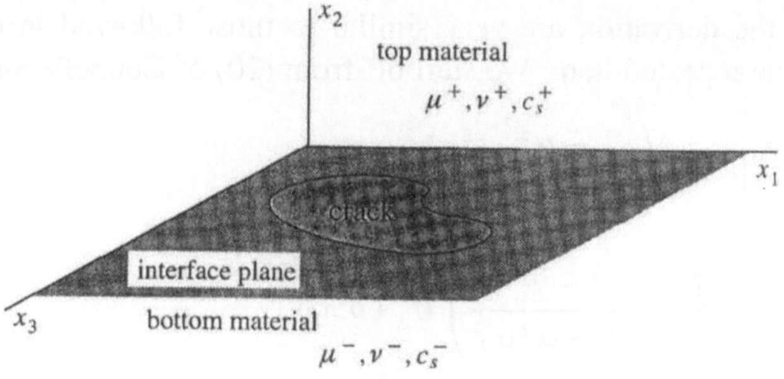

Figure 1. Three dimensional problem geometry of interfacial plane.

half spaces. A cartesian coordinate system is defined such that the interface plane, which is also the fracture plane, corresponds to $x_2 = 0$. Let $\sigma_{ij}(x_k, t)$* and $u_i(x_k, t)$ denote the stress and displacement components, respectively. The spectral scheme is a special form of the boundary integral relations existing between the traction stresses $\tau_j(x_1, x_3, t) = \sigma_{2j}(x_1, x_2 = 0, x_3, t)$ acting on the interfacial fracture plane and the resulting displacement components $u_j^{\pm}(x_1, x_3, t) = u_j(x_1, x_2 = 0^{\pm}, x_3, t)$**. More precisely, the formulation is expressed in the spectral domain between the time-dependent Fourier coefficients of the traction stresses and those of the displacements. The steps leading to the derivation of the bimaterial spectral formulation are, to a certain extent, similar to those used in the homogeneous case given by Geubelle and Rice (1995). As was the case there, the spectral formulation is first obtained in the simpler 2D in-plane situation. The 2D results are then combined with the bimaterial anti-plane shear relations described by Geubelle and Breitenfeld (1997) to obtain the fully 3D formulation.

2.1. Two-dimensional in-plane formulation

In this preliminary 2D step, we assume that the interfacial crack is infinite in the x_3-direction and that the displacements and stress fields depend solely on x_1 and x_2. The anti-plane shear solution has been presented by Geubelle and Breitenfeld (1997), and we focus now on the in-plane case. Let $T_\alpha(t : q)$ and $U_\alpha(t; q)$ denote the qth-mode Fourier coefficients of the in-plane traction stresses and displacements, respectively, as in

$$\begin{aligned} \sigma_{2\alpha}(x_1, 0^{\pm}, t) &= \tau_\alpha(x_1, t) = T_\alpha(t; q)\, \mathrm{e}^{iqx_1}, \\ u_\alpha(x_1, 0^{\pm}, t) &= u_\alpha^{\pm}(x_1, t) = U_\alpha^{\pm}(t; q)\, \mathrm{e}^{iqx_1}. \end{aligned} \tag{1}$$

* Conventional notation is adopted here, with Latin indices ranging over 1, 2, 3 and Greek indices taking the value of 1 or 2.

** As indicated in Figure 1, the superscripts '+' and '−' denote the top and bottom half spaces, respectively, and are often omitted for clarity purpose.

The initial steps of the derivation are very similar to those followed in Geubelle and Rice (1995) and will not be repeated here. We start off from (20) of Geubelle and Rice's paper,

$$\begin{aligned}
\widehat{T}_1(p;q) &= \mp\mu^{\pm}|q|\alpha_d^{\pm}\left(\frac{1-\alpha_s^{\pm^2}}{1-\alpha_s^{\pm}\alpha_d^{\pm}}\right)\widehat{U}_1^{\pm}(p;q) \\
&\quad +i\mu^{\pm}q\left(2-\frac{1-\alpha_s^{\pm^2}}{1-\alpha_s^{\pm}\alpha_d^{\pm}}\right)\widehat{U}_2^{\pm}(p;q), \\
\widehat{T}_2(p;q) &= -i\mu^{\pm}q\left(2-\frac{1-\alpha_s^{\pm^2}}{1-\alpha_s^{\pm}\alpha_d^{\pm}}\right)\widehat{U}_1^{\pm}(p;q) \\
&\quad \mp\mu^{\pm}|q|\frac{\alpha_s^{\pm}(1-\alpha_s^{\pm^2})}{1-\alpha_s^{\pm}\alpha_d^{\pm}}\widehat{U}_2^{\pm}(p;q),
\end{aligned} \tag{2}$$

where $\hat{f}(p)$ denotes the Laplace transform of $f(t)$, μ is the shear modulus, and

$$\alpha_d = \sqrt{1+\frac{p^2}{q^2c_d^2}}, \qquad \alpha_s = \sqrt{1+\frac{p^2}{q^2c_s^2}}. \tag{3}$$

In (3), c_d and c_s are the dilatation and shear wave speeds, respectively. Next, we extract the instantaneous response of the two half spaces

$$[\widehat{T}_1]_{\text{inst}} = \mp\frac{\mu^{\pm}}{c_s^{\pm}}p\widehat{U}_1^{\pm}(p;q), \qquad [\widehat{T}_2]_{\text{inst}} = \mp\frac{c_d^{\pm}}{c_s^{\pm^2}}\mu^{\pm}p\widehat{U}_2^{\pm}(p;q),$$

and rewrite (2) as

$$\begin{aligned}
\widehat{T}_1(p;q) &= -\frac{\mu^{\pm}}{c_s^{\pm}}p\widehat{U}_1^{\pm}(p;q) \mp \mu^{\pm}|q|\left[\frac{\alpha_d^{\pm}(1-\alpha_s^{\pm^2})}{1-\alpha_s^{\pm}\alpha_d^{\pm}} - \frac{p}{|q|c_s^{\pm}}\right]\widehat{U}_1^{\pm}(p;q) \\
&\quad +i\mu^{\pm}q\left[2-\frac{1-\alpha_s^{\pm^2}}{1-\alpha_s^{\pm}\alpha_d^{\pm}}\right]\widehat{U}_2^{\pm}(p;q), \\
\widehat{T}_2(p;q) &= \mp\frac{c_d^{\pm}}{c_s^{\pm^2}}\mu p\widehat{U}_2^{\pm}(p;q) \mp \mu^{\pm}|q|\left[\frac{\alpha_s^{\pm}(1-\alpha_s^{\pm^2})}{1-\alpha_s^{\pm}\alpha_d^{\pm}} - \frac{c_d^{\pm}}{c_s^{\pm}}\frac{p}{|q|c_s^{\pm}}\right]\widehat{U}_2^{\pm}(p;q) \\
&\quad -i\mu^{\pm}q\left[2-\frac{1-\alpha_s^{\pm^2}}{1-\alpha_s^{\pm}\alpha_d^{\pm}}\right]\widehat{U}_1^{\pm}(p;q).
\end{aligned} \tag{4}$$

Back in the space-time domain, the 2D elastodynamic relations between the traction components of the stress τ_α acting on the fracture plane and the resulting displacements $u_\alpha^{\pm}$ take the form

$$\begin{aligned}
\tau_1(x_1,t) &= \tau_1^o(x_1,t) \mp \frac{\mu^{\pm}}{c_s^{\pm}}\frac{\partial u_1^{\pm}(x_1,t)}{\partial t} + f_1^{\pm}(x_1,t), \\
\tau_2(x_1,t) &= \tau_2^o(x_1,t) \mp \frac{c_d^{\pm}}{c_s^{\pm}}\frac{\mu^{\pm}}{c_s^{\pm}}\frac{\partial u_2^{\pm}(x_1,t)}{\partial t} + f_2^{\pm}(x_1,t),
\end{aligned} \tag{5}$$

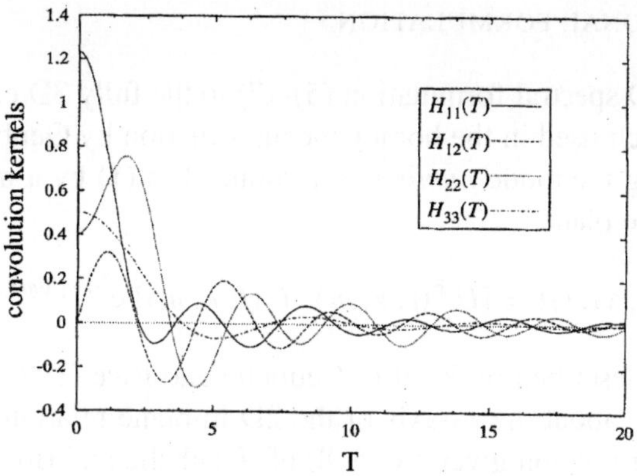

Figure 2. Convolution kernels ($\nu = 0.25$).

where $\tau_\alpha^o(x_1, t)$ are the externally applied traction stresses and $f_\alpha^\pm(x_1, t)$ represents the convolution terms corresponding to the last two terms of (4). The associated Fourier coefficients $F_\alpha(t; q)$ are givn by

$$\begin{aligned} F_1^\pm(t; q) &= \pm\mu^\pm|q| \int_0^t H_{11}(|q|c_s^\pm t')U_1^\pm(t - t'; q)|q|c_s^\pm \,\mathrm{d}t' + i(2 - \eta^\pm)\mu^\pm q U_2^\pm(t; q) \\ &\quad + i\mu^\pm q \int_0^t H_{12}(|q|c_s^\pm t')U_2^\pm(t - t', q)|q|c_s^\pm \,\mathrm{d}t', \\ F_2^\pm(t; q) &= \mp\mu^\pm|q| \int_0^t H_{22}(|q|c_s^\pm t')U_2^\pm(t - t'; q)|q|c_s^\pm \,\mathrm{d}t' - i(2 - \eta^\pm)\mu^\pm q U_1^\pm(t; q) \\ &\quad - i\mu^\pm q \int_0^t H_{12}(|q|c_s^\pm t')U_1^\pm(t - t', q)|q|c_s^\pm \,\mathrm{d}t', \end{aligned} \tag{6}$$

where $\eta = c_d/c_s$. Thc convolution kernels H_{11}, H_{12} and H_{22} correspond to the square bracketed terms in (4) and can be written as

$$\begin{aligned} H_{11}(T) &= \mathfrak{L}^{-1}\left[\frac{s^2\sqrt{s^2+\eta^2}}{\sqrt{s^2+\eta^2}\sqrt{s^2+1}-\eta} - s\right], \\ H_{12}(T) &= \mathfrak{L}^{-1}\left[\frac{\eta s^2}{\eta - \sqrt{s^2+\eta^2}\sqrt{1+s^2}} + \eta\right], \\ H_{22}(T) &= \mathfrak{L}^{-1}\left[\frac{s^2\sqrt{1+s^2}}{\sqrt{1+s^2/\eta^2}\sqrt{1+s^2}-1} - \eta s\right], \end{aligned} \tag{7}$$

where $s = p/|q|c_s$ is the nondimensional Laplace transform variable. A closed-form expression of the H_{11} and H_{22} kernels can be found in Appendix A. The Laplace inversion of $\widehat{H}_{12}$ has so far eluded out efforts, but can be readily performed numerically. The three convolution kernels are presented in Figure 2 for a Poisson's ration $\nu = 0.25$.

2.2. Three-dimensional formulation

The extension of the 2D spectral formulation (5)–(7) to the fully 3D case depicted in Figure 1 is similar to the approach used in the homogeneous situation by Geubelle and Rice (1995). It is obtained by replacing the mode *number* q introduced in (1) by a mode *vector* $\boldsymbol{q} = (k, m)$ which spans the fracture plane

$$[u_j^{\pm}(x_1, x_3, t), \tau_j(x_1, x_3, t)] = [U_j^{\pm}(t; k, m), T_j(t; k, m)]\, \mathrm{e}^{i(kx_1+mx_3)}. \tag{8}$$

Following the process described in detail in Geubelle and Rice (1995), the 3D formulation is obtained by a 'rotation' about the x_2-axis of the 2D in-plane relations derived in Section 2.1 and the anti-plane shear relation given by (5–8) of (Geubelle and Breitenfeld, 1996), yielding

$$\tau_j(x_1, x_3, t) = \tau_j^o(x_1, x_3, t) - V_{jk}^{\pm}\frac{\partial u_k^{\pm}(x_1, x_3, t)}{\partial t} + f_j^{\pm}(x_1, x_3, t), \tag{9}$$

where $[V_{ij}]$ is a diagonal matrix with

$$V_{11} = V_{33} = \frac{\mu}{c_s}, \qquad V_{22} = \eta\frac{\mu}{c_s}.$$

As was the case in the 2D formulation, the Fourier coefficients $F_j^{\pm}(t; k, m)$ of the $f_j^{\pm}$ (x_1,x_2,t) term in (9) involve a series of convolution integrals over the past displacement history

$$\begin{Bmatrix} F_1^{\pm}(t; k, m) \\ F_2^{\pm}(t; k, m) \\ F_3^{\pm}(t; k, m) \end{Bmatrix}$$

$$= -i\mu^{\pm}(2-\eta^{\pm})\begin{pmatrix} 0 & -k & 0 \\ k & 0 & m \\ 0 & -m & 0 \end{pmatrix}\begin{Bmatrix} U_1^{\pm}(t-t'; k, m) \\ U_2^{\pm}(t-t'; k, m) \\ U_3^{\pm}(t-t'; k, m) \end{Bmatrix}$$

$$-\mu^{\pm}q\int_0^t\left[i\frac{H_{12}(qc_s^{\pm}t')}{|q|}\begin{pmatrix} 0 & -k & 0 \\ k & 0 & m \\ 0 & -m & 0 \end{pmatrix} \pm H_{22}(qc_s^{\pm}t')\begin{pmatrix} 0 & 0 & 0 \\ 0 & 1 & 0 \\ 0 & 0 & 0 \end{pmatrix}\right.$$

$$\left.\pm\frac{H_{11}(qc_s^{\pm}t')}{q^2}\begin{pmatrix} k^2 & 0 & km \\ 0 & 0 & 0 \\ km & 0 & m^2 \end{pmatrix} \pm \frac{H_{33}(qc_s^{\pm}t')}{q^2}\begin{pmatrix} m^2 & 0 & -km \\ 0 & 0 & 0 \\ -km & 0 & k^2 \end{pmatrix}\right]$$

$$\times\begin{Bmatrix} U_1^{\pm}(t-t'; k, m) \\ U_2^{\pm}(t-t'; k, m) \\ U_3^{\pm}(t-t'; k, m) \end{Bmatrix}|q|c_s^{\pm}\,\mathrm{d}t', \tag{10}$$

where $q = \sqrt{k^2 + m^2}$. The in-plane convolution kernels $H_{11}(T)$, $H_{22}(T)$ and $H_{12}(T)$ are given by (7) and $H_{33}(T) = J_1(T)/T$ is the mode III kernel obtained in Geubelle and Breitenfeld (1997).

Note that, in contrast to the 3D homogeneous case in which the 'tensile' response (in the x_2-direction) is decoupled from the 'shear' response (in the x_1- and x_3-directions)(see Equations (29, 30) of (Geubelle and Rice, 1995)), the spectral formulation presented here for the bimaterial problem involves a coupling of all three fracture modes.

3. Numerical implementation

The implementation of the 3D spectral formulation (8)–(10) starts by expressing the $u_j^{\pm}$ and $f_j^{\pm}$ distributions on the fracture plane as a double Fourier series with periods X and Z in the x_1- and x_3-directions, respectively:

$$\left\{ \begin{array}{c} u_j^{\pm}(x_1, x_3, t) \\ f_j^{\pm}(x_1, x_2, t) \end{array} \right\} = \sum_{k=-K/2}^{K/2} \sum_{m=-M/2}^{M/2} \left\{ \begin{array}{c} U_j^{km\pm}(t) \\ F_j^{km\pm}(t) \end{array} \right\} e^{2\pi i \left(\frac{kx_1}{X} + \frac{mx_3}{Z} \right)}. \tag{11}$$

A conventional 2D FFT algorithm is used to link spatial and spectral representations, with the $K * M$ sampling points distributed uniformly over the $X * Z$ rectangular portion of the fracture plane. Once the convolution term is computed through (10) in the spectral domain and transformed back to the spatial domain, (9) is used to compute the updated velocity distribution $\dot{u}_k^{\pm}(x_1, x_3, t)$ which is then integrated in time with an explicit scheme to derive the displacement field

$$u_j^{\pm}(x_1, x_3, t + \Delta t) = u_j^{\pm}(x_1, x_3, t) + \Delta t \quad \dot{u}_j^{\pm}(x_1, x_3, t). \tag{12}$$

The time step Δt is chosen as a fraction of the time needed by the fastest shear wave speed to propagate the smallest spacing distance between the grid points defined on the fracture plane and used as sampling points for the FFT,

$$\Delta t = \beta \frac{\min(\Delta x, \Delta z)}{\max(c_s^+, c_s^-)}. \tag{13}$$

As will be discussed in Section 4.1, the user-defined parameter β plays a critical role in the stability and precision of the numerical scheme.

To complete the formulation, we still have to incorporate the continuity conditions along the interface plane, and introduce a cohesive failure model to allow for the spontaneous propagation of the interfacial crack. This failure model can take the general form

$$\tau_{n,s}^{str} = f_{n,s}(\delta_n, \delta_s, \dot{\delta}_n, \dot{\delta}_s, x_1, x_3, t), \tag{14}$$

where τ_n^{str} and τ_s^{str} are the normal and shear strengths of the interface; δ_n and δ_s are the normal and shear displacement discontinuities, respectively, and are defined by $\delta_n = \delta_2$ and $\delta_s = \sqrt{\delta_1^2 + \delta_3^2}$, where δ_j is the displacement continuity across the fracture plane in the x_j-direction

$$\delta_j(x_1, x_3, t) = u_j^{+}(x_1, x_2, t) - u_j^{-}(x_1, x_3, t). \tag{15}$$

In the remainder of this paper, we will adopt the rate-independent coupled failure model

$$\tau_{n,s}^{str} = \tau_{n,s}^{c} \left\langle 1 - \sqrt{(\delta_n/\delta_n^c)^2 + (\delta_s/\delta_s^c)^2} \right\rangle, \tag{16}$$

where $\langle a \rangle = a$ if $a > 0$ and $= 0$ otherwise; τ_n^c and τ_s^c denote the 'intact' normal and shear strengths of the interface; and δ_n^c and δ_s^c correspond to the normal and shear values of the crack opening displacement beyond which complete failure is achieved. It has to be noted that the general form (14) of the interface strength can be used not only to describe the interface failure process, but also to characterize frictional contact between the crack faces, as described below.

The sequence of operations performed at each time step can thus be summarized as follows:

Step (i) Update the displacement distribution $u_j^{\pm}$ with (12).

Step (ii) Update the applied load τ_j^o if, for example, it depends on the current extent of the crack.

Step (iii) Update the interface strength $\tau_{n,s}^{str}$ with (16).

Step (iv) Compute the convolution terms $f_j^{\pm}$ with (10) and (11) using a 2D FFT to link the spatial and spectral domains.

Step (v) Assuming that the interface does not undergo further failure and that the two half spaces move together ($\dot{u}_j^+ = \dot{u}_j^- = \dot{u}_j$), compute the resulting interface velocities $\dot{u}_j$ and tractions τ_j^{in}.

$$\frac{\dot{u}_1}{c_s^+} = \frac{1}{\mu^+}\left(\frac{f_1^+ - f_1^-}{1 + \frac{\xi}{\zeta}}\right), \qquad \tau_1^{in} = \tau_1^o + f_1^+ - \mu^+ \frac{\dot{u}_1}{c_s^+},$$

$$\frac{\dot{u}_2}{c_s^+} = \frac{1}{\mu^+}\left(\frac{f_2^+ - f_2^-}{\eta^+ + \frac{\xi}{\zeta}\eta^-}\right), \qquad \tau_2^{in} = \tau_2^o + f_2^+ - \mu^+\eta^+ \frac{\dot{u}_2}{c_s^+}, \tag{17}$$

$$\frac{\dot{u}_3}{c_s^+} = \frac{1}{\mu^+}\left(\frac{f_3^+ - f_3^-}{1 + \frac{\xi}{\zeta}}\right), \qquad \tau_3^{in} = \tau_3^o + f_3^+ - \mu^+ \frac{\dot{u}_3}{c_s^+},$$

where $\xi = c_s^+/c_s^-$ and $\zeta = \mu^+/\mu^-$ are the mismatch parameters.

Step (vi) Compare the normal and shear components of the interace tractions ($\tau_n^{in} = \tau_2^{in}$ and $\tau_s^{in} = \sqrt{(\tau_1^{in})^2 + (\tau_3^{in})^2}$) with the current normal and shear values of the strength (τ_n^{str} and τ_s^{str}) obtained in *Step (iii)*:

Step (vi.a) If no failure is detected, the solution found in *Step (v)* is valid.

Step (vi.b) If failure is detected, the top and bottom surfaces move at different speeds:

$$\frac{\dot{u}_2^+}{c_s^+} = \frac{1}{\mu^+}\frac{1}{\eta^+}(\tau_{d_2}^+ - \tau_n^{str}), \qquad \frac{\dot{u}_2^-}{c_s^+} = \frac{1}{\mu^+}\frac{\zeta}{\xi}\frac{1}{\eta^-}(\tau_n^{str} - \tau_{d_2}^-),$$

$$\dot{u}_\alpha^+ = \dot{u}_s^+ \frac{\tau_{d_\alpha}^+}{\tau_{d_s}^+}, \qquad \dot{u}_\alpha^- = \dot{u}_s^- \frac{\tau_{d_\alpha}^+}{\tau_{d_s}^+}, \qquad (\alpha = 1, 3) \tag{18}$$

where $\tau_{d_j} = \tau_j^o + f_j$ denotes the dynamic stress; $\tau_{d_s} = \sqrt{\tau_{d_1}^2 + \tau_{d_3}^2}$; and the shear component of the velocity $\dot{u}_s^\pm$ is given by

$$\frac{\dot{u}_s^+}{c_s^+} = \frac{1}{\mu^+}(\tau_{d_s}^+ - \tau_s^{str}), \qquad \frac{\dot{u}_s^-}{c_s^+} = \frac{1}{\mu^+}\frac{\zeta}{\xi}(\tau_s^{str} - \tau_{d_s}^-). \tag{19}$$

Step (vii) In the region where the crack surfaces move independently, check for possible overlapping by computing the predicted normal crack opening displacement (COD)

$$\delta_2^{\text{pred}} = u_2^+ - u_2^- + \Delta t(\dot{u}_2^+ - \dot{u}_2^-).$$

If δ_2^{pred} is negative, the local motion of the crack surfaces is modified to ensure a vanishing COD and a continuity of the normal traction, yielding

$$\dot{u}_2^+ = \frac{c_s^+}{\eta^+ + \frac{\xi\eta^-}{\zeta}}\left[\frac{\tau_{d_2}^+ - \tau_{d_2}^-}{\mu^+} - \frac{\xi\eta^-}{\zeta}\left(\frac{u_2^+ - u_2^-}{c_s^+\Delta t}\right)\right], \qquad \dot{u}_2^- = \dot{u}_2^+ + \frac{u_2^+ - u_2^-}{\Delta t},$$

while the resulting normal (compressive) contact traction stress $\tilde{\tau}_2$ is given by

$$\tilde{\tau}_2 = \tau_{d_2}^+ - \eta^+\mu^+\frac{\dot{u}_2^+}{c_s^+}. \tag{20}$$

Step (viii) Finally, the knowledge of the normal compressive traction acting on the contacting surfaces can then be used in conjunction with a Coulomb friction model to introduce a frictional resistance to the relative motion in shear ($\dot{u}_s^\pm$) of the two fracture surfaces. The procedure is similar to that used in *Steps (v)* and *(vi)*

Step (viii.a) First assume that the top and bottom crack surfaces stick together ($\dot{u}_{1,3}^+ = \dot{u}_{1,3}^- = \dot{u}_{1,3}$) and compute the velocities $\dot{u}_{1,3}$ and shear traction stresses $\tau_{1,3}^{in}$ with (17)

Step (viii.b) Next compare the resulting shear interface traction $\tau_s^{in} = \sqrt{(\tau_1^{in})^2 + (\tau_3^{in})^2}$ with the frictional 'resistance' $f_f|\tilde{\tau}_2|$ where f_f is the Coulomb coefficient of friction and $\tilde{\tau}_2$ has been computed in (20). If $\tau_s^{in} < f_f|\tilde{\tau}_2|$, the solution found in *Step (viii.a)* stands. Otherwise, relative slip is detected with velocities $\dot{u}_s^\pm$ given by (19) in which the shear strength has been replaced by $f_f|\tilde{\tau}_2|$. Finally, the shear velocities $\dot{u}_{1,3}^\pm$ of the slippling crack faces are given by (18).

This concludes the description of the proposed algorithm. A discussion of advantages and limitations of the numerical method can be found in (Geubelle, 1996). The stability and precision of the bimaterial implementation of the spectral scheme are assessed in the next section.

4. Validation of the numerical scheme

As indicated earlier, the spectral scheme described in the previous section is referred to as the *independent formulation* since the dynamic responses of the two half spaces are studied separately through (9) before being joined with the aid of the interface continuity conditions. The approach is quite different from the original *combined formulation* proposed by Geubelle and Rice (1995) in the homogeneous situation, in which the dynamic responses of the two half spaces were combined through a simple convolution relation involving the displacement discontinuities $\delta_j(x_1, x_3, t)$ defined in (15). The distinction between the combined formulation and the independent approach presented here has been discussed in detail in the 2D mode III case (Geubelle and Breitenfeld, 1997).

However, while the two approaches yield identical results and have similar numerical characteristics in the scalar (mode III) case, the vectorial (modes I/II) situation is different: as will become apparent hereafter, the stability and precision of the two formulations are quite different. Since the combined formulation has been derived only in the homogeneous case, the comparison described in the next section is performed for the classical homogeneous problem of a simple half space subjected to a sudden tensile and shear line load.

After assessing the stability and precision of the independent formulation, we validate the use of the bimaterial spectral scheme for interfacial dynamic fracture problems by investigating the classical 2D problem of a finite non-moving interfacial crack subjected to a sudden uniform loading.

4.1. Step line load on the surface of a half space

The case of a step line load suddenly applied on a linearly elastic half space is used here to compare the stability and precision of the combined and independent spectral formulations and to investigate the optimum value of β defined in (13). This particular application is chosen not only because it has a closed-form solution, but also because it is numerically challenging due to the presence of singularities in both time and space. Two separate cases are investigated: Case A and Case B refer to a vertical and horizontal line load applied at the midpoint of the discretized domain, respectively. The stability and precision of the numerical scheme are assessed by comparing the numerical value of the velocity components of a point located at a distance L from the load with the analytical expressions given by Eringen and Suhubi (1975, pp. 614–626).

The simulations consist of a constant normal or tangential line load applied at $t = 0$ at the midpoint of a domain discretized by 2048 elements. The observation point is located 97 grid spacings from the point of application of the step line load.

We first examine the dynamic problem referred to as Case A and illustrated in the insert of Figure 3. Figure 3 shows that, for the independent formulation, a lower β is better able to capture the singularity associated with the arrival of the Rayleigh wave. However, the solution at other points in time indicate that the spurious oscillations present in the numerical solution reaches a minimum at values of β approximately equal to 0.4 and increases at larger β. In addition, the independent formulation captures the small $\dot{u}_2$ velocity resulting from the arrival of the dilation wave very well for all values of β. In contrast, when the independent and the combined formulations are compared (Figure 4) for this tensile loading case, we observe that the combined method captures more precisely the singularity for a larger β but gives comparable results for dilation wave effects.

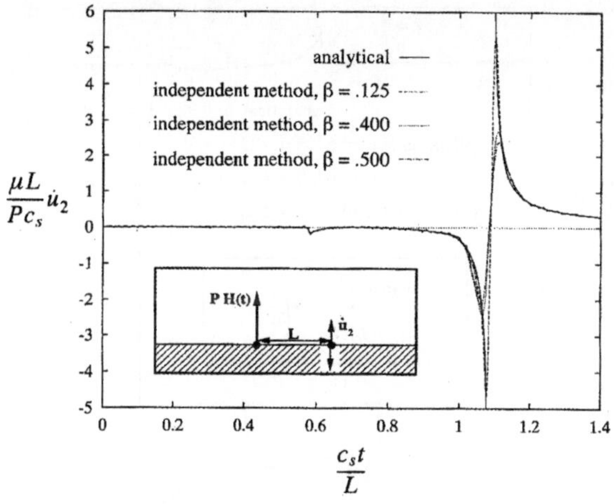

Figure 3. Normal velocity $\dot{u}_2$ computed with the independent formulation at a distance L from the point of application of a normal line load for various values of β.

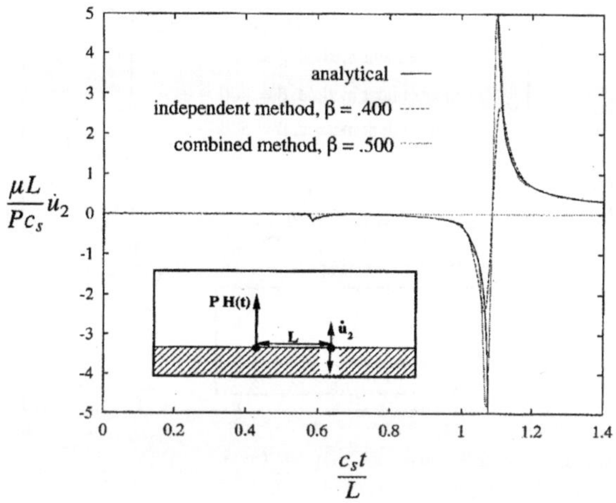

Figure 4. Comparison between the independent and combined formulations: evolution of the normal velocity $\dot{u}_2$ computed at a distance L from the point of application of a normal line load.

Unlike the combined formulation, the independent formulation does, in addition to the $\dot{u}_2$ velocity, provide the $\dot{u}_1$ velocity under this same Case A loading condition. Thus, the $\dot{u}_1$ velocity is plotted in Figure 5 which shows that the larger the time step, the more exaggerated are the effects of the dilation wave. Also note that the Rayleigh wave effects are better captured as β becomes smaller. In addition, small numerical oscillations are evident before the arrival of the dilation wave with the worst oscillations occurring for larger β.

As shown in Figure 6, the time step size has a similar effect on the stability and precision of the tangential velocity $\dot{u}_1$ obtained by the independent formulation for the case B loading. However, when the combined and independent formulations are compared (Figure 7), the combined formulation displays a large oscillatory nature that quickly leads to instability.

In conclusion, for the tensile (mode I) loading situation, the combined formulation (with $\beta = 0.5$) is the recommended scheme as long as the horizontal surface motion of the half space is not needed. However, under pure mode II or mixed-mode conditions, the combined formulation is much too unstable and requires the use of a very small time step. The in-

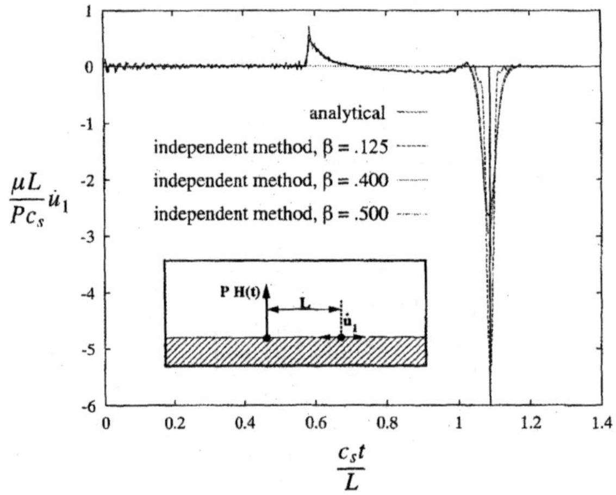

Figure 5. Evolution of the tangential velocity $\dot{u}_1$ computed with the independent formulation at a distance L from the point of application of a normal line load for various values of β.

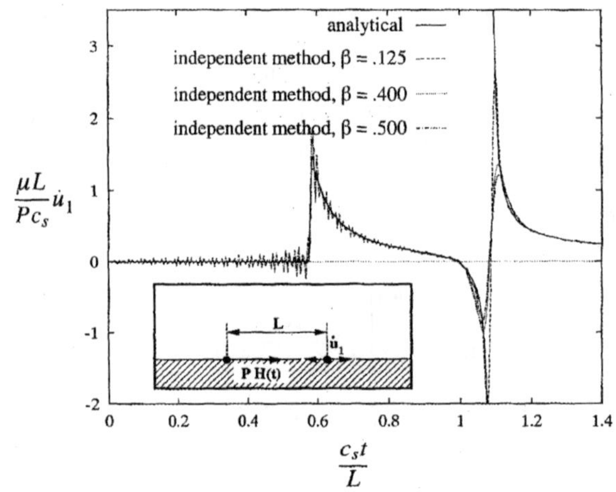

Figure 6. Same as in Figure 5, but for a suddenly applied tangential line load.

dependent formulation is therefore recommended in these cases, with a value of $\beta = 0.35$ to 0.4 which guarantees good stability, precision and efficiency. Since bimaterial problems are characterized by an intrinsic mixture of normal and tangential stresses and displacement discontinuities, the independent spectral scheme is adopted in all the simulations presented hereafter.

4.2. Nonpropagating Interface Crack under Sudden Uniform Mode I Loading

Let X denote the discretized portion of an uniformly spaced grid of 2048 elements, and let a pre-existing crack of length $2a = X/4$ be located at the center of the discretized domain. The material properties are chosen such that $\mu^+/\mu^- = 2$, $\rho^+/\rho^- = 1$ and $\nu^+ = \nu^- = 0.25$. The crack is suddenly subjected to a uniformly distributed in-plane tensile load τ_2^o and is prevented from extending by applying a very high interfacial strength τ_α^c in the uncracked region. Due to stability considerations mentioned in the previous section, the time step Δt is chosen as $0.4\Delta x/c_s^+$.

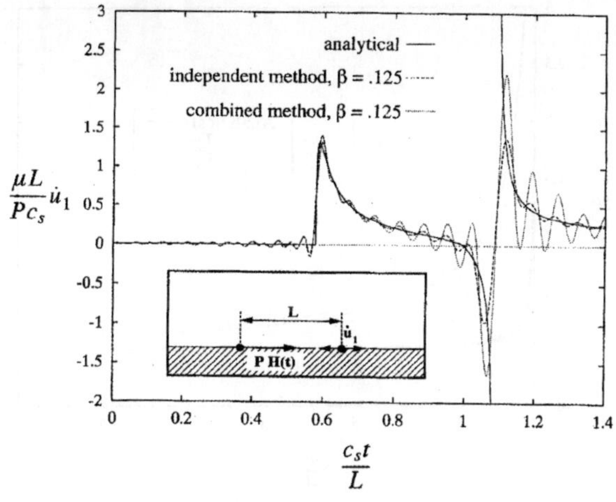

Figure 7. Comparison between the independent and combined formulations: evolution of the horizontal velocity $\dot{u}_1$ computed at a distance L from the point of application of a tangential line load.

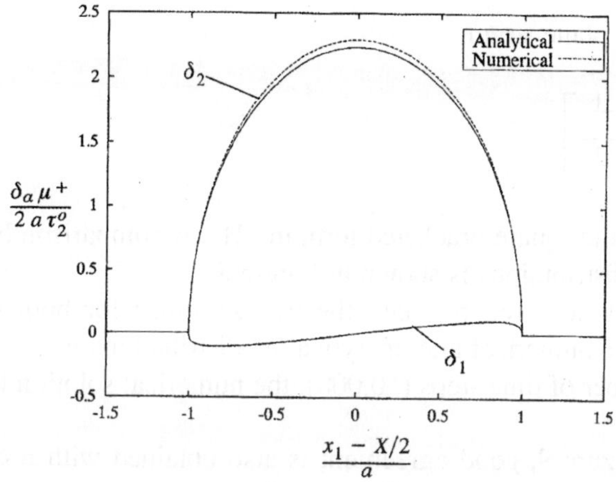

Figure 8. Long term (static) solution of the crack opening displacements δ_α for a center interfacial crack of length $2a$ subjected to a sudden uniform tensile loading τ_2^o.

The analytical solution of this dynamic problem is not known, except in its long-time (static) limit obtained by England (1965) who derived the crack opening displacements (COD) in the region $-a \leqslant x_1 \leqslant a$ as

$$\delta_2^{\text{static}}(x_1) = \frac{\tau_2^o\sqrt{\varphi}}{2(1+\varphi)}\left\{\frac{1+\kappa^+}{\mu^+} + \frac{1+\kappa^-}{\mu^-}\right\}\sqrt{a^2 - x_1^2}\cos\left(\epsilon \ln\left|\frac{x_1+a}{x_1-a}\right|\right),$$

$$\delta_1^{\text{static}}(x_1) = -\frac{\tau_2^o}{2(1+\varphi)}\left(\frac{1+\kappa^+}{\mu^+} + \frac{1+\kappa^-}{\mu^-}\right)\sqrt{\varphi(a^2 - x_1^2)}\sin\left(\epsilon \ln\left|\frac{x_1+a}{x_1-a}\right|\right),$$
$$-\frac{\tau_2^o}{2(1+\varphi)}x_1(1+\kappa^+ - \varphi(1+\kappa^-)),$$

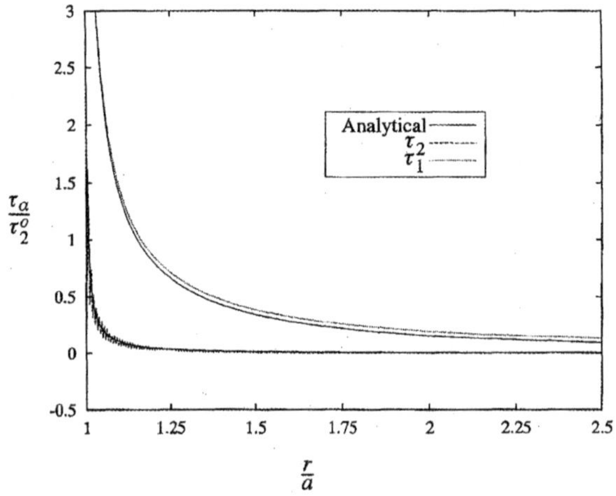

Figure 9. The long-time (static) traction stresses $\tau_\alpha^{\text{static}}$ along the interface for a center interfacial crack of length $2a$ subjected to a sudden uniform tensile loading τ_2^o.

where $\kappa = 3 - 4\nu$ for plane strain,

$$\epsilon = \frac{1}{2\pi} \ln \left[\frac{\frac{\kappa^+}{\mu^+} + \frac{1}{\mu^-}}{\frac{\kappa^-}{\mu^-} + \frac{1}{\mu^+}} \right], \tag{21}$$

and φ corresponds to the square bracketed term in (21). A comparison between the analytical and numerical COD distributions is shown in Figure 8.

Good agreement is achieved between the two solutions for both δ_1 and δ_2. The small difference between the numerical and analytical δ_2-distributions is attributed to the fact that, despite the large number of time steps (50,000), the numerical solution has not yet reached its final (static) state.

As indicated in Figure 9, good agreement is also obtained with regards to the long-time traction stresses τ_α along the interface. England (1965) gives the (static) expressions as

$$\tau_2^{\text{static}} = -\tau_2^o \left[1 \mp \frac{1}{\sqrt{x_1^2 - a^2}} \left\{ x_1 \cos\left(\left| \frac{x_1 + a}{x_1 - a} \right| \right) + 2a\epsilon \sin\left(\epsilon \ln \left| \frac{x_1 + a}{x_1 - a} \right| \right) \right\} \right],$$

$$\tau_1^{\text{static}} = \mp \frac{\tau_2^o}{\sqrt{x_1^2 - a^2}} \left[x_1 \sin\left(\epsilon \ln \left| \frac{x_1 + a}{x_1 - a} \right| \right) - 2a\epsilon \cos\left(\epsilon \ln \left| \frac{x_1 + a}{x_1 - a} \right| \right) \right],$$

where the upper and lower signs hold for $x_1 > a$ and $x_1 < a$, respectively. Except in the immediate vicinity of the crack tip (within a distance of a few spacings Δx) where the numerical solution is, as expected, unable to capture the stress singularity, and far from the crack tip where the presence of the neighboring cracks associated with the Fourier series representation of the displacement and traction distributions is felt (Geubelle, 1996), the spectral scheme provides a fairly accurate description of the traction stresses along and interface. In particular, it clearly captures the stress concentration appearing in the vicinity of the crack tips, which is essential when modeling the spontaneous motion of cracks and faults.

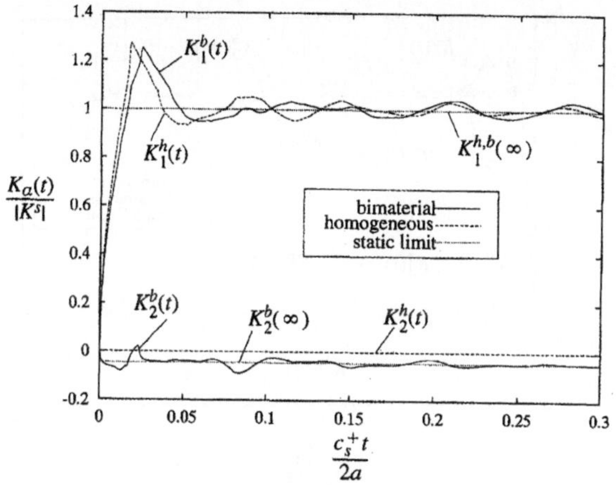

Figure 10. The evolution of the dynamic stress intensity factors $K_\alpha(t)$, normalized by the norm of the static limit $|K^s| = \sqrt{K_1^{s^2} + K_2^{s^2}}$, for the homogeneous (dashed curves, superscript h) and bimaterial (solid curves, superscript b) case of a pure tensile loading ($\tau_1^o = 0$).

5. Dynamic loading of a nonpropagating interface crack

5.1. 2D SIMULATIONS

By accurately capturing the stress and displacements distributions in the vicinity of the crack tip, the spectral scheme can be used to extract the dynamic stress intensity factors (SIF) $K_\alpha(t)$ characterizing the singular near-tip fields. As shown by Morrissey and Geubelle (1997), the SIF can be readily obtained from the displacement discontinuities δ_α immediately behind the crack tip. In the bimaterial situation, the relation between K_α and δ_α is (Rice, 1988)

$$L^{-i\epsilon}[K_1(t) + iK_2(t)] = \lim_{r\to 0} \frac{2\sqrt{2\pi}(1+2i\epsilon)\cosh(\pi\epsilon)(\delta_2 - i\delta_1)(r/L)^{-i\epsilon}}{(w^+ + w^-)\sqrt{r}},$$

where r is the distance measured from the crack tip, $w = 4(1-\nu)/\mu$, L is a characteristic length (chosen hereafter as the half-length a of the crack), and ϵ is the oscillatory index defined in (21).

As an illustration, let us reconsider the problem of a center interfacial crack of length $2a$ subjected to a sudden uniform mixed–mode loading (τ_1^o, τ_2^o). The expression of the static SIF is (Rice, 1988)

$$a^{i\epsilon}(K_1^s + iK_2^s) = a^{i\epsilon}(K_1(\infty) + iK_2(\infty)) = (\tau_2^o + i\tau_1^o)(1+2i\epsilon)(1/2)^{i\epsilon}\sqrt{\pi a}.$$

The evolution of the dynamic stress intensity factors $K_\alpha(t)$, normalized by the norm of their static limit $|K^s| = \sqrt{K_1^{s^2} + K_2^{s^2}}$ is presented in Figures 10 and 11 for both the homogeneous (dashed curves) and bimaterial (solid curves) cases. Figure 10 corresponds to a purely tensile loading ($\tau_1^o = 0$) while Figure 11 presents a mixed-mode situation ($\tau_1^o = 4\tau_2^o$). The maximum overshoot is always detected in the homogeneous case and is associated with the simultaneous arrival of the Rayleigh waves originating from the other crack tip and traveling along the top and bottom crack surfaces. In the bimaterial situation, the elastic waves travel at different

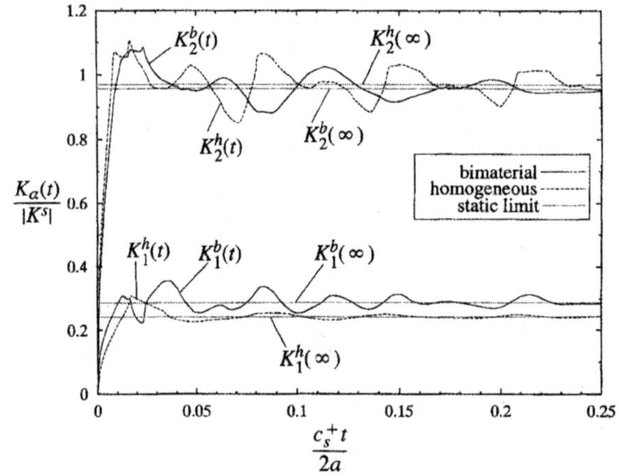

Figure 11. Same as in Figure 10 for a mixed-mode loading of the crack ($\tau_1^o = 4\tau_2^o$).

velocities along the top and bottom sides of the interface, and the resulting interface motion 'damps out' the dynamic overshoot.

5.2. 3D SIMULATIONS

To illustrate the capabilities of the 3D spectral scheme, we now investigate the 3D problem of a planar interfacial elliptical crack subjected to a uniform step loading of amplitude $\bar{\tau}$

$$\tau_1^o(x_1, x_3, t) = \bar{\tau} \sin \tfrac{1}{4}\pi H(t); \quad \tau_2^o(x_1, x_3, t) = \bar{\tau} \cos \tfrac{1}{4}\pi H(t); \quad \tau_3^o(x_1, x_3, t) = 0,$$

where $H(t)$ denotes the Heaviside step function. The major axis of the crack is aligned with the x_1-axis and its length is $3X/8$. The aspect ratio of the elliptical crack is 3/2. The portion $X * Z$ of the fracture plane is discretized with $512 * 512$ grid points and the time step is given by $\beta = 0.35$. The material mismatch parameters defined in (17) are $\xi = \sqrt{2}$ and $\zeta = 2$, and the Poisson's ratios of both materials are equal to 0.25. The homogeneous case is also shown for comparison purpose. Figures 12 and 13 illustrate the evolution of the deformed crack shape for the homogeneous and bimaterial cases, respectively. Figures 12a and 13a present 'snapshots' of the fracture problem during the early stages of the transient loading, showning the propagation of Rayleigh waves along the top and bottom crack surfaces. Unaware of its finite size, the center part of the crack moves uniformly at a constant velocity both horizontally and vertically, as illustrated by the flat center portion of the deformed shape. As apparent in Figures 13a,b, the waves travel faster along the top surface than along the bottom one: while the Rayleigh wave emanating from the sides are still converging toward the center of the bottom crack surface (Figure 13b), those propagating along the minor axis of the ellipse are about to cross in the middle of the top fracture surface (Figure 13a). The final shape of the homogeneous and interfacial cracks are shown in Figures 12b and 13c, clearly reflecting the mode mixity of the applied load and the asymmetry of the bimaterial problem.

6. Frictional contact instability

While it is generally accepted that frictional instability and stick/slip phenomena can be generated with rate-dependent Coulomb friction models (Gu et al., 1984), Adams (1995)

(a)

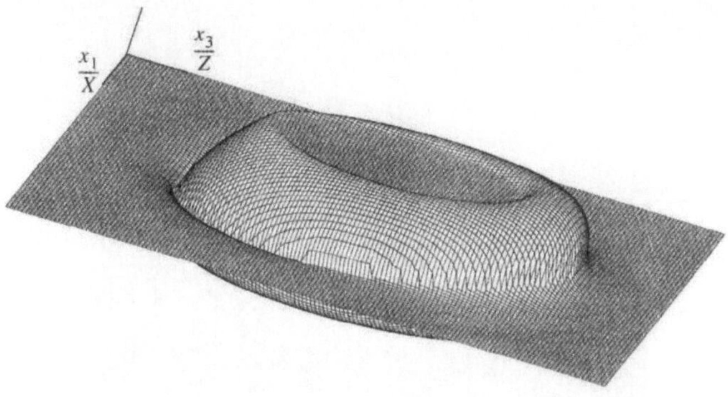

(b)

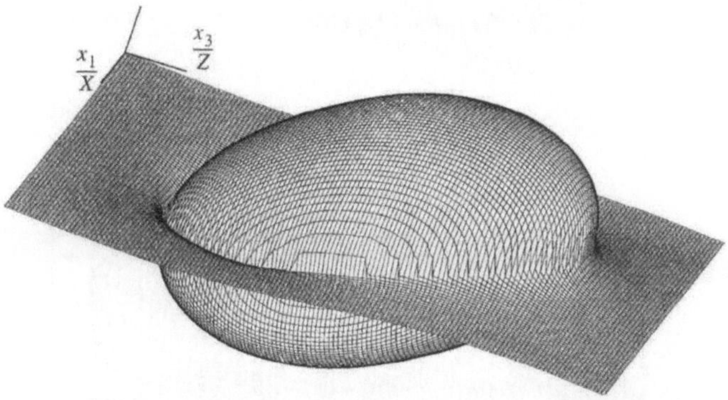

Figure 12. Sudden mixed-mode loading of a stationary elliptical crack embedded in a homogeneous material. 'Snapshots' of the deformed shape of the crack at the early *(a)* and final *(b)* stages of the transient problem.

recently showed that such instability can also occur in some bimaterial cases even when the friction coefficient is assumed to be constant. This unexpected result was obtained through a perturbation analysis of the elastodynamic response of two half spaces sliding at a constant average velocity with respect to each other. In this section, we investigate whether the proposed numerical scheme is able to capture this frictional instability.

Let p denote the normal compressive stress applied uniformly on the interface plane. In order to generate a uniform relative sliding of the two half spaces, a uniform tangential traction

$$\tau_1^o = \alpha f_f p \tag{22}$$

is also applied uniformly, where f_f denotes the constant coefficient of friction while α is an arbitrarily chosen positive constant determining the intensity of the relative sliding velocity

$$\frac{v_o}{c_s^+} = \frac{1}{\mu^+}\left(1 + \frac{\zeta}{\xi}\right)\langle \alpha - 1 \rangle f_f p,$$

where ζ and ξ have been defined in (17) and $\langle x \rangle = x$ if $x > 0$ and $= 0$ otherwise.

To investigate the stability of the sliding process, we introduce, at the center of the domain, a perturbation in the form of a step-like pulse of width $X/32$, duration $4\Delta t$ and amplitude $4\alpha f_f p$ in applied tangential traction, in addition to the uniform tangential traction given by (22). The material mismatch between the two half planes is set at $\mu^+/\mu^- = c_s^+/c_s^- = 2$.

(a)

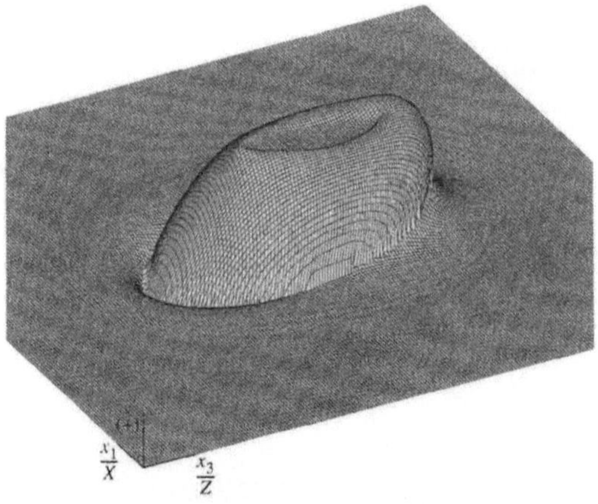

(b)

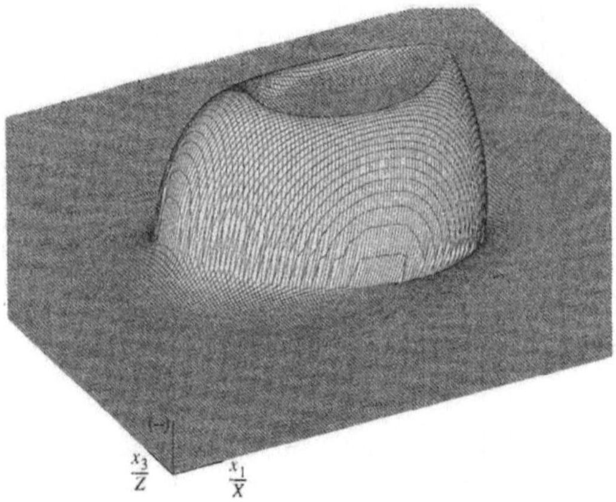

(c)

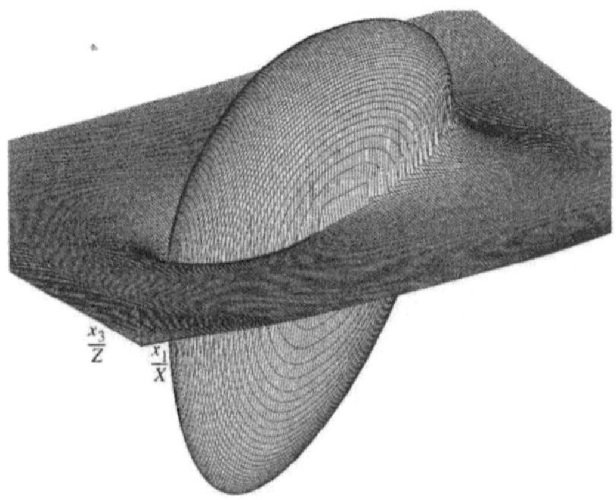

Figure 13. Same problem as in Figure 12 but for the bimaterial case (with $\mu^+/\mu^- = 2, \rho^+ = \rho^-$ and $\nu^+ = \nu^- = 0.25$). Figure 13a–b show the details of the top and bottom surfaces of the deformed crack at an early stage of the simulation, and Figure 13c presents the final shape of the interfacial crack.

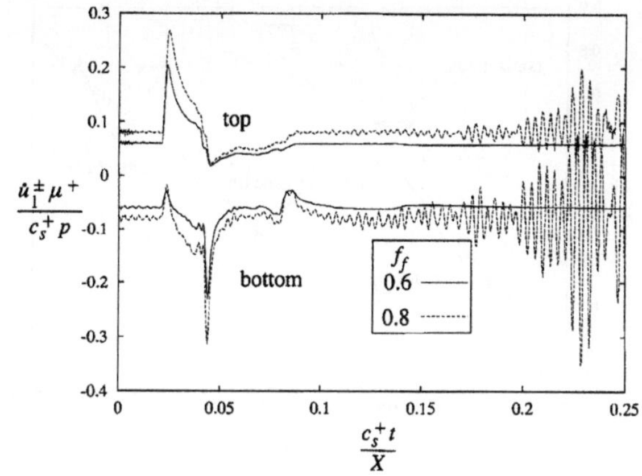

Figure 14. Evolution of the horizontal velocity computed at a point located along the interface between two elastic half spaces sliding at a constant average velocity with respect to each other, showing the effect of the friction coefficient f_f on the onset of instability.

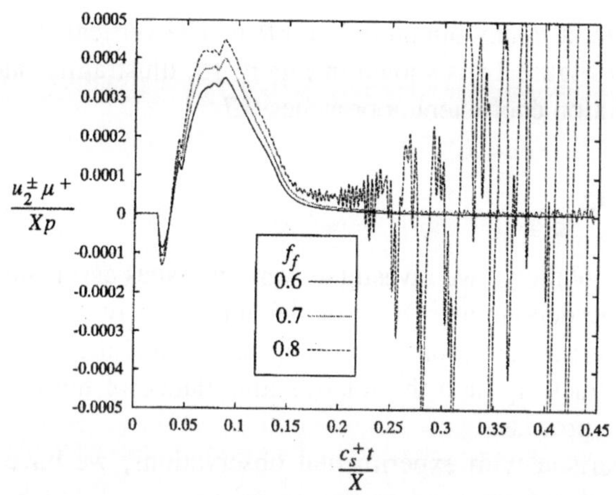

Figure 15. Frictional contact instability: Evolution of the vertical displacement of the contact interface for the perturbation problem presented in Figure 14, showing the coupling of normal and tangential motions characteristics of bimaterial problems.

The two half spaces have the same Poisson's ratio ($\nu^+ = \nu^- = 0.25$). The domain X is discretized by 1024 grid spacings, the time step defined in (13) corresponds to $\beta = 0.38$, and the parameter α introduced in (22) is set al 1.1. Under these conditions, Adams' analysis predicts the onset of instability to occur for values of the friction coefficient f_f in the vicinity of 0.7.

Figure 14 presents the evolution of the tangential velocities $\dot{u}_1^{\pm}$ of the top and bottom surfaces computed at a point located on the right half of the domain, 53 Δx away from the center of the domain. The arrival of the wave associated with the perturbation is clearly visible for both the top and bottom materials. But, while the perturbation eventually dies out and the two half spaces return to their uniform sliding motion for the smaller value of the friction coefficient ($f_f = 0.6$), instability is generated for the larger value of f_f (represented by dashed curves in Figure 14), in accordance with Adams' prediction. The origin of the instability is undoubtedly associated with the coupling between tangential and normal motions of the

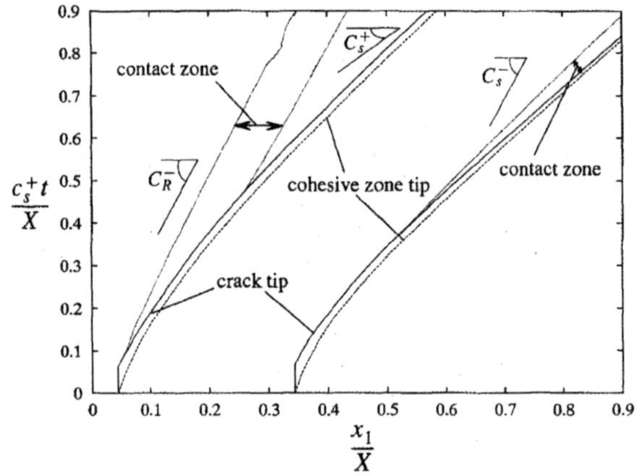

Figure 16. Spontaneous rapid debonding of a Aluminum/Homalite interface under mixed-mode loading. Location of the cohesive and contact zones for the case $\psi = 75°$ and $\psi = -75°$ with $f_f = 0.25$. For clarity purposes, the three curves associated with the $\psi = -75°$ case have been shifted to the right by 0.3.

interface, which is inherent in any bimaterial problem. The vertical motion of the interface at the same location as in Figure 14 is shown in Figure 15, illustrating once again the onset of instability when the friction coefficient approaches 0.7.

7. Dynamic debonding

In the final series of simulations, we investigate various issues associated with the mechanics of spontaneous dynamic debonding. Of particular interest here are the speed of the rapidly propagating interfacial crack, the spontaneous appearance and the behavior of a contact zone behind the advancing crack tip, and the relative importance of the frictional contact on the energetics of the failure process.

To allow for comparison with experimental observations, we have adopted the Homalite/Aluminum bimaterial system used in the experiments reported by Shukla and his co-workers (Singh and Shukla, 1996; Singh et al., 1997). The Young's modulus, Poisson's ratio and density of the top material (Aluminum) are $E^+ = 71\,\text{GPa}.\nu^+ = 0.33$ and $\rho^+ = 2770\,\text{kg/m}^3$, while the properties of the bottom material (Homalite) are $E^- = 5.3\,\text{GPa}.\nu^- = 0.35$ and $\rho^- = 1230\,\text{kg/m}^3$. The corresponding shear and Rayleigh wave speeds are $c_s^+ = 3100\,\text{m/s}$, $c_s^- = 1263\,\text{m/s}$, $c_R^+ = 2890\,\text{m/s}$, and $c_R^- = 1190\,\text{m/s}$, respectively. In the 2D simulations presented in this last section, a domain $X = 1\,\text{m}$ is discretized by 2048 equally spaced grid points and subjected to uniformly distributed normal and shear tractions

$$\tau_1^o(x_1, t) = \bar{\tau} \sin\psi, \qquad \tau_2^o(x_1, t) = \bar{\tau} \cos\psi, \tag{23}$$

with $\bar{\tau}$ chosen as 3.0 MPa. The cohesive failure model for the interface is given by (16) with $\tau_n^c = \tau_s^c = 5.0\,\text{MPa}$ and $\delta_n^c = \delta_s^c = 0.02\,\text{mm}$. The corresponding mode I and II fracture toughnesses are thus $G_{\text{Ic}} = G_{\text{IIc}} = 50\,\text{Pa m}$.

At time $t = 0$, a short (4.3 cm long) crack is introduced at the left edge of the discretized domain and, due to the stress concentration building up in the vicinity of the crack tips, starts to propagate to the right. Propagation in the left direction is prevented by introducing a region a very high fracture toughness immediately adjacent to the left crack tip. The time step is defined

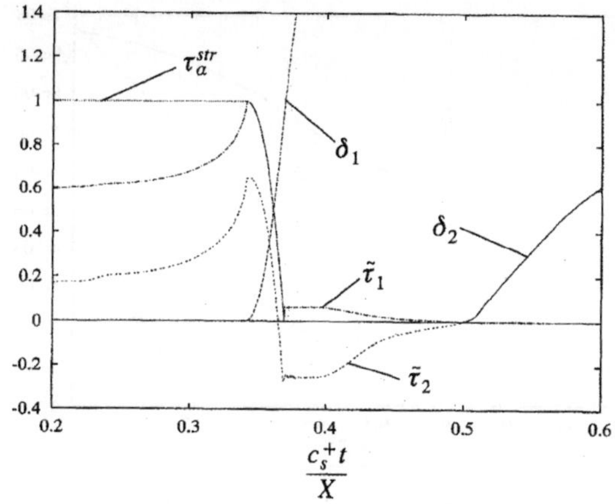

Figure 17. Evolution of the crack opening displacements δ_α (normalized by their respective critical value δ_α^c), of the tensile and shear strengths τ_α^{str} (normalized by their initial 'intact' value τ_α^c) and of the interface traction stresses $\tilde{\tau}_\alpha$ (normalized by τ_1^c) for a point located on the path of the advancing crack tip shown in Figure 16. The load conditions correspond to the case $\psi = 75°$.

by (13) with $\beta = 0.4$, and various values of the mode-mixity parameter ψ ranging from pure tensile to pure shear loading are considered. The value of the applied load and the critical values of the COD ensure a rapid accelaration of the crack tips while maintaining a sufficient discretization (at least 15 grid spacings) of the cohesive and contact zones throughout the simulation. Various values of the Coulomb friction coefficient f_f ranging from 0 to 0.5 are used to characterize the frictional contact between the crack faces which occurs under shear-dominated situations.

The evolution of the position of the cohesive zone tip (i.e., the right-most point on the fracture plane with a nonvanishing value of the COD), of the crack tip (i.e., the right-most point for which the interfacial strength vanishes) and of the contact zone limits is presented in Figure 16 for the case $\psi = \pm 75°$ and $f_f = 0.25$. As expected, in the initial stages of the simulation, the cohesive zone expands while the crack tip remains stationary while the crack progressively opens. When COD has been sufficiently accumulated, the crack tip (solid curves) quickly accelerates and 'catches-up' with the tip of the cohesive zone (dashed curves). After a little while, a contact zone (dotted curves) developes behind the crack tip.

Although the uniform and constant applied loading (23) does not capture the complexity of the loading conditions present in the experiments, we note a series of interesting similarities between the experimental observations reported by Singh et al. (1997) and the numerical results associated with the case $\psi = -75°$. Note that a negative shear component is needed in the experiments to maintain the crack along the interface: a positive applied shear (corresponding to a positive value of ψ) would make the crack kink into the Homalite. The first similarity between experimental and numerical results is the transonic speed of the crack tip and of the trailing edge of the contact zone. For the particular set of loading conditions and the interface properties, the crack tip rapidly moved in the transonic regime and approached the Rayleigh wave speed of Aluminum. Another similarity can be found in the appearance of the contact zone: just like in the experiments, no contact zone is detected in the subsonic regime, but it appears spontaneously only when the crack speed has exceeded the shear wave speed of Homalite. No steady-state solution is achieved in this particular simulations as the contact zone size increases monotonically.

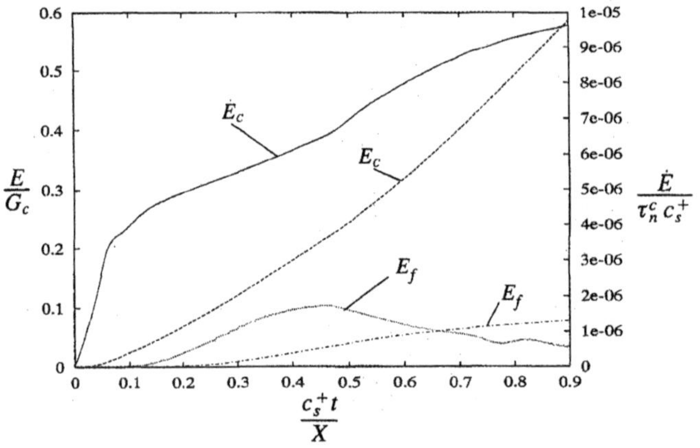

Figure 18. Evolution of the energy (E) and energy rate ($\dot{E}$) involved in the cohesive failure process (subscript 'c') and in the frictional contact (subscript 'f') for the spontaneous crack propagation shown in Figure 16 (with $\psi = -75°$), showing the relative importance of the friction in the crack energetics. The energy is normalized by $G_c = \tau_n^c \delta_n^c / 2$.

It is interesting to note that, while a similar 'transonic' solution is found for the cohesive zone and crack tips in the case $\psi = 75°$, the behaviour of the contact zone is radically different. After trailing the crack tip for a while, the contact zone eventually detaches from the transonically propagating crack tip and maintains a quasi-constant size. Its edges then propagate at the Rayleigh wave speed of Homalite. As indicated above, the experimental confirmation of this unexpected result is hard to achieve, unless the strength of the interface is sufficiently reduced to prevent crack kinking.

Details on the failure process at a point located along the crack path are presented in Figure 17, which shows the evolution of the normal and tangential crack opening displacements δ_α, of the tensile and shear strengths τ_α^{str}, and of the tensile and shear traction stresses $\tilde{\tau}_\alpha$ for a point located at $x_1 \simeq 0.2$ m. The loading conditions are the same as those used in the previous figure. For this particular case, the failure process, which corresponds to the time interval during which the strength progressively decreases from its original value to zero, takes place solely in shear: the crack actually opens only after the passage of the contact zone (i.e., for $c_s^+ t / X \gtrsim 0.5$).

The presence of frictional contact is expected to affect the energetics of crack propagation. The relative importance of the energy and energy rates dissipated in the cohesive failure and frictional processes associated with the right crack tip is shown in Figure 18. The initial expansion of the cohesive zone and the rapid acceleration of the crack can be detected from the evolution of the energy rate $\dot{E}_c$ associated with the failure process and defined by

$$\dot{E}_c(t) = \int_{\text{coh.zone}(t)} \tau_\alpha^{str}(x_1, t) \delta \dot{u}_\alpha(x_1, t) \, dx_1.$$

The energy rate $\dot{E}_f$ dissipated in the frictional contact, given by

$$\dot{E}_f(t) = \int_{\text{contact zone}(t)} f_f |\tilde{\tau}_2(x_1, t)| |\delta \dot{u}_1(x_1, t) \, dx_1,$$

where $\tilde{\tau}_2$ has been defined in (20). As apparent in Figure 18, the energy dissipated in the frictional contact remains, for this particular choice of loading conditions, interface properties

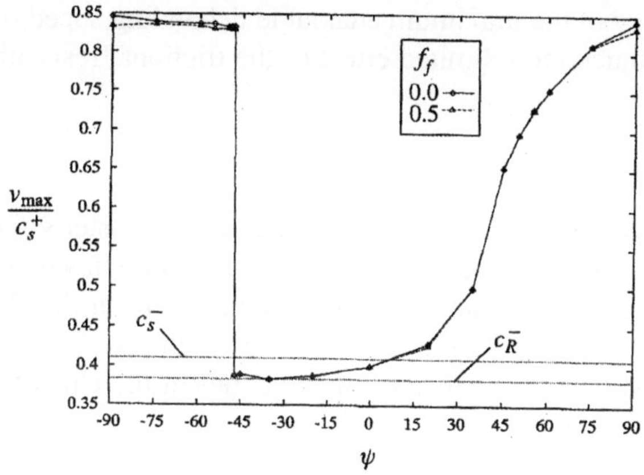

Figure 19. Effect of the coefficient of friction f_f and of the mode mixity parameter ψ on the maximum speed obtained for the right crack tip in spontaneous debonding simulations such as that presented in Figure 16, showing the appearance of transonic crack propagation under shear-dominated loading conditions.

and friction coefficient, a relatively small portion of the energy dissipated in the creation of new fracture surfaces. This relatively limited influence of the frictional contact on the crack propagation behavior is confirmed in Figure 19, which presents the variation of the maximum crack tip speed as a function of the mode-mixity parameter ψ introduced in (23) for two values of the friction coefficient $f_f (f_f = 0$ and $0.5)$. The values of the Rayleigh and shear wave speeds for the lower material (Homalite) are also indicated for reference purpose by horizontal dotted lines. Two interesting results can be emphasized: firstly, for the chosen cohesive failure model, transonic debonding seems to take place mostly under shear-dominated conditions. The maximum crack speed remains subsonic (i.e., close to c_R^-) for low absolute values of ψ, and exceeds both c_R^- and c_s^- as the external loading includes more and more shear. There also seems to be a sharp transition between transonic and subsonic crack propagation, especially in the negative shear loading regime. Secondly, as indicated above, the effect of f_f on the maximum debonding speed is relatively minor, and, as expected, appears only in the shear-dominated situations. Further studies are currently underway to better characterize the effect of frictional contact on the dynamic debonding process and their results will be reported in a subsequent paper.

8. Conclusion

A spectral scheme has been developed for 2D and 3D dynamic fracture problems involving planar interfacial cracks of arbitrary shapes and subjected to arbitrary space- and time-dependent loading conditions. The numerical scheme is based on an exact spectral representation of the elastodynamic integral relations describing the response of the two adjacent half spaces. A wide variety of cohesive failure models and friction laws can be incorporated in the algorithm to capture the dynamic phenomena associated with the spontaneous propagation of interfacial cracks and faults. Various problems have been examined in this paper, including the complex transient response of dynamically loaded stationary cracks, and the instability arising from the sliding motion of two elastic half spaces under frictional contact. Preliminary simulations of spontaneous debonding along Aluminum/Homalite interfaces have shown some important similarities with experimental observations of transonic crack propagation. They also indicated a strong dependency of the maximum attainable debonding speed on the mode-mixity of the applied load and a much more limited effect of the frictional resistance of the interface.

Acknowledgements

This paper has been written as part of M.S. Breitenfeld's Master's thesis work, supported partially by a grant from the Campus Research Board of the University of Illinois. Most of the simulations presented in this paper have been performed on the Power Challange array and the Origin 2000 supercomputers available at the National Center for Supercomputing Applications, located on the Urbana–Champaign campus of the University of Illinois.

Appendix A

As indicated in Section 2.1, the convolution kernels $H_{11}(T)$, $H_{12}(T)$ and $H_{22}(T)$ shown in Figure 2 can be computed numerically from their Laplace transform expressions (7). In this appendix, we summarize the steps leading to the closed form expression of $H_{11}(T)$ and $H_{22}(T)$. After some algebraic manipulations, we can rewrite $\widehat{H}_{11}(s)$ as the sum of four simpler functions

$$\widehat{H}_{11}(s) = \sum_{i=1}^{4} \hat{f}_i(s), \tag{A.1}$$

where

$$\hat{f}_1(s) = \frac{s^2}{(s+\sqrt{1+s^2})(s^2+1+\eta^2)}, \qquad \hat{f}_2(s) = \frac{\eta^2\sqrt{1+s^2}}{s^2+1+\eta^2},$$

$$\hat{f}_3(s) = \frac{\eta\sqrt{s^2+\eta^2}}{s^2+1+\eta^2}, \qquad \hat{f}_4(s) = -\frac{s(1+\eta^2)}{s^2+1+\eta^2}. \tag{A.2}$$

Using fundamental Laplace transform relations and the following transform property,

$$\text{if } \hat{g}(s) = \mathfrak{L}[g(T)] \text{ and } r = \sqrt{s^2+a^2},$$

$$\text{then } \mathfrak{L}^{-1}[\hat{g}(r)] = g(T) - a\int_0^T g[\sqrt{T^2-u^2}]J_1(au)\,du,$$

with $J_1(x)$ denoting the Bessel function, the four functions $\hat{f}_i(s)$ are inverted as

$$f_1(T) = \frac{J_1(T)}{T} - \sqrt{1+\eta^2}\int_0^T \frac{J_1(u)}{u}\sin(\sqrt{1+\eta^2}(T-u))\,du,$$

$$f_2(T) = \eta^2\cos\eta t - \int_0^T \eta^2\cos[\eta\sqrt{T^2-u^2}]J_1(u)\,du, \tag{A.3}$$

$$f_3(T) = \eta\cos t - \eta^2\int_0^T \cos[\eta\sqrt{T^2-u^2}]J_1(\eta u)\,du,$$

$$f_4(T) = -(1+\eta^2)\cos[\sqrt{1+\eta^2}T].$$

These terms can be further simplified by successive integration by parts

$$
\begin{aligned}
H_{11}(T) &= -\left(\tfrac{1}{2}+\eta^2\right)\cos[\sqrt{1+\eta^2}T] - \int_0^T \cos[\sqrt{1+\eta^2}(T-u)]\frac{J_2(u)}{u}\,\mathrm{d}u \\
&\quad +\eta^2 J_o(T) + \eta J_o(\eta t) - \eta^3 \int_0^T \sin[\eta u] J_o[\sqrt{T^2-u^2}]\,\mathrm{d}u \\
&\quad -\eta \int_0^T \sin u\, J_o[\eta\sqrt{T^2-u^2}]\,\mathrm{d}u. \qquad \text{(A.4)}
\end{aligned}
$$

Similarly, the convolution kernel $H_{22}(T)$ is shown to be

$$
\begin{aligned}
H_{22}(T) &= -\eta\left(1+\frac{\eta^2}{2}\right)\cos[\sqrt{1+\eta^2}T] - \eta^3 \int_0^T \cos[\sqrt{1+\eta^2}(T-u)]\frac{J_2(\eta u)}{u}\,\mathrm{d}u \\
&\quad +\eta^2 J_o(T) + \eta J_o(\eta t) - \eta^3 \int_0^T \sin[\eta u] J_o[\sqrt{T^2-u^2}]\,\mathrm{d}u \\
&\quad -\eta \int_0^T \sin u\, J_o[\eta\sqrt{T^2-u^2}]\,\mathrm{d}u. \qquad \text{(A.5)}
\end{aligned}
$$

The convolution integrals in (A.4) and (A.5) have to be computed numerically.

References

Adams, G.G. (1995). Self-excited oscillations of two elastic half-spaces sliding with a constant coefficient of friction. *Journal of Applied Mechanics* **62**(4), 867–872.

England, A.H. (1965). A crack between dissimilar media. *Journal of Applied Mechanics* **32**, 400–402.

Eringen, A.C. and Suhubi, E.S. (1975). *Elastodynamics: Vol. II – Linear Theory*. Academic Press. New York.

Geubelle, P.H. (1997). A numerical method for elastic and viscoelastic dynamic fracture problems in homogeneous and bimaterial systems. *Computational Mechanics* **20**(1/2), 20–25.

Geubelle, P.H. and Breitenfeld, M.S. (1997). Numerical analysis of dynamic debonding under anti-plane shear loading. *International Journal of Fracture* **85**, 265–282.

Geubelle, P.H. and Rice, J.R. (1995). A spectral method for 3D elastodynamic fracture problems. *Journal of Mechanics and Physics of Solids* **43**, 1791–1824.

Gu, J.C., Rice, J.R., Ruina, A.L. and Tse, S.T. (1984). Slip motion and stability of a single degree of freedom elastic system with rate and state dependent friction. *Journal of Mechanics and Physics of Solids* **32**(3), 167–196.

Lambros, J.M. and Rosakis, A.L. (1995). Shear dominated transonic interface crack growth in a bimaterial – I. Experimental observations. *Journal of Mechanics and Physics of Solids* **43**(2), 169–188.

Morrissey, J.W. and Geubelle, P.H. (1997). A numerical scheme for mode III dynamic fracture problems. *International Journal of Numerical Methods in Engineering* **40**, 1181–1196.

Rice, J.R. (1988). Elastic fracture mechanics concepts for interfacial cracks. *Journal of Applied Mechanics* **55**, 98–103.

Singh, R.P. and Shukla, A. (1996). Subsonic and intersonic crack growth along a bimaterial interface. *Journal of Applied Mechanics* **63**(4), 919–924.

Singh, R.P., Lambros, J., Shukla, A. and Rosakis, A.J. (1997). Investigation of the mechanics of intersonic crack propagation along a bimaterial interface using coherent gradient sensing and photoelasticity. *Proceedings of the Royal Society of London* **A453**, 2649–2667.

Williams, M.L. (1959). The stress around a fault or crack in dissimilar media. *Bulletin of the Seismological Society of America* **49**, 199–204.

Yang, W., Suo, Z. and Shih, C.F. (1991). Mechanics of dynamic debonding. *Proceedings of the Royal Society of London*, **A433**, 679–697.

International Journal of Fracture **93:** 39–50, 1998.

Dynamic ductile fracture of aluminum SEN specimens an experimental-numerical analysis

JONGHEE LEE[1], MATTHEW T. KOKALY[2] and ALBERT S. KOBAYASHI[2]
[1]*Lt. Colonel, Republic of Korea Air Force, Headquarters, Chungnam 320-910, South Korea*
[2]*University of Washington,Department of Mechanical Engineering, Seattle, Washington 98195-2600, U.S.A.*
e-mail: ask@u.washington.edu

Received 5 September 1997; accepted in revised form 3 September 1998

Abstract. Hybrid experimental-numerical and experimental analyses were used to explore possible dynamic ductile fracture parameters associated with rapid crack propagation in 7075-T6 and 2024-T3 aluminum alloy, single edge notched (SEN) specimens of 1.6 mm thickness. Dynamic Moiré interferometry was used to record the displacement field, which was used either to drive a dynamic elasto-plastic finite element (FE) model of the fracturing SEN specimen or by itself, to determine the crack-tip J-integral, the T_{ε}^{*}-integral and the crack tip opening angle (CTOA). The near-field J vanished but the near-field T_{ε}^{*} reached a constant value with crack propagation. The CTOA associated with a low crack velocity also remained constant during crack propagation but fluctuated at higher crack velocity. The results of this preliminary study suggest that either the T_{ε}^{*} or the CTOA criteria proposed for stable crack growth could be a suitable parameter for characterizing rapid crack propagation in these thin aluminum specimens.

Key words: Dynamic ductile fracture, aluminum SEN specimen, Moiré interferometry, dynamic finite element analysis, J-integral, T_{ε}^{*}-integral, crack tip opening angle (CTOA).

1. Introduction

While dynamic fracture in the presence of small scale yielding has been amply characterized experimentally, by the dynamic fracture toughness K_{ID}, no comparable fracture parameter has been identified for dynamic fracture in the presence of large scale yielding. The J-integral, which is widely and justifiably used to identify the onset of ductile fracture, loses its path independency with crack extension due to the large scale unloading in the trailing wake of the propagating crack. Cottrell and Atkins (1996) have shown that the far field J-resistance curve represents an increase screening of the energy release rate by the plastic zone and thus is not a fundamental ductile fracture parameter. In addition, the near-field J-integral has been shown to vanish after reaching a peak value at the initiation of crack extension and thus cannot be used as a fracture parameter. The resultant inadequacy of the ASTM procedure (ASTM, 1987) for determining the J resistance curve for stable crack growth with a trailing unloaded region has been documented by Brocks and Yuan (1989) and May and Kobayashi (1995).

2. Ductile fracture criteria and dynamic fracture

2.1. T_ε^* INTEGRAL CRITERION

For a stably growing crack, Brust et al. (1985) showed that the T^* integral, which is based on the incremental theory of plasticity, reaches a steady state value during stable crack growth and could be an effective stable crack growth parameter. This integral incorporates the unloading effect in the region trailing the crack tip by stretching the integration contour with the extending crack. For an integration contour close to the crack, T^* represents the energy inflow to the wake as well as to the frontal crack tip region (Okada and Atluri, 1997). In contrast, the near-field J-integral represents the energy inflow to a closed contour which moves with the crack tip and does not account for the energy dissipated in the wake of the crack tip (Cotterell and Atkins, 1996). *Like the J-integral, the T^*-integral is a path dependent integral for an extending ductile crack and vanishes when the integration contour is collapsed to the crack line.* Nevertheless, the J- or the T^*-integral must be evaluated along a contour Γ close to the crack tip[1] if either integral is to be considered a fracture parameter. This near-field T^* is designated as the T_ε^*-integral.

Following the definition of Stonesifer and Atluri (1982), T_ε^* become

$$T_\varepsilon^* = \int_{\Gamma_\varepsilon} \left(W n_1 - t_i \frac{\partial u_i}{\partial x_1} \right) \mathrm{d}\Gamma, \tag{1}$$

where t_i are the surface tractions on the contour Γ_ε, W is the strain energy density and n_1 is the first component of the normal to the curve. Γ_ε is an arbitrary small contour immediately surrounding the crack and more importantly it elongates together with crack extension. T^*, as defined by Equation (1), is identical in form to the J-integral and therefore T_ε^* coincides with J where the deformation theory of plasticity prevails, such as prior to crack extension of a monotonically loaded fracture specimen. In terms of the incremental theory of plasticity, T_ε^*-integral at the end of the Nth load steps is the sum of ΔT_ε^*, which is the incremental change of T_ε^* over a load step, or

$$T_\varepsilon^* = \sum^N \Delta T_\varepsilon^*, \tag{2a}$$

where

$$\Delta T = \int_{\Gamma_\varepsilon} \left[\Delta W n_1 - (t_i + \Delta t_i) \frac{\partial \Delta u_i}{\partial x_1} - \Delta t_i \frac{\partial u_i}{\partial x_1} \right] \mathrm{d}\Gamma. \tag{2b}$$

The current T_ε^*, as defined by (2), is thus dependent on the prior loading history, a property that is essential *for elastic-plastic analysis with crack growth.* Although such incremental analysis can be routinely conducted with finite element (FE) analysis, it is not practical in experimental analysis as the cumulative experimental errors per load step will eventually swamp the sought data. Fortunately Pyo et al. (1995) have shown, through numerical experiments involving elastic-plastic FE analysis, that the total T_ε^* integral, as represented by (1), computed directly by *using the stresses and strains based on the incremental theory of plasticity*, was for all practical purpose, equal to the summed ΔT_ε^* of (2). Thus (1) can be used, as an approximation, for

[1] The J-integral evaluated along a far-field contour as per ASTM procedure, thus is not a material parameter.

crack growth study without the cumbersome incremental procedure indicated by (2) provided the states of stress and strain used in the contour integration are computed on the basis of the incremental theory of plasticity. Since the near-field J-integral can also be computed by using the same elastic-plastic FE results along a similar small contour, the distinction between the near field J- and T_ε^*-integrals now reduces to the shape of the contours used, namely the moving and elongating contours, respectively. To reiterate, *the near field J-integral accounts for the plastic work in the confined contour moving with the crack tip while the T_ε^*-integral accounts for the total plastic work in the oblong contour which elongates with crack extension.*

Through another series of numerical experiments, again based on elastic-plastic FE analysis, Okada and Atluri (1996) showed that the integrand in (1) vanished along the trailing contour very close to the traction free crack surface. Thus the numerical integration on a trailing elongated contour along a traction free crack can be ignored. The magnitude of ε is governed by a characteristic dimension which assures that a plane stress state exists along the integration contour of Γ_ε. This distance ε is equated to the plate thickness after Narashimhan and Rosakis (1990)[2]. For a plane strain state, this characteristic distance could be several times the crack tip radius. This simplification of the T_ε^*-integral further distinguishes it from the J-integral in that *the T_ε^*-integral is evaluated along a partial near-crack tip contour while the J-integral is evaluated along a completely closed near-crack contour*. The effect of this subtle difference in the contours for these two integrals will be shown in the following.

2.2. CTOA CRITERION

The critical CTOA criterion assumes that stable crack growth occurs when an angle made by a point on the upper surface of a crack, the crack tip, and a point on the lower surface reaches a critical angle. For convenience, a point 1 mm behind the crack tip has been used (Dawicke et al., 1995; Omori et al., 1997). Extensive experimental results from thin aluminum fracture specimens have shown that after an initial transient period, the CTOA remains constant throughout Mode I stable crack growth (Dawicke et al., 1995; Dawicke et al., 1997). Moreover, a two-dimension, elastic-plastic FE simulation of stable crack growth based on the CTOA criterion correctly predicted the load-crack opening displacement (COD) relations and the Mode I crack extension histories of fracture specimens. A similar constant CTOA in dynamic fracture was sought in this study.

2.3. DYNAMIC FRACTURE

As for experimental dynamic ductile fracture studies, Kobayashi et al. (1967a, b) used a single frame, ultra-high speed photography and geometric Moiré to show that the transient strain fields in fracturing magnesium, 7075-T6 and 7178-T6 aluminum alloy, central-notched (CN) specimens varied with a strain singularity between 0.4–0.6 while the corresponding static strain was about 0.6–0.8. These results suggested that a propagating ductile crack can be modeled, as a first approximation, by dynamic linear elastic fracture mechanics (LEFM) and thus justified the earlier and extensive LEFM approach to dynamic fracture as documented by Kanninen and Popelar (1985). The above pseudo-elastic response was associated with a

[2] The three-dimensional elastic-plastic finite element analysis of a flat crack in a plate by Narashimhan and Rosakis (1990) showed that the plane stress state prevailed at one half of the plate thickness from the crack tip. Since a 100 percent shear lip is the norm in ductile fracture, the minimum distance ε was conservatively picked as the plate thickness

relatively high crack velocity, which was generated by an over-driven crack initiating from a blunt notch, of 10 to 20 percent of the Rayleigh wave velocity. In contrast, most dynamic ductile fracture experiments are associated with a lower crack velocity of about 5 to 10 percent of the Rayleigh wave velocity. This paper explores the dynamic ductile fracture response associated with a crack propagating at this quasi-static crack velocity.

3. Method of approach

3.1. Hybrid experimental-numerical approach

For a rapidly propagating crack in a ductile material considered in this paper, the inertia effect is expected to be minimal and thus the problem will be treated as a quasi-static stable crack growth problem. Initially Moiré interferometry was used to determine only the transient displacement field perpendicular to the running crack, i.e., the v-field in fatigue-precracked 7075-T6 and 2024-T3 aluminum alloy, single edge notched (SEN) specimens. The missing displacement field parallel to the crack, i.e., the u-field, was computed through the use of a hybrid experimental-numerical analysis where the measured transient displacement field together with the crack propagation history were used to drive a dynamic, elasto-plastic FE model of the SEN specimen. The u- and v-displacement fields obtained through this FE analysis were then used to extract dynamic ductile fracture parameters, such as the J- and the T_ε^*-integrals and the CTOA, of the propagating ductile crack by the procedure described in Section 5.1. The results obtained through this hybrid experimental-numerical procedure are designated as the 'first series' of tests.

3.2. Experimental approach

Results and experience gained from the above FE analysis, particularly the negligible inertia effect associated with dynamic ductile fracture of aluminum fracture specimens, led to the use of a purely experimental procedure involving only Moiré interferometry. In this procedure, the displacements perpendicular and parallel to the running crack, v- and u-displacements, respectively, were recorded separately by using a multitude of machine-notched SEN specimens of identical geometry. The results of these tests were then assembled as a combined transient u- and v-displacement records which were used to compute the T_ε^* directly by the procedure described in Section 4.2. This purely experimental procedure was previously validated by the excellent agreement between the experimentally and numerically determined crack opening and crack sliding displacements in rapidly fracturing concrete specimens (Yu et al., 1995). The results obtained experimentally are designated as the 'second series' of tests.

4. Experimental procedure

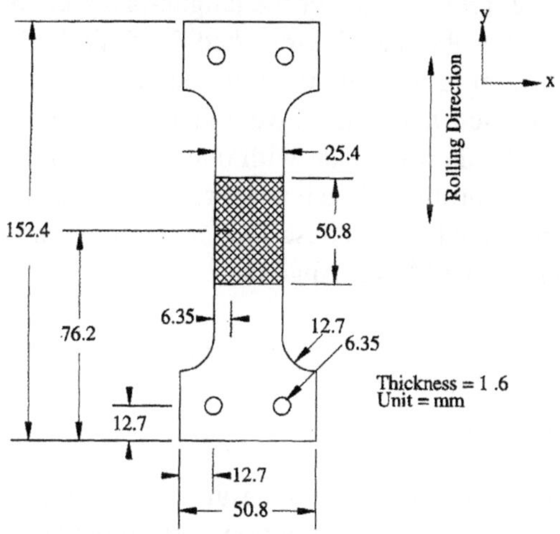

Figure 1. Aluminum SEN specimen.

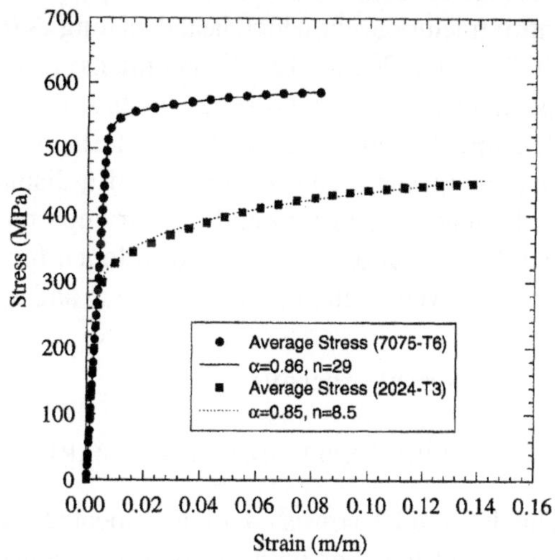

Figure 2. Uniaxial stress-strain relations of aluminum.

Figure 1 shows the SEN specimen and Figure 2 shows the static uniaxial stress versus strain relations for the 7075-T6 and 2024-T3 aluminum plates used in this investigation. The latter were obtained by averaging the uniaxial stress-strain data in the direction parallel and perpendicular to the rolling direction. The use of static stress-strain relations in this dynamic analysis was justified since the estimated maximum strain rate during the dynamic fracture event was only 539 sec^{-1}. For the elasto-plastic FE analysis, the following power hardening relation was fitted to the experimentally determined stress-strain relations.

$$\frac{\varepsilon}{\varepsilon_0} = \frac{\sigma}{\sigma_0} + \alpha \left[\frac{\sigma}{\sigma_0} \right]^n , \tag{3}$$

where σ_c is a reference yield stress and $\varepsilon_0 = \sigma_0/E$ is the associated elastic strain, and α and n are the material constants. The two coefficients, α and n, for the power hardening relations for each aluminum are listed next to their respective stress-strain curves in Figure 2.

The starter cracks in the first series of tests involving 7075-T6 and 2024-T3 SEN specimens were fatigue-precracked while the starter crack in the second series of tests involving only 7075-T6 SEN specimens were machine-notched. The machine notch delayed the initiation of the dynamic event and the crack propagated at a higher velocity. Excess ductility in the machine-notched 2024-T6 SEN specimens precluded Moiré analysis using the 40 lines/mm grating. Thus only the machine-notched 7075-T6 SEN specimens provided the Moiré data at the higher crack velocity.

A new Moiré interferometry technique by Wang et al. (1994), which combines the advantages of geometric Moiré and traditional Moiré interferometry, was used to measure the large strains in the vicinity of a running crack tip. The method uses a steep grating of low-spatial frequency, i.e., 40 lines/mm in this study, on a mirror finished specimen surface to achieve high contrast Moiré fringes on the specimen surface. Four frames of the Moiré fringe patterns corresponding to the dominant vertical displacements were recorded by an IMACON 790 camera. This limited number of frames and the fixed framing rates, i.e., 10,000 (first series of experiment) or 100,000 (second series of experiment) frames per second, required multiple recordings of identically loaded SEN specimens at different delay timing in order to capture the entire fracture event which lasted as long as 0.8 and 12 milliseconds for the fatigue-precracked 7075-T6 and 2024-T3 SEN specimens, respectively. For the machine-notched 7075-T6 SEN specimens in the second series of tests, the fracture event lasted only 0.1 millisecond. Despite all efforts to generate reproducible tests, no two dynamic fracture tests were identical. Thus a final composite record of the u- and v-displacement fields was made with due consideration of the load-time history and the varying crack opening profile of each fracture event. The J- and T_ε^*-evaluation procedures developed for numerical analysis, as described in the following sections, were then used to evaluate J and T_ε^* based on the Moiré fringe data.

5. Numerical procedure

5.1. Hybrid experimental-numerical approach

The FE model consisted of a truncated SEN specimen where the measured vertical displacement together with an assumed vanishing tangential surface traction were prescribed on a horizontal cross section, 10 mm from the crack line. This assumption was verified by comparing the FE and experimental results of a controlled 7075-T6 SEN specimen in which the u- and v-displacement fields at the onset of fracture were recorded simultaneously. The tangential surface tractions, which were obtained by the FE analysis of this test, were less than 0.1 percent of the corresponding normal surface tractions thus justifying the assumption.

In all FE analysis, symmetry about the crackline was assumed and therefore only half of the specimen was modeled. This FE model of the truncated SEN specimen consisted of graded finite elements with the smallest element being 0.25 mm square along the crack. A dynamic elasto-plastic FE analysis was conducted by driving the FE model with the measured time-varying normal displacement and vanishing tangential surface traction boundary conditions along a horizontal cross section 10 mm from the crackline, together with the instantaneous crack tip location.

The FE results were then used to compute the J- and T_ε^*-integrals following the procedure developed by Okada and Atluri (1996). The algorithm for dynamic J computation was developed in-house since the dynamic elasto-plastic FE code used in this study (ABAQUS Version 5.4-1) did not contain such a subroutine. Preliminary FE analysis showed that the area integral involving the inertia term in evaluating the dynamic J-integral was less than three percent of the maximum J-integral values of the 2024-T3 and 7075-T6 SEN specimens and thus the dynamic J- and T_ε^*-integrals were reduced to their static counterparts. J-integral was computed along a square contour which remained fixed in size and moved with the crack tip. The T_ε^*-integral was computed along a partial contour Γ_ε which also moved together with the crack tip, thus simplifying the T_ε^* computation by neglecting the contour integral behind the propagating crack. $\varepsilon = 2$ mm from the crack was used in the J- and the T_ε^*-integration processes. Figure 3 shows the completely closed moving and the partial Γ_ε contours used for evaluating J- and T_ε^*-integrals, respectively. *The difference between the J- and T_ε^*-integrals was then reduced to the lack of the downward closing line segment in Γ_ε of the latter.*

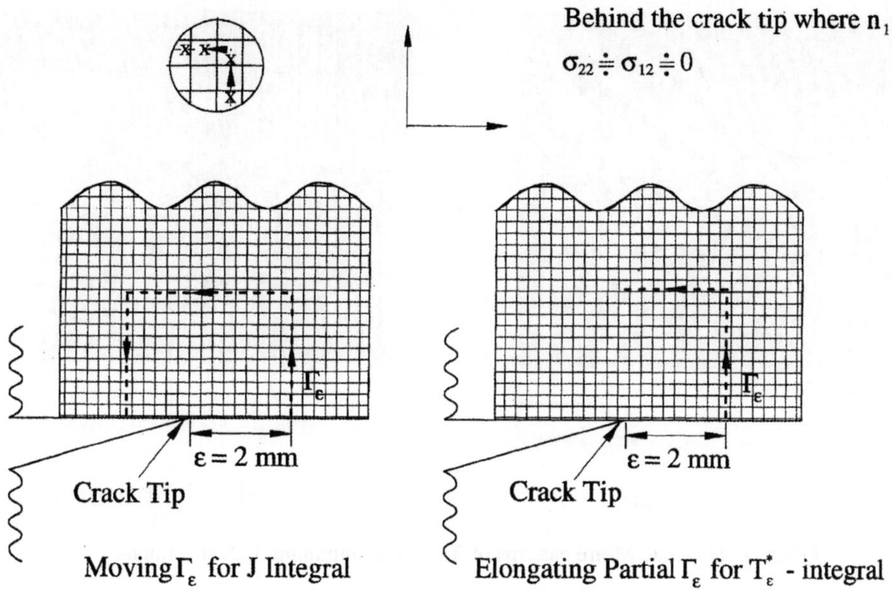

Figure 3. Integration contours.

5.2. Experimental approach

The above numerical procedure for computing T^*_ε can also be used to determine T^*_ε directly from the Moiré data. The *u*- and *v*-displacement fields provided the strains from which the stress can be computed. The state of deformation along the moving partial Γ_ε contour continually increases with crack extension and thus the deformation theory of plasticity can be used to compute the stresses from the current state of strains. With the strains and stresses known along the partial Γ_ε contour, T^*_ε was evaluated directly, without the time-consuming FE analysis, from the Moiré data of the machine-notched (second series) 7075-T6 SEN specimens.

6. Results

Figure 4 shows typical dynamic Moiré fringe patterns associated with propagating cracks in 7075-T6 and 2024-T3 SEN specimens. While the cracks in the fatigue-precracked (first series) and machine-notched (second series) 7075-T6 SEN specimens accelerated to crack velocities of 35 and 300 m/s, respectively, the cracks in the 2024-T3 SEN specimens gradually decelerated from a high of 2.5 m/s and eventually arrested.

Figure 5 shows the variations of CTOA in the 7075-T6 and 2024-T3 specimens. The CTOA reached a steady state value after an initial high value for the first series of tests and is similar to that observed by Dawicke et al. (1995, 1997). The close matches between the computed and measured CTOA's and between the computed and measured crack opening profiles in Figure 6, validated the FE modeling of the dynamic ductile fracture experiments. The effect of the crack tip blunting prior to crack extension is retained throughout the crack extension history of the 2024-T3 SEN specimen.

The computed and measured normal strains, ε_{yy}, distribution ahead of the propagating crack are shown in Figure 7. The latter was obtained by differentiating the displacement data

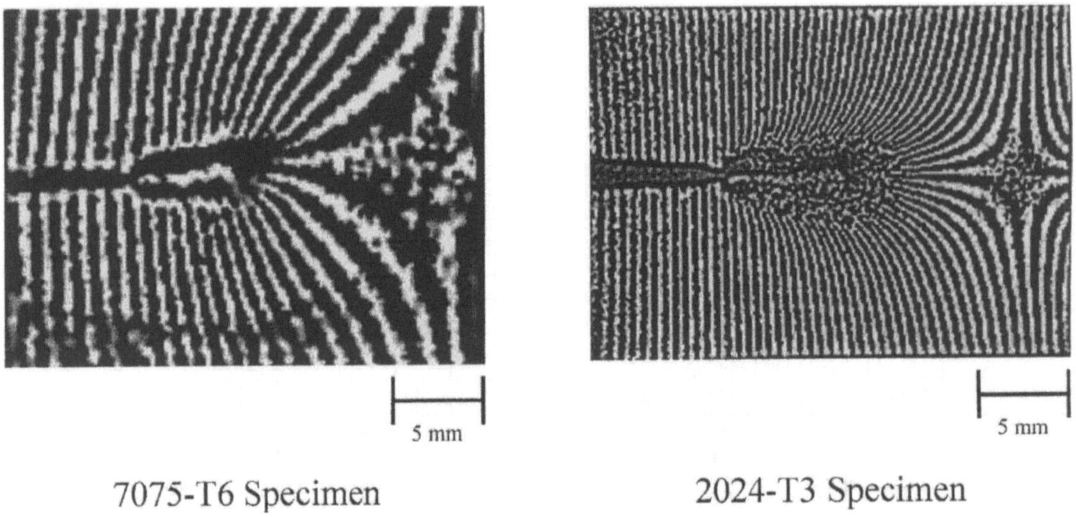

Figure 4. Dynamic Moiré patterns of fracturing aluminum SEN specimens.

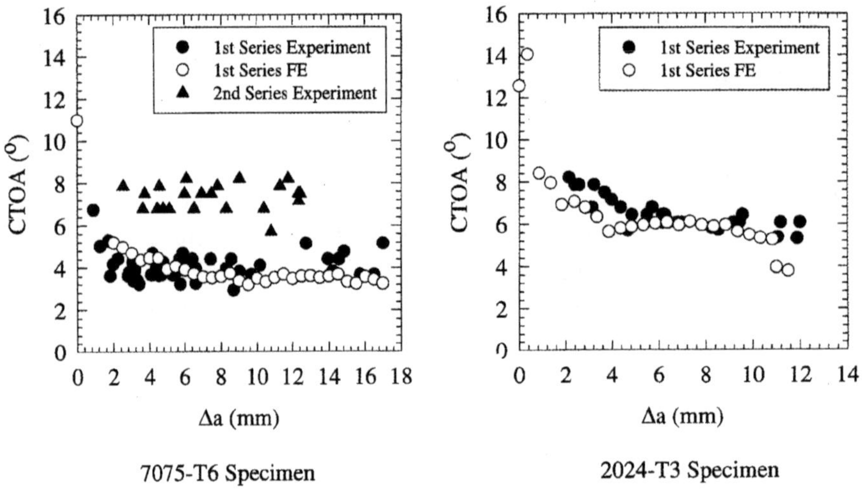

Figure 5. CTOA variations with rapid crack extension in aluminum SEN specimens.

obtained from the Moiré fringes. Unlike the CTOA, which remained constant, the strain ε_{yy} at the crack tip continuously increased with crack extension. The popular notion that CTOA is representative of the crack tip strain is thus not applicable for dynamic crack propagation. The elastic-plastic boundary, as surmised from the yield strains in Figure 2, is nearly 4 mm and 8 mm ahead of the propagating crack tip in the 7075-T6 and 2024-T3 SEN specimens, respectively. Again the good agreement between the computed and measured strains verifies the FE modeling.

The FE analysis was then used to compute the dynamic J- and the T_ε^*-integral values shown in Figure 8. While the near field J continually decreased after reaching a maximum value, the T_ε^* for the identical $\varepsilon = 2$ mm reached a steady state value during dynamic crack propagation.

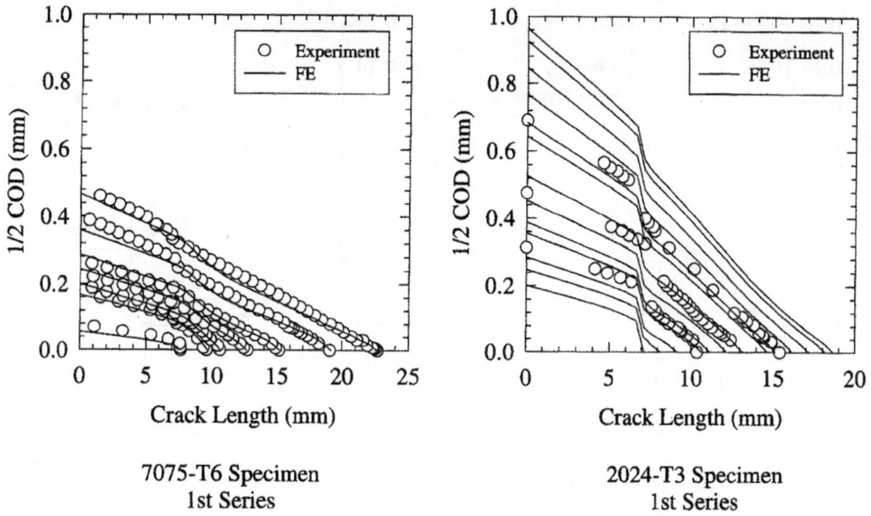

Figure 6. Variations in crack opening profiles in Aluminum SEN specimens.

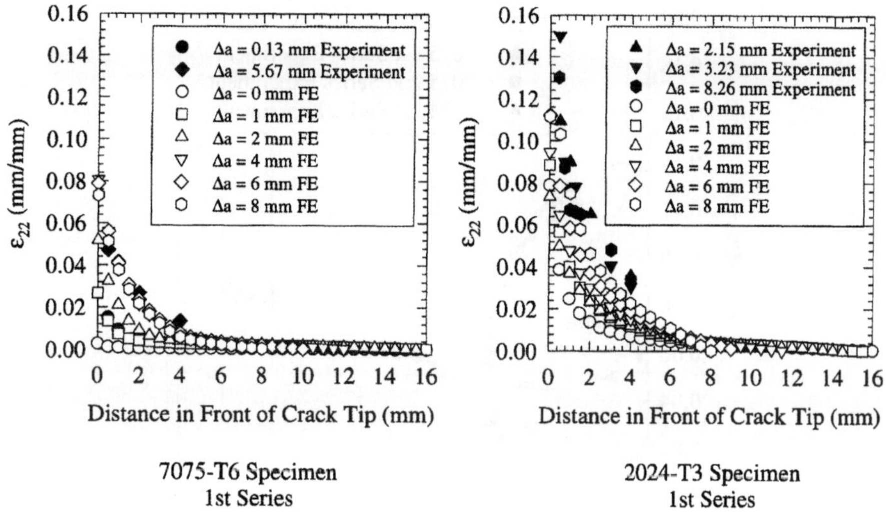

Figure 7. ε_{yy} distribution ahead of crack tip.

Figure 9 shows the T_ε^* versus crack velocity relations of the 7075-T6 and 2024-T3 SEN specimens. Initially T_ε^* increased with increased crack velocity and eventually leveled off at the terminal velocity of about 300 m/s in 7075-T6 SEN specimens. In contrast, the crack did not reach its terminal velocity in the fatigue precracked 2024-T3 SEN specimens and arrested at a T_ε^* higher than its initiation value.

The variation in dynamic stress intensity factor K_{ID} with respect to crack velocity in the 7075-T6 SEN specimens, like other somewhat brittle materials, was expected to exhibit the characteristic gamma-shaped curve. To test this postulate, the crack tip opening displacement (COD) at a crack tip distance, $r = 1$ mm, was used to compute K_{ID}, and hence the strain energy release rate G_{ID} based on LEFM of both the fatigue-precracked (first series) and the machine-notched (second series) 7075-T6 SEN specimens. The same procedure was used to compute G_{ID} of the blunt-notched 7075-T6 SEN specimens of 1967 (Kobayashi and Eng-

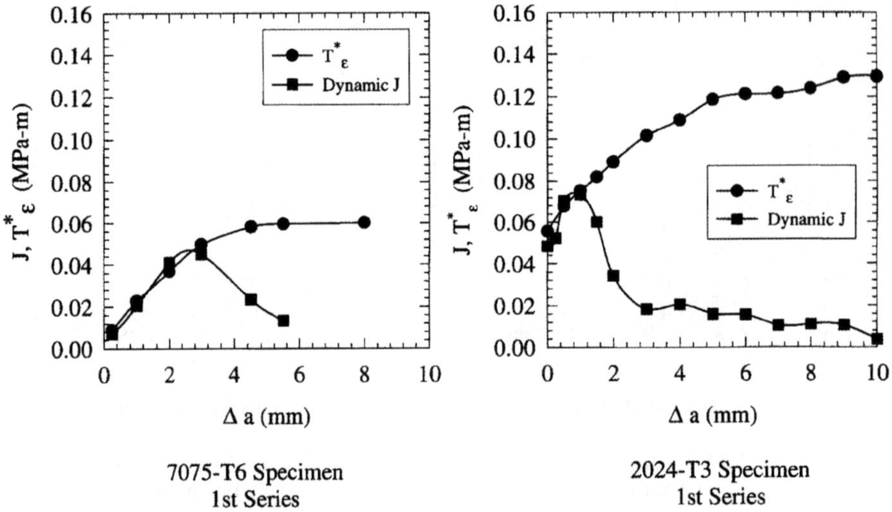

Figure 8. T_ε^* and J variations in aluminum SEN specimens $\varepsilon = 2.0\,\text{mm}$.

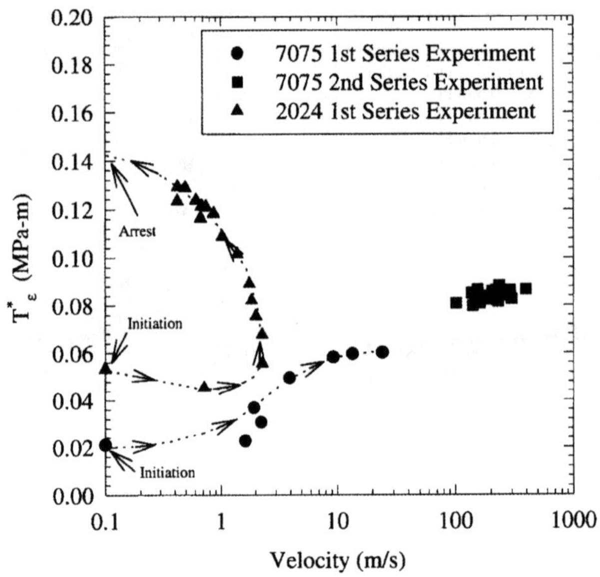

Figure 9. T_ε^* versus crack velocity.

strom, 1967b). The G_{ID} versus crack velocity relation in Figure 10 exhibits this characteristic gamma shaped curve.

7. Discussions

Both the T_ε^* and CTOA are being considered for stable crack growth criteria and likewise possibilities exist as dynamic ductile fracture criteria. CTOA, by definition, is a local crack tip parameter which exhibited a precipitous drop at the initial phase of rapid crack propagation in the fatigue-precracked 2024-T3 and 7075-T6 SEN specimens. While Dawicke et al. (1995, 1997) attributed this drop to the crack front tunneling prior to crack extension, the drop has

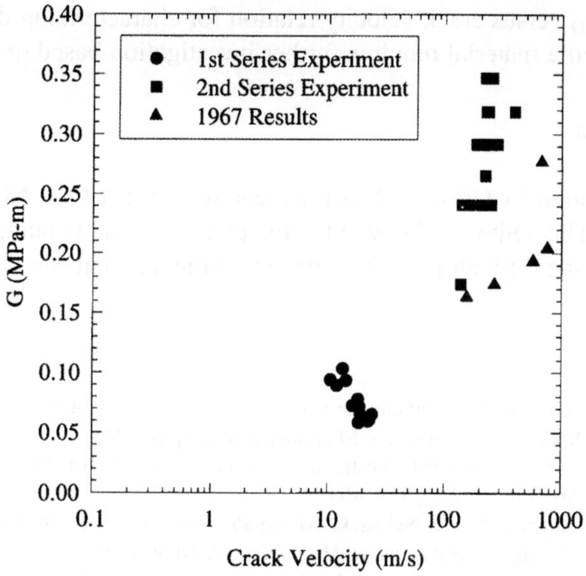

Figure 10. G versus crack velocity.

been duplicated by plane stress FE analysis without tunneling. The blunted machine-notched 7075-T6 SEN specimens did not exhibit an initially high CTOA. Thus the initial high value in CTOA in the fatigue-precracked SEN specimens is probably due to crack tip blunting prior to crack extension. Also the constant CTOA's with crack extension in Figure 5 does not correlate with the increasing crack tip ε_{yy} with crack extension as shown in Figure 7. Thus CTOA no longer represents the state of crack tip strain of a propagating crack.

Both CTOA and T_{ε}^{*} remained constant despite the increasing velocity in the 7075-T6 SEN specimens and the decreasing crack velocity in the 2024-T3 specimens. CTOA in the 2024-T3 SEN specimens at crack arrest is lower than both CTOA's at the initiation of rapid crack propagation and during crack propagation. On the other hand, T_{ε}^{*} at crack arrest in the 2024-T3 SEN specimens is higher than its constant value during crack propagation as shown in Figure 9.

The distinct difference in the G and the T_{ε}^{*} representation of the dynamic responses, as shown in Figures 9 and 10, suggests that the traditional practice of characterizing dynamic fracture of somewhat ductile material through the K_{ID} versus crack velocity relation based on LEFM could be misleading. The LEFM approach results in a terminal velocity, which is insensitive to the driving force K_{ID} while the T_{ε}^{*} approach, based on elastic-plastic fracture mechanics (EPFM), suggests that the terminal crack velocity is a consequence of the saturation of the dissipated plastic energy.

8. Conclusions

Both the T_{ε}^{*}-integral and the CTOA remained constant while J-integral diminished during the dynamic fracture event involving crack acceleration and deceleration. This study shows that the T_{ε}^{*} or the CTOA criteria proposed for stable crack growth could also be used to characterize rapid crack propagation in ductile material.

The traditional K_{ID} versus crack velocity relation for characterizing dynamic ductile fracture of somewhat ductile material requires further investigation based on EPFM.

9. Acknowledgment

This study was supported by Office of Naval Research Contract Nos. N00014-89-J-1276 and N00014-96-1-1218. The authors acknowledge the patience and encouragement of Dr. Y.D.S. Rajapakse, ONR, during difficult periods of the experimental analysis.

References

ASTM E1152-87 (1987). Standard test methods for determining J–R curves, *Annual Book of ASTM Standards*, Vol. 03.01, American Society for Testing and Materials, Philadelphia, PA.

Brocks, W. and Yuan, H. (1989). Numerical investigations on the significance of J for large stable crack growth. *Engineering Fracture Mechanics* **32**(3), 459–468.

Brust, F.W., Nishioka, T., Atluri, S.N. and Nakagaki, M. (1985). Further studies on elastic-plastic stable fracture utilizing the T^*-integral. *Engineering Fracture Mechanics* **22**, 1079–1103.

Cottrell, B. and Atkins, A.G. (1996). A review of the J- and I-integrals and their implications for crack growth resistance and toughness in ductile fracture. *International Journal of Fracture* **81**, 357–372.

Dawicke, D.S., Sutton, M., Newman, J.C. and Bigelow, C.A. (1995). Measurement and analysis of critical CTOA for thin-sheet aluminum alloy materials. *Fracture Mechanics 25th Volume, ASTM STP 1220* (Edited by F. Erdogan and J. Hartranft), American Society for Testing and Materials, 358–379.

Dawicke, D.S., Piascik, R.S. and Newman, J.C., Jr. (1997). Prediction of stable tearing and fracture of a 2000-series aluminum alloy plate using a CTOA criterion. *Fatigue and Fracture Mechanics: 27th Volume, ASTM STP 1296* (Edited by R.S. Piascik, J.C. Newman, Jr., and N.E. Dowling), American Society for Teating and Materials, 90–104.

Kanninen, M.F. and Popelar, C.H. (1985). Chapter 4, Dynamic fracture mechanics. *Advanced Fracture Mechanics*, 192–280.

Kobayashi, A.S., Harris, D.O. and Engstrom, W.L. (1967a). Transient analysis in a fracturing magnesium plate. *Experimental Mechanics* **7**(10), 434–440.

Kobayashi, A.S. and Engstrom, W.L. (1967b). Transient analysis in fracturing aluminum plates, *Proceedings 1967 JSME Semi-International Symposium*, 172–181.

May, G.B. and Kobayashi, A.S. (1995). Plane stress stable crack growth and J-integral/HRR. *Journal of Solids and Structures* **31**, 857–881.

Narasimhan, R. and Rosakis, A.J. (1990). Three-dimensional effects near a crack tip in a ductile three-point bend specimen: Part I – A numerical investigation. *ASME Journal of Applied Mechanics* **57**, 607–617.

Okada, H. and Atluri, S.N. (1996). Further study on the near tip integral parameter T_ε^* in stable crack propagation in thin ductile plate. *Proceedings of Aerospace Division, AD* **52** (Edited by J.C.L. Chang et al.), American Society of Mechanical Engineers, 251–260.

Omori, Y., Okada, H., Ma.L., Atluri, S.N. and Kobayashi, A.S. (1997). T_ε^*-integral under plane stress crack growth. *Fracture Mechanics, 27th Volume,* ASTM STP **296**, 61–71.

Pyo, C.R., Okada, H. and Atluri, S.N. (1995). An elastic-plastic finite element alternating method for analyzing wide spread fatigue damage in aircraft structures. *Computational Mechanics* **16**, 62–68.

Stonesifer, R.C. and Atluri, S.N. (1982). On the study of the $(\Delta \underline{T})_c$ and C^*-integrals for fracture analysis under non-steady creep. *Engineering Fracture Mechanics* **16**(5), 625–643.

Wang, F.X., May, G.B. and Kobayashi, A.S. (1994). Low-spatial frequency steep geometric grating for use in Moiré interferometry. *Optical Engineering* **33**, 1125–1131.

Yu, C.T., Kobayashi, A.S. and Hawkins, N.M. (1995). Dynamic fracture of concrete under mixed mode loading. *Fracture Mechanics: 25th Volume, ASTM STP 1220* (Edited by F. Erdogan and J. Hartranft), American Society for Testing and Materials, 283–293.

International Journal of Fracture **93:** 51–61, 1998.

Transient effects during rapid crack propagation

J.G. WILLIAMS
Imperial College of Science, Technology and Medicine, Mechanical Engineering Department, London, SW7 2BX, U.K.

Received 28 October 1997; accepted in revised form 5 March 1998

Abstract. An analysis is described for rapid crack propagation in an axially loaded strip and for a driven wedge. Each has a dynamic steady state, but this is usually not achieved because the initial boundary conditions of crack length and speed are violated. It is shown that the induced transient behaviour may be described using a perturbation analysis. In the strip case there is a nonlinear oscillating solution, while for the wedge a damped harmonic motion results.

Key words: Dynamic fracture, steady state, perturbation analysis.

1. Introduction

There is a considerable interest in determining the fracture resistance of materials to rapid crack propagation (Blackman et al., 1996; Ritchie and Leevers, 1993; Greenshields and Leevers, 1996). Some recent studies have considered high speed delamination of composites where the interest arises from attempts to predict impact damage (Blackman et al., 1996). From practical considerations it is easiest to test the materials using rapidly loaded double cantilever (DCB) specimens. During these tests the crack was recorded using a high speed camera and it was expected that the crack would run at its steady state condition. The steady state in these cases is determined from the equation of motion for the crack which is derived from the energy balancc. The steady state solution, analogous to the particular integral for the equation, is a monotonic variation of crack length with time of the form $a\alpha t^n$ with $a = 0$ at $t = 0$. For the DCB specimen $n = \frac{1}{2}$ and $\dot{a}\alpha t^{-(1/2)}$. In the experiments such a decrease in speed was observed, but the speed oscillated about this condition. Finite element analyses (Williams, 1994, 1995; Wang and Williams, 1994) also revealed a similar phenomena. Originally this was thought to be an artefact of the test, but the consistent nature of the effect led to a more careful consideration. An analysis was made considering a small perturbation p from the steady state and it was found for the DCB that this had the equation of motion (Blackman et al., 1996)

$$0 = \alpha p + \tfrac{1}{2}\dot{p}t + \ddot{p}t^2,$$

where α is a constant, which gave solution of the form

$$p = t^{1/4}(A \sin X + B \cos X)$$

where $X = \text{const.}(\ln t)$. The transient effect is triggered by the crack starting at rest and not the steady state. The speed $\dot{p}\alpha t^{-(3/4)}$ and thus converges on the steady state and the nonlinear

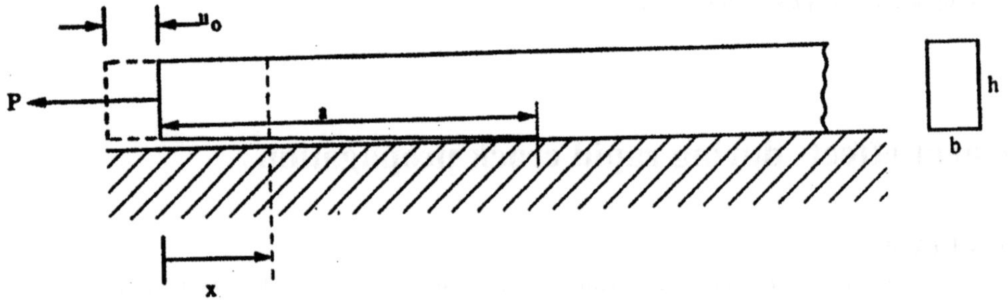

Figure 1. The axially loaded strip.

nature of the oscillation arise from the varying stiffness at the crack grows. It was found that the equation gave a very good fit to the observations (Blackman et al., 1996).

The scheme of considering perturbation has subsequently proved useful in analysing other test configurations and two such are described here which give different characteristic equations. It is felt that the methodology is useful in understanding dynamic fracture.

2. The axially loaded strip

This simple configuration is shown in Figure 1 and would be in mode II. It was analysed in some detail elsewhere (Williams, 1993) where it was shown that the static G is given by,

$$G_s = \frac{Eh}{2}\left(\frac{u_0}{a}\right)^2 \tag{1}$$

and that the solution is the same for the torsion loaded strip for which E is replaced by $\frac{1}{3}(b/d)^2\mu$, μ is the shear modulus and d is the distance between the load points and for the shear beam in which E is replaced by $\frac{2}{3}\mu$. The displacement distribution in the strip is

$$u = u_0\left(1 - \frac{x}{a}\right) \tag{2}$$

and hence for a crack speed $\dot{a}$ and when $u_0 = Vt$

$$\dot{u} = V\left[1 - \left(1 - \frac{\dot{a}t}{a}\right)\frac{x}{a}\right]. \tag{3}$$

The dynamic G is,

$$G = G_s - \frac{1}{b}\frac{\mathrm{d}U_k}{\mathrm{d}a},$$

where U_k is the kinetic energy given by

$$U_k = \tfrac{1}{2}\rho bh \int_0^a \dot{u}^2\,\mathrm{d}x, \quad \text{i.e.}$$

$$U_k = \rho bha\frac{V^2}{6}\left[1 + \frac{\dot{a}t}{a} + \left(\frac{\dot{a}t}{a}\right)^2\right],$$

where ρ is the density. Differentiating again we have

$$\frac{\mathrm{d}U_k}{b\,\mathrm{d}a} = \frac{Eh}{6}\cdot\left(\frac{V}{c}\right)^2\left[2+2\frac{\dot{a}t}{a}-\left(\frac{\dot{a}t}{a}\right)^2+\frac{\ddot{a}t}{\dot{a}}+2\frac{\ddot{a}t^2}{a}\right] \tag{4}$$

and $c^2 = E/\rho$ the wave speed.

We may use the expression for G assuming that $G = G_0$, a constant, for propagation. There are three cases of interest.

2.1. Initiation

For the period up to initiation $G < G_0$ with $\dot{a} = \ddot{a} = 0$ and at initiation

$$G = G_1 = \frac{Eh}{2}\cdot\left(\frac{Vt_0}{a_0}\right)^2 - \frac{Eh}{3}\left(\frac{V}{c}\right)^2,$$

where t_0 is the initiation time and a_0 the initial crack length, i.e.

$$\left(\frac{ct_0}{a_0}\right)^2 = \frac{2G_1}{Eh}\left(\frac{c}{V}\right)^2 + \frac{2}{3}. \tag{5}$$

2.2. Steady state propagation

A constant static G_s is achieved for constant speed crack growth, i.e.

$$a = \dot{a}_0 t \tag{6}$$

for which $\ddot{a} = 0$ and $\dot{a}t/a = 1$ and hence for propagation

$$G_0 = \frac{Eh}{2}\cdot\left(\frac{V}{\dot{a}_0}\right)^2 - \frac{Eh}{2}\left(\frac{V}{c}\right)^2 \quad \text{and}$$
$$\left(\frac{c}{\dot{a}_0}\right)^2 = \frac{2G_0}{Eh}\left(\frac{c}{V}\right)^2 + 1. \tag{7}$$

Note that as $V \to \infty$, and $G_0 \to 0$, $\dot{a}_0 \to c$ and that for a finite initial crack length the boundary conditions of this solution are violated since at $t = t_0$

$$\left(\frac{ct_0}{a}\right)^2 = \frac{2G_0}{Eh}\left(\frac{c}{V}\right)^2 + 1$$

and thus from (5), $a \neq a_0$. Also $\dot{a} = 0$ at initiation and not $\dot{a}_0$ thus stimulating transients in the motion of a.

2.3. TRANSIENT SOLUTION

The full equation of motion for constant G is somewhat intractable, but a useful solution is obtained by considering a small perturbation, p, from the steady state

$$a = \dot{a}_0 t + p, \tag{8}$$

substituting in (4) we have,

$$\frac{1}{b}\frac{\mathrm{d}U_k}{\mathrm{d}a} = \frac{Eh}{2}\left(\frac{V}{c}\right)^2\left(1 + \frac{\ddot{p}t}{\dot{a}_0}\right)$$

and hence

$$G_0 = \frac{Eh}{2}\left(\frac{V}{\dot{a}_0}\right)^2\left(1 - \frac{2p}{\dot{a}_0 t}\right) - \frac{Eh}{2}\left(\frac{V}{c}\right)^2\left(1 + \frac{\ddot{p}t}{\dot{a}_0}\right), \quad \text{i.e.,}$$
$$\left(\frac{c}{\dot{a}_0}\right)^2 = \left[\frac{2G_0}{Eh}\left(\frac{c}{V}\right)^2 + 1\right] + \frac{1}{\dot{a}_0 t}\left[2\left(\frac{c}{\dot{a}_0}\right)^2 p + \ddot{p}t^2\right]. \tag{9}$$

Thus for small perturbations the equation of motion for p is,

$$0 = \alpha p + \ddot{p}t^2 \tag{10}$$

where

$$\alpha = 2\left(\frac{c}{\dot{a}_0}\right)^2 = 2\left[\frac{2G_0}{Eh}\left(\frac{c}{V}\right)^2 + 1\right].$$

Equation (10) has a solution of the form

$$p = \text{const.}\ t^{(1/2)\pm i\varepsilon}$$

where

$$\varepsilon = \tfrac{1}{2}\sqrt{4\alpha - 1} = \frac{1}{2}\sqrt{8\left(\frac{c}{\dot{a}_0}\right)^2 - 1} \tag{11}$$

which has a real part of

$$p = \left(\frac{t}{t_0}\right)^{1/2}\left[B_1 \sin\left(\varepsilon \ln\frac{t}{t_0}\right) + B_2 \cos\left(\varepsilon \ln\frac{t}{t_0}\right)\right]. \tag{12}$$

The constants B_1 and B_2 can be found from the mismatch in a and $\dot{a}$ at $t = t_0$ discussed previously, i.e., at

$$t = t_0, \qquad \dot{p}_0 = -\dot{a}_0 \quad \text{and} \quad p_0 = a_0(1 - \beta),$$

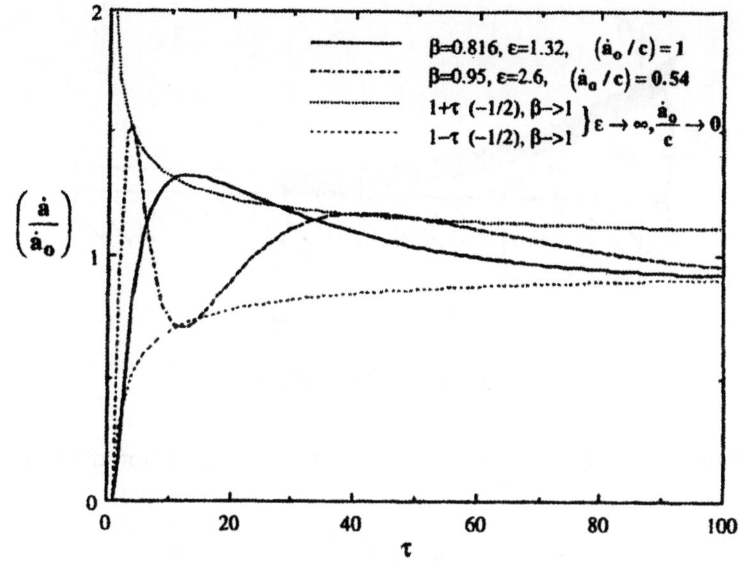

Figure 2. Variations of velocity with time for the parallel strip.

where

$$\beta^2 = \left(\frac{\dot{a}_0 t_0}{a_0}\right)^2 = \frac{\frac{2}{3} + \gamma\left(\frac{2G_0}{Eh}\right)\left(\frac{c}{V}\right)^2}{1 + \left(\frac{2G_0}{Eh}\right)\left(\frac{c}{V}\right)^2} = \gamma - (\gamma - \tfrac{2}{3})\left(\frac{\dot{a}_0}{c}\right)^2$$

and

$$\varepsilon^2 = \frac{21\gamma - 16 + 3\beta^2}{12(\gamma - \beta^2)} \quad \text{with} \quad \gamma = \frac{G_1}{G_0}$$

the ratio of initiation to propagation toughness. We have finally

$$\begin{aligned} \frac{a}{a_0} &= \beta\tau + \tau^{1/2}\left[(1-\beta)\cos X - \frac{(1+\beta)}{2\varepsilon}\sin X\right], \\ \frac{\dot{a}}{\dot{a}_0} &= 1 - \tau^{-(1/2)}\left[\cos X + \frac{1}{\beta}\left(\frac{(1+\beta)}{4\varepsilon} + (1-\beta)\varepsilon\right)\sin X\right], \end{aligned} \tag{13}$$

where $\tau = t/t_0$ and $X = \varepsilon \ln \tau$.

The form of the perturbations is shown in Figure 2 where $\dot{a}/\dot{a}_0$ is shown as a function of τ. For low crack speeds $\beta \to 1$ and $\varepsilon \to \infty$ giving very rapid oscillations between bounds of $\pm\tau^{-(1/2)}$ as shown. As the crack speed increases the frequency decreases as ε decreases and a typical curve for $\beta = 0.95$, $\varepsilon = 2.6$, $\dot{a}_0/c = 0.54$ is shown which illustrates the decreasing frequency with time. The upper bound as $V \to \infty$ or $G_0 \to 0$ is also shown for which $\beta^2 = \frac{2}{3}$, $\varepsilon = 1.32$, $\dot{a}_0/c = 1$.

3. The driven wedge

A further example which yields a different form of transient behaviour is shown in Figure 3. In this case a wedge of angle α is driven by a force F and this yields an exact dynamic solution

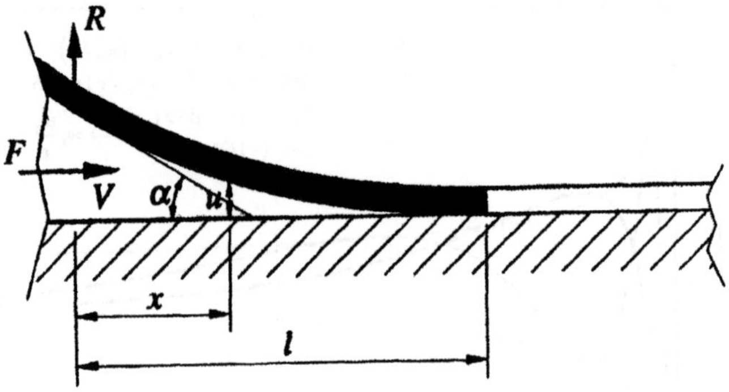

Figure 3. The driven wedge.

for the steady-state case (Williams, 1993; Freund, 1990). The equation of motion of the beam is

$$\frac{d^4u}{dx^4} + \frac{1}{\lambda^2}\frac{d^2u}{dx^2} = 0, \quad \text{where} \quad \frac{1}{\lambda^2} = \frac{A}{I}\cdot\left(\frac{V}{c}\right)^2, \qquad c^2 = \frac{E}{\rho} \tag{14}$$

giving the solution

$$u = A_0 + A_1 x + A_2 \sin\frac{x}{\lambda} + A_3 \cos\frac{x}{\lambda}.$$

The boundary conditions are

$$u = \frac{du}{dx} = 0 \quad \text{at} \quad x = \ell \quad \text{and} \quad \frac{du}{dx} = -\alpha, \qquad \frac{d^2u}{dx^2} = 0, \qquad \frac{d^3u}{dx^3} = \frac{R}{EI}$$

at $x = 0$ which gives the result

$$u = \frac{\alpha\lambda}{1 - \cos\frac{\ell}{\lambda}}\left[\sin\frac{\ell}{\lambda} - \frac{\ell}{\lambda}\cos\frac{\ell}{\lambda} + \frac{x}{\lambda}\cos\frac{\ell}{\lambda} - \sin\frac{x}{\lambda}\right]. \tag{15}$$

The reaction at $x = 0$ is

$$R = EI\frac{d^3u}{dx^3} = \frac{EI\alpha}{\lambda^2\left(1 - \cos\frac{\ell}{\lambda}\right)}$$

and the horizontal force $F = R\alpha$ so that the rate of external work input is

$$\dot{U}_{\text{ext}} = F.V. = \frac{EI\alpha^2.V}{\lambda^2\left(1 - \cos\frac{\ell}{\lambda}\right)}. \tag{16}$$

There is no change in the strain energy in this case (i.e. $\dot{U}_s = 0$) but there is an outflow of kinetic energy at the contact point ($x = 0$) given by

$$\dot{U}_{k0} = \frac{EA}{2}\left(\frac{\alpha V}{c}\right)^2.V = \frac{EI\alpha^2}{2\lambda^2}.V. \tag{17}$$

In the steady state the debonding point ($x = \ell$) also moves at velocity V so we have

$$bGV = \dot{U}_{\text{ext}} - \dot{U}_s - \dot{U}_k, \quad \text{i.e.}$$

$$G = \frac{EI\alpha^2}{b\lambda^2}\left(\frac{1}{1-\cos\chi} - \frac{1}{2}\right)$$

where

$$\chi = \ell\sqrt{\frac{A}{I}} \cdot \frac{V}{c}.$$

The applied force on the wedge is

$$F = \frac{EI\alpha^2}{\lambda^2(1-\cos\chi)} \tag{18}$$

and we may define this as the applied G, i.e.

$$G_s = \frac{F}{b} = \frac{EI\alpha^2}{b\lambda^2(1-\cos\chi)}. \tag{19}$$

For a constant propagation G value of G_0 the motion of the crack is given by

$$\left(\frac{bG_0}{F}\right) = \delta = \tfrac{1}{2}(1+\cos\chi)$$

and the velocity may be expressed as

$$\left(\frac{V}{c}\right)^2 = \left(\frac{1-\delta}{\delta}\right)\left(\frac{2bG_0}{EA\alpha^2}\right). \tag{20}$$

At low speeds

$$G_0 = \frac{2EI\alpha^2}{b\ell_0^2},$$

where ℓ_0 is the loaded beam length and the length at other speeds is given by

$$\frac{\ell}{\ell_0} = \frac{1}{2}\sqrt{\frac{\delta}{1-\delta}}.\cos^{-1}(2\delta - 1). \tag{21}$$

The time characterising the fracture is

$$t = \left(\frac{\ell}{V}\right) = \frac{1}{2}\left(\frac{\delta}{1-\delta}\right)\cos^{-1}(2\delta-1)\cdot\bar{t} \tag{22}$$

where

$$\bar{t} = \frac{\ell_0^2}{2c}\sqrt{\frac{A}{I}} = \frac{\sqrt{AI}}{c}\cdot\frac{E\alpha^2}{bG_0}. \tag{23}$$

3.1. Transient solution

A perturbation from the steady state in this case is given by a small change in ℓ and the perturbed velocity is

$$\dot{u} = -V\frac{\partial u}{\partial x} + \dot{p}\frac{\partial u}{\partial \ell}.$$

From (15) we have

$$\frac{\partial u}{\partial x} = -\frac{\alpha}{1-\cos\chi}(\cos\chi\varsigma - \cos\chi), \qquad \varsigma = \frac{x}{\ell}$$

and noting that

$$\frac{\mathrm{d}\chi}{\mathrm{d}\ell} = \frac{1}{\lambda},$$

$$\frac{\partial u}{\partial \ell} = \frac{\alpha\sin\chi}{(1-\cos\chi)^2}[\chi(1-\varsigma) - (\sin\chi - \sin\chi\varsigma)].$$

We may now derive the perturbed energy rate changes and hence,

$$\dot{U}_{\text{ext}} = \frac{EI\alpha^2 V}{\lambda^2(1-\cos\chi)}\left[1 - \left(\frac{\sin\chi}{1-\cos\chi}\right)\frac{p}{\lambda} + \frac{\sin\chi(\chi-\sin\chi)}{(1-\cos\chi)^2}\cdot\frac{\dot{p}}{V}\right]. \tag{24}$$

There is now a change in U_s via the change in ℓ and since

$$U_s = \tfrac{1}{2}EI\int_0^\ell \left(\frac{\partial^2 u}{\partial x^2}\right)^2 \mathrm{d}x,$$

we have

$$\dot{U}_s = -\frac{EI\alpha^2 V}{2\lambda^2(1-\cos\chi)}\cdot\left[\frac{\sin\chi(\chi-\sin\chi)}{(1-\cos\chi)^2}\right]\cdot\frac{\dot{p}}{V}. \tag{25}$$

There is a kinetic energy change at the load point

$$\dot{U}_{k0} = -\frac{EI\alpha^2 V}{2\lambda^2}\left[1 + \frac{2\sin\chi(\chi-\sin\chi)}{(1-\cos\chi)^2}\cdot\frac{\dot{p}}{V}\right] \tag{26}$$

and also a change via ℓ derived from

$$U_{k1} = \frac{EA}{2}\int_0^\ell \left(\frac{\dot{u}}{c}\right)^2 \mathrm{d}x, \quad \text{i.e.}$$

$$\dot{U}_{k1} = \frac{EI\alpha^2}{2\lambda^2}\cdot V\left(I_1\frac{\dot{p}}{V} + I_2\frac{\lambda\ddot{p}}{V^2}\right), \tag{27}$$

where

$$I_1 = \frac{\sin\chi}{(1-\cos\chi)^3}[2(\sin\chi - \chi\cos\chi) - (\chi - \sin\chi\cos\chi)] \quad \text{and}$$

$$I_2 = \frac{\sin\chi}{(1-\cos\chi)^3}[4(1-\cos\chi) + 2\chi\sin\chi\cos\chi - 3\sin^2\chi - \chi^2\cos\chi].$$

The equation of motion is now

$$\begin{aligned}\delta\left(1+\frac{\dot p}{V}\right) &= \left[1-\left(\frac{\sin\chi}{1-\cos\chi}\right)\frac{p}{\lambda}+\tfrac{3}{2}I_3\frac{\dot p}{V}\right]\\ &\quad -\left(\frac{1-\cos\chi}{2}\right)\left[1+2I_3\frac{\dot p}{V}+I_1\frac{\dot p}{V}+I_2\frac{\lambda\ddot p}{V^2}\right],\end{aligned} \tag{28}$$

where

$$I_3 = \frac{\sin\chi}{(1-\cos\chi)^2}(\chi - \sin\chi).$$

For $p = 0$ we retrieve the steady state result and the perturbation equation is a damped harmonic of the form

$$\ddot p + \frac{2}{\tau}\cdot\dot p + \omega_0^2 p = 0,$$

with

$$\omega_0^2 = \frac{2\sin\chi}{(1-\cos\chi)^2}\frac{1}{I_2}\cdot\left(\frac{V}{\lambda}\right)^2,$$

$$\frac{1}{\tau} = \frac{1}{2I_2}\left[\left(\frac{1+\cos\chi}{1-\cos\chi}\right) - \left(\frac{1+2\cos\chi}{1-\cos\chi}\right)I_3 + I_1\right]\left(\frac{V}{\lambda}\right),$$

with

$$\frac{V}{\lambda} = \left(\frac{1-\cos\chi}{1+\cos\chi}\right)\frac{2}{\bar t}.$$

The final forms are

$$(\omega_0\bar t)^2 = \frac{(1-\cos\chi)^3}{(1+\cos\chi)^2}\cdot\frac{8}{\phi_2} \quad \text{and} \quad \frac{\bar t}{\tau} = \frac{(1-\cos\chi)}{(1+\cos\chi)}\cdot\frac{2\phi_1}{\phi_2}, \tag{29}$$

where

$$\phi_1 = 2(\sin\chi - \chi\cos\chi) - (\chi - \sin\chi\cos\chi) \quad \text{and}$$

$$\phi_2 = 4(1-\cos\chi) + 2\chi\sin\chi\cos\chi - 3\sin^2\chi - \chi^2\cos\chi.$$

The frequency is

$$\omega\bar t = \sqrt{(\omega_0\bar t)^2 - \left(\frac{\bar t}{\tau}\right)^2}$$

and the perturbation is

$$p = \mathrm{e}^{-(t/\tau)}(A \sin \omega t + B \cos \omega t).$$

The approximate boundary conditions for a crack starting at rest are $p = 0$ and $\dot{p} = -V$ at $t = 0$ giving

$$\dot{a} = V\left[1 - \mathrm{e}^{-(t/\tau)}\left(\cos \omega t - \frac{\sin \omega t}{\omega \tau}\right)\right]. \tag{30}$$

For low speeds $\phi_1 \to \chi^5/15$ and $\phi_2 \to 3/72\chi^6$ giving

$$(\omega_0 \bar{t})^2 = \tfrac{18}{7} \quad \text{and} \quad \left(\frac{\bar{t}}{\tau}\right) = \tfrac{24}{35} \cdot \cos^{-1}(2\delta - 1)$$

and the frequency tends to $1.6/\bar{t}$ with zero damping. As the speed increases so does both the frequency and the damping. At $\chi = \frac{1}{2}\pi$, i.e.

$$\delta = 0.5, \qquad \omega = \frac{2.7}{\bar{t}} \quad \text{and} \quad \frac{1}{\tau} = \frac{0.86}{\bar{t}}.$$

Thus the system is always dynamically stable with the crack speed converging on the steady state. The higher speed solutions for $\chi > \pi$ are unstable since $1/\tau < 0$.

4. Conclusions

The two examples of the perturbation analysis given show that the method works well for the different configurations. All give damped oscillations and the differences arise mainly from the stiffness variations as the crack grows. For the DCB the stiffness goes as a^{-3}, for the strip as a^{-1} and for the wedge it is constant. The last case gives damped harmonic equations and clearly damping leads to a convergence on the steady state. Both the previous examples have velocities which converge on the steady state and are thus stable. Under some circumstances the damping is positive leading to unstable behaviour.

It was pointed out many years ago (Bilek and Burns, 1974) that steady states are difficult to achieve with finite initial crack lenghts, since the steady state conditions are initially violated which leads to the transient phenomena. The scheme used here enables this to be described and gives a useful basis for understanding dynamic fracture tests and other phenomena.

Dedication

Part of this paper (Section 2) was given at a session of ICFII in 1997 in honour of Professor M.L. Williams. The work has subsequently been extended in this more complete paper. The solutions of the characteristic equations for the DCB and loaded strip case are similar in form to those obtained for complex singularities by MLW (Williams, 1959) and would probably not have been recognised by this author, without a knowledge of that work.

References

Bilek, Z.J. and Burns, S.J. (1974). Crack propagation in wedged double cantilevered beam specimens. *Journal of the Mechanics and Physics of Solids* **22**, 85–95.

Blackman, B.R.K., Kinloch, A.J., Wang, Y. and Williams, J.G. (1996). The failure of fibre-composites and adhesively-bonded fibre-composites under high rates of test. Part II: Mode I loading – Dynamic effects. *Journal of Materials Science* **31**, 4451–4466.

Freund, L.B. (1990). *Dynamic Fracture Mechanics*. Cambridge University Press.

Greenshields, C.J. and Leevers, P.S. (1996). Rapid crack propagation in plastic water pipes: Measurement of dynamic fracture resistance. *International Journal of Fracture* **79**, 85–95.

Ritchie, S.J.K. and Leevers, P.S. (1993). The high speed double torsion test impact and dynamic fracture of polymers and composites. *ESIS 19* (Edited by J.G. Williams and A. Pavan), Mechanical Engineering Publications, Bury St. Edmunds (1995) 137–146.

Wang, Y. and Williams, J.G. (1994). A numerical study of dynamic crack growth in isotropic DCB specimens. *Composites* **25**(5), 323–331.

Williams, J.G. (1993). A review of the determination of energy release rates for strips in tension and bending. Part II – Dynamic solutions. *Strain Analysis for Engineering Design* **28**(4), 247–256.

Williams, J.G. (1994). High speed delamination of composites. *European Journal of Mechaics A/Solids* **13**(4), suppl., 227–241.

Williams, J.G. (1995). Dynamic fracture analysis of the DCB specimen. *Proc. 1995 ASME/JSME Pressure Vessels and Piping Conference*.

Williams, M.L. (1959). The stresses around a fault or crack in a dissimilar media. *Bulletin of the Seismological Society of America* **49**, 199–204.

International Journal of Fracture **93:** 63–86, 1998.

Crack detection by the electric method: Uniqueness and approximation

Dedicated to Max and Mel Williams on the occasion of their retirement from the editorship of this Journal.

DANG DINH ANG[1]* and DANG DUC TRONG[2]**

[1]*Department of Mathematics and Computer Science, Hochiminh City National University, 227 Nguyen Van Cu, Q5, Hochiminh City, Vietnam*

[2]*Address for Editorial Correspondence: Dang Dinh Ang, 162 Dien Bien Phu, Q3, Hochiminh City, Vietnam*

Received 10 September 1997; accepted in revised form 3 September 1998

Abstract. Let U be a bounded domain in $\mathbb{R}^2$ the boundary ∂U of which is a closed nonself-intersecting curve. Let ω be a hole (a crack) in the interior of U the boundary of which is a closed nonself-intersecting curve. Let U and ω be star-shaped. It is shown that the problem of finding a harmonic function u and its domain $\Omega = U \setminus \bar{\omega}$ ($\bar{\omega} \equiv \omega \cup \partial\omega$), subject to given Cauchy conditions on an open subset of ∂U and to the condition $\partial u/\partial n = 0$ on $\partial\omega$ has at most one solution. This uniqueness result shows that an insulating crack (e.g. one made up of air) can be identified by the electric method. The problem of determining $\partial\omega$ is an ill-posed problem; it is here regularized using finite dimensional approximations.

Key words: Crack detection, electric method, domain identification, Laplace equation, finite dimensional regularization.

We shall consider the problem of identifying the location and shape of a crack in a solid by the electric method, from the measured potential and flux on a portion of the outer boundary. Assuming the electric conductivity to be 1, the flux is simply $\partial u/\partial n$, the normal derivative to the boundary.

In the electric method, we have to solve the Laplace equation for the potential u

$$\Delta u = 0$$

subject to conditions on the potential u and on the flux $\partial u/\partial n$ on the outer boundary and a condition on the surface of the crack, that can be

$$u = 0$$

or

$$\frac{\partial u}{\partial n} = 0$$

* Professor.

** Dr.

the latter condition corresponding to an insulating crack (like air).

The case $u = 0$ (a conductive crack) was treated in Ang and Vy (1995). We are concerned here with an insulating crack. The latter case was discussed, e.g., in Bui (1994). Particular cases of the identification problem by the electric or electromagnetic method are considered in Kubo (1990), McIver (1991) and the references therein. It is noteworthy that cracks in elastic solids can also be detected by mechanical methods, in other words, the location and shape of a crack in an elastic solid can be determined from the stresses and displacements measured on a portion of the outer surface of the elastic body cf., Ang, Trong and Yamamoto (1996, 1997).

Before giving a mathematical formulation of the problem, we first set some notations and definitions.

$L^2(\Omega)$ is the set of functions measurable on $\Omega \subset \mathbb{R}^n$, $n \in \mathbb{N}$ such that

$$\|f\|^2_{L^2(\Omega)} \equiv \int_\Omega |f(x)|^2 \, \mathrm{d}x < \infty.$$

$H^1(\Omega)$ ($\Omega \subset \mathbb{R}^2$) is the set of measurable functions $f(= f(x_1, x_2))$ in $L^2(\Omega)$ having the generalized derivatives $\frac{\partial f}{\partial x_1}, \frac{\partial f}{\partial x_2}$ such that

$$\|f\|^2_{H^1(\Omega)} \equiv \int_\Omega \left\{ |f(x_1, x_2)|^2 + \left|\frac{\partial f}{\partial x_1}(x_1, x_2)\right|^2 + \left|\frac{\partial f}{\partial x_2}(x_1, x_2)\right|^2 \right\} \mathrm{d}x_1 \, \mathrm{d}x_2 < \infty.$$

$H^1(a, b)$, $a < b$, is the set of functions f in $L^2(a, b)$ having the generalized derivative f' such that

$$\|f\|^2_{H^1(a,b)} \equiv \int_a^b \{|f(x)|^2 + |f'(x)|^2\} \, \mathrm{d}x < \infty.$$

$H_0^1(0, 1)$: the set of functions f in $H^1(0, 1)$ satisfying $f(0) = f(1) = 0$.

$H^2(a, b)$, $a < b$, is the set of functions f in $L^2(a, b)$ having the generalized derivatives f', f'' such that

$$\|f\|^2_{H^2(a,b)} \equiv \|f\|^2_{L^2(a,b)} + \|f'\|^2_{L^2(a,b)} + \|f''\|^2_{L^2(a,b)} < \infty.$$

$C^k(\Omega)$ is the set of functions on Ω having continuous derivatives up to order k.

$L^2(\Gamma_0)$: the set of measurable functions defined on the curve Γ_0 such that

$$\|f\|^2_{L^2(\Gamma_0)} = \int_{\Gamma_0} |f|^2 \, \mathrm{d}s < \infty,$$

where $\mathrm{d}s$ is the arc length element of Γ_0.

We shall also refer to $H^{1/2}(\Gamma_0)$, the definition and properties of which can be found in Lions and Magenes (1972), page 34, or Rodrigues (1987), page 65.

1. Mathematical formulation and summary

Let U be a bounded domain in $\mathbb{R}^2$, the boundary ∂U of which is a closed nonself-intersecting curve. Let ω be a hole (a crack) in the interior of U the boundary of which is a closed nonself-intersecting curve. Denoting by $\bar{\omega}$ the set $\omega \cup \partial\omega$ (the closure of ω), we assume that $\bar{\omega} \subset U$. We assume known the domain U. The problem is to determine the subdomain ω from electric potential and flux measured on a portion of ∂U and under the condition of zero flux on $\partial\omega$. Mathematically, the problem is to identify the domain of a function satisfying the Laplace equation

$$\Delta u = 0 \quad \text{on} \quad \Omega = U \setminus \bar{\omega}, \tag{1}$$

subject to the conditions

$$u|_{\Gamma_0} = f, \qquad \left.\frac{\partial u}{\partial n}\right|_{\Gamma_0} = g, \tag{2}$$

$$\left.\frac{\partial u}{\partial n}\right|_{\partial\omega} = 0, \tag{3}$$

where Γ_0 is an open subset of ∂U.

It should be noted that exact solutions of Problem (1)–(3) usually do not exist. In fact, the set of the (f, g)'s for which the system (1)–(3) has no solution is dense in $L^2(\Gamma_0) \times L^2(\Gamma_0)$. If (1)–(3) has a solution (u, Ω) with $u \in H^1(\Omega)$ (which is a natural solution space), then $f \in H^{1/2}(\Gamma_0)$ (cf. Rodrigues (1987), Theorem 5.6, Chapter 3, page 74). Hence if f, g are step functions, then (1)–(3) has no solution in $H^1(\Omega)$. Now, in practice f, g come from experimental measurements and thus are given as finite sets of points that are conveniently patched into L^2-functions. In this case, (1)–(3) usually has no solution. If we denote by (f_0, g_0) the (probably unknown) exact Cauchy data and by (f_n, g_n) the nth approximation, then it is natural to assume that

$$(f_n, g_n) \to (f_0, g_0) \quad \text{in } L^2(\Gamma_0) \times L^2(\Gamma_0) \quad \text{as } n \to \infty \tag{4}$$

and from (f_n, g_n) we construct the regularized solutions of (1)–(3).

In Ang and Vy (1995), the problem of identifying the domain of a harmonic function satisfying $u|_{\partial\omega} = 0$ instead of (3) was studied. The exact solution was approximated by the solutions of a sequence of minimization problems in finite-dimensional spaces. In the present paper we also use the minimization method. However, we do not assume that the unknown boundary $\partial\omega$ is C^1. As is known, the boundary of a crack $\partial\omega$ is in general not smooth, it can have corners. We shall study crack surfaces consisting of two C^1-curves. Note that our analysis does not exclude the possibility of a smooth crack surface (corresponding to the case that the union of the two C^1-curves is a C^1 closed curve) or of a crack surface with just one corner (corresponding to the case that the union of the two C^1-curves is smooth everywhere except at one point, the corner). On the other hand, we do not have the values of u or of $\partial u/\partial n$ on all of the outer boundary, but only on a portion Γ_0. The main results of our paper are the uniqueness of the exact problem and the convergence of regularized solutions to the exact solution.

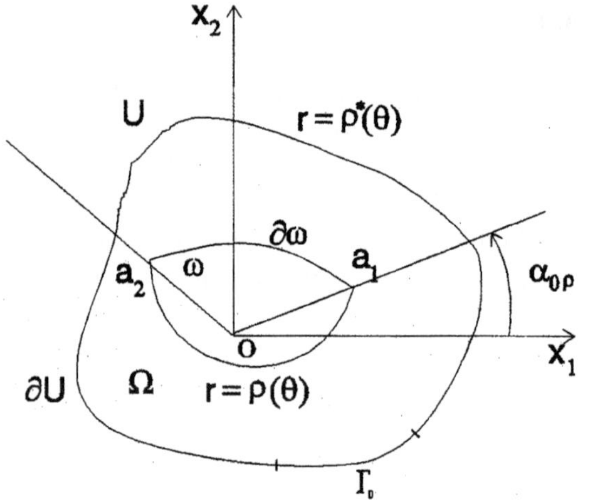

Figure 1.

The remainder of the paper consists of three sections. Section 2 gives a uniqueness result, i.e., a proof of crack identifiability by the electric method. In Section 3, the problem is transformed into one on a fixed known domain and is given a variational formulation. In Section 4 we give a parametrization of the crack, in fact, the crack is represented by a point in some function space. The final Section 5 deals with finite dimensional approximations and regularization.

2. Uniqueness result

Let U, ω be as in Section 1. We shall assume that there is a point $\xi_0 \in \mathbb{R}^2$ such that U, ω are star-shaped with respect to ξ_0. By using the translation $y = x - \xi_0$, we can reduce the problem to the case $\xi_0 = 0$. The set ω can be seen as a crack of $\Omega = U \setminus \bar{\omega}$. We shall assume that the known outer boundary ∂U of Ω can be represented in the polar coordinate system by the equation

$$r = \rho^*(\theta), \quad 0 < \theta < 2\pi, \tag{5}$$

where ρ^* is 2π-periodic and piecewise C^1. The unknown inner boundary has the equation

$$r = \rho(\theta), \quad 0 < \rho(\theta) < \rho^*(\theta), \quad \theta \in [0, 2\pi], \tag{6}$$

where ρ is 2π-periodic and piecewise C^1 (cf. Figure 1).

We consider solutions (u, Ω) with $u \in H^1(\Omega)$ and C^1 on $\Omega \cup \Gamma_0 \cup \partial\omega$ except possibly at a_1 and a_2, i.e., $u \in H^1(\Omega) \cap C^1(\Omega \cup (\partial\omega \setminus \{a_1, a_2\}) \cup \Gamma_0)$. Note that if the number of corners is at most equal to n, then the following analysis will also apply, only with a higher dimensional space E (cf. (28) for the definition of E).

THEOREM 1. *If $f \not\equiv$ const or $g \not\equiv 0$ on Γ_0, then the problem* (1)–(3) *has at most one solution* (u, Ω),

$$u \in H^1(\Omega) \cap C^1(\Omega \cup (\partial\omega \setminus \{a_1, a_2\}) \cup \Gamma_0)$$

for which the inner boundary $\partial\omega$ is in the family of curves satisfying (6).

Proof. Suppose that (u_1, Ω_1), (u_2, Ω_2) satisfy (1)–(3) with

$$u_i \in H^1(\Omega_i) \cap C^1(\Omega_i \cup (\partial\omega_i \setminus \{a_{i1}, a_{i2}\}) \cup \Gamma_0), \quad i = 1, 2,$$

where a_{i1}, a_{i2} may be corners on the crack surface $\partial\omega_i$, $i = 1, 2$. We prove that $\Omega_1 = \Omega_2$.

We first have that $\Omega_1 \cap \Omega_2$ is connected. Indeed, one has $\partial\omega_i (i = 1, 2)$ (the boundary of crack) represented by the equation

$$r = \rho_i(\theta), \quad i = 1, 2,$$

with

$$0 < \rho_i(\theta) < \rho^*(\theta), \quad \forall\theta \in [0, 2\pi], \quad i = 1, 2.$$

Hence

$$\Omega_1 \cap \Omega_2 = \{(r, \theta) : 0 \leqslant \theta \leqslant 2\pi,\ \max(\rho_1(\theta), \rho_2(\theta)) < r < \rho^*(\theta)\}.$$

Now, it can be checked directly that $\Omega_1 \cap \Omega_2$ is connected.

Putting

$$w = u_1 - u_2 \quad \text{in} \quad \Omega_1 \cap \Omega_2,$$

one has

$$\Delta w = 0 \quad \text{on} \quad \Omega_1 \cap \Omega_2,$$

$$w|_{\Gamma_0} = 0, \qquad \left.\frac{\partial w}{\partial n}\right|_{\Gamma_0} = 0.$$

By uniqueness of Cauchy's problem for harmonic functions, the above equations give:

$$u_1 - u_2 = w = 0 \quad \text{on} \quad \Omega_1 \cap \Omega_2. \tag{7}$$

Suppose by contradiction that $\Omega_1 \neq \Omega_2$ or equivalently that

$$\rho_1 \neq \rho_2. \tag{8}$$

Then, we can find a point $\eta \in [0, 2\pi]$ such that

$$\rho_1(\eta) \neq \rho_2(\eta).$$

Without loss of generality, we assume that

$$0 < \rho_1(\eta) < \rho_2(\eta) < \rho^*(\eta). \tag{9}$$

We have two cases

(i) $\partial\omega_1 \cap \partial\omega_2 \neq \emptyset$

(ii) $\partial\omega_1 \cap \partial\omega_2 = \emptyset$.

If (i) holds, let (ξ_1, ξ_2) be the maximal interval of R satisfying $\eta \in (\xi_1, \xi_2)$ and such that

$$0 < \rho_1(\xi) < \rho_2(\xi) < \rho^*(\xi), \quad \forall \xi \in (\xi_1, \xi_2).$$

Put

$$U_0 = \{(r, \theta) : \theta \in (\xi_1, \xi_2),\ 0 < \rho_1(\theta) < r < \rho_2(\theta)\}.$$

From the definition one has

$$\partial U_0 = C_1 \cup C_2, \tag{10}$$

where

$$C_1 = \{(r, \theta) : \theta \in [\xi_1, \xi_2],\ r = \rho_1(\theta)\},$$

$$C_2 = \{(r, \theta) : \theta \in [\xi_1, \xi_2],\ r = \rho_2(\theta)\}.$$

Note that

$$U_0 \subset \Omega_1, \qquad C_2 \subset \partial\omega_2 \cap \partial(\Omega_1 \cap \Omega_2).$$

Hence, from (7) one has

$$\frac{\partial u_1}{\partial n} = \frac{\partial u_2}{\partial n} = 0 \quad \text{on} \quad C_2 \setminus \{a_{21}, a_{22}\}. \tag{11}$$

On the other hand, since $C_1 \subset \partial\omega_1$, it follows that

$$\frac{\partial u_1}{\partial n} = 0 \quad \text{on} \quad C_1 \setminus \{a_{11}, a_{12}\} \tag{12}$$

In view of (10)–(12) one has

$$\frac{\partial u_1}{\partial n} = 0 \quad \text{on} \quad \partial U_0 \setminus \{a_{11}, a_{12}, a_{21}, a_{22}\}. \tag{13}$$

Now, multiplying by u_1 the Laplace equation in u_1 and integrating the result thus obtained on U_0, one has in view of (13)

$$-\int_{U_0} |\nabla u_1|^2 \, dx = \int_{U_0} u_1 \Delta u_1 \, dx = 0.$$

Hence $\nabla u_1 = 0$ a.e. on U_0. Since U_0 is connected, this equality gives

$$u_1 = c \text{ (constant) on } U_0.$$

Since $U_0 \subset \Omega_1$ and since Ω_1 is connected it follows that

$$u_1 = c \quad \text{on} \quad \Omega_1.$$

Thus

$$g = \left.\frac{\partial u_1}{\partial n}\right|_{\Gamma_0} = 0, \quad f|_{\Gamma_0} = c.$$

This is a contradiction. Hence, (i) does not hold.

Now, if (ii) holds then we have

$$U_0 \equiv \{(r, \theta) : \theta \in [0, 2\pi], \ 0 < \rho_1(\theta) < r < \rho_2(\theta)\} = \Omega_1 \setminus \bar{\Omega}_2$$

In this case

$$\partial U_0 = \partial\omega_1 \cup \partial\omega_2, \quad \partial\omega_2 \subset \Omega_1.$$

Using the same technique as for the case (i), we get a contradiction. This completes the proof of Theorem 1.

3. Transformation of (1)–(3) into a variational problem on the rectangle $Q = (\frac{1}{2}, 1) \times (0, 2\pi)$

Let Ω be a domain as in Figure 1. We shall first give a variational formulation of the problem on the (unknown) domain Ω (Subsection 3.1). Since it is rather unwieldy working on an unknown domain, we shall transform the problem into one of finding the coefficients of a variational equation on a fixed known domain. There are many choices for such a domain. We chose to map Ω onto the rectangle $Q = (\frac{1}{2}, 1) \times (0, 2\pi)$. Accordingly, the problem reduces the more standard problem of finding the coefficients of a variational equation on the rectangle Q. This is the object of Subsections 3.2, 3.3 and 3.4. The complete formulation of the problem will be given in Subsection 3.4.

3.1. Variational formulation on Ω

Let v_1 be in $H^1(\Omega)$. Multiplying Equation (1) by v_1 and integrating the result thus obtained on Ω, we get in view of the divergence theorem

$$\int_\Omega \nabla u \nabla v_1 \, dx - \int_{\partial\Omega} v_1 \frac{\partial u}{\partial n} \, ds = 0,$$

where ds is the element of arc length. Since

$$\partial\Omega = \partial U \cup \partial\omega = (\partial U \setminus \Gamma_0) \cup \Gamma_0 \cup \partial\omega$$

we get in view of (2), (3) that

$$\int_{\partial\Omega} v_1 \frac{\partial u}{\partial n} \, ds = \int_{\partial U \setminus \Gamma_0} v_1 \frac{\partial u}{\partial n} \, ds + \int_{\Gamma_0} v_1 g \, ds.$$

Hence

$$\int_\Omega \nabla u \nabla v_1 \, dx = \int_{\Gamma_0} v_1 g \, ds, \quad \forall v_1 \in H^1(\Omega), \ v_1|_{\partial U \setminus \Gamma_0} = 0. \tag{14}$$

The system (1)–(3) is equivalent to the problem of finding (u, Ω) satisfying (14) and

$$u|_{\Gamma_0} = f. \tag{15}$$

Note that we are working on an unknown domain Ω. For reasons mentioned earlier, we shall map Ω onto the rectangle $Q = (\frac{1}{2}, 1) \times (0, 2\pi)$ and derive a variational equation on Q.

3.2. TRANSFORMATION OF Ω ONTO $Q=(\frac{1}{2},1)\times(0,2\pi)$

Consider the transformation

$$T\colon (r, \theta) \longmapsto (R, \theta)$$

with

$$R = \tfrac{1}{2}(\rho^*(\theta) - \rho(\theta))^{-1}(r + \rho^*(\theta) - 2\rho(\theta)),$$

where, we recall, $r = \rho^*(\theta), r = \rho(\theta)$ are the polar equations of the outer and inner boundaries of Ω respectively. We can verify directly that the above transformation is one-to-one continuous from Ω onto the annulus with inner radius $\frac{1}{2}$ and outer radius 1. The polar representation of this annulus is the domain $Q = (\frac{1}{2}, 1) \times (0, 2\pi)$. The images of the outer and the inner boundaries are $\{1\} \times (0, 2\pi)$ and $\{\frac{1}{2}\} \times (0, 2\pi)$ resp., i.e.,

$$T(\rho^*(\theta), \theta) = (1, \theta), \qquad T(\rho(\theta), \theta) = (\tfrac{1}{2}, \theta), \quad \forall\theta \in (0, 2\pi).$$

3.3. VARIATIONAL FORMULATION ON $Q=(\frac{1}{2},1)\times(0,2\pi)$

For definiteness, we assume that

$$\Gamma_0 = \{(\rho^*(\theta) \cos\theta, \rho^*(\theta) \sin\theta)\colon \delta_1 < \theta < \delta_2\},$$

where δ_1, δ_2 are two constants satisfying $-\pi \leqslant \delta_1 < \delta_2 \leqslant \pi$.

Accordingly,

$$\partial U \setminus \Gamma_0 = \{(\rho^*(\theta) \cos\theta, \rho^*(\theta) \sin\theta)\colon \delta_2 \leqslant \theta \leqslant 2\pi - \delta_1\}.$$

Hence if v_1 is a function in $H^1(\Omega)$ satisfying $v_1|_{\partial U\setminus\Gamma_0} = 0$ then

$$v_1(\rho^*(\theta) \cos\theta, \rho^*(\theta) \sin\theta) = 0, \quad \forall\delta_2 \leqslant \theta \leqslant 2\pi - \delta_1.$$

We now transform (14) into a variational equation on Q. Let (u, ρ) be the solution of (14)–(15). From the definition of the transformation T, one has the following relation between the two variables r and R

$$r = 2(\rho^*(\theta) - \rho(\theta))R - \rho^*(\theta) + 2\rho(\theta).$$

Put

$$z(R, \theta) = u(r \cos\theta, r \sin\theta),$$

where r is calculated from R, θ as above.

Similarly, for $v_1 \in H^1(\Omega)$ such that $v_1|_{\partial U \setminus \Gamma_0} = 0$, we set

$$v(R, \theta) = v_1(r \cos \theta, r \sin \theta).$$

From the properties of T, one has

$$v \in H^1(Q), \qquad v(R, 0) = v(R, 2\pi) \quad \forall R \in (\tfrac{1}{2}, 1).$$

Since $v_1|_{\partial U \setminus \Gamma_0} = 0$, one has

$$v(1, \theta) = v_1(\rho^*(\theta) \cos \theta, \rho^*(\theta) \sin \theta) = 0, \qquad \delta_2 \leqslant \theta \leqslant 2\pi - \delta_1.$$

Thus $v_1 \in H^1(\Omega)$ satisfies $v_1|_{\partial U \setminus \Gamma_0} = 0$ if and only if $v \in V$, where

$$V = \{v \in H^1(Q) \colon v(R, 0) = v(R, 2\pi) \ \forall R \in (\tfrac{1}{2}, 1),$$
$$v(1, \theta) = 0 \ \forall \delta_2 \leqslant \theta \leqslant 2\pi - \delta_1\}. \tag{16}$$

By a change of variables, we have

$$\begin{aligned}\int_\Omega \nabla u \nabla v_1 \, \mathrm{d}x &= \int_0^{2\pi} \int_{1/2}^1 \bigg\{ a_{1\rho}(R, \theta) \frac{\partial z}{\partial R} \cdot \frac{\partial v}{\partial R} \\ &\quad + \frac{2(\rho^*(\theta) - \rho(\theta))}{r_\rho(R, \theta)} \left(\frac{\partial z}{\partial \theta} + a_{2\rho}(R, \theta) \frac{\partial z}{\partial R} \right) \\ &\quad \times \left(\frac{\partial v}{\partial \theta} + a_{2\rho}(R, \theta) \frac{\partial v}{\partial R} \right) \bigg\} \, \mathrm{d}R \, \mathrm{d}\theta = A_\rho(z, v),\end{aligned} \tag{17}$$

where

$$r_\rho(R, \theta) = 2R(\rho^*(\theta) - \rho(\theta)) - \rho^*(\theta) + 2\rho(\theta)$$

$$a_{1\rho}(R, \theta) = \tfrac{1}{2} r_\rho(R, \theta)(\rho^*(\theta) - \rho(\theta))^{-1}$$

$$a_{2\rho}(R, \theta) = \frac{\rho(\theta)(\rho^*(\theta))' - \rho'(\theta)\rho^*(\theta) - r_\rho(R, \theta)((\rho^*(\theta))' - \rho'(\theta))}{2(\rho^*(\theta) - \rho(\theta))^2}$$

and

$$\begin{aligned}\int_{\Gamma_0} v_1 g \, \mathrm{d}s &= \int_{\delta_1}^{\delta_2} v_1(\rho^*(\theta) \cos \theta, \rho^*(\theta) \sin \theta) g(\theta) \\ &\quad \times \left\{ |\rho^*(\theta)|^2 + \left| \frac{\mathrm{d}\rho^*}{\mathrm{d}\theta}(\theta) \right|^2 \right\}^{1/2} \mathrm{d}\theta \\ &= \int_{\delta_1}^{\delta_2} v(1, \theta) g(\theta) \left\{ |\rho^*(\theta)|^2 + \left| \frac{\mathrm{d}\rho^*}{\mathrm{d}\theta}(\theta) \right|^2 \right\}^{1/2} \mathrm{d}\theta.\end{aligned}$$

Hence the system (14)–(15) is equivalent to

$$A_\rho(z, v) = L_g v \quad \forall v \in V \tag{18}$$

$$z(1, \theta) = f(\theta) \quad \forall \theta \in (\delta_1, \delta_2), \tag{19}$$

where

$$L_g v = \int_{\delta_1}^{\delta_2} v(1, \theta) g(\theta) \left\{ |\rho^*(\theta)|^2 + \left| \frac{\mathrm{d}\rho^*}{\mathrm{d}\theta}(\theta) \right|^2 \right\}^{1/2} \mathrm{d}\theta, \quad \forall v \in V$$

and thus, the problem of finding (u, ρ) satisfying (14)–(15) is equivalent to that of finding $(z, \rho) \in H^1(Q) \times C[0, 2\pi]$ with $z(R, 0) = z(R, 2\pi)\ \forall R \in (\frac{1}{2}, 1)$, $\rho(0) = \rho(2\pi)$, satisfying (18)–(19).

3.4. The problem will first be considered on the space V (defined in (16)). In order to find (z, ρ) satisfying (18)–(19), we could use the method of variational inequality as in Ang and Vy (1995). However, it is more advantageous for calculations if we apply the Galerkin method and the Lax–Milgram theorem (cf. Brezis (1987), Chapter 5, page 84 or Rodrigues (1987), Chapter 4, page 95). To this end, we shall make a decomposition of z into a sum of a function w in V and a function Φ in $H^1(0, 2\pi)$ (identified with the set of the functions $y(R, \theta)$ in $H^1(Q)$, that are independent from R). Anticipating somewhat on our strategy, we first solve the variational equation (21) for w (treating ρ and Φ as known) using the Lax–Milgram theorem, which applies since the bilinear operator $A_\rho(w, v)$ (defined in (17)) is coercive on the space V. It remains to find ρ and Φ, which will be handled by minimization methods combined with Galerkin approximations. Returning to our decomposition of $z \in H^1(\Omega)$, we choose a Φ in $H^1(0, 2\pi)$ that is 2π-periodic and such that

$$\Phi(\theta) = z(1, \theta) \qquad \forall \theta \in (\delta_2, 2\pi - \delta_1) \tag{20}$$

and we let

$$w(R, \theta) = z(R, \theta) - \Phi(\theta).$$

Then

$$z(R, \theta) = w(R, \theta) + \Phi(\theta).$$

Since z and Φ are 2π-periodic with respect to θ, the function w is also 2π-periodic with respect to θ. Since $z \in H^1(Q)$ and since $\Phi \in H^1(0, 2\pi)$ we have $w \in H^1(Q)$. Finally, one has in view of (20)

$$w(1, \theta) = z(1, \theta) - \Phi(\theta) = 0, \quad \forall \theta \in (\delta_2, 2\pi - \delta_1).$$

Thus we have $w \in V$. It follows that

$$z = \Phi + w, \quad w \in V, \Phi \in H^1(0, 2\pi).$$

Substituting the latter relation into (18)–(19) gives

$$A_\rho(w, v) = -A_\rho(\Phi, v) + L_g v, \quad \forall v \in V, \tag{21}$$

$$w(1, \theta) = -\Phi(\theta) + f(\theta), \quad \forall \theta \in (\delta_1, \delta_2). \tag{22}$$

We see that our problem is equivalent to that of finding a triplet (w, Φ, ρ) satisfying (21) and (22) with $w \in V$, $\Phi \in H^1(0, 2\pi)$, $\rho \in C[0, 2\pi]$, w, Φ and ρ 2π-periodic with respect to θ and ρ satisfying the double inequality

$$0 < \rho(\theta) < \rho^*(\theta) \quad \forall \theta \in [0, 2\pi].$$

Consider (21). The bilinear operator $A_\rho(w, v)$ is coercive on V since Poincaré's inequality holds in V (cf. Rodrigues (1987), Theorem 5.8, Chapter 3, page 75). Hence Lax–Milgram's theorem can be applied to (21) and thus for each (ρ, Φ), there exists, by the latter theorem, a unique solution $w(.; \rho, \Phi) \in V$ of (21). The problem becomes one of finding (ρ, Φ) such that

$$w(1, \theta; \rho, \Phi) = -\Phi(\theta) + f(\theta) \quad \forall \theta \in (\delta_1, \delta_2)$$

which is solved by minimization methods combined with finite dimensional approximations.

The remainder of the paper consists of Section 4 giving a representation of the crack by a point in a function space and Section 5 dealing with finite dimensional approximations and regularization, culminating in a convergence theorem, Theorem 2, which is one of the two main results of this paper.

4. Representation of the crack

4.1. Description of the crack

From now on, we shall study cracks with a boundary made up of two C^1-curves and two edges a_1, a_2 (see Figure 1). In this case, the function ρ is completely defined by the knowledge of a_1, a_2 and of two C^1-functions representing two C^1-curves. The locations of a_1, a_2 (cf. Figure 1) are defined by the angles between the rays Oa_1, Oa_2 with the horizontal axis Ox_1 respectively and by the distances p_0, p_1 from a_1, a_2 to the origin O. More specifically, the crack $r = \rho(\theta)$ can be represented by six elements: two angles α_0, α_1 two distances p_0, p_1 and two functions ρ_1, ρ_2 such that

$$\begin{aligned} \rho(\theta) &= \rho_1(\theta) \quad \alpha_0 < \theta < \alpha_1 \\ &= \rho_2(\theta) \quad \alpha_1 < \theta < \alpha_0 + 2\pi, \end{aligned}$$

where $\rho_1 \in C^1[\alpha_0, \alpha_1]$, $\rho_2 \in C^1[\alpha_1, \alpha_0 + 2\pi]$ and

$$\rho_1(\alpha_0) = \rho_2(\alpha_0 + 2\pi), \qquad \rho_1(\alpha_1) = \rho_2(\alpha_1).$$

If we put for $0 \leqslant t \leqslant 1$

$$\tilde{\rho}_1(t) = \rho_1(\alpha_0 + t(\alpha_1 - \alpha_0)) - (p_0 + t(p_1 - p_0)),$$

$$\tilde{\rho}_2(t) = \rho_2(\alpha_1 + t(\alpha_0 + 2\pi - \alpha_1)) - (p_1 + t(p_0 - p_1)),$$

then

$$\tilde{\rho}_1, \tilde{\rho}_2 \in C^1[0,1], \qquad \tilde{\rho}_1(0) = \tilde{\rho}_1(1) = \tilde{\rho}_2(0) = \tilde{\rho}_2(1) = 0. \tag{23}$$

In terms of the vector $\boldsymbol{\rho} = (\alpha_0, \alpha_1, p_0, p_1, \tilde{\rho}_1, \tilde{\rho}_2)$ satisfying (23) and the double inequality

$$0 < \alpha_0 < \alpha_1 < 2\pi \tag{24}$$

we can define the 2π-periodic function $\rho = S(\boldsymbol{\rho})$ as follows

$$\begin{aligned} S(\boldsymbol{\rho})(\theta) &= \tilde{\rho}_1\left(\frac{\theta - \alpha_0}{\alpha_1 - \alpha_0}\right) + p_0 + \frac{\theta - \alpha_0}{\alpha_1 - \alpha_0}(p_1 - p_0) \quad \alpha_0 < \theta < \alpha_1 \\ &= \tilde{\rho}_2\left(\frac{\theta - \alpha_1}{\alpha_0 + 2\pi - \alpha_1}\right) + p_1 \\ &\quad + \frac{\theta - \alpha_1}{\alpha_0 + 2\pi - \alpha_1}(p_0 - p_1) \quad \alpha_1 < \theta < \alpha_0 + 2\pi. \end{aligned} \tag{25}$$

4.2. CONDITIONS ON $(\alpha_0, \alpha_1, p_0, p_1, \tilde{\rho}_1, \tilde{\rho}_2)$

It is clear from (6) and (25) that the following conditions hold

$$0 < \tilde{\rho}_1(t) + p_0 + t(p_1 - p_0) < \rho^*(\alpha_0 + t(\alpha_1 - \alpha_0)), \tag{26}$$

$$0 < \tilde{\rho}_2(t) + p_1 + t(p_0 - p_1) < \rho^*(\alpha_1 + t(\alpha_0 + 2\pi - \alpha_1)). \tag{27}$$

It results from the foregoing discussion that the function ρ is completely defined by the vector $(\alpha_0, \alpha_1, p_0, p_1, \tilde{\rho}_1, \tilde{\rho}_2)$ satisfying (23)–(27). Define

$$E = \{\boldsymbol{\rho} = (\alpha_0, \alpha_1, p_0, p_1, \tilde{\rho}_1, \tilde{\rho}_2) : (23)\text{–}(27) \text{ hold}\}. \tag{28}$$

Then, instead of identifying ρ we propose to identify vectors $\boldsymbol{\rho}$ in E. Note that E is a subset of the normed space $\mathbb{R}^4 \times (C^1[0,1])^2$, and thus, E can be seen as a metric space with the metric induced by the norm of $\mathbb{R}^4 \times (C^1[0,1])^2$.

The exact values of the angles α_0, α_1, (see Figure 1) are not known. However, their upper bounds and lower bounds are usually available, i.e. there are two known intervals

$$I_1 = [\gamma_1, \gamma_2], \qquad I_2 = [\gamma_3, \gamma_4], \quad 0 < \gamma_1 < \gamma_2 < \gamma_3 < \gamma_4 < 2\pi,$$

such that

$$\alpha_0 \in I_1, \quad \alpha_1 \in I_2.$$

Similarly, we can find an $\alpha > 0$ such that

$$\alpha < \rho(\theta) < \rho^*(\theta) - \alpha \quad \forall \theta \in [0, 2\pi]$$

or equivalently

$$\alpha \leqslant \tilde{\rho}_1(t) + p_0 + t(p_1 - p_0) \leqslant \rho^*(\alpha_0 + t(\alpha_1 - \alpha_0)) - \alpha, \tag{29}$$

$$\alpha \leqslant \tilde{\rho}_2(t) + p_1 + t(p_0 - p_1) \leqslant \rho^*(\alpha_1 + t(\alpha_0 + 2\pi - \alpha_1)) - \alpha, \tag{30}$$

for all $t \in [0, 1]$.

Finally, we shall assume that the unknown vector $\boldsymbol{\rho}$ lies in a compact subset of E. We consider $\tilde{\rho}_1, \tilde{\rho}_2$ in $H^2(0, 1)$ and assume that there is an $M > 0$ such that

$$|p_0| + |p_1| + \|\tilde{\rho}_1\|_{H^2(0,1)} + \|\tilde{\rho}_2\|_{H^2(0,1)} \leqslant M. \tag{31}$$

These remarks being made, we shall identify the unknown vector $\boldsymbol{\rho}$ in the following compact subset of E

$$\begin{aligned}\mathcal{K}(I_1, I_2, M) &= \{\boldsymbol{\rho} = (\alpha_0, \alpha_1, p_0, p_1, \tilde{\rho}_1, \tilde{\rho}_2) \in E: \tilde{\rho}_1, \tilde{\rho}_2 \in H^2(0, 1),\\ &\qquad \alpha_0 \in I_1, \alpha_1 \in I_2 \text{ and (29)–(31) hold}\}.\end{aligned}$$

5. Finite dimensional approximations and regularization

This section is devoted to a regularization of Problem (1)–(3) by finite dimensional approximations. The section is divided into three subsections. In Subsection 5.1, we give a sketch of the method. In Subsection 5.2 we construct finite dimensional approximations of the function spaces considered in the problem. The final subsection 5.3 deals with a discussion of our minimization method and establishes our main result on regularization.

5.1. A DESCRIPTION OF THE REGULARIZATION METHOD

We assume that the system (1)–(3) with respect to the exact Cauchy data (f_0, g_0) has the (unique) solution (u_0, Ω_0), i.e.,

$$\Delta u_0 = 0 \quad \text{on} \quad \Omega$$

$$u_0|_{\Gamma_0} = f_0, \qquad \left.\frac{\partial u_0}{\partial n}\right|_{\Gamma_0} = g_0, \qquad \left.\frac{\partial u_0}{\partial n}\right|_{\partial\omega_0} = 0,$$

where ω_0 is the crack of Ω_0 and $\partial\omega_0$ has the polar representation $r = \rho_0(\theta)$. We note that (f_0, g_0) is probably unknown. We only know a sequence $\{(f_n, g_n)\}$ such that (4) holds, i.e.,

$$(f_n, g_n) \to (f_0, g_0) \quad \text{in} \quad L^2(\Gamma_0) \times L^2(\Gamma_0) \quad \text{as} \quad n \to \infty.$$

Now, from (f_n, g_n) we shall construct a sequence $\{\rho_{0n}\} \subset C[0, 2\pi]$ such that

$$\rho_{0n} \to \rho_0 \quad \text{in} \quad C[0, 2\pi] \quad \text{as} \quad n \to \infty.$$

The ρ_{0n}'s will be constructed in 5.3 as minima of certain functionals.

Note that the variational formulation (21)–(22) reads as follows

$$\begin{aligned} &A_\rho(w, v) = -A_\rho(\Phi, v) + L_{g_0} v \quad \forall v \in V, \\ &w(1, \theta) = -\Phi(\theta) + f_0(\theta) \quad \forall \theta \in (\delta_1, \delta_2). \end{aligned} \tag{32}$$

To construct $\{\rho_{0n}\}$ converging to ρ_0, we shall use finite dimensional approximations of the foregoing problem (with (f_0, g_0) replaced by (f_n, g_n)). We shall construct finite dimensional approximations of the elements of V and of the functions ρ, Φ.

5.2. Finite dimensional approximations of function spaces

Let $\{V_n\} \subset V$, $L_n \subset H^2(0, 2\pi)$ $\{K_n\} \subset H^2(0,1) \cap H_0^1(0,1)$ be sequences of finite dimensional subspaces of 2π-periodic functions satisfying

$$\lim_{n\to\infty} \inf_{\phi \in V_n} \|v - \phi\|_{H^1(Q)} = 0 \quad \forall v \in V,$$

$$\lim_{n\to\infty} \inf_{\psi \in L_n} \|\Phi - \psi\|_{H^2(0,2\pi)} = 0 \quad \forall \Phi \in H^2(0, 2\pi)$$

and let $\{K_n\} \subset H^2(0,1) \cap H_0^1(0,1)$ be sequences of finite dimensional subspaces of 2π-periodic functions satisfying

$$\lim_{n\to\infty} \inf_{\varphi \in K_n} \|\tilde{\rho} - \varphi\|_{H^2(0,1)} = 0 \quad \forall \tilde{\rho} \in H^2(0,1) \cap H_0^1(0,1).$$

There are many ways to construct V_n, K_n, L_n. For example, to construct K_n, L_n, we put

$$\begin{aligned} K_n \;=\; & \Bigg\{ \eta \colon \eta(x) = \sum_{j=0}^{n} (a_j \varphi_1(nx - j) + b_j \varphi_2(nx - j)), \\ & \qquad a_0 = a_n = 0, a_j, b_j \in \mathbb{R}, 0 \leqslant j \leqslant n \Bigg\} \end{aligned}$$

$$\begin{aligned} L_n \;=\; & \Bigg\{ \zeta \colon \zeta(\theta) = \sum_{j=0}^{n} (\alpha_j \varphi_1(2n\pi\theta - j) + \beta_j \varphi_2(2n\pi\theta - j)), \\ & \qquad \alpha_0 = \alpha_n, \alpha_j, \beta_j \in \mathbb{R}, 0 \leqslant j \leqslant n \Bigg\}, \end{aligned}$$

where φ_1, φ_2 are spline functions such that

$$\varphi_i \in C^2(\mathbb{R}),\ \operatorname{supp} \varphi_i \subset [-1, 1], \quad i = 1, 2,$$

$$\varphi_1(0) = 1, \qquad \varphi_1'(0) = 0,$$

$$\varphi_2(0) = 0, \qquad \varphi_2'(0) = 1$$

(see e.g., Marchouk–Agochkov 1985, Chapter 2, page 161).

We now turn to a construction of our finite dimensional subspace V_n of V, recalling that

$$V = \{v \in H^1(Q)\colon v(R,\theta) = v(R, 2\pi)\forall R \in (\tfrac{1}{2}, 1)$$
$$v(1,\theta) = 0 \ \forall\theta \in (\delta_2, 2\pi - \delta_1)\}.$$

We can consider the problem with Cauchy data on a subset of Γ_0 with the variable θ in $(-\pi/m, \pi/m)$ (for m sufficiently large). For notational simplicity we take

$$\Gamma_0 = \{(\rho^*(\theta)\cos\theta, \rho^*(\theta)\sin\theta)\colon -\pi/m < \theta < \pi/m\}$$

i.e., $\delta_1 = -\pi/m, \ \delta_2 = \pi/m$.

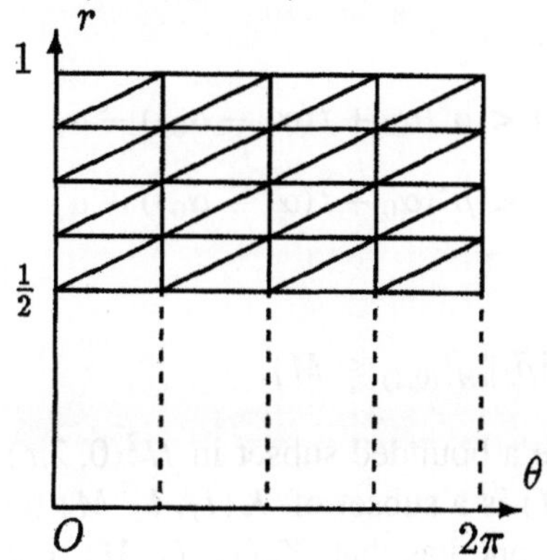

Figure 2.

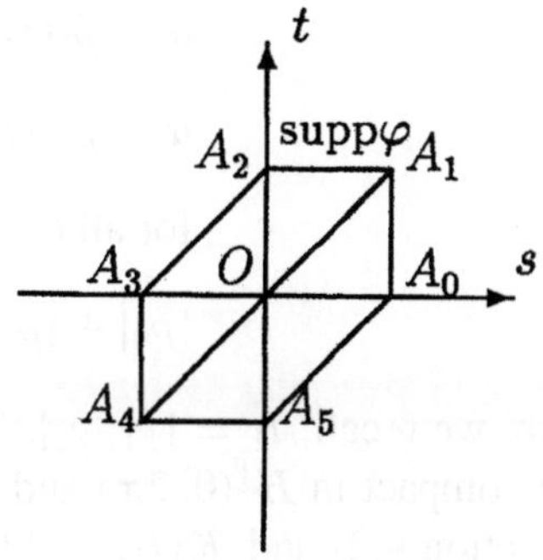

Figure 3.

We consider the triangulation of the rectangle $Q = (\frac{1}{2}, 1) \times (0, 2\pi)$ by the points

$$x_{kl} = \left(\frac{1}{2} + \frac{k}{4nm}, \frac{l\pi}{nm}\right) \qquad 0 \leqslant k, l \leqslant 2nm$$

(cf. Figure 2).

Put

$$V_n = \left\{\xi\colon \xi(R,\theta) = \sum_{k,l=0}^{2nm} c_{kl}\varphi_{kl}(R,\theta), \quad c_{kl} \in \mathbb{R}, 0 \leqslant k, l \leqslant 2nm\right.$$
$$c_{k0} = c_{k(2nm)} \quad 0 \leqslant k \leqslant 2nm$$
$$\left. c_{(2nm)l} = 0 \quad 2n \leqslant l \leqslant n(2m-1)\right\}$$

with φ_{kl} defined by

$$\varphi_{kl}(R,\theta) = \varphi(2(R - \tfrac{1}{2})nm - k, 2nm\theta/\pi - l),$$

where $\varphi \in C(\mathbb{R}^2)$ is a function satisfying

(a) supp $\varphi \subset$ polygon $A_0A_1A_2A_3A_4A_5$ (cf. Figure 3)

(b) φ is an affine function on each triangle $OA_iA_{i+1}(i = 0, 1, 2, 3, 4)$ and OA_5A_0
(c) $\varphi(0) = 1$

(see, e.g., Marchouk–Agochkov 1985, Chapter 2, page 111 for an explicit expression for φ).

Now put

$$\begin{aligned}
\mathcal{L}(M) &= \{\Phi \in H^2(0, 2\pi)\colon \Phi(0) = \Phi(2\pi),\ \|\Phi\|_{H^2(0,2\pi)} \leqslant M\} \\
\mathcal{L}_n(M) &= \{\Phi \in L_n \subset H^2(0, 2\pi)\colon \|\Phi\|_{H^2(0,2\pi)} \leqslant M\} \\
\mathcal{K}_n(I_1, I_2, M) &= \{\boldsymbol{\rho} = (\alpha_0, \alpha_1, p_0, p_1, \tilde{\rho}_1, \tilde{\rho}_2) : \\
&\qquad \alpha_0 \in I_1,\ \alpha_1 \in I_2, \tilde{\rho}_1, \tilde{\rho}_2 \in K_n, \\
&\qquad \alpha < \tilde{\rho}_1(t) + p_0 + t(p_1 - p_0) < \rho^*(\alpha_0 + t(\alpha_1 - \alpha_0)) - \alpha, \\
&\qquad \alpha < \tilde{\rho}_2(t) + p_1 + t(p_0 - p_1) < \rho^*(\alpha_0 + t(\alpha_1 - \alpha_0)) - \alpha \\
&\qquad \text{for all } t \in [0, 1], \text{ and} \\
&\qquad |p_0| + |p_1| + \|\tilde{\rho}_1\|_{H^2(0,1)} + \|\tilde{\rho}_2\|_{H^2(0,1)} \leqslant M\},
\end{aligned}$$

where, we recall, $I_1 = [\gamma_1, \gamma_2]$, $I_2 = [\gamma_3, \gamma_4]$. Since a bounded subset in $H^2(0, 2\pi)$ is relatively compact in $H^1(0, 2\pi)$ and since $\mathcal{K}_n(I_1, I_2, M)$ is a subset of $\mathcal{K}(I_1, I_2, M)$ (defined in Subsection 4.2) and $\mathcal{K}(I_1, I_2, M)$ is compact in E, one has that $\mathcal{K}_n(I_1, I_2, M) \times \mathcal{L}_n(M)$ is compact in $E \times H^1(0, 2\pi)$. Here, we recall

$$\begin{aligned}
E = \{&\boldsymbol{\rho} = (\alpha_0, \alpha_1, p_0, p_1, \tilde{\rho}_1, \tilde{\rho}_2)\colon 0 < \alpha_0 < \alpha_1 < 2\pi, \\
&\tilde{\rho}_1, \tilde{\rho}_2 \in C^1[0, 1],\ \tilde{\rho}_1(0) = \tilde{\rho}_1(1) = \tilde{\rho}_2(0) = \tilde{\rho}_2(1) = 0, \\
&0 < \tilde{\rho}_1(t) + p_0 + t(p_1 - p_0) < \rho^*(\alpha_0 + t(\alpha_1 - \alpha_0)), \\
&0 < \tilde{\rho}_2(t) + p_1 + t(p_0 - p_1) < \rho^*(\alpha_0 + t(\alpha_1 - \alpha_0)) \\
&\text{for all } t \in [0, 1]\},
\end{aligned}$$

$$\begin{aligned}
\mathcal{K}(I_1, I_2, M) &= \{\boldsymbol{\rho} = (\alpha_0, \alpha_1, p_0, p_1, \tilde{\rho}_1, \tilde{\rho}_2) \in E\colon \alpha_0 \in I_1, \alpha_1 \in I_2, \\
&\qquad \alpha < \tilde{\rho}_1(t) + p_0 + t(p_1 - p_0) < \rho^*(\alpha_0 + t(\alpha_1 - \alpha_0)) - \alpha, \\
&\qquad \alpha < \tilde{\rho}_2(t) + p_1 + t(p_0 - p_1) < \rho^*(\alpha_0 + t(\alpha_1 - \alpha_0)) - \alpha \\
&\qquad \text{for all } t \in [0, 1], \text{ and} \\
&\qquad |p_0| + |p_1| + \|\tilde{\rho}_1\|_{H^2(0,1)} + \|\tilde{\rho}_2\|_{H^2(0,1)} \leqslant M\}.
\end{aligned}$$

5.3. The minimization problem and a result on regularization

We consider the following Galerkin approximation of the problem in (32)

$$A_\rho(w, v) = -A_\rho(\Phi, v) + L_{g_n} v \quad \forall v \in V_n, \tag{33}$$

where $\Phi \in \mathcal{L}_n(M)$, $\rho = S(\boldsymbol{\rho})$ (see (25)) with $\boldsymbol{\rho} \in \mathcal{K}_n(I_1, I_2, M)$.

By the Lax–Milgram theorem, for each $(\boldsymbol{\rho}, \Phi)$ in $\mathcal{K}_n(I_1, I_2, M) \times \mathcal{L}_n(M)$, the problem (33) has a unique solution

$$w_n(.; \boldsymbol{\rho}, \Phi) \in V_n.$$

Note that $w_n(.; \boldsymbol{\rho}, \Phi)$ depends continuously on $(\boldsymbol{\rho}, \Phi)$ (cf. Lemma 1 below).

Put

$$J_n(\boldsymbol{\rho}, \Phi) = \int_{\delta_1}^{\delta_2} |(w_n(1, \theta; \boldsymbol{\rho}, \Phi) + \Phi(1, \theta) - f_n(\theta)|^2 \, d\theta.$$

By the compactness of the set $\mathcal{K}_n(I_1, I_2, M) \times \mathcal{L}_n(M)$, the minimization problem

$$\min J_n(\boldsymbol{\rho}, \Phi) \qquad (\boldsymbol{\rho}, \Phi) \in \mathcal{K}_n(I_1, I_2, M) \times \mathcal{L}_n(M) \tag{34}$$

has a solution $(\boldsymbol{\rho}_{0n}, \Phi_{0n})$. We claim that

$$S(\boldsymbol{\rho}_{0n}) \to \rho_0 \quad \text{in} \quad C[0, 2\pi] \text{ as } n \to \infty.$$

In fact, we have

THEOREM 2. *Let $M > 0$, $0 < \gamma_1 < \gamma_2 < \gamma_3 < \gamma_4 < 2\pi$, let $(f_0, g_0) \in (L^2(\Gamma_0))^2$ satisfy $f_0 \not\equiv c$ (constant) or $g_0 \not\equiv 0$ and assume that the problem*

$$\Delta u = 0 \quad \text{on} \quad \Omega$$

$$u|_{\Gamma_0} = f_0, \qquad \left.\frac{\partial u}{\partial n}\right|_{\Gamma_0} = g_0, \qquad \left.\frac{\partial u}{\partial n}\right|_{\partial\omega} = 0$$

has the (unique) solution (u_0, Ω_0) such that:

(i) $u_0|_{\partial U} \in H^2(0, 2\pi)$, $\qquad \|u_0|_{\partial U}\|^2_{H^2(0,1)} < M$,

(ii) *there is a vector $\boldsymbol{\rho}_0 \in \mathcal{K}(I_1, I_2, M)$ such that the polar representation $r = \rho_0(\theta)$ of the boundary $\partial\omega_0$ of the crack $\omega_0 \subset \Omega_0$ satisfies*

$$\rho_0 = S(\boldsymbol{\rho}_0),$$

where $I_1 = [\gamma_1, \gamma_2]$, $I_2 = [\gamma_3, \gamma_4]$ and $\mathcal{K}(I_1, I_2, M)$ is as in Subsection 5.2.
Then one has

$$S(\boldsymbol{\rho}_{0n}) \to \rho_0 \quad \text{in} \quad C[0, 2\pi] \quad \text{as} \quad n \to \infty,$$

where

$$(\boldsymbol{\rho}_{0n}, \Phi_{0n}) \in \mathcal{K}_n(I_1, I_2, M) \times \mathcal{L}_n(M)$$

is solution of the minimization problem (34).

The proof of Theorem 2 is based on the following lemma (the proof of which will be given later).

LEMMA 1. *Let* $(\boldsymbol{\rho}_\infty, \Phi_\infty) \in E \times H^1(0, 2\pi)$ *and let* $(\boldsymbol{\rho}_k, \Phi_k)$ *be a sequence in* $E \times H^1(0, 2\pi)$ *such that*

$$(\boldsymbol{\rho}_k, \Phi_k) \to (\boldsymbol{\rho}_\infty, \Phi_\infty) \text{ in } E \times H^1(0, 2\pi) \quad \text{as} \quad k \to \infty.$$

If $w_n(.; \boldsymbol{\rho}_k, \Phi_k)$ *and* $w_n(.; \boldsymbol{\rho}_\infty, \Phi_\infty)$ *are solutions of problem* (33) *with* $(\boldsymbol{\rho}, \Phi) = (\boldsymbol{\rho}_k, \Phi_k)$ *and* $(\boldsymbol{\rho}, \Phi) = (\boldsymbol{\rho}_\infty, \Phi_\infty)$ *respectively, then*

$$w_n(.; \boldsymbol{\rho}_k, \Phi_k) \to w_n(.; \boldsymbol{\rho}_\infty, \Phi_\infty) \quad in \ H^1(Q) \quad as\, k \to \infty. \tag{35}$$

Moreover, if there exists a sequence $n_1 < n_2 < \cdots < n_k < \cdots$ *of natural integers such that*

$$(\boldsymbol{\rho}_k, \Phi_k) \in \mathcal{K}_{n_k}(I_1, I_2, M) \times \mathcal{L}_{n_k}(M)$$

and if $w(.; \boldsymbol{\rho}_\infty, \Phi_\infty)$ *is solution of problem* (32) *with respect to* $(\rho, \Phi) = (S(\boldsymbol{\rho}_\infty), \Phi_\infty)$*, then*

$$w_{n_k}(.; \boldsymbol{\rho}_k, \Phi_k) \to w(.; \boldsymbol{\rho}_\infty, \Phi_\infty) \quad in\, H^1(Q) \quad as\, k \to \infty. \tag{36}$$

Proof of Theorem 2. Since (35) holds, it results from the properties of the set $\mathcal{L}_n(M)$ that there is a sequence $\bar{\Phi}_{0n} \in \mathcal{L}_n(M)$ such that

$$\bar{\Phi}_{0n} \to \Phi_0 \equiv u_0|_{\partial U} \quad \text{in } H^2(0, 2\pi) \quad \text{as } n \to \infty. \tag{37}$$

Let the vector $\boldsymbol{\rho}_0$ be of the following form

$$\boldsymbol{\rho}_0 = (\alpha_{00}, \alpha_{01}, p_{00}, p_{01}, \tilde{\rho}_{01}, \tilde{\rho}_{02}).$$

We choose

$$(\rho_{n1}, \rho_{n2}) \in K_n$$

such that

$$(\rho_{n1}, \rho_{n2}) \to (\tilde{\rho}_{01}, \tilde{\rho}_{02}) \quad \text{in } H^2(0, 1) \text{ as } n \to \infty. \tag{38}$$

Put

$$\boldsymbol{\rho}_n = (\alpha_{00}, \alpha_{01}, p_{00}, p_{01}, \rho_{n1}, \rho_{n2}) \in \mathcal{K}_n(I_1, I_2, M).$$

By (39), one has:

$$\boldsymbol{\rho}_n \to \boldsymbol{\rho}_0 \quad \text{in } E \text{ as } n \to \infty. \tag{39}$$

By Lemma 1, one has

$$\begin{aligned} J_n(\boldsymbol{\rho}_n, \bar{\Phi}_{0n}) &\equiv \int_{\delta_1}^{\delta_2} |w_n(1, \theta; \boldsymbol{\rho}_n, \bar{\Phi}_{0n}) + \bar{\Phi}_{0n}(\theta) - f_n(\theta)|^2 \, \mathrm{d}\theta \\ &\to \int_{\delta_1}^{\delta_2} |w(1, \theta; \boldsymbol{\rho}_0, \Phi_0) + \Phi_0(\theta) - f_0(\theta)|^2 \, \mathrm{d}\theta = 0. \end{aligned} \tag{40}$$

On the other hand, $(\boldsymbol{\rho}_{0n}, \Phi_{0n})$ is solution of the minimization problem (34). Hence

$$J_n(\boldsymbol{\rho}_{0n}, \Phi_{0n}) \leqslant J_n(\boldsymbol{\rho}_n, \bar{\Phi}_{0n}). \tag{41}$$

In view of (41), (42), one has

$$\lim_{n\to\infty} J_n(\boldsymbol{\rho}_{0n}, \Phi_{0n}) = 0. \tag{42}$$

We claim that

$$\rho_{0n} \to \rho_0 \quad \text{in } C[0, 2\pi] \text{ as } n \to \infty,$$

where

$$\rho_{0n} = S(\boldsymbol{\rho}_{0n}), \qquad \rho_0 = S(\boldsymbol{\rho}_0).$$

By the compactness of $\mathcal{K}(I_1, I_2, M) \times \mathcal{L}(M)$ in $E \times H^1(Q)$ and the continuity of S, it is sufficient to prove that ρ_0 is the limit of all subsequences $S(\boldsymbol{\rho}_{0n_k})$ where $(\boldsymbol{\rho}_{0n_k}, \Phi_{0n_k})$ is a convergent subsequence of $(\boldsymbol{\rho}_{0n}, \Phi_{0n})$. Assume that

$$(\boldsymbol{\rho}_{0n_k}, \Phi_{0n_k}) \to (\boldsymbol{\rho}_0^\infty, \Phi_0^\infty) \in E \times H^1(Q) \quad \text{as } k \to \infty.$$

We prove that

$$S(\boldsymbol{\rho}_0^\infty) = \rho_0.$$

From Lemma 1, one has

$$w_{n_k}(.; \boldsymbol{\rho}_{0n_k}, \Phi_{0n_k}) \to w(.; \boldsymbol{\rho}_0^\infty, \Phi_0^\infty) \quad \text{in } H^1(Q) \text{ as } k \to \infty. \tag{43}$$

Combining (43), (44) gives

$$\int_{\delta_1}^{\delta_2} |w(1, \theta; \boldsymbol{\rho}_0^\infty, \Phi_0^\infty) + \Phi_0^\infty(\theta) - f_0(\theta)|^2 \, \mathrm{d}\theta = 0$$

that is

$$w(1, \theta; \boldsymbol{\rho}_0^\infty, \Phi_0^\infty) + \Phi_0^\infty(\theta) = f_0(\theta) \quad \delta_1 \leqslant \theta \leqslant \delta_2.$$

By (25), using $\boldsymbol{\rho}_0^\infty$ we can write the equation of the boundary of a crack as follows

$$r = \rho_0^\infty(\theta) \equiv S(\boldsymbol{\rho}_0^\infty).$$

We shall denote this crack by ω_0^∞. Put

$$\Omega_0^\infty = U \setminus \bar{\omega}_0^\infty$$

and

$$u_0^\infty(r, \theta) = w(R_{\rho_0^\infty}(r, \theta); \boldsymbol{\rho}_0^\infty, \Phi_0^\infty) + \Phi_0^\infty(\theta),$$

where

$$R_{\rho_0^\infty}(r,\theta) = \tfrac{1}{2}(\rho^*(\theta) - \rho_0^\infty(\theta))^{-1}(r + \rho^*(\theta) - 2\rho_0^\infty(\theta)).$$

Then u_0^∞ is solution of the following problem (see Section 3)

$$\begin{aligned} &\Delta u_0^\infty = 0 \quad \text{on } \Omega_0^\infty, \\ &u_0^\infty|_{\Gamma_0} = f_0, \quad \left.\frac{\partial u_0^\infty}{\partial n}\right|_{\Gamma_0} = g_0, \\ &\left.\frac{\partial u_0^\infty}{\partial n}\right|_{\partial\omega_0^\infty} = 0. \end{aligned} \tag{44}$$

Note that, we also have

$$\begin{aligned} &\Delta u_0 = 0 \quad \text{on } \Omega_0 \\ &u_0|_{\Gamma_0} = f_0, \quad \left.\frac{\partial u_0}{\partial n}\right|_{\Gamma_0} = g_0, \\ &\left.\frac{\partial u_0}{\partial n}\right|_{\partial\omega_0} = 0. \end{aligned} \tag{45}$$

By Theorem 1, we get in view of (45), (46) that

$$(u_0^\infty, \Omega_0^\infty) = (u_0, \Omega_0) \tag{46}$$

which gives

$$\rho_0^\infty(\theta) = \rho_0(\theta) \quad 0 \leqslant \theta \leqslant 2\pi. \tag{47}$$

The proof of Theorem 2 will be completed once Lemma 1 is proved.

Proof of Lemma 1. We shall give the proof of (37) only. Since the proof of (36) is similar to (in fact, easier than) that of (37), it is omitted. We first show that there exists a $C_0 > 0$ satisfying

$$A_{\rho_k}(v,v) \geqslant C_0\|v\|^2_{H^1(Q)} \quad \forall v \in V_k, \quad k = 1,2, \tag{48}$$

where $\rho_k = S(\boldsymbol{\rho}_k)$ as in (25). Denote by r_{ρ_k} the function

$$r_{\rho_k}(R,\theta) = 2R(\rho^*(\theta) - \rho_k(\theta)) - \rho^*(\theta) + 2\rho_k(\theta), \quad R \in (\tfrac{1}{2}, 1).$$

It is obvious that

$$0 < \rho_k(\theta) < r_{\rho_k}(R,\theta) < \rho^*(\theta) \quad \forall (R,\theta) \in Q. \tag{49}$$

Since $\boldsymbol{\rho}_k \in \mathcal{K}(I_1, I_2, M)$, one has

$$\rho^*(\theta) - \alpha \geqslant \rho_k(\theta) \geqslant \alpha > 0 \quad \forall \theta \in [0, 2\pi], \quad k = 1,2, \tag{50}$$

(see Subsection 4.3).

Now from the definition of A_ρ (see Section 3), one gets in view of (50)–(51) that

$$A_{\rho_k}(v, v) \geqslant \frac{\alpha}{2M_1} \int_Q \left| \frac{\partial v}{\partial R} \right|^2 \mathrm{d}R\, \mathrm{d}\theta \quad \forall v \in V_{n_k}, \tag{51}$$

where

$$Q = (\tfrac{1}{2}, 1) \times (0, 2\pi) \quad \text{and } M_1 = \|\rho^*\|_{W^{1,\infty}(Q)}.$$

Similarly, from the definition of A_ρ, we can derive directly the inequality

$$A_{\rho_k}(v, v) \geqslant \frac{2\alpha}{M_1} \int_Q \left(\frac{\partial v}{\partial \theta} + a_{2\rho_k}(R, \theta) \frac{\partial v}{\partial R} \right)^2 \mathrm{d}R\, \mathrm{d}\theta$$

which implies

$$A_{\rho_k}(v, v) + \frac{2\alpha}{M_1} \int_Q |a_{2\rho_k}(R, \theta)|^2 \left(\frac{\partial v}{\partial R} \right)^2 \mathrm{d}R\, \mathrm{d}\theta \geqslant \frac{2\alpha}{M_1} \int_Q \left(\frac{\partial v}{\partial \theta} \right)^2 \mathrm{d}R\, \mathrm{d}\theta.$$

But, since $\boldsymbol{\rho}_k \in \mathcal{K}(I_1, I_2, M)$, we can find an $M_2 > 0$ such that

$$|\rho_k'| \leqslant M_2 \quad \text{a.e. } \theta \in [0, 2\pi] \tag{52}$$

and by direct computations, one has

$$|a_{2\rho_k}(R, \theta)| \leqslant M_1(M_1 + M_2)\alpha^{-2}.$$

Hence

$$A_{\rho_k}(v, v) + \frac{2(M_1 + M_2)}{\alpha} \int_Q \left(\frac{\partial v}{\partial R} \right)^2 \mathrm{d}R\, \mathrm{d}\theta \geqslant \frac{2\alpha}{M_1} \int_Q \left(\frac{\partial v}{\partial \theta} \right)^2 \mathrm{d}R\, \mathrm{d}\theta. \tag{53}$$

Combining (53), (54) gives

$$A_{\rho_k}(v, v) \geqslant C_0 \|v\|^2_{H^1(Q)},$$

where

$$C_0 = \min \left\{ \frac{\alpha}{2M_1}, \frac{2\alpha^3}{M_1(\alpha^2 + 4M_1(M_1 + M_2))} \right\}.$$

Now, we show that

$$w_{n_k}(.; \boldsymbol{\rho}_k, \Phi_k) \to w(.; \boldsymbol{\rho}_\infty, \Phi_\infty) \quad \text{in } H^1(Q) \text{ as } k \to \infty.$$

For simplicity, we put

$$\begin{aligned} w_k &= w_{n_k}(.; \boldsymbol{\rho}_k, \Phi_k), & w_0 &= w(.; \boldsymbol{\rho}_\infty, \Phi_\infty), \\ A_k &= A_{\rho_k}, & A_0 &= A_\rho, \\ L_k &= L_{g_{n_k}}, & L_0 &= L_{g_0}, \end{aligned}$$

where L_g is as in Subsection 3.3.

From the assumption about $\{V_n\}$, there exists a sequence $\{\bar{w}_k\}$, $\bar{w}_k \in V_{n_k}$ such that

$$\bar{w}_k \to w_0 \quad \text{in } V \text{ as } k \to \infty. \tag{54}$$

One has

$$A_k(w_k, w_k - \bar{w}_k) = L_k(w_k - \bar{w}_k) - A_k(\Phi_k, w_k - \bar{w}_k),$$

$$A_0(w_0, w_k - \bar{w}_k) = L_0(w_k - \bar{w}_k) - A_0(\Phi_\infty, w_k - \bar{w}_k).$$

From (49), it follows that

$$A_k(w_k - \bar{w}_k, w_k - \bar{w}_k) \geqslant C_0\|w_k - \bar{w}_k\|_V^2. \tag{55}$$

But

$$\begin{aligned} &A_k(w_k - \bar{w}_k, w_k - \bar{w}_k) \\ &\quad = -(A_k - A_0)(w_0, w_k - \bar{w}_k) - (A_k - A_0)(\Phi_\infty, w_k - \bar{w}_k) \\ &\qquad - A_k(\bar{w}_k - w_0, w_k - \bar{w}_k) + A_k(\Phi_\infty - \Phi_k, w_k - \bar{w}_k) \\ &\qquad + (L_k - L_0)(w_k - \bar{w}_k). \end{aligned}$$

The last three terms in the right-hand side of the foregoing equality tend to 0 as $k \to \infty$ (on account of (55), (4) and of the assumption $\Phi_k \to \Phi$ in $H^1(0, 2\pi)$). Now, if

$$(A_k - A_0)(w_0, w_k - \bar{w}_k) \to 0 \quad \text{as } k \to \infty, \tag{56}$$

and

$$(A_k - A_0)(\Phi_\infty, w_k - \bar{w}_k) \to 0 \quad \text{as } k \to \infty, \tag{57}$$

then

$$A_k(w_k - \bar{w}_k, w_k - \bar{w}_k) \to 0 \quad \text{as } k \to \infty. \tag{58}$$

Combining (55), (56), (59) gives

$$w_k \to w_0 \quad \text{in } H^1(Q) \text{ as } k \to \infty$$

which is the desired result. Hence, the proof of the Lemma will be completed once (57), (58) are established. Thus, we show that indeed (57) and (58) hold. We prove (57) (the proof of (58) which is similar is omitted).

Put

$$a_{11}(\rho)(R, \theta) = \frac{2(\rho^*(\theta) - \rho(\theta))}{r_\rho(R, \theta)},$$

$$a_{12}(\rho)(R, \theta) = \frac{2a_{2\rho}(R, \theta)(\rho^*(\theta) - \rho(\theta))}{r_\rho(R, \theta)},$$

$$a_{21}(\rho)(R,\theta) = a_{12}(\rho)(R,\theta),$$

$$a_{22}(\rho)(R,\theta) = \frac{2a_{2\rho}^2(R,\theta)(\rho^*(\theta) - \rho(\theta))}{r_\rho(R,\theta)} + a_{1\rho}(R,\theta),$$

where $a_{1\rho}, a_{2\rho}, r_\rho$ are defined Subsection 3.3.

By direct computations, one has

$$|(A_k - A_0)(w_0, w_k - \bar{w}_k)|^2$$
$$\leqslant C\|w_k - \bar{w}_k)\|_V^2 \int_Q \left[\left(\frac{\partial w_0}{\partial R}\right)^2 + \left(\frac{\partial w_0}{\partial \theta}\right)^2\right]$$
$$\times \sum_{i,j=1}^{2} (a_{ij}(\rho_\infty) - a_{ij}(\rho_k))^2 \,\mathrm{d}R\,\mathrm{d}\theta. \tag{59}$$

Using (51), (53) one can prove that

$$|a_{ij}(\rho_\infty) - a_{ij}(\rho_k)| \leqslant C(M) \quad k = 1, 2, \ldots, i, j = 1, 2, \tag{60}$$

where $C(M)$ depends only on M.

On the other hand, we have

$$a_{ij}(\rho_k)(R,\theta) \to a_{ij}(\rho_\infty)(R,\theta) \text{ a.e. } (R,\theta) \in Q \quad i, j = 1, 2. \tag{61}$$

Hence in view of (61) and (62), we have by Lebesgue's dominated convergence theorem

$$\int_Q \left[\left(\frac{\partial w_0}{\partial R}\right)^2 + \left(\frac{\partial w_0}{\partial \theta}\right)^2\right] \sum_{i,j=1}^{2} (a_{ij}(\rho_\infty) - a_{ij}(\rho_k))^2 \,\mathrm{d}R\,\mathrm{d}\theta \to 0 \quad \text{as } k \to \infty.$$

In view of (60) and of the latter relation, (57) is proved. This establishes Lemma 1 and thus completes the proof of Theorem 2.

Acknowledgments

The authors wish to express their heart-felt thanks and appreciation to the referees for the criticisms of the first version and for the most valuable suggestions leading to the present improved version of our paper.

References

Ang, D.D. and Vy, L.K. (1995). Domain identification for harmonic function. *Acta Applicandae Mathematicae* **38**, 217–238.

Ang, D.D., Trong, D.D. and Yamamoto, M. (1996). Unique continuation and identification of boundary of an elastic body. *Journal of Inverse and Ill-posed Problem* **3**(6), 417–428.

Ang, D.D., Trong, D.D. and Yamamoto, M. (1997). *Crack Identification in Plane Nonhomogeneous Isotropic Elasticity* (to appear).

Bui, H.D. (1994). *Inverse Problems in the Mechanics of Materials: an Introduction*, CRC Press Inc.

Brezis, H. (1987). *Analyse Fonctionelle, Théorie et Applications,* Masson.

Kubo, S. (1990). Requirements for uniqueness crack identification from electric potential distribution, in *Inverse Problems in Engineering Sciences*, Springer-Verlag.

Lions, J.L. and Magenes, E. (1972). *Non-homogeneous Boundary Value Problems and Applications,* **I**, Springer-Verlag.

Marchouk, G. and Agochkov, V. (1985). *Introduction aux Méthodes des Élément Finis*, Edition Mir, Moscou.

McIver, M. (1991). An inverse problem in electro-magnetic crack detection, *IMA Journal on Applied Mathematics* **47**, 127–145.

Rodrigues, J.F. (1987). *Obstacle Problem in Mathematical Physics,* North-Holland, Amsterdam.

International Journal of Fracture **93:** 87–114, 1998.

Singularity-reduced integral equations for displacement discontinuities in three-dimensional linear elastic media

SONGSHAN LI and MARK E. MEAR
Department of Aerospace Engineering and Engineering Mechanics, The University of Texas at Austin, Austin, TX 78712, USA; e-mail: mmear@max.ae.utexas.edu

Received 21 October 1997; accepted in revised form 13 July 1998

Abstract. A systematic procedure is followed to develop singularity-reduced integral equations for displacement discontinuities in homogeneous linear elastic media. The procedure readily reproduces and generalizes, in a unified manner, various integral equations previously developed by other means, and it leads to a new stress relation from which a general weakly-singular, weak-form traction integral equation is established. An isolated discontinuity is treated first (including, as special cases, cracks and dislocations) after which singularity-reduced integral equations are obtained for cracks in a finite domain. The first step in the development is to regularize Somigliana's identity by utilizing a stress function for the stress fundamental solution to effect an integration by parts. The resulting integral equation is valid irrespective of the choice of stress function (as guaranteed by a certain 'closure condition' established for the integral operator), but certain particular forms of the stress function are introduced and discussed, including one which admits an interpretation as a 'line discontinuity'. A singularity-reduced integral equation for the displacement gradients is then obtained by utilizing a relation between the stress function and the stress fundamental solution along with the closure condition. This construction does not rely upon a particular choice of stress function, and the final integral equation (which is a generalization of Mura's (1963) formula) has a kernel which is a simple function of the stress fundamental solution. From this relation, singularity-reduced integral equations for the stress and traction are easily obtained. The key step in the further development is the construction of an alternative stress integral equation for which a differential operator has been 'factored out' of the integral. This is accomplished by, in essence, establishing a stress function for the stress field induced by the discontinuity. A weak-form traction integral equation is then readily obtained and involves a kernel which is only weakly-singular. The nonuniqueness of this kernel is discussed in detail and it is shown that, at least in a certain sense, the kernel which is given is the simplest possible. The results for an isolated discontinuity are then adapted to treat cracks in a finite domain. In doing so, emphasis is given to the development of weakly-singular, weak-form displacement and traction integral equations since these form the basis of an effective numerical procedure for fracture analysis (Li et al., 1998), and such equations are presented for both elastostatics and elastodynamics. A noteworthy aspect of the development is that there is no need to introduce Cauchy principal value integrals much less Hadamard finite part integrals. Finally, the utility of the systematic procedure presented here for use in obtaining singularity-reduced integral equations for other unbounded media (viz. the half-space and bi-material) is indicated.

Key words: Discontinuities, cracks, integral equations, boundary elements.

1. Introduction

Displacement discontinuities in linear elastic media are conveniently modeled using singular integral equations. A widely applied method of deriving such equations utilizes Betti's reciprocal theorem, along with the solution for a concentrated force in an unbounded domain, to obtain an expression for the displacement field which is known as Somigliana's identity. From this result integral relations for other field quantities of interest can be established, but the integral relations obtained in the most direct fashion are typically not as convenient as 'singularity-reduced integral representations' obtained by additional manipulations involving, in essence, integration by parts.

In the analysis of dislocations, the relative displacement across the surface of the discontinuity is constant and specified, and interest is in determining the fields induced by the dislocation. Direct utilization of Somigliana's identity results in relations which involve integrals over the surface of the discontinuity, but a regularization of the type described above can be used to obtain line integral representations for the quantities of interest. For dislocations in a whole space such line integral representations for stress, displacement gradient and energy have been extensively studied, and various forms for the results have been obtained (e.g. Hirth and Lothe, 1982). It should be noted that since the displacement field is discontinuous across the surface of the dislocation, it is not determined entirely by the dislocation loop but rather information about the surface of the discontinuity is required. As discussed below, however, such information can be cast in terms of a line integral of a certain type which involves a 'multi-valued' kernel.

For cracks, the relative crack-face displacement is usually not known in advance but instead is to be determined such that traction boundary conditions are satisfied on the crack surface. The quantities of primary interest are the stress intensity factors which characterize the amplitude of the singular stress field in the neighborhood of the crack front, and these quantities can be evaluated once the relative crack-face displacement has been determined.

Apart from a small number of special cases, the relative crack-face displacement cannot be established in closed form and resort must be made to numerical methods. Boundary element methods based upon singular integral equations are attractive for this purpose since typically only the boundary of the body and the surface of the crack require discretization rather than the domain containing the crack. However, when applied to fracture problems the standard boundary element method based on Somigliana's identity suffers from an indeterminacy associated with the geometric coincidence of the 'upper' and 'lower' crack surfaces (e.g. Cruse, 1988).

To circumvent this difficulty an integral equation for the traction on the crack surface must be developed, but the traction integral equation obtained directly from Somigliana's identity involves a hypersingular kernel which requires a special interpretation and which is challenging to evaluate (e.g. Guiggiani et al., 1991; Balakrishna et al., 1994). Further, for the integral equation to be valid the gradients of the relative crack-face displacements must be continuous, and this requirement is both difficult and computationally expensive to implement (e.g. Guiggiani et al., 1991; Martin and Rizzo 1996; Muci-Küchle and Rudolphi, 1994).

A singularity-reduced form of the traction integral equations which involves only Cauchy singular kernels has been established by Bui (1977) and Weaver (1977) for planar cracks and by Sládek and Sládek (1982) for curved cracks. The essential step in deriving these equations is to shift the differentiation from the kernel to relative crack face displacements through an integration by parts (hence the kernel involves gradients of relative crack face displacements rather than the displacements themselves). It is important to note, however, that even though the singularity of the integral kernels has been reduced to Cauchy type, there remains the requirement that the derivatives of the relative crack-face displacements are continuous (e.g. Martin and Rizzo, 1996; Bonnet and Bui, 1993).

To reduce this regularity requirement on the crack-face displacement data, Galerkin solution strategies based upon 'weak-form' statements of the traction integral equation have been pursued. The straightforward Galerkin approach applied to the hypersingular traction integral equation gives rise to a double surface integral with the inner integral still hypersingular, hence treatment of strongly singular integrands is still necessary. An alternative weak-form traction integral equation which contains only weakly singular kernels has been developed by Gu and Yew (1988) for the restricted case of mode-I planar cracks in an unbounded domain. Xu and Ortiz (1993) established a weakly-singular formulation for curved cracks in an unbounded domain by modeling the crack as a distribution of dislocation loops. Their formulation is for the strain-energy of the body (rather than being a weak form statement of the traction integral equation) and forms the basis of a numerical procedure obtained through a variational statement of the boundary value problem.

In this paper, we follow a systematic procedure to develop singularity-reduced integral representations for displacement discontinuities in three-dimensional linear elastic media. With the procedure we readily reproduce and generalize, in a unified fashion, various of the singularity-reduced integral equations originally obtained by other means, and we developed a new stress relation from which a general weakly-singular, weak form traction integral equation is established. The starting point for the development is Somigliana's identity which, in its most general form, involves two kernels: one which corresponds to the displacement field and the other which corresponds to the stress field induced by a unit concentrated force in an unbounded medium. Of course, the stress fundamental solution is the most singular of the two, and the term involving it is also the more difficult to regularize. For this reason attention is initially focused upon this kernel by considering a discontinuity in an unbounded domain for which the displacement kernel is absent.

A regularization of Somigliana's identity is achieved through an integration by parts facilitated by the introduction of a stress function for the stress fundamental solution. The kernel of the resulting integral equation is, naturally, the stress function itself which is necessarily 'multi-valued' in a certain sense. That the choice of the (nonunique) stress function does not effect the validity of the integral equation is guaranteed by a 'closure condition' established for the integral operator; this closure condition plays a central role in much of the development. A singularity-reduced integral equation for the displacement gradients is then obtained from this displacement relation by utilizing a connection between the stress function and the stress fundamental solution along with the closure condition. It should be noted that the construction which is followed does not hinge upon a particular choice of stress function, it applies for general anisotropy, and the final integral equation has a kernel which is a simple function of the stress fundamental solution. From the displacement gradient relation, singularity-reduced integral equations for the stress and traction are readily obtained.

Next, an alternative stress integral equation is established for which a differential operator has been 'factored out' of the integral such that the remaining kernel is only weakly singular. This is accomplished by, in essence, constructing a stress function for the stress field induced by the discontinuity. While the existence of this alternative stress relation is guaranteed for an anisotropic medium, the kernel of the integral relation is explicitly established only for elastic isotropy. From this alternative stress relation a weakly-singular, weak-form traction integral equation follows from a straightforward process involving an integration by parts.

The development given for discontinuities in an unbounded domain provides a basis for an extension to treat cracks in finite domains, and a suite of singularity-reduced integral equations are also established for this case. In particular, a pair of weakly-singular, weak-form displacement and traction integral equations are constructed; these two equations can be utilized to obtain a 'symmetric' Galerkin formulation well suited for the numerical treatment of cracks in finite domains (Li et al., 1998). Finally, a straightforward generalization of the weakly-singular, weak-form integral equations to elastodynamics is provided.

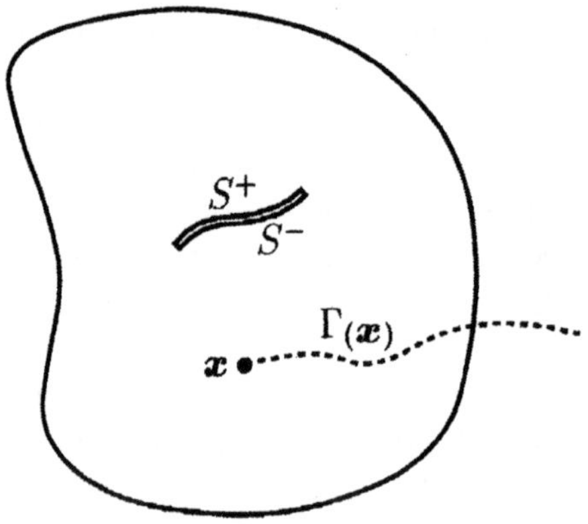

Figure 1. Schematic of a discontinuity in an unbounded domain.

2. Discontinuities in a whole space

Consider a discontinuity embedded in an unbounded homogeneous linear elastic medium which is free of remote loading, Figure 1. Let the discontinuity have 'upper' surface S^+ and (geometrically coincident) 'lower' surface S^-, and assume that the tractions acting on the discontinuity are such that[1] $\Sigma t_j \equiv t_j^+ + t_j^- = 0$. Specializing Somigliana's identity to this case gives an integral relation for the displacement at a point $\boldsymbol{x}$ in the domain as

$$u_p(\boldsymbol{x}) = -\int_{S^+} n_i^+(\boldsymbol{\xi}) S_{ij}^p(\boldsymbol{\xi} - \boldsymbol{x}) \Delta u_j(\boldsymbol{\xi}) \, \mathrm{d}s(\boldsymbol{\xi}), \tag{1}$$

in which S_{ij}^p is the stress fundamental solution (see Appendix A), n_i^+ is the unit normal to the 'upper' surface (pointing 'into' the discontinuity) and $\Delta u_j \equiv u_j^+ - u_j^-$ is the relative displacement across the discontinuity.

For a dislocation Δu_j is a prescribed constant and interest is in evaluating the displacement, stress and energy induced by the dislocation, whereas for a crack typically the tractions acting on the crack faces are prescribed and interest is in evaluating Δu_j (from which the stress intensity factors can be extracted). In the latter case, Somigliana's identity is clearly degenerate in that it contains no information about the crack-face tractions, and for this reason a traction integral equation must be established. However, the traction integral equation derived directly from (1) involves a hypersingular integral which, as discussed previously, is difficult to treat.

To obtain integral equations which are more convenient for analytical and numerical treatment, we seek to reduce the singularity involved in the integral representation rather than to retain the kernels resulting directly from use of Somigliana's identity. Toward this end we consider first the development of singularity-reduced integral equations for the displacement field.

[1] Here and in what follows the superscripts + and − are used to indicate that the quantity is associated with the 'upper' and 'lower' surface of the discontinuity, respectively.

2.1. Integral expressions for displacements

As a means to reduce the singularity for the kernel in (1) we introduce a stress function $\psi^p_{mj}(\boldsymbol{\xi}, \boldsymbol{x})$ for the stress fundamental solution $S^p_{ij}(\boldsymbol{\xi} - \boldsymbol{x})$ as

$$S^p_{ij}(\boldsymbol{\xi} - \boldsymbol{x}) = \varepsilon_{ikm} \frac{\partial \psi^p_{mj}(\boldsymbol{\xi}, \boldsymbol{x})}{\partial \xi_k}, \tag{2}$$

where ε_{ikm} is the usual permutation symbol. Clearly the stress function is not unique (although any two possibilities differ only by a gradient of a function), but an essential feature of all such functions is that they are (at the very least) singular along a path, or 'cut', which contains the source point $\boldsymbol{\xi} = \boldsymbol{x}$ (see the discussion given in Appendix B).

As shown in Appendix B, one such stress function for the stress fundamental solution is given by

$$\psi^p_{mj}(\boldsymbol{\xi}, \boldsymbol{x}) = \int_{\Gamma_{(\boldsymbol{x})}} \varepsilon_{mrt} S^p_{rj}(\boldsymbol{\xi} - \boldsymbol{z})\, \mathrm{d}z_t, \tag{3}$$

in which $\Gamma_{(\boldsymbol{x})}$ is a path which originates at $\boldsymbol{x}$ and extends to infinity in $\boldsymbol{\xi}$-space. This expression for the stress function proves convenient for certain aspects of the development to follow, but further below alternative expressions are discussed and a simple explicit expression is given for isotropic materials. We remark that for the stress function (2), the quantity $u_p = a_{mj}\psi^p_{mj}$ corresponds to a displacement field (with zero body force) for arbitrary choices for the constants[2] a_{mj}; this result follows immediately from the fact that the stress fundamental solution S^p_{mj} possesses this property.

To proceed, we note that

$$\begin{aligned} n^+_i(\boldsymbol{\xi}) S^p_{ij}(\boldsymbol{\xi} - \boldsymbol{x}) \Delta u_j(\boldsymbol{\xi}) &= D_m \psi^p_{mj}(\boldsymbol{\xi}, \boldsymbol{x}) \Delta u_j(\boldsymbol{\xi}) \\ &= D_m[\psi^p_{mj}(\boldsymbol{\xi}, \boldsymbol{x}) \Delta u_j(\boldsymbol{\xi})] - \psi^p_{mj}(\boldsymbol{\xi}, \boldsymbol{x}) D_m \Delta u_j(\boldsymbol{\xi}), \end{aligned} \tag{4}$$

where

$$D_m = n^+_i \varepsilon_{ikm} \frac{\partial}{\partial \xi_k}$$

is a tangential operator. To verify that D_m is in fact a tangential operator, let η_1 and η_2 be surface coordinates such that the Cartesian coordinates of points on the surface S^+ (relative to a fixed coordinate system) are given by $\xi_j = \xi_j(\eta_1, \eta_2)$. Then the unit normal vector $\boldsymbol{n}^+$ to S^+ is given by $\boldsymbol{n}^+ = (\boldsymbol{s} \times \boldsymbol{t})/\mathcal{J}$ where $\boldsymbol{s} = \partial\boldsymbol{\xi}/\partial\eta_1$ and $\boldsymbol{t} = \partial\boldsymbol{\xi}/\partial\eta_2$ are vectors in the tangent plane to the surface and where $\mathcal{J} = | \boldsymbol{s} \times \boldsymbol{t} |$. It then follows that

$$D_m = \frac{1}{\mathcal{J}} \left(\frac{\partial}{\partial \eta_1} \frac{\partial \xi_m}{\partial \eta_2} - \frac{\partial}{\partial \eta_2} \frac{\partial \xi_m}{\partial \eta_1} \right), \tag{5}$$

[2] Indeed, the function ψ^p_{mj} admits an interpretation as a 'line discontinuity' along the (specified) path $\Gamma_{(\boldsymbol{x})}$ and further, a semi-infinite discontinuity strip in $\boldsymbol{\xi}$-space is generated by interpreting $a_{mj}(\boldsymbol{x})$ as a density and integrating $a_{mj}(\boldsymbol{x})\psi^p_{mj}(\boldsymbol{\xi}, \boldsymbol{x})$, with respect to $\boldsymbol{x}$, along a prescribed arc.

so that $D_m \Delta u_j$ can be evaluated in terms of data Δu_j given only on the surface of the discontinuity. We also note that, in terms of the tangential operator D_m, Stokes' theorem on the surface S^+ takes the form

$$\int_{S^+} D_m f(\boldsymbol{\xi}) \, \mathrm{d}s(\boldsymbol{\xi}) = \oint_{\partial S^+} f(\boldsymbol{\xi}) \, \mathrm{d}\xi_m, \tag{6}$$

in which $f(\boldsymbol{\xi})$ is a piece-wise smooth function on S^+.

Now, substituting (4) into (1) and effecting an integration by parts via Stokes' theorem, we obtain a singularity reduced integral equation for the displacements as

$$u_p(\boldsymbol{x}) = -\oint_{\partial S^+} \psi^p_{mj}(\boldsymbol{\xi}, \boldsymbol{x}) \Delta u_j(\boldsymbol{\xi}) \, \mathrm{d}\xi_m + \int_{S^+} \psi^p_{mj}(\boldsymbol{\xi}, \boldsymbol{x}) D_m \Delta u_j(\boldsymbol{\xi}) \, \mathrm{d}s(\boldsymbol{\xi}). \tag{7}$$

In developing this expression, $\boldsymbol{x}$ has been taken to represent a point in the domain exterior to the discontinuity, but the final relation (7) also applies for points which are on S^+ (or, in fact, on S^-). We emphasize that in establishing this equation it has been assumed that the path $\Gamma_{(\boldsymbol{x})}$ does not intersect the surface S^+; if it intersects the surface then the integration by parts gives rise to an extra contribution associated with the intersection point which must be retained[3].

Of course, for (7) to be meaningful its value can not depend upon the choice of the stress function (and, in particular, upon the choice of the curve $\Gamma_{(\boldsymbol{x})}$) for (3)). That it does not is a consequence of a 'closure condition' for the operator D_m (readily verified by use of Stokes' theorem along with the fact that $D_m[\partial F/\partial \xi_m] = 0$) which states that

$$\int_{S^+} \frac{\partial F(\boldsymbol{\xi})}{\partial \xi_m} D_m \Delta u_j(\boldsymbol{\xi}) \, \mathrm{d}s(\boldsymbol{\xi}) - \oint_{\partial S^+} \frac{\partial F(\boldsymbol{\xi})}{\partial \xi_m} \Delta u_j(\boldsymbol{\xi}) \, \mathrm{d}\xi_m = 0, \tag{8}$$

for every suitably differentiable function $F(\boldsymbol{\xi})$. This closure condition, in conjunction with the fact that possible stress functions differ from each other only by a gradient of a function, guarantees the validity of (7) for all ψ^p_{mj}. When (8) is expressed in parameterized form it becomes

$$\int_{S^+} \left[\frac{\partial \Delta u_j(\boldsymbol{\xi})}{\partial \eta_1} \frac{\partial F(\boldsymbol{\xi})}{\partial \eta_2} - \frac{\partial \Delta u_j(\boldsymbol{\xi})}{\partial \eta_2} \frac{\partial F(\boldsymbol{\xi})}{\partial \eta_1} \right] \mathrm{d}\eta_1 \, \mathrm{d}\eta_2 - \oint_{\partial S^+} \Delta u_j(\boldsymbol{\xi}) \, \mathrm{d}F(\boldsymbol{\xi}) = 0, \tag{9}$$

from which it is evident that, as expected, only data associated with the surface is involved in its evaluation.

2.2. Integral expressions for stress and traction

An expression for the displacement gradients $\partial u_p/\partial x_q$ (valid for points $\boldsymbol{x}$ off of the surface S^+) results by forming the gradient of the right hand side of (7). Using the properties of the stress function as well as the closure condition, the derivatives with respect to x_q which appear in this expression can be exchanged for derivatives with respect to $-\xi_q$ to obtain (see Appendix C)

$$\frac{\partial u_p(\boldsymbol{x})}{\partial x_q} = \oint_{\partial S^+} \frac{\partial \psi^p_{mj}(\boldsymbol{\xi}, \boldsymbol{x})}{\partial \xi_q} \Delta u_j(\boldsymbol{\xi}) \, \mathrm{d}\xi_m - \int_{S^+} \frac{\partial \psi^p_{mj}(\boldsymbol{\xi}, \boldsymbol{x})}{\partial \xi_q} D_m \Delta u_j(\boldsymbol{\xi}) \, \mathrm{d}s(\boldsymbol{\xi}). \tag{10}$$

---

[3] With the path restricted not to intersect the surface, the jump in displacement across the discontinuity manifests itself in the requirement that the path be taken 'away' from the surface.

A simplified expression for the displacement gradients can then be obtained by exploiting a relation between the stress function $\psi^p_{mj}(\boldsymbol{\xi}, \boldsymbol{x})$ and the stress fundamental solution $S^p_{ij}(\boldsymbol{\xi} - \boldsymbol{x})$. Indeed, forming $\varepsilon_{sqm} S^p_{sj}(\boldsymbol{\xi} - \boldsymbol{x})$ and using (2) along with the delta-epsilon identity we have

$$\varepsilon_{sqm} S^p_{sj}(\boldsymbol{\xi} - \boldsymbol{x}) = \frac{\partial \psi^p_{mj}(\boldsymbol{\xi}, \boldsymbol{x})}{\partial \xi_q} - \frac{\partial \psi^p_{qj}(\boldsymbol{\xi}, \boldsymbol{x})}{\partial \xi_m} \tag{11}$$

so that, by virtue of the closure condition, (10) can be reduced to

$$\frac{\partial u_p(\boldsymbol{x})}{\partial x_q} = \oint_{\partial S^+} \varepsilon_{sqm} S^p_{sj}(\boldsymbol{\xi} - \boldsymbol{x}) \Delta u_j(\boldsymbol{\xi})\, \mathrm{d}\xi_m - \int_{S^+} \varepsilon_{sqm} S^p_{sj}(\boldsymbol{\xi} - \boldsymbol{x}) D_m \Delta u_j(\boldsymbol{\xi})\, \mathrm{d}s(\boldsymbol{\xi}). \tag{12}$$

For a dislocation (i.e. for Δu_j constant) this relation specializes to a well-known result established by Mura (1963). Note that the development of this relation does not rest upon a particular choice of the stress function ψ^p_{mj}.

An integral relation for the stress then follows from (12) and Hooke's law with the result

$$\sigma_{lk}(\boldsymbol{x}) = \oint_{\partial S^+} \Lambda^{lk}_{mj}(\boldsymbol{\xi} - \boldsymbol{x}) \Delta u_j(\boldsymbol{\xi})\, \mathrm{d}\xi_m - \int_{S^+} \Lambda^{lk}_{mj}(\boldsymbol{\xi} - \boldsymbol{x}) D_m \Delta u_j(\boldsymbol{\xi})\, \mathrm{d}s(\boldsymbol{\xi}), \tag{13}$$

in which

$$\Lambda^{lk}_{mj} \equiv E_{lkpq} \varepsilon_{sqm} S^p_{sj}, \tag{14}$$

where E_{lkpq} are the elastic moduli. For future reference we note that, as a consequence of the closure condition, the validity of (13) is unaltered if Λ^{lk}_{mj} is replaced by

$$\hat{\Lambda}^{lk}_{mj} \equiv \Lambda^{lk}_{mj} + \frac{\partial \Phi^{lk}_j}{\partial \xi_m}, \tag{15}$$

where $\Phi^{lk}_j(\boldsymbol{\xi}; \boldsymbol{x})$ is an arbitrary function (suitably well behaved on S^+).

A Cauchy singular traction integral equation for points on the discontinuity can be obtained from (13) through an appropriate limiting process. For the special case of a crack the resulting integral relation is identical to that obtained through a different approach by Sládek and Sládek (1982). Since this equation involves a Cauchy singular kernel, continuity of the derivatives of the relative displacements on S^+ is required for the expression to be valid. To relax this requirement, a weak-form statement of the traction integral equation is required.

2.3. Alternative expression for stress. Weak-form traction equation

Toward obtaining a weak-form traction integral equation, an alternative integral relation for the stress is first developed. To do so we seek a function Φ^{lk}_j such that the kernel $\hat{\Lambda}^{lk}_{mj}$ defined by (15) admits a representation as

$$\hat{\Lambda}^{lk}_{mj}(\boldsymbol{\xi} - \boldsymbol{x}) = \varepsilon_{lrt} \frac{\partial C^{tk}_{mj}(\boldsymbol{\xi} - \boldsymbol{x})}{\partial \xi_r}, \tag{16}$$

in which the function C^{tk}_{mj} is singular only at $\boldsymbol{\xi} = \boldsymbol{x}$. For such a function to exist, $\hat{\Lambda}^{lk}_{mj}$ must be divergence free in the domain exterior to $\boldsymbol{\xi} = \boldsymbol{x}$, it must itself be singular only at $\boldsymbol{\xi} = \boldsymbol{x}$, and it must be 'free of source' at $\boldsymbol{\xi} = \boldsymbol{x}$. That is, $\Phi^{lk}_j(\boldsymbol{\xi} - \boldsymbol{x})$ is sought such that

$$\frac{\partial \hat{\Lambda}^{lk}_{mj}(\boldsymbol{\xi} - \boldsymbol{x})}{\partial \xi_l} = E_{lkpq}\varepsilon_{sqm}\frac{\partial S^p_{sj}(\boldsymbol{\xi} - \boldsymbol{x})}{\partial \xi_l} + \frac{\partial^2 \Phi^{lk}_j(\boldsymbol{\xi} - \boldsymbol{x})}{\partial \xi_m \partial \xi_l} = 0, \tag{17}$$

for all $\boldsymbol{\xi} \neq \boldsymbol{x}$ *and* such that

$$\int_C \hat{\Lambda}^{lk}_{mj}(\boldsymbol{\xi} - \boldsymbol{x}) n_l(\boldsymbol{\xi})\, \mathrm{d}s(\boldsymbol{\xi}) = 0, \tag{18}$$

in which C is an arbitrary closed surface surrounding the point $\boldsymbol{x}$. As shown in Appendix D, a particular solution for Φ^{lk}_j is

$$\Phi^{lk}_j(\boldsymbol{\zeta}) = \tfrac{1}{2}E_{nkpq}\varepsilon_{sqm}\left[\delta_{ln}\zeta_m S^p_{sj}(\boldsymbol{\zeta}) - \delta_{mn}E_{sjal}U^p_a(\boldsymbol{\zeta})\right], \tag{19}$$

where U^p_a is the displacement fundamental solution (Appendix A). While the existence of the function C^{tk}_{mj} is guaranteed, there remains to solve the differential equation (16).

An integral of (16) for general anisotropy is difficult to establish, but for isotropic materials it is relatively straightforward to obtain an explicit expression for C^{tk}_{mj}. To do so, we first perform a direct integration of (17) to obtain a solution for Φ^{lk}_j as[4]

$$\Phi^{lk}_j(\boldsymbol{\zeta}) = -\frac{\mu}{8\pi(1-\nu)r}\left[(1-2\nu)\varepsilon_{lkj} - \varepsilon_{lks}\frac{\zeta_j\zeta_s}{r^2}\right], \tag{20}$$

in which $r^2(\boldsymbol{\zeta}) = \zeta_i\zeta_i$, ν is Poisson's ratio and μ is the shear modulus. Using this result, a direct integration of (16) then yields

$$C^{tk}_{mj}(\boldsymbol{\zeta}) = \frac{\mu}{4\pi(1-\nu)r}\left[(1-\nu)\delta_{kt}\delta_{jm} + 2\nu\delta_{km}\delta_{jt} - \delta_{kj}\delta_{tm} - \frac{\zeta_k\zeta_j}{r^2}\delta_{tm}\right]. \tag{21}$$

Further below it is shown that, at least in a certain sense, (21) is the simplest of all possible kernels for an isotropic material.

Returning to (13) with Λ^{lk}_{mj} replaced by $\hat{\Lambda}^{lk}_{mj}$, and using (16) along with the fact that $\partial C^{tk}_{mj}/\partial \xi_r = -\partial C^{tk}_{mj}/\partial x_r$, an alternative expression for the stress is obtained as

$$\sigma_{lk}(\boldsymbol{x}) = \varepsilon_{lrt}\frac{\partial \Upsilon_{tk}(\boldsymbol{x})}{\partial x_r}, \tag{22}$$

where the quantity

$$\Upsilon_{tk}(\boldsymbol{x}) = -\oint_{\partial S^+} C^{tk}_{mj}(\boldsymbol{\xi} - \boldsymbol{x})\Delta u_j(\boldsymbol{\xi})\, \mathrm{d}\xi_m + \int_{S^+} C^{tk}_{mj}(\boldsymbol{\xi} - \boldsymbol{x}) D_m \Delta u_j(\boldsymbol{\xi})\, \mathrm{d}s(\boldsymbol{\xi}) \tag{23}$$

[4] This form for Φ^{lk}_j and that given by (19) differ by a gradient of a function.

involves only weakly-singular integrands and admits an interpretation as a stress function for stress field induced by the defect. To obtain an integral equation for the traction $t_k^+(\boldsymbol{y}) = n_l^+(\boldsymbol{y})\sigma_{lk}(\boldsymbol{y})$ acting at a point $\boldsymbol{y}$ on the surface of the discontinuity, we form $n_l^+(\boldsymbol{y})\sigma_{lk}(\boldsymbol{x})$ and take a limit as $\boldsymbol{x} \to \boldsymbol{y}$ with the result

$$t_k^+(\boldsymbol{y}) = n_l^+(\boldsymbol{y})\varepsilon_{lrt}\frac{\partial \Upsilon_{tk}(\boldsymbol{y})}{\partial y_r} = D_t\Upsilon_{tk}(\boldsymbol{y}). \tag{24}$$

Finally, to obtain a weak-form traction integral equation we multiply the above relation by a continuous test function $\Delta u_k^*(y)$, integrate over S^+, and employ Stokes' theorem with the result

$$\int_{S^+} t_k^+(\boldsymbol{y})\Delta u_k^*(\boldsymbol{y})\,\mathrm{d}s(\boldsymbol{y}) = \oint_{\partial S^+} \Upsilon_{tk}(\boldsymbol{y})\Delta u_k^*(\boldsymbol{y})\,\mathrm{d}y_t - \int_{S^+} \Upsilon_{tk}(\boldsymbol{y})D_t\Delta u_k^*(\boldsymbol{y})\,\mathrm{d}s(\boldsymbol{y}). \tag{25}$$

This general relation can be specialized to the case of a crack by taking $\Delta u_k^*(\boldsymbol{y}) = 0$ on ∂S^+ (consistent with the fact that $\Delta u_k(\boldsymbol{y})$ vanishes along the crack front) to obtain

$$\int_{S^+} t_k^+(\boldsymbol{y})\Delta u_k^*(\boldsymbol{y})\,\mathrm{d}s(\boldsymbol{y}) = -\int_{S^+}\int_{S^+} C_{mj}^{tk}(\boldsymbol{\xi} - \boldsymbol{y})D_m\Delta u_j(\boldsymbol{\xi})D_t\Delta u_k^*(\boldsymbol{y})\,\mathrm{d}s(\boldsymbol{\xi})\,\mathrm{d}s(\boldsymbol{y}). \tag{26}$$

For the special case of a planar crack under mode I loading this expression reduces to that obtained by Gu and Yew (1988), and when the test function Δu_k^* is taken as $1/2\Delta u_k$ (hence giving an expression for the strain energy of the body containing the crack) it is related to the result obtained by Xu and Ortiz (1993).

2.4. DISCUSSION OF KERNELS ψ_{mj}^p AND C_{mj}^{tk}

Before discussing further the application of the singularity-reduced integral equations to modeling discontinuities, it is convenient to explore certain properties of the kernels ψ_{mj}^p and C_{mj}^{tk}. In particular, we examine the singular structure of ψ_{mj}^p associated with the path involved in its definition, and we discuss in detail the nonuniqueness associated with the kernel C_{mj}^{tk}.

2.4.1. *Kernel* ψ_{mj}^p

To examine the singular nature of ψ_{mj}^p we first construct a particular solution for this function which explicitly reveals the role of the path. Toward this end, we decompose the stress fundamental solution into a part associated with the point source at $\boldsymbol{\xi} = \boldsymbol{x}$ and a remainder which is divergence free and free of source. This is accomplished by first noting that the function $H_{ij}^p(\boldsymbol{\xi} - \boldsymbol{x})$ defined in terms of the Newtonian potential as

$$H_{ij}^p(\boldsymbol{\xi} - \boldsymbol{x}) \equiv -\frac{1}{4\pi}\delta_{pj}\frac{\partial}{\partial \xi_i}\left(\frac{1}{r}\right) = \frac{1}{4\pi}\frac{(\xi_i - x_i)\delta_{pj}}{r^3} \tag{27}$$

is divergence free for $\boldsymbol{\xi} \neq \boldsymbol{x}$ and contains a source at $\boldsymbol{\xi} = \boldsymbol{x}$ of strength δ_{pj} in the sense that

$$\int_C H_{ij}^p(\boldsymbol{\xi} - \boldsymbol{x})n_i(\boldsymbol{\xi})\,\mathrm{d}s(\boldsymbol{\xi}) = \delta_{pj}, \tag{28}$$

in which C is an arbitrary closed surface surrounding the point $\boldsymbol{x}$. The function

$$F_{ij}^{p}(\boldsymbol{\xi}-\boldsymbol{x}) \equiv S_{ij}^{p}(\boldsymbol{\xi}-\boldsymbol{x}) + H_{ij}^{p}(\boldsymbol{\xi}-\boldsymbol{x}) \tag{29}$$

is then both divergence free and source free. It follows that there exists a function $G_{ij}^{p}(\boldsymbol{\xi}-\boldsymbol{x})$ (singular only at $\boldsymbol{\xi}=\boldsymbol{x}$) such that $F_{ij}^{p} = \varepsilon_{ikm}\partial G_{mj}^{p}(\boldsymbol{\xi}-\boldsymbol{x})/\partial\xi_k$, hence a representation for the stress fundamental solution is

$$S_{ij}^{p}(\boldsymbol{\xi}-\boldsymbol{x}) = \varepsilon_{ikm}\frac{\partial G_{mj}^{p}(\boldsymbol{\xi}-\boldsymbol{x})}{\partial\xi_k} - H_{ij}^{p}(\boldsymbol{\xi}-\boldsymbol{x}). \tag{30}$$

We note that for an isotropic material a particular solution for G_{mj}^{p} is given by

$$G_{mj}^{p}(\boldsymbol{\zeta}) = \frac{1}{8\pi(1-\nu)r}\left[(1-2\nu)\varepsilon_{mpj} + \varepsilon_{mtj}\frac{\zeta_t\zeta_p}{r^2}\right]. \tag{31}$$

To proceed, we substitute (30) into (3) to find

$$\psi_{mj}^{p}(\boldsymbol{\xi},\boldsymbol{x}) = G_{mj}^{p}(\boldsymbol{\xi}-\boldsymbol{x}) - \delta_{pj}\phi_m(\boldsymbol{\xi},\boldsymbol{x}) - \frac{\partial}{\partial\xi_m}\int_{\Gamma(\boldsymbol{x})} G_{tj}^{p}(\boldsymbol{\xi}-\boldsymbol{z})\,\mathrm{d}z_t, \tag{32}$$

in which ϕ_m is defined by

$$\phi_m(\boldsymbol{\xi},\boldsymbol{x}) \equiv \frac{1}{4\pi}\int_{\Gamma(\boldsymbol{x})} \varepsilon_{mrt}\frac{\xi_r - z_r}{r^3(\boldsymbol{\xi}-\boldsymbol{z})}\,\mathrm{d}z_t \tag{33}$$

and satisfies

$$\varepsilon_{ikm}\frac{\partial\phi_m}{\partial\xi_k} = \frac{1}{4\pi}\frac{\xi_i - x_i}{r^3}, \tag{34}$$

(hence H_{ij}^{p} is given as the curl of $\delta_{pj}\phi_m$). Upon noting that the last term in (32) is just a gradient of a function, we obtain a relatively simple stress function for S_{ij}^{p} as[5]

$$\psi_{mj}^{p}(\boldsymbol{\xi},\boldsymbol{x}) = G_{mj}^{p}(\boldsymbol{\xi}-\boldsymbol{x}) - \delta_{pj}\phi_m(\boldsymbol{\xi},\boldsymbol{x}). \tag{35}$$

Since G_{mj}^{p} is singular only at $\boldsymbol{\xi}=\boldsymbol{x}$, it is clear that it is the term $\phi_m(\boldsymbol{\xi},\boldsymbol{x})$ which gives rise to the singularity in ψ_{mj}^{p} along the 'cut' $\Gamma(\boldsymbol{x})$. To examine the nature of this singularity, consider the special case in which $\Gamma(\boldsymbol{x})$ is a straight line from $\boldsymbol{x}$ to infinity. With the unit directional vector along the line given as e_k and with $g_k \equiv (x_k - \xi_k)/r$, (33) reduces to

$$\phi_m(\boldsymbol{\xi},\boldsymbol{x}) = -\frac{1}{4\pi}\frac{\varepsilon_{mkt}g_k e_t}{1+g_l e_l}\frac{1}{r}. \tag{36}$$

Without loss of generality, now let $\boldsymbol{x}$ be the origin and orient the positive ξ_3-axis along the cut. Expressing the coordinates of a point near the cut as $\boldsymbol{\xi} = \{\rho\cos\theta, \rho\sin\theta, h\}$ with $h > 0$, we

[5] Note that whereas the stress functions given by (3) and (32) correspond to a displacement solution for any given pair of indicies $\{m, j\}$, the stress function defined by (35) does not possess this property.

have $r = \sqrt{h^2 + \rho^2}$, $\boldsymbol{e} = \{0, 0, 1\}$ and $\boldsymbol{g} = \{-\rho\cos\theta, -\rho\sin\theta, -h\}/r$. Substituting these results into (36), and forming the limit as $\rho \to 0$ with h fixed, we obtain

$$\lim_{\rho\to 0} \rho\phi_1 = \frac{\sin\theta}{8\pi}, \qquad \lim_{\rho\to 0} \rho\phi_2 = -\frac{\cos\theta}{8\pi}, \qquad \phi_3 = 0. \tag{37}$$

From this relation, it is apparent that as $\boldsymbol{\xi}$ approaches the cut along a line perpendicular to the cut, both ϕ_1 and ϕ_2 exhibit a $1/\rho$ singularity with the intensity of the singularity dependent upon the direction in which the cut is approached. In this sense ϕ_m is singular and multi-valued along the cut.

2.4.2. *Kernel* C_{mj}^{tk}

To establish the kernel C_{mj}^{tk}, a function Φ_j^{lk} was sought such that $\hat{\Lambda}_{mj}^{lk}$ defined by (15) is both divergence free and free of source. Clearly there is not a unique choice for Φ_j^{lk} which achieves this objective and further, for a given Φ_j^{lk} and associated $\hat{\Lambda}_{mj}^{lk}$ the function C_{mj}^{tk} defined by (16) is not unique. Indeed, while kernels which are functions of $(\boldsymbol{\xi} - \boldsymbol{x})$ and singular only at $\boldsymbol{\xi} = \boldsymbol{x}$ have been sought for the convenience which they afford, this property is not essential[6].

The nonuniqueness associated with the kernel C_{mj}^{tk} is evident from its role in the stress function $\Upsilon_{tk}(\boldsymbol{x})$ defined by (23) and (16). Clearly terms of the form $\partial A_j^{tk}(\boldsymbol{\xi}, \boldsymbol{x})/\partial\xi_m$ and $\partial B_{mj}^{k}(\boldsymbol{\xi}, \boldsymbol{x})/\partial x_t$ can be added to the kernel without violating these equations; terms of the first form can be added by virtue of the closure condition, while terms of the second form can added as a consequence of the curl operation associated with (16).

In addition to being nonunique, the kernel in (23) satisfies a duality property which states that the validity of the integral relation for the stress is unaltered if a given kernel $C_{mj}^{tk}(\boldsymbol{\xi}, \boldsymbol{x})$ is replaced by $C_{tk}^{mj}(\boldsymbol{x}, \boldsymbol{\xi})$. A consequence of this property is that it is always possible to establish a kernel which possesses the symmetry $C_{mj}^{tk}(\boldsymbol{\xi}, \boldsymbol{x}) = C_{tk}^{mj}(\boldsymbol{x}, \boldsymbol{\xi})$. We remark that (21) is the simplest possible kernel for an isotropic material in the sense that it involves no 'extra' gradients with respect to ξ_m and x_t (associated with the nonuniqueness described above) and it satisfies the latter symmetry condition.

To establish the duality property for the kernel, we follow a construction in which an 'auxiliary' discontinuity is introduced in the presence of the actual discontinuity. It should be noted from the outset that both the surface of the auxiliary discontinuity $\tilde{S}^+$ and the associated relative displacements $\Delta\tilde{u}_j$ are arbitrary, and that it is this arbitrariness which is exploited to establish the desired result. The presentation of the development is facilitated by defining the integral operator

$$\mathcal{I}_m[C, F(\boldsymbol{\xi}), \Delta v(\boldsymbol{\xi})] \equiv -\oint_{\partial C} F(\boldsymbol{\xi})\Delta v(\boldsymbol{\xi})\, \mathrm{d}\xi_m + \int_C F(\boldsymbol{\xi}) D_m \Delta v(\boldsymbol{\xi})\, \mathrm{d}s(\boldsymbol{\xi}), \tag{38}$$

in which $F(\boldsymbol{\xi})$ is the kernel of the integral operator and where (the data) $\Delta v(\xi)$ is defined over the surface C. By Stokes' theorem, (38) can be re-expressed as

$$\mathcal{I}_m[C, F(\boldsymbol{\xi}), \Delta v(\boldsymbol{\xi})] = -\int_C n_i(\boldsymbol{\xi})\varepsilon_{ikm}\frac{\partial F(\boldsymbol{\xi})}{\partial \xi_k}\Delta v(\boldsymbol{\xi})\, \mathrm{d}s(\boldsymbol{\xi}). \tag{39}$$

---

6 For example, a stress function for the self-equilibrated stress field $E_{lkpq}\partial\psi_{mj}^{P}/\partial\xi_q$ (in which ψ_{mj}^{P} is given by (3) with a particular choice of path $\Gamma_{(\boldsymbol{x})}$) is a valid kernel, yet it is necessarily singular along the path associated with the 'line discontinuity'.

Now, from (22) and (38) the stress induced by the actual discontinuity in the *absence* of the auxiliary discontinuity is given by

$$\sigma_{lk}(\boldsymbol{x}) = \varepsilon_{lrt}\frac{\partial}{\partial x_r}\mathcal{I}_m[S^+, C^{tk}_{mj}(\boldsymbol{\xi}, \boldsymbol{x}), \Delta u_j(\boldsymbol{\xi})] \tag{40}$$

and, similarly, the stress induced by the auxiliary discontinuity in the absence of the actual discontinuity is

$$\tilde{\sigma}_{lk}(\boldsymbol{x}) = \varepsilon_{lrt}\frac{\partial}{\partial x_r}\mathcal{I}_m[\tilde{S}^+, C^{tk}_{mj}(\boldsymbol{\eta}, \boldsymbol{x}), \Delta\tilde{u}_j(\boldsymbol{\eta})]. \tag{41}$$

From a consideration of the 'interaction energy' between the discontinuities, along with an application of the reciprocal theorem, it can be shown that

$$\int_{\tilde{S}^+} \tilde{n}^+_l(\boldsymbol{\eta})\sigma_{lk}(\boldsymbol{\eta})\Delta\tilde{u}_k(\boldsymbol{\eta})\,\mathrm{d}s(\boldsymbol{\eta}) = \int_{S^+} n^+_l(\boldsymbol{\xi})\tilde{\sigma}_{lk}(\boldsymbol{\xi})\Delta u_k(\boldsymbol{\xi})\,\mathrm{d}s(\boldsymbol{\xi}), \tag{42}$$

in which, for emphasis, the unit normal to $\tilde{S}^+$ has been denoted as $\tilde{n}^+_k$. Using (40) and (39), the right side of this expression can be re-written as

$$\int_{S^+} n^+_l(\boldsymbol{\xi})\tilde{\sigma}_{lk}(\boldsymbol{\xi})\Delta u_k(\boldsymbol{\xi})\,\mathrm{d}s(\boldsymbol{\xi}) = -\mathcal{I}_t[S^+, \mathcal{I}_m[\tilde{S}^+, C^{tk}_{mj}(\boldsymbol{\eta}, \boldsymbol{\xi}), \Delta\tilde{u}_j(\boldsymbol{\eta})], \Delta u_k(\boldsymbol{\xi})], \tag{43}$$

or, upon an interchange of the order of integration, as

$$\int_{S^+} n^+_l(\boldsymbol{\xi})\tilde{\sigma}_{lk}(\boldsymbol{\xi})\Delta u_k(\boldsymbol{\xi})\,\mathrm{d}s(\boldsymbol{\xi}) = -\mathcal{I}_t[\tilde{S}^+, \mathcal{I}_m[S^+, C^{mj}_{tk}(\boldsymbol{\eta}, \boldsymbol{\xi}), \Delta u_j(\boldsymbol{\xi})], \Delta\tilde{u}_k(\boldsymbol{\eta})]. \tag{44}$$

Using this result along with (42) and (39) we then find

$$\int_{\tilde{S}^+} \tilde{\beta}_{lk}(\boldsymbol{\eta})\sigma_{lk}(\boldsymbol{\eta})\,\mathrm{d}s(\boldsymbol{\eta}) = \int_{\tilde{S}^+} \tilde{\beta}_{lk}(\boldsymbol{\eta})\,\varepsilon_{lrt}\frac{\partial\mathcal{I}_m[S^+, C^{mj}_{tk}(\boldsymbol{\eta}, \boldsymbol{\xi}), \Delta u_j(\boldsymbol{\xi})]}{\partial\eta_r}\,\mathrm{d}s(\boldsymbol{\eta}), \tag{45}$$

in which $\tilde{\beta}_{lk}(\boldsymbol{\eta}) \equiv \tilde{n}^+_l(\boldsymbol{\eta})\Delta\tilde{u}_k(\boldsymbol{\eta})$.

The latter expression applies for any auxiliary discontinuity and, in particular, for one which is arbitrarily small. Without loss, now assume that this discontinuity is a flat dislocation which is 'centered' at a point $\boldsymbol{x}$ and which is of infinitesimal size. Since both the orientation $\tilde{n}_l$ of the dislocation and the constants $\Delta\tilde{u}_k$ can be selected arbitrarily (i.e. $\tilde{\beta}_{lk}$ is an arbitrary constant tensor), it follows from (45) that

$$\sigma_{lk}(\boldsymbol{x}) = \varepsilon_{lrt}\frac{\partial\mathcal{I}_m[S^+, C^{mj}_{tk}(\boldsymbol{x}, \boldsymbol{\xi}), \Delta u_j(\boldsymbol{\xi})]}{\partial x_r}. \tag{46}$$

Comparing this relation with (41) we deduce that for any given kernel $C^{tk}_{mj}(\boldsymbol{\xi}, \boldsymbol{x})$ the quantity $C^{mj}_{tk}(\boldsymbol{x}, \boldsymbol{\xi})$ is also a valid kernel for the stress relation (22). It then follows that the function $1/2[C^{tk}_{mj}(\boldsymbol{\xi}, \boldsymbol{x}) + C^{mj}_{tk}(\boldsymbol{x}, \boldsymbol{\xi})]$ is a valid kernel and as a consequence it is always possible to construct a kernel which is symmetric in the sense that $C^{tk}_{mj}(\boldsymbol{\xi}, \boldsymbol{x}) = C^{mj}_{tk}(\boldsymbol{x}, \boldsymbol{\xi})$.

3. Dislocations in an unbounded domain

The singularity-reduced integral equations developed above specialize to the case of a dislocation simply by specifying the relative displacement as constant over the surface of the discontinuity. In this section we discuss relations for the displacement field, the stress field and the energy which results from this specialization. The results take simple and explicit forms and serve as alternatives to corresponding integral relations which exist in the literature.

3.1. Line integral expression for displacement and stress

Consider a dislocation which has Burgers vector $b_j = \Delta u_j$. From (1) and (30), a variant of Burger's equation for the displacement field induced by such a dislocation is

$$u_p(\boldsymbol{x}) = b_j \oint_{\partial S^+} G^p_{mj}(\boldsymbol{\xi} - \boldsymbol{x})\, \mathrm{d}\xi_m - \frac{b_p}{4\pi}\Omega(\boldsymbol{x}), \tag{47}$$

in which

$$\Omega(\boldsymbol{x}) = \int_{S^+} n_i(\boldsymbol{\xi}) \frac{(\xi_i - x_i)}{r^3}\, \mathrm{d}s(\boldsymbol{\xi}) \tag{48}$$

is the solid angle to the surface and where the function G^p_{mj} is defined by (30) (and, for an isotropic material, given by (31)). Using (34), Ω can be converted into a line integral expression as

$$\Omega(\boldsymbol{x}) = 4\pi \oint_{\partial S^+} \phi_m(\boldsymbol{\xi}, \boldsymbol{x})\, \mathrm{d}\xi_m, \tag{49}$$

in which ϕ_m is defined by (33). Using the latter result, the displacement relation (47) can be re-expressed as

$$u_p(\boldsymbol{x}) = b_j \oint_{\partial S^+} \left[G^p_{mj} - \delta_{jp}\phi_m \right] \mathrm{d}\xi_m. \tag{50}$$

This line integral relation for the displacement differs in an important respect from that established by Peach and Koehler (1950). The critical difference is that in their expression the integral kernel associated with the term Ω is singular along two orthogonal lines, and it is often difficult if not impossible to select the orientation of these lines such that they do not intersect the dislocation surface. In contrast, the integral kernel ϕ_m involved in (50) is singular only along a half infinite curve and further, this curve can always be chosen such that it does not intersect the dislocation surface. We remark that in applying this result the special choice of a piece-wise linear curve may prove convenient since for this case the line integral can be expressed in closed form.

A line integral representation for the stress can be obtained by applying Hooke's law to Mura's formula for the displacement gradient (see (13)). An alternative expression for the stress field follows from (22) and (23) as

$$\sigma_{lk}(\boldsymbol{x}) = -b_j \varepsilon_{lrt} \frac{\partial}{\partial x_r} \oint_{\partial S^+} C^{tk}_{mj}(\boldsymbol{\xi} - \boldsymbol{x})\, \mathrm{d}\xi_m, \tag{51}$$

in which C_{mj}^{tk} is defined by (16) and, for an isotropic material, given by (21).

3.2. Line integral expression for energy

For purpose of generality, we now assume that two dislocation loops are present: a loop ∂S_1^+ with Burgers vectors b_j^1 and a loop ∂S_2^+ with Burgers vectors b_j^2. The energy induced by these two loops can be calculated from (25) and, upon observing that the two 'cross terms' which arise are equal as a consequence of the fact that $C_{mj}^{tk}(\boldsymbol{\xi} - \boldsymbol{x})$ can be replaced by $C_{tk}^{mj}(\boldsymbol{x} - \boldsymbol{\xi})$, the result can be expressed as

$$W = W_{11} + W_{12} + W_{22}, \tag{52}$$

where

$$\begin{aligned} W_{11} &= -\frac{1}{2}\oint_{\partial S_1^+}\oint_{\partial S_1^+} C_{mj}^{tk}(\boldsymbol{\xi} - \boldsymbol{\eta})\, b_k^1 b_j^1 \,\mathrm{d}\eta_t \,\mathrm{d}\xi_m, \\ W_{12} &= -\oint_{\partial S_1^+}\oint_{\partial S_2^+} C_{mj}^{tk}(\boldsymbol{\xi} - \boldsymbol{\eta})\, b_k^1 b_j^2 \,\mathrm{d}\eta_t \,\mathrm{d}\xi_m, \\ W_{22} &= -\frac{1}{2}\oint_{\partial S_2^+}\oint_{\partial S_2^+} C_{mj}^{tk}(\boldsymbol{\xi} - \boldsymbol{\eta})\, b_k^2 b_j^2 \,\mathrm{d}\eta_t \,\mathrm{d}\xi_m. \end{aligned} \tag{53}$$

The terms W_{11} and W_{22} correspond to the 'self-energy' (i.e. the energy induced by each dislocation acting separately) whereas W_{12} is the 'interaction energy' of one loop with respect to the other[7]. There exist several well known expressions for the self-energy and the interaction energy for dislocation loops (see Hirth and Lothe, 1982) and the present expressions represent but an alternative; the discussion given previously for the nonuniqueness of the kernel C_{mj}^{tk} serves to clarify the nature of the differences between any such possibilities. We remark that one feature of the result (**??**) which distinguishes it from previous results is that the same kernel is naturally associated with both the self-energy and the interaction-energy.

4. Cracks in a finite domain

We now turn attention to a finite domain which contains an embedded or surface-breaking crack as shown schematically in Figure 2. For purposes of modeling the crack it is still necessary to utilize a traction integral equation, but the regular boundary can be treated using either a displacement or a traction integral equation or, in fact, a combination of such equations. Here attention is focused on the development of a system of weak-form, weakly-singular integral equations, and both traction and displacement integral equations are established.

The development of the integral equations rests upon Somigliana's identity which gives the displacement u_p at a point $\boldsymbol{x}$ within the domain in terms of the displacement and traction data $\{u_i, t_i\}$ at points $\boldsymbol{\xi}$ on its boundary as

$$u_p(\boldsymbol{x}) = \int_S U_j^p(\boldsymbol{\xi} - \boldsymbol{x}) t_j(\boldsymbol{\xi}) \,\mathrm{d}s(\boldsymbol{\xi}) - \int_S n_i(\boldsymbol{\xi}) S_{ij}^p(\boldsymbol{\xi} - \boldsymbol{x}) u_j(\boldsymbol{\xi}) \,\mathrm{d}s(\boldsymbol{\xi}), \tag{54}$$

[7] The self energy is logarithmically unbounded. To render it finite a 'cut-off' parameter is typically introduced (see Hirth and Lothe, 1982).

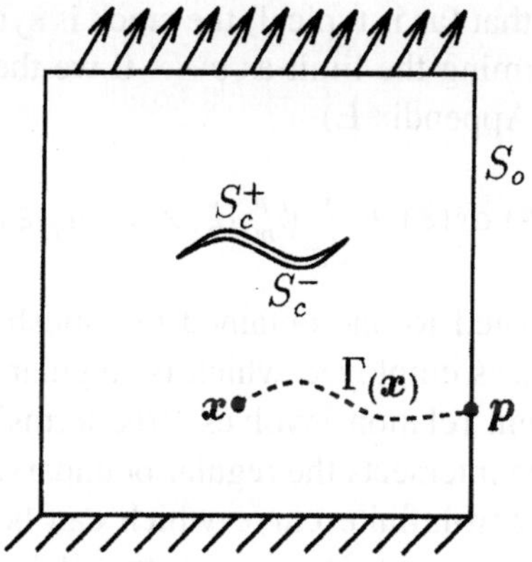

Figure 2. Schematic of a crack in a finite domain.

in which $n_i(\xi)$ denotes the unit outward normal to the surface S. Note that in this expression the surfaces of the crack are regarded as an extension of the regular boundary in the sense that S denotes the total (closed) boundary given as $S = S_o + S_c^+ + S_c^-$ where S_o is the regular boundary and S_c^+ and S_c^- are the upper and lower surfaces of the crack, respectively. Of course, since the normal to the 'upper' crack surface is related to the normal to the 'lower' crack surface by $n_i^+(\xi) = -n_i^-(\xi)$, (54) can be reduced to an expression involving only integrals over the regular boundary and a single crack surface, say the 'upper' surface, as

$$u_p(x) = \int_{S_o} U_j^p(\xi - x)t_j(\xi)\,\mathrm{d}s(\xi) + \int_{S^+} U_j^p(\xi - x)\Sigma t_j(\xi)\,\mathrm{d}s(\xi)$$

$$-\int_{S_o} n_i(\xi)S_{ij}^p(\xi - x)u_j(\xi)\,\mathrm{d}s(\xi) - \int_{S^+} n_i^+(\xi)S_{ij}^p(\xi - x)\Delta u_j(\xi)\,\mathrm{d}s(\xi), \tag{55}$$

in which $\Sigma t_j = t_j^+ + t_j^-$ is the sum of the tractions acting on the crack surfaces and $\Delta u_j = u_j^+ - u_j^-$ is the relative crack-face displacement. It is convenient, however, to reserve this reduction until after the development of the integral equations has been completed.

4.1. Displacement integral equation for regular boundary

Consider first the development of a displacement integral equation suitable for treating the regular boundary. In a fashion analogous to that pursued for the isolated crack, a singularity-reduced statement of Somigliana's identity (54) can be established by employing the stress function $\psi_{mj}^p(\xi, x)$ defined by (2). To do so, assume that the path $\Gamma_{(x)}$ associated with the stress function intersects the regular boundary at a single point p and let S_ε be a small part of S which contains p and which has 'size' ε. Then using (2) in (54) and noting that $\psi_{mj}^p(\xi, x)$ is differentiable on $S - S_\varepsilon$, Stokes' theorem can be applied with the result

$$u_p(x) = \int_{S_o} U_j^p(\xi - x)t_j(\xi)\,\mathrm{d}s(\xi) + \int_{S_\varepsilon} n_i(\xi)S_{ij}^p(\xi - x)u_j(\xi)\,\mathrm{d}s(\xi)$$

$$+\int_{S-S_\varepsilon} \psi_{mj}^p(\xi, x)D_m u_j(\xi)\,\mathrm{d}s(\xi) - \oint_{\partial(S-S_\varepsilon)} \psi_{mj}^p(\xi, x)u_j(\xi)\,\mathrm{d}\xi_m, \tag{56}$$

in which it has been assumed that (as is typical) the crack is symmetrically loaded in the sense that $\Sigma t_i = t_i^+ + t_i^- = 0$. Forming the limit as $\varepsilon \to 0$ we then obtain a singularity-reduced displacement relation as (see Appendix E)

$$u_p(\boldsymbol{x}) = \int_{S_o} U_j^p(\boldsymbol{\xi} - \boldsymbol{x}) t_j(\boldsymbol{\xi}) \, \mathrm{d}s(\boldsymbol{\xi}) + \int_S \psi_{mj}^p(\boldsymbol{\xi}, \boldsymbol{x}) D_m u_j(\boldsymbol{\xi}) \, \mathrm{d}s(\boldsymbol{\xi}) + u_p(\boldsymbol{p}). \tag{57}$$

This relation is closely related to one obtained by Ghosh and Mukerjee (1987). In their development, a stress function is employed which is singular along an infinite line, and as a consequence their displacement relation involves 'free terms' associated with (at least) two points at which the infinite line intersects the regular boundary. In the relation developed here, the path $\Gamma_{(\boldsymbol{x})}$ is taken as a semi-infinite curve which can be chosen such that it intersects the regular boundary at a single point. We remark that if $\Gamma_{(\boldsymbol{x})}$ intersects the surface of the discontinuity or if it intersects the regular boundary at more than one location, then (57) must be modified to account for the contributions from these intersection points (Appendix E).

The displacement relation (56) forms the basis for the development of the traction integral equation to follow. Its limit form (57) could also be used as the basis of a weakly-singular, weak-form displacement integral equation for the regular boundary, but it suffers from the undesirable characteristics that the kernel ψ_{mj}^k is multi-valued and that there is a 'free term' associated with the intersection of $\Gamma_{(\boldsymbol{x})}$ with the boundary.

A singularity-reduced displacement integral equation without these drawbacks can be established by several means. One technique is to utilize a subtraction procedure often motivated by consideration of a rigid body motion (e.g. Brebbia and Dominguez, 1992), while another is to employ an appropriate decomposition of the fundamental solution. Here we follow the latter approach, and for this purpose we employ the particular decomposition of the stress fundamental solution given by (30).

Incorporating the decomposition (30) into (54), and utilizing Stokes' theorem to perform an integration by parts, we obtain

$$\begin{aligned} u_p(\boldsymbol{x}) = & \int_{S_o} t_i(\boldsymbol{\xi}) U_i^p(\boldsymbol{\xi} - \boldsymbol{x}) \, \mathrm{d}s(\boldsymbol{\xi}) + \int_S G_{mj}^p(\boldsymbol{\xi} - \boldsymbol{x}) D_m u_j(\boldsymbol{\xi}) \, \mathrm{d}s(\boldsymbol{\xi}) \\ & + \int_S n_i(\boldsymbol{\xi}) H_{ij}^p(\boldsymbol{\xi} - \boldsymbol{x}) u_j(\boldsymbol{\xi}) \, \mathrm{d}s(\boldsymbol{\xi}). \end{aligned} \tag{58}$$

To obtain a form of this equation applicable for points $z \in S_o$, we assume (without loss for the weak-form equation to follow) that the boundary is smooth at the point z of interest. The limit $\boldsymbol{x} \to z$ can be taken directly for the first two integrals in (58) (since the kernels involved are weakly singular), whereas for the third integral we have (e.g. Kellog, 1953)

$$\lim_{\boldsymbol{x} \to z} \int_S n_i(\boldsymbol{\xi}) H_{ij}^p(\boldsymbol{\xi} - \boldsymbol{x}) u_j(\boldsymbol{\xi}) \, \mathrm{d}s(\boldsymbol{\xi}) = \tfrac{1}{2} u_p(z) + \int_S n_i(\boldsymbol{\xi}) H_{ij}^p(\boldsymbol{\xi} - z) u_j(\boldsymbol{\xi}) \, \mathrm{d}s(\boldsymbol{\xi}). \tag{59}$$

Using this result, a relation for the displacement on the boundary is obtained as

$$\begin{aligned} \tfrac{1}{2} u_p(z) = & \int_{S_o} t_i(\boldsymbol{\xi}) U_i^p(\boldsymbol{\xi} - z) \, \mathrm{d}s(\boldsymbol{\xi}) + \int_S G_{mj}^p(\boldsymbol{\xi} - z) D_m u_j(\boldsymbol{\xi}) \, \mathrm{d}s(\boldsymbol{\xi}) \\ & + \int_S n_i(\boldsymbol{\xi}) H_{ij}^p(\boldsymbol{\xi} - z) u_j(\boldsymbol{\xi}) \, \mathrm{d}s(\boldsymbol{\xi}). \end{aligned} \tag{60}$$

Note that since the boundary is presumed smooth at z, the term $n_i(\boldsymbol{\xi})(\xi_i - z_i)/r$ is $O(r)$ as $r \to 0$ and the kernel

$$n_i(\boldsymbol{\xi})H_{ij}^p(\boldsymbol{\xi} - z) = \frac{1}{4\pi} n_i(\boldsymbol{\xi}) \frac{(\xi_i - z_i)}{r^3} \delta_{jp} \tag{61}$$

is in fact weakly singular (i.e. $O(1/r)$ as $r \to 0$).

Multiplying (60) by a test function $t_p^*(z)$ and integrating the result over the regular boundary, a weak-form displacement integral equation for the regular boundary is obtained as

$$\begin{aligned}\int_{S_o} \tfrac{1}{2} u_p(z) t_p^*(z)\, \mathrm{d}s(z) \; = \; & \int_{S_o} t_p^*(z) \int_{S_o} t_i(\boldsymbol{\xi}) U_i^p(\boldsymbol{\xi} - z)\, \mathrm{d}s(\boldsymbol{\xi})\, \mathrm{d}s(z) \\ & + \int_{S_o} t_p^*(z) \int_S G_{mj}^p(\boldsymbol{\xi} - z) D_m u_j(\boldsymbol{\xi})\, \mathrm{d}s(\boldsymbol{\xi})\, \mathrm{d}s(z) \\ & + \int_{S_o} t_p^*(z) \int_S n_i(\boldsymbol{\xi}) H_{ij}^p(\boldsymbol{\xi} - z) u_j(\boldsymbol{\xi})\, \mathrm{d}s(\boldsymbol{\xi})\, \mathrm{d}s(z).\end{aligned} \tag{62}$$

It is important to observe that the test function which appears in this relation need not be continuous; this property can be exploited to treat displacement boundary conditions at corners or edge points where other integral formulation usually provide insufficient information to determine the unknown tractions.

4.2. Traction integral equations for crack and boundary

Consider next the development of a weakly-singular, weak-form traction integral equation applicable to the regular boundary as well as to the crack. As discussed previously, it is necessary to have such a traction integral equation for purposes of modeling the crack, whereas for the regular boundary it serves as an alternative to the displacement integral equation (62).

To develop the traction integral equation, we return to (56) and obtain a relation for the displacement gradients (valid at internal points $\boldsymbol{x}$) as

$$\begin{aligned} u_{p,q}(\boldsymbol{x}) \; = \; & - \int_{S_o} \frac{\partial U_j^p(\boldsymbol{\xi} - \boldsymbol{x})}{\partial \xi_q} t_j(\boldsymbol{\xi})\, \mathrm{d}s(\boldsymbol{\xi}) + \frac{\partial}{\partial x_q} \int_{S_\varepsilon} n_i(\boldsymbol{\xi}) S_{ij}^p(\boldsymbol{\xi} - \boldsymbol{x}) u_j(\boldsymbol{\xi})\, \mathrm{d}s(\boldsymbol{\xi}) \\ & - \int_{S - S_\varepsilon} \frac{\partial \psi_{mj}^p(\boldsymbol{\xi}, \boldsymbol{x})}{\partial \xi_q} D_m u_j(\boldsymbol{\xi})\, \mathrm{d}s(\boldsymbol{\xi}) + \oint_{\partial(S - S_\varepsilon)} \frac{\partial \psi_{mj}^p(\boldsymbol{\xi}, \boldsymbol{x})}{\partial \xi_q} u_j(\boldsymbol{\xi})\, \mathrm{d}\xi_m, \end{aligned} \tag{63}$$

where, for all but the integral over S_ϵ, the derivative with respect to $\boldsymbol{x}$ have been exchanged for a derivative with respect to $\boldsymbol{\xi}$ along with a change in sign. This exchange is obvious for the first terms since U_j^p is a function of $(\boldsymbol{\xi} - \boldsymbol{x})$, and for the last two terms involving the kernel ψ_{mj}^p it is justified by the result of Appendix C.

Now, $\psi_{mj}^p(\boldsymbol{\xi}, \boldsymbol{x})$ is differentiable on $S - S_\varepsilon$, and it follows from (11) and the closure condition (8) that, for the last two terms in (63), $\partial \psi_{mj}^p(\boldsymbol{\xi}, \boldsymbol{x})/\partial \xi_q$ can be replaced by $\varepsilon_{sqm} S_{sj}^p(\boldsymbol{\xi} - \boldsymbol{x})$. Making this substitution in (63) and forming the limit as ε tends to zero gives an integral equation for the displacement gradients as

$$u_{p,q}(\boldsymbol{x}) = - \int_{S_o} \frac{\partial U_j^p(\boldsymbol{\xi} - \boldsymbol{x})}{\partial \xi_q} t_j(\boldsymbol{\xi})\, \mathrm{d}s(\boldsymbol{\xi}) - \int_S \varepsilon_{sqm} S_{sj}^p(\boldsymbol{\xi} - \boldsymbol{x}) D_m u_j(\boldsymbol{\xi})\, \mathrm{d}s(\boldsymbol{\xi}). \tag{64}$$

An integral relation for the stress at an internal point can then be obtained by using Hooke's law and the fact that $U_j^p(\boldsymbol{\xi}-\boldsymbol{x}) = U_p^j(\boldsymbol{\xi}-\boldsymbol{x})$ with the result

$$\sigma_{lk}(\boldsymbol{x}) = -\int_{S_o} S_{lk}^j(\boldsymbol{\xi}-\boldsymbol{x})t_j(\boldsymbol{\xi})\,\mathrm{d}s(\boldsymbol{\xi}) - \int_S \Lambda_{mj}^{lk}(\boldsymbol{\xi}-\boldsymbol{x})D_m u_j(\boldsymbol{\xi})\,\mathrm{d}s(\boldsymbol{\xi}) \tag{65}$$

in which, as before, $\Lambda_{mj}^{lk}(\boldsymbol{\xi}-\boldsymbol{x}) = E_{lkpq}\varepsilon_{iqm}S_{ij}^p(\boldsymbol{\xi}-\boldsymbol{x})$.

To proceed, we use relation (16) along with the closure condition to re-express the last term in (65) as

$$\int_S \Lambda_{mj}^{lk}(\boldsymbol{\xi}-\boldsymbol{x})D_m u_j(\boldsymbol{\xi})\,\mathrm{d}s(\boldsymbol{\xi}) = -\varepsilon_{lrt}\frac{\partial}{\partial x_r}\int_S C_{mj}^{tk}(\boldsymbol{\xi}-\boldsymbol{x})D_m u_j(\boldsymbol{\xi})\,\mathrm{d}s(\boldsymbol{\xi}), \tag{66}$$

and we utilize the decomposition (30) to re-express the first term in (65) as

$$\begin{aligned}\int_{S_o} S_{lk}^j(\boldsymbol{\xi}-\boldsymbol{x})t_j(\boldsymbol{\xi})\,\mathrm{d}s(\boldsymbol{\xi}) &= -\varepsilon_{lsm}\frac{\partial}{\partial x_s}\int_{S_o} G_{mk}^j(\boldsymbol{\xi}-\boldsymbol{x})t_j(\boldsymbol{\xi})\,\mathrm{d}s(\boldsymbol{\xi}) \\ &\quad -\int_{S_o} H_{lk}^j(\boldsymbol{\xi}-\boldsymbol{x})t_j(\boldsymbol{\xi})\,\mathrm{d}s(\boldsymbol{\xi}).\end{aligned} \tag{67}$$

Substituting these results into (65), and forming $n_l(z)\sigma_{lk}(\boldsymbol{x})$ in which $n_l(z)$ is the unit out normal to the boundary S at a point z of interest, we have

$$\begin{aligned}n_l(z)\sigma_{lk}(\boldsymbol{x}) &= n_l(z)\varepsilon_{lrt}\frac{\partial}{\partial x_r}\int_S C_{mj}^{tk}(\boldsymbol{\xi}-\boldsymbol{x})D_m u_j(\boldsymbol{\xi})\,\mathrm{d}s(\boldsymbol{\xi}) \\ &\quad + n_l(z)\varepsilon_{lsm}\frac{\partial}{\partial x_s}\int_{S_o} G_{mk}^j(\boldsymbol{\xi}-\boldsymbol{x})t_j(\boldsymbol{\xi})\,\mathrm{d}s(\boldsymbol{\xi}) \\ &\quad + \int_{S_o} n_l(z)H_{lk}^j(\boldsymbol{\xi}-\boldsymbol{x})t_j(\boldsymbol{\xi})\,\mathrm{d}s(\boldsymbol{\xi}).\end{aligned} \tag{68}$$

Assuming now that the boundary is smooth at z and that the traction is continuous at this location, the limit $\boldsymbol{x} \to z$ (in which z can be either on the regular boundary or the crack surfaces) can readily be taken. Indeed, using a result similar to (59) we find

$$\begin{aligned}c(z)t_k(z) &= n_l(z)\varepsilon_{lrt}\frac{\partial}{\partial z_r}\int_S C_{mj}^{tk}(\boldsymbol{\xi}-z)D_m u_j(\boldsymbol{\xi})\,\mathrm{d}s(\boldsymbol{\xi}) \\ &\quad + n_l(z)\varepsilon_{lsm}\frac{\partial}{\partial z_s}\int_{S_o} G_{mk}^j(\boldsymbol{\xi}-z)t_j(\boldsymbol{\xi})\,\mathrm{d}s(\boldsymbol{\xi}) \\ &\quad + \int_{S_o} n_l(z)H_{lk}^j(\boldsymbol{\xi}-z)t_j(\boldsymbol{\xi})\,\mathrm{d}s(\boldsymbol{\xi}),\end{aligned} \tag{69}$$

in which $c = \frac{1}{2}$ for $z \in S_o$ whereas $c = 1$ for $z \in S_c^+$ or $z \in S_c^-$. We note that the kernel $n_l(z)H_{lk}^j(\boldsymbol{\xi}-z)$ is weakly-singular; the proof of this fact is straightforward and similar to the proof that $n_l(\boldsymbol{\xi})H_{lk}^j(\boldsymbol{\xi}-z)$ is weakly-singular. We also note that the traction integral equation (69) is valid for either crack surface and it holds without modification for both embedded and surface-breaking cracks.

Finally, multiplying (69) by a *continuous test* function $u_k^*(z)$ and employing Stokes' theorem, we obtain a weak-form, weakly-singular traction integral equation as

$$\begin{aligned}\int_S c(z)t_k u_k^*(z)\,\mathrm{d}s(z) &= -\int_S D_t u_k^*(z)\int_S C_{mj}^{tk}(\xi - z)D_m u_j(\xi)\,\mathrm{d}s(\xi)\,\mathrm{d}s(z)\\ &\quad -\int_S D_m u_k^*(z)\int_{S_o} G_{mk}^i(\xi - z)t_i(\xi)\,\mathrm{d}s(\xi)\,\mathrm{d}s(z)\\ &\quad +\int_S u_k^*(z)\int_{S_o} n_l(z)H_{lk}^j(\xi - z)t_j(\xi)\,\mathrm{d}s(\xi)\,\mathrm{d}s(z).\end{aligned} \tag{70}$$

4.3. Discussion of weak-form equations

In the development of the weakly-singular, weak-form displacement and traction integral equations it was convenient to retain both crack surfaces despite the fact that they are geometrically coincident. Using the fact that $n_k^+ = -n_k^-$ at a given location on the crack, the final relations can now be reduced to involve the regular boundary S_o and the upper (or lower) crack surface alone. When this is done, the displacements of the individual crack surfaces no longer appear, but rather the displacement data for the crack is in terms of the relative crack-face displacement $\Delta u_j = u_j^+ - u_j^-$. For example, the term

$$\int_S D_t u_k^*(z)\int_S C_{mj}^{tk}(\xi - z)D_m u_j(\xi)\,\mathrm{d}s(\xi)\,\mathrm{d}s(z), \tag{71}$$

which appears in (70) can be expressed as

$$\begin{aligned}&\int_{S_o} D_t u_k^*(z)\int_{S_o} C_{mj}^{tk}(\xi - z)D_m u_j(\xi)\,\mathrm{d}s(\xi)\,\mathrm{d}s(z)\\ &\quad +\int_{S_o} D_t u_k^*(z)\int_{S_c^+} C_{mj}^{tk}(\xi - z)D_m \Delta u_j(\xi)\,\mathrm{d}s(\xi)\,\mathrm{d}s(z)\\ &\quad +\int_{S_c^+} D_t \Delta u_k^*(z)\int_{S_o} C_{mj}^{tk}(\xi - z)D_m u_j(\xi)\,\mathrm{d}s(\xi)\,\mathrm{d}s(z)\\ &\quad +\int_{S_c^+} D_t \Delta u_k^*(z)\int_{S_c^+} C_{mj}^{tk}(\xi - z)D_m \Delta u_j(\xi)\,\mathrm{d}s(\xi)\,\mathrm{d}s(z).\end{aligned} \tag{72}$$

It is important to recognize that after this reduction has been made there is but a single test function Δu_k^* associated with the crack.

As already emphasized, all the kernels associated with the weak-form equations (62) and (70) are weakly singular (i.e. $O(1/r)$ as $r \to 0$). An important consequence of this feature is that numerical implementations of these equations can utilize C^o elements for all the displacement data including the relative crack face displacements.

Now, either the weak-form displacement equation or the weak-form traction equation can be adopted to treat the regular boundary, but the weak-form traction integral equation is required for treatment of the crack. We remark that it is not difficult to also establish a weak-form displacement integral equation for the crack, but this equation is not required to solve the fracture problem if only the relative crack-face displacements are desired. Such an equation would be required, however, to evaluate the displacements of each individual crack-face (also see Chang and Mear, 1995; Li et al., 1998).

Naturally, the formulation also applies to bodies which do not contain cracks. In this case the weakly-singular, weak-form equations developed here are closely related to those established through a different approach by Bonnet (1995). As part of his development, Bonnet employs a subtraction technique to regularize the Cauchy-singular kernel associated with the standard displacement integral equation and he utilizes a result of Nedelec (1982) to effect a regularization of the hyper-singular kernel associated with the standard traction integral equation. The differences between the kernels appearing in his final weak-form equations and the corresponding kernels given here can be traced to the different treatments of the integral equations and, in particular, to the nonuniqueness associated with the kernel C^{tk}_{mj}.

4.4. Extension to Elastodynamics

As is well known, the fundamental solution for elastodynamics can be decomposed into a static part corresponding to Kelvin's solution and a time-dependent remainder which is nonsingular. As a consequence, the results developed above for static problems can be directly utilized to obtain singularity-reduced integral equations for elastodynamics. This can be done in the time domain or by use of integral transforms; here the development is carried through only for the time domain since an extension of the results to the Fourier or Laplace domain can be accomplished in a straightforward fashion.

Consider a time varying concentrated force acting at a location $\boldsymbol{x}$ in the whole space. Let $\mathcal{U}^p_i[\boldsymbol{\xi}-\boldsymbol{x}, t\mid]$ and $\mathcal{S}^p_{ij}[\boldsymbol{\xi}-\boldsymbol{x}, t\mid]$ be operators defined such that, for a force $f(t)$ acting in the p-th coordinate direction, the displacement and stress fields are given by $\mathcal{U}^p_i[\boldsymbol{\xi}-\boldsymbol{x}, t\mid f(t)]$ and $\mathcal{S}^p_{ij}[\boldsymbol{\xi}-\boldsymbol{x}, t\mid f(t)]$, respectively (see Appendix A). Further, define the operator $\mathcal{R}^p_{ij}[\boldsymbol{\xi}-\boldsymbol{x}, t\mid]$ such that

$$\mathcal{R}^p_{ij}[\boldsymbol{\xi}-\boldsymbol{x}, t\mid f(t)] = \mathcal{S}^p_{ij}[\boldsymbol{\xi}-\boldsymbol{x}, t\mid f(t)] - f(t)S^p_{ij}(\boldsymbol{\xi}-\boldsymbol{x}), \tag{73}$$

in which $S^p_{ij}(\boldsymbol{\xi}-\boldsymbol{x})$ is the static stress fundamental solution (i.e. Kelvin's solution). The key result to be exploited here is that, for an arbitrary twice differentiable function $f(t)$, we have $\mathcal{R}^p_{ij}[\boldsymbol{\xi}-\boldsymbol{x}, t\mid f(t)] = \ddot{f}(t)O(1)$ as $r\to 0$ hence the 'remainder kernel' $\mathcal{R}^p_{ij}$ is nonsingular (see e.g. Bonnet and Bui, 1993).

In terms of these operators, the generalization of Somigliana's identity to elastodynamics gives the (time varying) displacement at an internal point $\boldsymbol{x}$ of a body in terms of the traction and displacement acting on the surface of the body as

$$\begin{aligned} u_p(\boldsymbol{x}, t) &= \int_{S_o} \mathcal{U}^p_i[\boldsymbol{\xi}-\boldsymbol{x}, t\mid t_i(\boldsymbol{\xi}, t)]\,\mathrm{d}s(\boldsymbol{\xi}) \\ &\quad - \int_S n_i(\boldsymbol{\xi})\mathcal{S}^p_{ij}[\boldsymbol{\xi}-\boldsymbol{x}, t\mid u_j(\boldsymbol{\xi}, t)]\,\mathrm{d}s(\boldsymbol{\xi}). \end{aligned} \tag{74}$$

Exploiting the decomposition (73) for the operator $\mathcal{S}^p_{ij}$, and following a procedure analogous to that for the static case, a weakly-singular, weak-form displacement integral equation for elastodynamics is obtained as

$$\int_{S_o} \tfrac{1}{2}u_p(z, t)t^*_p(z)\,\mathrm{d}s(z)$$

$$= \int_{S_o} t_p^*(z) \int_{S_o} \mathcal{U}_i^p[\xi - z, t \mid t_i(\xi, t)] \, ds(\xi) \, ds(z)$$
$$+ \int_{S_o} t_p^*(z) \int_S G_{mj}^p(\xi - z) D_m u_j(\xi, t) \, ds(\xi) \, ds(z)$$
$$+ \int_{S_o} t_p^*(z) \int_S n_i(\xi) H_{ij}^p(\xi - z) u_j(\xi, t) \, ds(\xi) \, ds(z)$$
$$+ \int_{S_o} t_p^*(z) \int_S n_i(\xi) \mathcal{R}_{ij}^p[\xi - z, t \,| u_j(\xi, t)] \, ds(\xi) \, ds(z). \tag{75}$$

Likewise, a weakly-singular, weak-form traction integral equation can be established as

$$\int_S c(z) t_k(z, t) u_k^*(z) \, ds(z)$$
$$= - \int_S D_t u_k^*(z) \int_S C_{mj}^{tk}(\xi - z) D_m u_j(\xi, t) \, ds(\xi) \, ds(z)$$
$$- \int_S D_m u_k^*(z) \int_{S_o} G_{mk}^i(\xi - z) t_i(\xi, t) \, ds(\xi) \, ds(z)$$
$$+ \int_S u_k^*(z) \int_{S_o} n_l(z) H_{lk}^j(\xi - z) t_j(\xi, t) \, ds(\xi) \, ds(z)$$
$$+ \int_S u_k^*(z) n_l(z) \int_S n_i(\xi) E_{lkpq} \frac{\partial R_{ij}^p[\xi - z, t \mid u_j(\xi, t)]}{\partial \xi_q} \, ds(\xi) \, ds(z)$$
$$+ \int_S u_k^*(z) n_l(z) \int_{S_o} R_{lk}^i[\xi - z, t \mid t_i(\xi, t)] \, ds(\xi) \, ds(z). \tag{76}$$

5. Conclusions

A systematic procedure has been followed to develop singularity-reduced integral equations for discontinuities in linear elastic media. In the procedure, a regularization of the integral equations is achieved through an integration by parts which is facilitated by the introduction of certain stress functions (one for the stress fundamental solution and another, in essence, for the stress field induced by a discontinuity in an unbounded domain). The development is applicable to arbitrary material anisotropy, and for isotropic materials all of the relevant kernels have been established in closed form.

General discontinuities in an unbounded domain have been considered in detail, and singularity-reduced integral equations have been developed for all field quantities of interest. For the special case of a dislocation, the line integral expressions obtained for the displacement, stress and energy serve as alternatives to corresponding relations which exist in the literature. For the special case of a crack, the weakly-singular, weak-form traction integral equation which has been developed is of particular significance because of its usefulness for the numerical treatment of fracture problems.

Singularity-reduced integral equations have also been established for finite domains containing cracks and, in particular, a pair of weakly-singular, weak-form displacement and traction integral equations have been developed. Recently, Li et al. (1998) used these weak-form equations as the basis of a numerical procedure suitable for treating finite domains containing arbitrary cracks. In their work, the weak-form integral equations are first combined to obtain a 'symmetric formulation' which is then treated using Galerkin's method. An important aspect of the work is the construction of a special crack tip element which allows the stress intensity factors (for all three fracture modes) to be extracted directly and accurately as functions of arc length along the crack front. The procedure is demonstrated for various example cases (including both embedded and surface-breaking cracks), and it is shown that highly accurate stress intensity factors can be obtained using relatively coarse meshes.

Finally, we note that the systematic procedure presented here for the development of singularity-reduced integral equations in the whole space admits a generalization to treat discontinuities in other unbounded domains for which the fundamental solution exists in closed form. Indeed, we have recently carried out this extension for isotropic half-spaces and bi-materials. For both of these cases, singularity-reduced integral equations have been derived for all field quantities of interest, and a weakly-singular, weak-form traction integral equations analogous to that for the whole-space has also been established (Li, 1996; Li and Mear, 1998a,b,c). When specialized to dislocations, the results provide line integral representations for displacement, displacement gradient, stress and energy. For cracks, the weakly singular, weak-form traction integral equation is of particular interest because it provides a convenient basis for the numerical treatment of fractures in the half-space and bi-material.

Appendix A

For a homogeneous and isotropic whole-space, the displacement and stress fields induced by a (static) unit concentrated force acting in the pth coordinate direction are given by (e.g. Gurtin, 1972)

$$U_j^p(\boldsymbol{\zeta}) = \frac{1}{16\pi(1-\nu)\mu r}\left[(3-4\nu)\delta_{pj} + \frac{\zeta_p\zeta_j}{r^2}\right],$$

and

$$S_{ij}^p(\boldsymbol{\zeta}) = \frac{1}{8\pi(1-\nu)r^2}\left[\frac{(1-2\nu)}{r}(\zeta_p\delta_{ij} - \zeta_i\delta_{pj} - \zeta_j\delta_{pi}) - \frac{3\zeta_p\zeta_i\zeta_j}{r^3}\right],$$

in which ν is Poisson's ratio, μ is the shear modulus, $\boldsymbol{\zeta}$ is the position vector from the point of application of the load and $r^2 = \zeta_i\zeta_i$. For a time-varying concentrated force $f(t)$, the displacement and stress fields are given by (e.g. Gurtin, 1972; Bonnet and Bui, 1993)

$$\mathcal{U}_i^p[\boldsymbol{\zeta}, t\,|f(t)] = \frac{1}{4\pi\mu r}[Br_{,i}r_{,k} + (A + f(t - r/c_2))\delta_{ip}],$$

and

$$\mathcal{S}_{ij}^p[\boldsymbol{\zeta}, t\,|f(t)] = \frac{1}{4\pi r^2}[2Cr_{,p}r_{,i}r_{,j} + \delta_{ij}r_{,p}D + (\delta_{jp}r_{,i} + \delta_{ip}r_{,j})E],$$

in which

$$A = c_2^2 \int_{1/c_2}^{1/c_1} \lambda f(t - \lambda r)\, \mathrm{d}\lambda,$$

$$B = -3A - [f(t - r/c_2) - \gamma^2 f(t - r/c_1)],$$

$$C = -5B + \left[f\left(t - \frac{r}{c_2}\right) + \frac{r}{c_2}\dot{f}\left(t - \frac{r}{c_2}\right)\right] - \gamma^2 \left[f\left(t - \frac{r}{c_1}\right) + \frac{r}{c_1}\dot{f}\left(t - \frac{r}{c_1}\right)\right],$$

$$D = 2B - (1 - 2\gamma^2)[f(t - r/c_1) + \frac{r}{c_1}\dot{f}(t - r/c_1)],$$

$$E = 2B - [f(t - r/c_2) + \frac{r}{c_2}\dot{f}(t - r/c_2)],$$

and

$$c_1 = \sqrt{\frac{2\mu(1-\nu)}{(1-2\nu)\rho}}, \qquad c_2 = \sqrt{\frac{\mu}{\rho}}, \qquad \gamma = \frac{c_2}{c_1},$$

where ρ is the density.

Appendix B

The fact that the stress function $\psi^p_{mj}(\boldsymbol{\xi}, \boldsymbol{x})$ necessarily has a 'cut' along which it is singular can be motivated as follows. First we note that for a domain whose boundary consists of more than one connected component (i.e. more than one distinct closed surface), a stress function of the form (2) which is nonsingular throughout the domain will exist only when the resultant force on each connected component vanishes individually[8] (e.g. Gurtin, 1972). Consider then a modified domain in which a 'tunnel' extending from an outer surface and surrounding the point $\boldsymbol{\xi} = \boldsymbol{x}$ has been removed, and let the tractions which act on this surface be derived from the stress field $S^p_{ij}(\boldsymbol{\xi} - \boldsymbol{x})$. Then since the boundary of the domain has a single connected component and the body is in equilibrium, there does exists a stress function which is nonsingular throughout the domain. However, it cannot be expected that the extension of this function into the tunnel will be regular, for if it were then the body could be altered to one for which the boundary consists of more than one connected component yet with a resultant force acting on each component. As the original unbounded domain is recovered by taking the limit as the 'size' of the tunnel tends to zero and the outer boundary tends to infinity, the resulting stress function will be singular along the path about which the tunnel is shrunk. Note that in this construction a tunnel which extends across the domain could be considered (giving rise to an infinite path), but more simply the tunnel can be taken to terminate around the point $\boldsymbol{x}$ hence giving rise to a semi-infinite path. We remark that this path is analogous to the branch cut associated with a stress function (defined such that the stress is obtained

[8] The concentrated force solution can be considered as the limit of a sequence of problems in which tractions are prescribed on an interior closed surface surrounding the source point $\boldsymbol{\xi} = \boldsymbol{x}$.

via a single curl operation) for the two-dimensional stress fundamental solution, and that for the two-dimensional case the stress function corresponds to the displacement solution for a dislocation (see Chang and Mear, 1995). We also note that it is possible to construct stress functions which are singular on a surface rather than on a curve; within the context of the current discussion these can be interpreted as resulting from a 'cut' which is a surface containing the point $\boldsymbol{x}$.

To now establish that (3) is a stress function for the fundamental solution, it suffices to consider a vector valued function $v_i(\boldsymbol{\xi})$ which is nonsingular and divergence free for all $\boldsymbol{\xi} \neq \mathbf{0}$ but which may be singular at $\boldsymbol{\xi} = \mathbf{0}$. What is then sought is a function $w_t(\boldsymbol{\xi})$ such that

$$v_i = \varepsilon_{irt} w_{t,r}.$$

Assuming that v_i vanishes at infinity, it is readily verified that a particular solution for w_t is given in terms of a line integral as

$$w_t(\boldsymbol{\xi}) = \int_{\Gamma(\mathbf{0})} \varepsilon_{taj} v_a(\boldsymbol{\xi} - \boldsymbol{y})\, \mathrm{d}y_j,$$

in which $\Gamma_{(\mathbf{0})}$ is a semi-infinite curve originating at the origin. Indeed,

$$\begin{aligned}
\varepsilon_{irt} w_{t,r} &= \int_{\Gamma(\mathbf{0})} \varepsilon_{irt}\varepsilon_{taj} \frac{\partial v_a(\boldsymbol{\xi} - \boldsymbol{y})}{\partial \xi_r}\, \mathrm{d}y_j \\
&= -\int_{\Gamma(\mathbf{0})} (\delta_{ia}\delta_{rj} - \delta_{ij}\delta_{ra}) \frac{\partial v_a(\boldsymbol{\xi} - \boldsymbol{y})}{\partial y_r}\, \mathrm{d}y_j \\
&= -\int_{\Gamma(\mathbf{0})} \frac{\partial v_i(\boldsymbol{\xi} - \boldsymbol{y})}{\partial y_j}\, \mathrm{d}y_j = v_i(\boldsymbol{\xi}).
\end{aligned}$$

This result is readily generalized to tensor valued functions and, in particular, a consideration for the stress fundamental solution leads to (3).

While it is not pertinent to the current study, we note that if $v_i(\boldsymbol{\xi})$ is a homogeneous function of degree n with $n \neq -2$, then $\xi_r v_{i,r} = n v_i$ from which it can be established that a solution for $w_t(\boldsymbol{\xi})$ is

$$w_t = -\frac{1}{n+2} \varepsilon_{abt} \xi_a v_b.$$

This result could, for example, be used to generate stress functions for 'point-force dipoles'.

Appendix C

To prove (10) we first note that, by definition,

$$\frac{\partial u_p(\boldsymbol{x})}{\partial x_q} = \lim_{\Delta x_q \to 0} \frac{u_p(\boldsymbol{x} + \Delta \boldsymbol{x}) - u_p(\boldsymbol{x})}{\Delta x_q}. \tag{C.1}$$

From (7), $u_p(\boldsymbol{x})$ and $u_p(\boldsymbol{x} + \Delta \boldsymbol{x})$ are given as

$$u_p(\boldsymbol{x}) = \oint_{\partial S^+} \psi^p_{mj}(\boldsymbol{\xi}, \boldsymbol{x}) \Delta u_j(\boldsymbol{\xi})\, \mathrm{d}\xi_m - \int_{S^+} \psi^p_{mj}(\boldsymbol{\xi}, \boldsymbol{x}) D_m \Delta u_j(\boldsymbol{\xi})\, \mathrm{d}s(\boldsymbol{\xi}), \tag{C.2}$$

$$u_p(\boldsymbol{x}+\Delta\boldsymbol{x}) = \oint_{\partial S^+} \psi^p_{mj}(\boldsymbol{\xi}, \boldsymbol{x}+\Delta\boldsymbol{x})\Delta u_j(\boldsymbol{\xi})\, \mathrm{d}\xi_m$$

$$- \int_{S^+} \psi^p_{mj}(\boldsymbol{\xi}, \boldsymbol{x}+\Delta\boldsymbol{x}) D_m \Delta u_j(\boldsymbol{\xi})\, \mathrm{d}s(\boldsymbol{\xi}), \qquad \text{(C.3)}$$

in which

$$\psi^p_{mj}(\boldsymbol{\xi}, \boldsymbol{x}) = \int_{\Gamma(\boldsymbol{x})} \varepsilon_{mrt} S^p_{rj}(\boldsymbol{\xi} - \boldsymbol{z})\, \mathrm{d}z_t, \qquad \text{(C.4)}$$

$$\psi^p_{mj}(\boldsymbol{\xi}, \boldsymbol{x}+\Delta\boldsymbol{x}) = \int_{\Gamma'_{(\boldsymbol{x}+\Delta\boldsymbol{x})}} \varepsilon_{mrt} S^p_{rj}(\boldsymbol{\xi} - \boldsymbol{z})\, \mathrm{d}z_t. \qquad \text{(C.4)}$$

The closure condition guarantees that $\Gamma'_{(\boldsymbol{x}+\Delta\boldsymbol{x})}$ can be any semi-infinite curve originating at $\boldsymbol{x}+\Delta\boldsymbol{x}$ provided only that it does not intersect S^+, and for convenience we now choose it to be a translation of $\Gamma_{(\boldsymbol{x})}$ such that

$$\Gamma'_{(\boldsymbol{x}+\Delta\boldsymbol{x})} = \Delta\boldsymbol{x} + \Gamma_{(\boldsymbol{x})}.$$

From (C.5) we then have

$$\psi^p_{mj}(\boldsymbol{\xi}, \boldsymbol{x}+\Delta\boldsymbol{x}) = \int_{\Delta\boldsymbol{x}+\Gamma_{(\boldsymbol{x})}} \varepsilon_{mrt} S^p_{rj}(\boldsymbol{\xi} - \boldsymbol{z})\, \mathrm{d}z_t$$

$$= \int_{\Gamma_{(\boldsymbol{x})}} \varepsilon_{mrt} S^p_{rj}(\boldsymbol{\xi} - \boldsymbol{z} - \Delta\boldsymbol{x})\, \mathrm{d}z_t.$$

This result in conjunction with (C.4) leads to

$$\psi^p_{mj}(\boldsymbol{\xi}, \boldsymbol{x}+\Delta\boldsymbol{x}) = \psi^p_{mj}(\boldsymbol{\xi} - \Delta\boldsymbol{x}, \boldsymbol{x}),$$

which allows (C.3) to be re-written as

$$u_p(\boldsymbol{x}+\Delta\boldsymbol{x}) = \oint_{\partial S^+} \psi^p_{mj}(\boldsymbol{\xi} - \Delta\boldsymbol{x}, \boldsymbol{x})\Delta u_j(\boldsymbol{\xi})\, \mathrm{d}\xi_m$$

$$- \int_{S^+} \psi^p_{mj}(\boldsymbol{\xi} - \Delta\boldsymbol{x}, \boldsymbol{x}) D_m \Delta u_j(\boldsymbol{\xi})\, \mathrm{d}s(\boldsymbol{\xi}).$$

Substituting this relation and (C.2) into (C.1), and noting that

$$\lim_{\Delta x_q \to 0} \frac{\psi^p_{mj}(\boldsymbol{\xi} - \Delta\boldsymbol{x}, \boldsymbol{x}) - \psi^p_{mj}(\boldsymbol{\xi}, \boldsymbol{x})}{\Delta x_q} = -\frac{\partial \psi^p_{mj}(\boldsymbol{\xi}, \boldsymbol{x})}{\partial \xi_q},$$

we finally have

$$\frac{\partial u_p(\boldsymbol{x})}{\partial x_q} = -\oint_{\partial S^+} \frac{\partial \psi^p_{mj}(\boldsymbol{\xi}, \boldsymbol{x})}{\partial \xi_q} \Delta u_j(\boldsymbol{\xi})\, \mathrm{d}\xi_m$$

$$+ \int_{S^+} \frac{\partial \psi^p_{mj}(\boldsymbol{\xi}, \boldsymbol{x})}{\partial \xi_q} D_m \Delta u_j(\boldsymbol{\xi})\, \mathrm{d}s(\boldsymbol{\xi}).$$

Now, while this result has been established by appealing to the particular stress function (2), the closure condition guarantees that it applies for all possible stress functions (provided that the 'cut' associated with the stress function does not intersect the surface of the discontinuity).

Appendix D

We seek a function $\Phi_j^{lk}(\boldsymbol{\xi} - \boldsymbol{x})$ which is singular only at $\boldsymbol{\xi} = \boldsymbol{x}$ and which satisfies both (17) and (18). Since S_{sj}^p is homogeneous of degree -2 we infer that $\partial\Phi_j^{lk}(\boldsymbol{\xi} - \boldsymbol{x})/\partial\xi_l$ is also homogeneous of degree -2, and it then follows that

$$(\xi_m - x_m)\frac{\partial^2\Phi_j^{lk}(\boldsymbol{\xi} - \boldsymbol{x})}{\partial\xi_m\partial\xi_l} = -2\frac{\partial\Phi_j^{lk}(\boldsymbol{\xi} - \boldsymbol{x})}{\partial\xi_l} = -(\xi_m - x_m)E_{lkpq}\varepsilon_{sqm}\frac{\partial S_{sj}^p(\boldsymbol{\xi} - \boldsymbol{x})}{\partial\xi_l}.$$

Re-expressing this equation as

$$\begin{aligned}\frac{\partial\Phi_j^{lk}(\boldsymbol{\xi} - \boldsymbol{x})}{\partial\xi_l} &= \tfrac{1}{2}(\xi_m - x_m)E_{lkpq}\varepsilon_{sqm}\frac{\partial S_{sj}^p(\boldsymbol{\xi} - \boldsymbol{x})}{\partial\xi_l}\\ &= \tfrac{1}{2}E_{lkpq}\varepsilon_{sqm}\frac{\partial[(\xi_m - x_m)S_{sj}^p(\boldsymbol{\xi} - \boldsymbol{x})]}{\partial\xi_l}\\ &\quad -\tfrac{1}{2}E_{mkpq}\varepsilon_{sqm}S_{sj}^p(\boldsymbol{\xi} - \boldsymbol{x})\\ &= \tfrac{1}{2}E_{lkpq}\varepsilon_{sqm}\frac{\partial[(\xi_m - x_m)S_{sj}^p(\boldsymbol{\xi} - \boldsymbol{x})]}{\partial\xi_l}\\ &\quad -\tfrac{1}{2}E_{mkpq}\varepsilon_{sqm}E_{sjal}\frac{\partial U_a^p(\boldsymbol{\xi} - \boldsymbol{x})}{\partial\xi_l},\end{aligned}$$

it is evident that a particular solution for Φ_j^{lk} is given by

$$\Phi_j^{lk}(\boldsymbol{\zeta}) = \tfrac{1}{2}E_{nkpq}\varepsilon_{sqm}[\delta_{ln}\zeta_m S_{sj}^p(\boldsymbol{\zeta}) - \delta_{mn}E_{sjal}U_a^p(\boldsymbol{\zeta})],$$

with $\boldsymbol{\zeta} = \boldsymbol{\xi} - \boldsymbol{x}$. When this relation is specialized to isotropic materials it can be directly verified that the associated function $\hat{\Lambda}_{mj}^{lk}$ is, as desired, free of source in the sense that (18) is satisfied, and we expect that this also holds for general anisotropy.

We remark that any particular solution of (17), say $\tilde{\Phi}_j^{lk}$, which *does not* satisfy (18) can readily be 'corrected' such that this condition is satisfied. To demonstrate this, let $\tilde{\Lambda}_{mj}^{lk} \equiv E_{lkpq}\varepsilon_{sqm}S_{sj}^p + \partial\tilde{\Phi}_j^{lk}/\partial\xi_m$ and let

$$q_{mj}^k \equiv -\int_{|\boldsymbol{\zeta}|=1}\tilde{\Lambda}_{mj}^{lk}(\boldsymbol{\zeta})n_l(\boldsymbol{\zeta})\,\mathrm{d}s(\boldsymbol{\zeta}).$$

Then upon noting that the function

$$\hat{\Phi}_j^{lk}(\boldsymbol{\zeta}) \equiv -\frac{q_{nj}^k}{8\pi r}\left[\delta_{ln} + \frac{\zeta_l\zeta_n}{r^2}\right]$$

is such that $\partial\hat{\Phi}/\partial\xi_m$ is divergence free (i.e. $\partial[\partial\hat{\Phi}/\partial\xi_m]/\partial\xi_l = 0$) for all $\boldsymbol{\xi} \neq \boldsymbol{x}$ and that it has a source of strength $-q^k_{mj}$ at $\boldsymbol{\xi} = \boldsymbol{x}$, it is evident that the function

$$\Phi^{lk}_j \equiv \tilde{\Phi}^{lk}_j + \hat{\Phi}^{lk}_j$$

is a particular solution satisfying both (17) and (18).

Appendix E

Let

$$I_p \equiv -\oint_{\partial(S-S_\varepsilon)} \psi^p_{mj}(\boldsymbol{\xi}, \boldsymbol{x}) u_j(\boldsymbol{\xi})\, \mathrm{d}\xi_m, \tag{E.1}$$

in which S_ε is a small part of S which contains $\boldsymbol{p}$ and which has 'size' ε. What is to be shown is that $\lim_{\epsilon\to 0} I_p = u_p(\boldsymbol{p})$. Toward this end, write

$$I_p = -\oint_{\partial(S-S_\varepsilon)} \psi^p_{mj}(\boldsymbol{\xi}, \boldsymbol{x})[u_j(\boldsymbol{\xi}) - u_j(\boldsymbol{p})]\, \mathrm{d}\xi_m - u_j(\boldsymbol{p}) \oint_{\partial(S-S_\varepsilon)} \psi^p_{mj}\, \mathrm{d}\xi_m \tag{E.2}$$

and note that the first term in this expression vanishes as $\epsilon \to 0$ (see the discussion of the singular structure of ψ^p_{mj} given in Section 2.4.1). To treat the second term, Stokes' theorem can be used along with the definition of the stress function (2) to obtain

$$J_{pj} \equiv \oint_{\partial(S-S_\varepsilon)} \psi^p_{mj}\, \mathrm{d}\xi_m = \int_{S^*} n_i \varepsilon_{ikm} \frac{\partial \psi^p_{mj}}{\partial \xi_k}\, \mathrm{d}s(\boldsymbol{\xi}) = \int_{S^*} n_i S^p_{ij}\, \mathrm{d}s(\boldsymbol{\xi}), \tag{E.3}$$

in which S^* is any surface which has edge $\partial(S - S_\varepsilon)$ and which does not intersect the path $\Gamma_{(\boldsymbol{x})}$ (hence it necessarily surrounds the source point $\boldsymbol{\xi} = \boldsymbol{x}$). In the limit $\epsilon \to 0$ the surface S^* tends to a closed surface surrounding the source point $\boldsymbol{\xi} = \boldsymbol{x}$, hence $\lim_{\epsilon\to 0} J_{pj} = -\delta_{pj}$ and so $\lim_{\epsilon\to 0} I_p = u_p(\boldsymbol{p})$.

In developing this result, consideration has been limited to the case in which the path $\Gamma_{(\boldsymbol{x})}$ intersects the regular boundary S_o at a single location. If instead it intersects this boundary at more than one location (or, in fact, if it intersects the crack), then the construction followed above can be applied to separately evaluate the contribution from each intersection point $\boldsymbol{p}$. When this is done it is found that at each location at which the path 'exits' the body there is a contribution $u_p(\boldsymbol{p})$, whereas at each location where it 're-enters' the body there is a contribution $-u_p(\boldsymbol{p})$.

Acknowledgments

This work was supported by the NASA Airframe Structural Integrity Program (contract NAG-1-1121) under the direction of Dr. C.E. Harris and Dr. J.C. Newman. We are grateful to Charlie Harris and Jim Newman for the continued support and encouragement which they have provided.

References

Balakrishna, C., Gray, L.J. and Kane, J.H. (1994). Efficient analytical integration of symmetric Galerkin boundary integrals over curved elements; elasticity formulation. *Computer Methods in Applied Mechanics and Engineering* **117**, 157–179.

Bonnet, M. (1995). Regularized direct and indirect symmetric variational BIE formulations for three-dimensional elasticity. *Engineering Analysis with Boundary Elements* **15**, 93–102.

Bonnet, M. and Bui, H.D. (1993). Regularization of the displacement and traction BIE for 3D elastodynamics using indirect methods. *Advances in Boundary Element Techniques*. (Edited by J.H. Kane and M. Maier), Springer-Verlag.

Brebbia, C.A. and Dominguez, J. (1992). *Boundary Elements, An Introductory Course*, 2nd edn. Computational Mechanics Publications.

Chang, C.C. and Mear, M.E. (1995). A boundary element method for two dimensional linear elastic fracture analysis. *International Journal of Fracture* **74**, 219–251.

Cruse, T.A. (1988). *Boundary Element Analysis in Computational Fracture Mechanics*, Kluwer Academic Publishers.

Eringen, A.C. and Suhubi, E.S. (1975). *Elastodynamics, vol. II*, Academic Press.

Ghosh, N. and Mukherjee, S. (1987). A new boundary element method formulation for three-dimensional problems in linear elasticity. *Acta Mechanica* **67**, 107–119.

Gu, H.D. and Yew, C.H. (1988). Finite element solution of a boundary integral equation for mode-I embedded three-dimensional fractures. *International Journal for Numerical Methods in Engineering* **26**, 1525–1540.

Guiggiani, M., Krishnasamy, G., Rizzo, F.J. and Rudolphi, T.J. (1991). Hypersingular boundary integral equations. A new approach to their numerical treatment. *Boundary Integral Methods, Theory and Applications* (Edited by L. Morino, R. Piva and R.G. Piva), Berlin, 211–220.

Gurtin, M.E. (1972). The linear theory of elasticity. *Handbuch der Physik* (Edited by C. Truesdell), VIa/2, Springer-Verlag, Berlin and New York.

Hirth, P.J. and Lothe, J. (1982). *Theory of Dislocations*. W.J. Publishing.

Kellog, O.D. (1953). *Foundations of Potential Theory*, Dover, New York.

Li, S. (1996) *Singularity-reduced Integral Equations for Discontinuities in Three Dimensional Elastic Media*. Ph.D. dissertation, the University of Texas at Austin.

Li, S. and Mear, M.E. (1998a). *Singularity-reduced Integral Equations for Discontinuities in an Isotropic Half-space*. In preparation.

Li, S. and Mear, M.E. (1998b). *Singularity-reduced Integral Equations for Discontinuities in a Bi-material for which the Poisson's Ratio are Equal*. In preparation.

Li, S. and Mear, M.E. (1998c). *Singularity-reduced Integral Equations for Discontinuities in a General Isotropic, Linear Elastic Bi-material*. In preparation.

Li, S., Mear, M.E. and Xiao, L. (1998). Symmetric weak-form integral equation method for three dimensional fracture analysis. *Computer Methods in Applied Mechanics and Engineering*, **151**, 435–459.

Martin, P.A. and Rizzo, F.J. (1996). Hypersingular integrals: how smooth must the density be? *International Journal for Numerical Methods in Engineering* **39**, 687–704.

Muci-Küchler, K.H. and Rudolphi, T.J. (1994). Application of tangent derivative boundary integral equations to the formulation of higher order boundary elements. *International Journal of Solids and Structures* **31**, 1565–1584.

Mura, T. (1963). Continuous distribution of moving dislocations. *Philosophical Magazine* **8**, 843–857.

Nedelec, J.C. (1982). Integral equations with nonintegrable kernels. *Integral Equations and Operator Theory* **5**, 563–572.

Peach, M. and Koehler, J. (1956). The forces exerted on dislocations and the stress fields produced by them. *Physical Review* **80**, 436–439.

Sládek, V., and Sládek, J. (1982). Three-dimensional crack analysis for an anisotropic body. *Applied Mathematical Modelling* **6**, 374–380.

Weaver, J. (1977). Three-dimensional crack analysis. *International Journal of Solids and Structures* **13**, 321–330.

Xu, G. and Ortiz, M. (1993). A variational boundary integral method for the analysis of 3-D cracks of arbitrary geometry modelled as continuous distributions of dislocation loops. *International Journal for Numerical Methods in Engineering* **36**, 3675–3701.

International Journal of Fracture **93:** 115–144, 1998.

A single-domain dual-boundary-element formulation incorporating a cohesive zone model for elastostatic cracks

B. YANG and K. RAVI-CHANDAR
Department of Mechanical Engineering, University of Houston, Houston, Tx 77204-4792, U.S.A.
e-mail: ravi@uh.edu

Received 11 December 1997; accepted in revised form 19 June 1998

Abstract. A cracked elastostatic structure is artificially divided into subdomains of simpler topology such that the well-developed classic dual integral equations can be applied appropriately to each domain. Applying the continuity and equilibrium conditions along artificial boundaries and properties of the integral kernels a single-domain dual-boundary-integral equation formulation is derived for a cracked elastic structure. A cohesive zone model is used to model the crack tip processes and is coupled with the single-domain dual-boundary-integral equation formulation; the resulting nonlinear equations are solved using the iterative method of successive-over-relaxation. The constitutive law used for a crack includes three parts: a law relating cohesive force to crack displacement difference when a crack is opening, a characterization of tangential interaction between crack surfaces when the crack surfaces are in contact, and a maximum principal stress criterion of crack advance. Incorporation of local unloading effect of the cohesive zone material has enabled a simulation of fracture with initial damage, partial development of the failure process zone at structural instability and multiple crack interaction. Some of the features of the method are demonstrated by considering three examples. The first problem is a single-edge-cracked specimen that exhibits a snap-back instability. The second example is the development of wing cracks from an angled crack under compression. The last example demonstrates the capability to consider mixed-mode crack growth and interaction of cracks. Thus, the problem of crack growth has been reduced to the determination of the cohesive model for the fracture process.

Key words: Cohesive zone model, boundary integral method.

1. Introduction

Boundary element method (BEM) has received much attention recently, especially in application to fracture mechanics (FM) (see Cruse, 1996 and Aliabadi, 1997 for reviews). The method is attractive because it involves discretization of the boundary alone; the dimensionality of the stiffness matrix formed in BEM is then reduced by one in comparison to a domain method, such as finite element method (FEM), although the stiffness matrix is full and asymmetric in general. A particularly attractive advantage of BEM in application to crack problems over FEM is that domain remeshing is not necessary when a crack grows; only one more element of a crack is added with all the already existing elements untouched. However, as a crack is modeled mathematically with the two crack surfaces being coincident, the classic boundary integral equation (BIE) can not be applied directly or the resulting stiffness matrix formed will be ill-conditioned. A great deal of effort has been expended in dealing with this difficulty; many methods – such as the crack Green's function method (Snyder and Cruse, 1975), the displacement discontinuity method (Crouch, 1976 and Wen, 1996), the subdomains method (Blandford, Ingraffea and Liggett, 1981), the dual boundary element method (DBEM) (Hong and Chen, 1988, Portela, Aliabadi and Rooke, 1992, and Chen and Chen,

1995), the single-domain traction boundary element method (Young, 1996), and some hybrid methods optimizing the advantages of some of the above mentioned methods (Ameen and Raghuprasad, 1994) – have been proposed. It is not our intent to review these methods here; one may see Chen and Chen (1995) for a discussion of some of these methods. Among the methods mentioned above the method of subdomains eliminates the discrepancy noted above by cutting a cracked structure into pieces of simpler topology such that BIE can be applied properly to each of domain separately. However, this formulation increases the computational effort as a result of the additional artificial boundaries that appear from the cutting process; moreover, when a crack advances, remeshing is still needed in general. Nevertheless, the idea of the method of subdomains is valuable and is adopted in the present paper to derive a mathematically rigorous and simple formulation of the *single-domain dual-boundary-integral equations* (SDDBIEs) of a cracked structure. The classic dual integral equations for a simple structure is, of course, well established in the literature. These SDDBIEs were given by Young (1996), and applied to solve a traction-free crack problem using continuous elements.

The present paper is organized as follows: In Section 2, we summarize the classic dual integral equations for a simple elastostatic structure, and derive the single-domain dual-integral-equations for an elastostatic structure containing a mathematically modeled crack. By applying continuity conditions and properties of the integral kernels across the cut, the integrals along the artificial boundaries that appear due to cutting are eliminated. In most crack problems in which tractions on two opposite crack surfaces are equal in magnitude and opposite in directions, the single-domain dual-boundary-integral equations can be further simplified with integrals of crack tractions eliminated. In Section 3, we consider the problem of a cohesive crack. A line spring model is used to characterize interaction between two cohesive crack surfaces opening or sliding. Material softening with accumulated damage is also incorporated into the cohesive model. This model seems most appropriate to a crazing zone in polymers, in which the bridging long chain molecules are fully orientated and flexible, but can also be applied to most bridged crack problems, such as in ceramics and fiber or particle reinforced materials.

Coulomb friction law is applied to crack surfaces when they come in contact due to normal pressure. This line spring model may contribute to tangential interaction too if not broken completely in that case. The maximum principal stress criterion is applied for the marching of an existing crack. A physical crack tip would follow along the same path as a natural result of complete debonding of the cohesive zone material behind the fictitious crack tip. Note that the incorporation of local unloading effect of the cohesive zone material has enabled a simulation of fracture with an initial process zone, the partial development at structural instability and most importantly arbitrary crack interaction. In Section 4, we introduce an iterative boundary element method of successive over-relaxation to approach numerically the problem of a cohesive crack formulated in Section 3. An iterative process is necessary in the present problem since behavior of the cohesive zone material is nonlinear and history dependent. Crack opening and contact are both programmed simply in the iterative process. No special iterative procedure is needed for mixed crack opening and contact problem. In Section 5, we simulate a two-dimensional rectangular specimen with three different crack configurations. Problems involving structural instability, mixed-mode crack growth, crack surface compression and friction and multiple crack interactions are all explored as examples demonstrating the power of the cohesive zone formulation of the boundary element method.

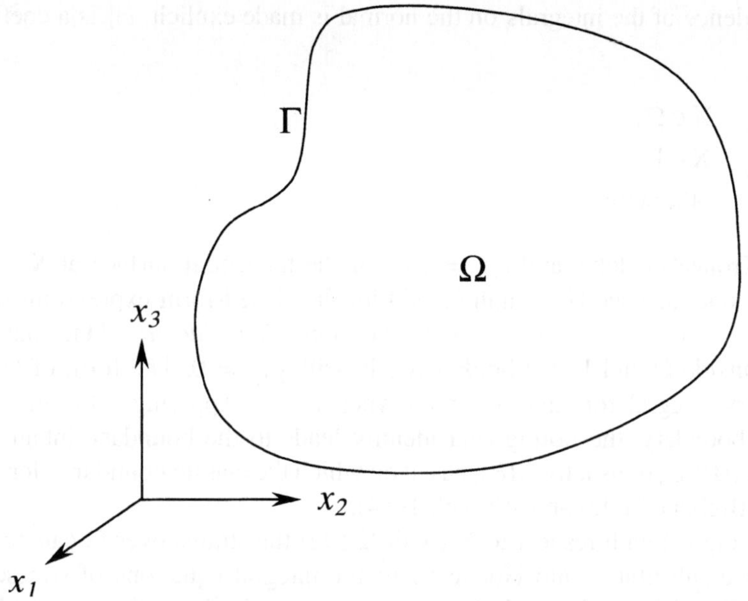

Figure 1. An isotropic, homogeneous, linear elastic domain Ω with boundary Γ.

2. Single-domain dual boundary integral equations of a cracked structure in elastostatics

2.1. SUMMARY OF THE DUAL INTEGRAL EQUATIONS FOR A SIMPLE STRUCTURE

Consider a homogeneous, isotropic, linearly elastic domain Ω with piecewise smooth boundary Γ as shown in Figure 1. A Cartesian (Lagrangian) coordinate system[1] is used along with standard indicial notation. The displacement components $u_i(\mathbf{X})$, at a point $\mathbf{X}$, can be represented in the following form

$$\underset{\sim}{c}_{ij}(\mathbf{X})u_j(\mathbf{X}) = \int_\Gamma \{u^*_{ij}(\mathbf{X},\mathbf{x})p_j(\mathbf{x}) - p^*_{ij}(\mathbf{X},\mathbf{x},\mathbf{n})u_j(\mathbf{x})\}\,\mathrm{d}\Gamma(\mathbf{x}) + \int_\Omega u^*_{ij}(\mathbf{X},\mathbf{x})b_j(\mathbf{x})\,\mathrm{d}\Omega(\mathbf{x}), \tag{1}$$

where p_j are traction components, b_j are body force components. u^*_{ij} and p^*_{ij} are the fundamental solutions representing the displacements and tractions respectively in the jth direction at a field point $\mathbf{x}$ due to a unit force acting in the ith direction at a source point $\mathbf{X}$. Note that p^*_{ij} are taken along the outward normal vector, $\mathbf{n}$, of Γ at $\mathbf{x}$ and that with the inclusion of $\mathbf{n}$ in

[1] Standard index notation is used. The range for Latin subscript indices is 3 and the range for Greek subscript indices is 2. Summation over repeated subscripts over their range is implied unless suspended explicitly. Superscripts do not follow this range and summation convention; their range would be indicated explicitly and summation is indicated by the summation symbol.

p^*_{ij}, the dependence of the integrals on the normal is made explicit. c_{ij} is a coefficient matrix given by

$$\underline{c}_{ij} = \begin{cases} \delta_{ij} & \mathbf{X} \in \Omega, \\ c_{ij} & \mathbf{X} \in \Gamma, \\ 0 & \text{otherwise}, \end{cases} \tag{2}$$

where δ_{ij} is Kronecker delta, and $c_{ij} = \delta_{ij}/2$ if the tangential surface at $\mathbf{X}$ is smooth; if it is not the case one may see Hartmann (1980) for the closed-form expressions of this matrix. When $\mathbf{X} \in \Omega$, (1) with $\underline{c}_{ij} = \delta_{ij}$ is called the *Somigliana identity*. One may easily prove that, if $\mathbf{X}$ is outside Ω and Γ, (1) holds trivially with $\underline{c}_{ij} = 0$. The form of (1) most useful in the boundary integral formulation arises when $\mathbf{X} \in \Gamma$. Applying a limiting process as $\mathbf{X}$ approaches a boundary, the Somigliana identity leads to the boundary integral equation of displacements (BIED, or usually BIE), based on which the classic boundary element technique is developed (Brebbia, Telles and Wrobel, 1984).

Differentiating (1) with respect to $\mathbf{X}$ (with $\mathbf{X} \in \Omega$) the strains over Ω can be calculated. If these strains are substituted into Hooke's law, the integral equations of stresses at a source point $\mathbf{X}$ (with $\mathbf{X} \in \Omega$) are obtained. A limiting process, similar to the one used to obtain the BIE, can be applied to the resulting integral equation as $\mathbf{X}$ approaches a boundary, leading to the boundary integral equation of stresses. These integral equations are summarized below

$$\begin{aligned} \underline{C}_{ik}(\mathbf{X})\sigma_{kj}(\mathbf{X}) &= \int_\Gamma \{U^*_{ijk}(\mathbf{X}, \mathbf{x})p_k(\mathbf{x}) - P^*_{ijk}(\mathbf{X}, \mathbf{x}, \mathbf{n})u_k(\mathbf{x})\}\,\mathrm{d}\Gamma(\mathbf{x}) \\ &\quad + \int_\Omega U^*_{ijk}(\mathbf{X}, \mathbf{x})b_k(\mathbf{x})\,\mathrm{d}\Omega(\mathbf{x}), \end{aligned} \tag{3}$$

where U^*_{ijk} and P^*_{ijk} are linear combinations of derivatives of u^*_{ij} and p^*_{ij}, with respect to $\mathbf{X}$. As in the case of p^*_{ij}, P^*_{ijk} are taken along the outward normal vector, $\mathbf{n}$, of Γ at $\mathbf{x}$. $\underline{C}_{ij}$ is a coefficient matrix given by

$$\underline{C}_{ij} = \begin{cases} \delta_{ij} & \mathbf{X} \in \Omega, \\ \delta_{ij}/2 & \mathbf{X} \in \Gamma, \\ 0 & \text{otherwise}, \end{cases} \tag{4}$$

which is identical to $\underline{c}_{ij}$ provided that corner points where tractions are not well defined are excluded. If (3) are multiplied on both sides by the outward normal vector at $\mathbf{X}$ as $\mathbf{X}$ in Γ, then, the boundary integral equations of tractions (BIET) are obtained, which may be employed to formulate a boundary element method in a similar way as BIE.

The fundamental solutions used in the above formulation are due to Kelvin (Love, 1944), and are the basic singular solutions to a point load in an infinite medium. The components u^*_{ij} and p^*_{ij} are given below for two (plane-strain) and three-dimensional problems:

$$u^*_{ij}(\mathbf{X}, \mathbf{x}) = \frac{1}{16\pi\mu(1-\nu)r}[(3-4\nu)\delta_{ij} + r_{,i}r_{,j}], \qquad \text{(3-D)}, \tag{5}$$

$$u^*_{ij}(\mathbf{X}, \mathbf{x}) = \frac{1}{8\pi\mu(1-\nu)}\left[(3-4\nu)\ln\left(\frac{1}{r}\right)\delta_{ij} + r_{,i}r_{,j}\right], \qquad \text{(2-D; plane-strain)}, \tag{6}$$

$$p_{ij}^*(\mathbf{X}, \mathbf{x}, \mathbf{n})$$

$$= \frac{-1}{4\overline{\alpha}\pi(1-\nu)r^{\overline{\alpha}}}\left[\frac{\partial r}{\partial n}((1-2\nu)\delta_{ij} + \overline{\beta}r_{,i}r_{,j}) - (1-2\nu)(r_{,i}n_j - r_{,j}n_i)\right], \tag{7}$$

where μ is shear modulus, and ν is Poisson ratio; $\overline{\alpha} = 2$, $\overline{\beta} = 3$ for three-dimensional problems and $\overline{\alpha} = 1$, $\overline{\beta} = 2$ for two-dimensional plane-strain problems. Moreover, $r = r(\mathbf{X}, \mathbf{x})$ represents the distance between the points $\mathbf{X}$ and $\mathbf{x}$, and its derivatives are taken with respect to $\mathbf{x}$. U_{ijk}^* and P_{ijk}^* are given by

$$U_{ijk}^*(\mathbf{X}, \mathbf{x}) = \frac{1}{4\overline{\alpha}\pi(1-\nu)r^{\overline{\alpha}}}[(1-2\nu)(r_{,j}\delta_{ik} + r_{,i}\delta_{jk} - r_{,k}\delta_{ij}) + \overline{\beta}r_{,i}r_{,j}r_{,k}], \tag{8}$$

$$P_{ijk}^*(\mathbf{X}, \mathbf{x}, \mathbf{n})$$

$$= \frac{\mu}{2\overline{\alpha}\pi(1-\nu)r^{\overline{\beta}}}\left\{\overline{\beta}\frac{\partial r}{\partial n}[(1-2\nu)r_{,k}\delta_{ij} + \nu(r_{,i}\delta_{jk} + r_{,j}\delta_{ik}) - \overline{\gamma}r_{,i}r_{,j}r_{,k}]\right.$$
$$+\overline{\beta}\nu(r_{,i}n_j r_{,k} + n_i r_{,j} r_{,k})$$
$$\left. +(1-2\nu)(\overline{\beta}n_k r_{,i} r_{,j} + n_j\delta_{ik} + n_i\delta_{jk}) - (1-4\nu)n_k\delta_{ij}\right\}, \tag{9}$$

where $\overline{\alpha}$ and $\overline{\beta}$ are as given above, and $\overline{\gamma} = 5$ for three-dimensional problems and $\overline{\gamma} = 4$ for two-dimensional problems. The plane strain expressions are valid for plane stress provided that ν is replaced by $\overline{\nu} = \nu/(1+\nu)$. As a source point $\mathbf{X}$ and a field point $\mathbf{x}$ coincide on a boundary, the integral kernels of u_{ij}^*, p_{ij}^*, U_{ijk}^*, and P_{ijk}^* may be singular and even hypersingular. The corresponding integrals in (1) and (3) are taken in the sense of Cauchy principal value if singular, and are taken in the sense of Hadamard principal value if hypersingular.

2.2. The single-domain dual integral equations for a cracked structure

A structure containing a mathematically sharp crack degenerates the boundary integral formulation due to coincidence of the two crack surfaces. In that case, one cannot apply either BIED or BIET to produce a boundary element method directly in general. This problem has been addressed in a number of ways: partitioning the domain into multi-domains, using crack Green's functions, the displacement discontinuity technique, and so on. In this section, we describe the formulation of a single-domain dual-boundary-integral method using Kelvin's solutions. We cut a cracked elastostatic structure into subdomains of simple topology so that the dual integral equations for a simple structure can be applied to each domain appropriately. In a further step, using the continuity and equilibrium conditions along artificial boundaries, as well as some properties of the integral kernels, we eliminate entirely the integrals involving any artificial boundaries. In order to illustrate the derivation of the single-domain integral equations, we shall consider a structure containing one crack; however, the same approach can be applied to any number of cracks in the structure. Furthermore, for simplicity, we shall ignore the body forces; if necessary, they may be added back into the final equations without any difficulty.

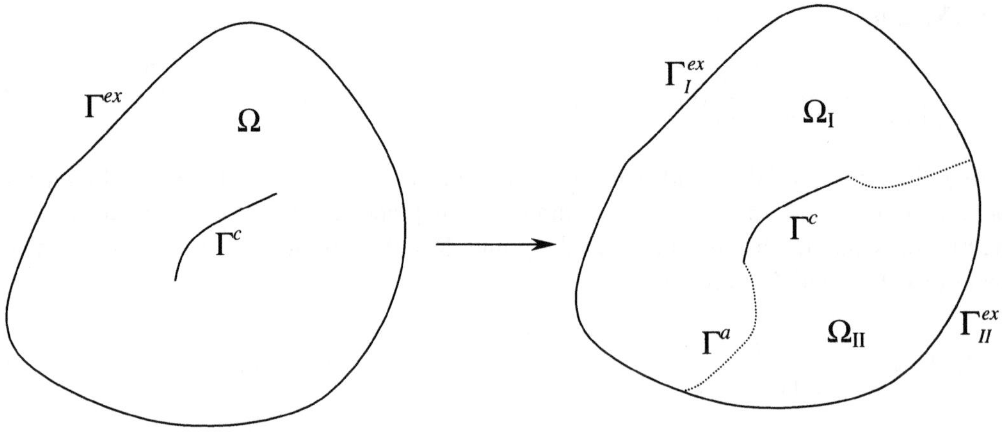

Figure 2. A domain Ω containing a crack Γ^c is cut into two subdomains Ω_{I} and Ω_{II} along the dotted line path Γ^a.

As shown in Figure 2, a structure containing one crack is cut into two subdomains with a path that passes through the crack. We denote the two subdomains without a crack by Ω_{I} and Ω_{II} respectively. Their boundaries are denoted by Γ_{I} and Γ_{II} respectively. The topology of the subdomains is simple such that (1) and (3) may be applied to them. This results in the following sets of equations for the displacement and traction components in the two subdomains. For the domain Ω_{I}, we have

$$\underline{c}^{\mathrm{I}}_{ij}(\mathbf{X})u^{\mathrm{I}}_j(\mathbf{X}) = \int_{\Gamma_{\mathrm{I}}} \{u^*_{ij}(\mathbf{X},\mathbf{x})p^{\mathrm{I}}_j(\mathbf{x}) - p^*_{ij}(\mathbf{X},\mathbf{x},\mathbf{n}^{\mathrm{I}})u^{\mathrm{I}}_j(\mathbf{x})\}\,\mathrm{d}\Gamma(\mathbf{x}), \tag{10}$$

with

$$\underline{c}^{\mathrm{I}}_{ij} = \begin{cases} \delta_{ij} & \mathbf{X}\in\Omega_{\mathrm{I}}, \\ c^{\mathrm{I}}_{ij} & \mathbf{X}\in\Gamma_{\mathrm{I}}, \\ 0 & \text{otherwise} \end{cases} \tag{11}$$

and

$$\underline{C}^{\mathrm{I}}_{ik}(\mathbf{X})\sigma^{\mathrm{I}}_{kj}(\mathbf{X}) = \int_{\Gamma_{\mathrm{I}}} \{U^*_{ijk}(\mathbf{X},\mathbf{x})p^{\mathrm{I}}_k(\mathbf{x}) - P^*_{ijk}(\mathbf{X},\mathbf{x},\mathbf{n}^{\mathrm{I}})u^{\mathrm{I}}_k(\mathbf{x})\}\,\mathrm{d}\Gamma(\mathbf{x}), \tag{12}$$

with

$$\underline{C}^{\mathrm{I}}_{ik} = \begin{cases} \delta_{ik} & \mathbf{X}\in\Omega_{\mathrm{I}}, \\ \delta_{ik}/2 & \mathbf{X}\in\Gamma_{\mathrm{I}}, \\ 0 & \text{otherwise.} \end{cases} \tag{13}$$

For the domain Ω_{II}, we also have

$$\underline{c}^{\mathrm{II}}_{ij}(\mathbf{X})u^{\mathrm{II}}_j(\mathbf{X}) = \int_{\Gamma_{\mathrm{II}}} \{u^*_{ij}(\mathbf{X},\mathbf{x})p^{\mathrm{II}}_j(\mathbf{x}) - p^*_{ij}(\mathbf{X},\mathbf{x},\mathbf{n}^{\mathrm{II}})u^{\mathrm{II}}_j(\mathbf{x})\}\,\mathrm{d}\Gamma(\mathbf{x}), \tag{14}$$

with

$$\underline{c}^{\mathrm{II}}_{ij} = \begin{cases} \delta_{ij} & \mathbf{X} \in \Omega_{\mathrm{II}}, \\ c^{\mathrm{II}}_{ij} & \mathbf{X} \in \Gamma_{\mathrm{II}}, \\ 0 & \text{otherwise}, \end{cases} \tag{15}$$

and

$$\underline{C}^{\mathrm{II}}_{ik}(\mathbf{X})\sigma^{\mathrm{II}}_{kj}(\mathbf{X}) = \int_{\Gamma_{\mathrm{II}}} \{U^*_{ijk}(\mathbf{X},\mathbf{x})p^{\mathrm{II}}_k(\mathbf{x}) - P^*_{ijk}(\mathbf{X},\mathbf{x},\mathbf{n}^{\mathrm{II}})u^{\mathrm{II}}_k(\mathbf{x})\}\,\mathrm{d}\Gamma(\mathbf{x}), \tag{16}$$

with

$$\underline{C}^{\mathrm{II}}_{ik} = \begin{cases} \delta_{ik} & \mathbf{X} \in \Omega_{\mathrm{II}}, \\ \delta_{ik}/2, & \mathbf{X} \in \Gamma_{\mathrm{II}}, \\ 0, & \text{otherwise}. \end{cases} \tag{17}$$

All of the functions are defined for all $\mathbf{X}$ and are single-valued. Thus, we add (10) and (14), and (12) and (16) algebraically, and rearrange them in the following forms

$$\begin{aligned} &\underline{c}^{\mathrm{I}}_{ij}(\mathbf{X})u^{\mathrm{I}}_j(\mathbf{X}) + \underline{c}^{\mathrm{II}}_{ij}(\mathbf{X})u^{\mathrm{II}}_j(\mathbf{X}) \\ &\quad= \int_{\Gamma^{\mathrm{ex}}} \{u^*_{ij}(\mathbf{X},\mathbf{x})p_j(\mathbf{x}) - p^*_{ij}(\mathbf{X},\mathbf{x},\mathbf{n})u_j(\mathbf{x})\}\,\mathrm{d}\Gamma(\mathbf{x}) \\ &\qquad+ \int_{\Gamma^{c+a}} \{u^*_{ij}(\mathbf{X},\mathbf{x})[p^{\mathrm{I}}_j(\mathbf{x}) + p^{\mathrm{II}}_j(\mathbf{x})] \\ &\qquad-[p^*_{ij}(\mathbf{X},\mathbf{x},\mathbf{n}^{\mathrm{I}})u^{\mathrm{I}}_j(\mathbf{x}) + p^*_{ij}(\mathbf{X},\mathbf{x},\mathbf{n}^{\mathrm{II}})u^{\mathrm{II}}_j(\mathbf{x})]\,\mathrm{d}\Gamma(\mathbf{x}), \end{aligned} \tag{18}$$

and

$$\begin{aligned} &\underline{C}^{\mathrm{I}}_{ik}(\mathbf{X})\sigma^{\mathrm{I}}_{kj}(\mathbf{X}) + \underline{C}^{\mathrm{II}}_{ik}(\mathbf{X})\sigma^{\mathrm{II}}_{kj}(\mathbf{X}) \\ &\quad= \int_{\Gamma^{\mathrm{ex}}} \{U^*_{ijk}(\mathbf{X},\mathbf{x})p_k(\mathbf{x}) - P^*_{ijk}(\mathbf{X},\mathbf{x},\mathbf{n})u_k(\mathbf{x})\}\,\mathrm{d}\Gamma(\mathbf{x}) \\ &\qquad+ \int_{\Gamma^{c+a}} \{U^*_{ijk}(\mathbf{X},\mathbf{x})[p^{\mathrm{I}}_k(\mathbf{x}) + p^{\mathrm{II}}_k(\mathbf{x})] \\ &\qquad-[P^*_{ijk}(\mathbf{X},\mathbf{x},\mathbf{n}^{\mathrm{I}})u^{\mathrm{I}}_k(\mathbf{x}) + P^*_{ijk}(\mathbf{X},\mathbf{x},\mathbf{n}^{\mathrm{II}})u^{\mathrm{II}}_k(\mathbf{x})]\,\mathrm{d}\Gamma(\mathbf{x}), \end{aligned} \tag{19}$$

where the superscript ex denotes an external regular boundary, the superscript c denotes a crack boundary, and the superscript a denotes an artificial boundary generated by the cutting process. Moreover, the integrals over $\Gamma^{\mathrm{ex}}_{\mathrm{I}}$ and $\Gamma^{\mathrm{ex}}_{\mathrm{II}}$ have been put together with $\Gamma^{\mathrm{ex}} = \Gamma^{\mathrm{ex}}_{\mathrm{I}} + \Gamma^{\mathrm{ex}}_{\mathrm{II}}$; the superscripts I and II have been dropped since $\Gamma^{\mathrm{ex}}_{\mathrm{I}}$ and $\Gamma^{\mathrm{ex}}_{\mathrm{II}}$ do not overlap. Equilibrium of the whole structure indicates that the tractions on two coincident opposite artificial boundaries are equal in magnitude and opposite in direction; i.e.,

$$p^{\mathrm{I}}_i(\mathbf{x}) = -p^{\mathrm{II}}_i(\mathbf{x}), \quad \text{as } \mathbf{x} \in \Gamma^a. \tag{20}$$

Also, due to the requirement of continuity of the displacement components at the coincident points on the artificial boundaries, it holds that

$$u_i^{\mathrm{I}}(\mathbf{x}) = u_i^{\mathrm{II}}(\mathbf{x}), \quad \text{as } \mathbf{x} \in \Gamma^a. \tag{21}$$

Moreover, the integral kernels in (7) and (9) have the following properties

$$\begin{aligned} &p_{ij}^*(\mathbf{X}, \mathbf{x}, -\mathbf{n}) = -p_{ij}^*(\mathbf{X}, \mathbf{x}, \mathbf{n}), \\ &\quad \text{i.e., } p_{ij}^*(\mathbf{X}, \mathbf{x}, \mathbf{n}^{\mathrm{I}}) = -p_{ij}^*(\mathbf{X}, \mathbf{x}, \mathbf{n}^{\mathrm{II}}), \quad \text{as } \mathbf{x} \in \Gamma^{c+a}, \end{aligned} \tag{22}$$

$$\begin{aligned} &P_{ijk}^*(\mathbf{X}, \mathbf{x}, -\mathbf{n}) = -P_{ijk}^*(\mathbf{X}, \mathbf{x}, \mathbf{n}), \\ &\quad \text{i.e., } P_{ijk}^*(\mathbf{X}, \mathbf{x}, \mathbf{n}^{\mathrm{I}}) = -P_{ijk}^*(\mathbf{X}, \mathbf{x}, \mathbf{n}^{\mathrm{II}}), \quad \text{as } \mathbf{x} \in \Gamma^{c+a}. \end{aligned} \tag{23}$$

In deriving the above, the reversal of the outward normal $\mathbf{n}$ at the two coincident points on the crack and artificial boundaries has been applied

$$\mathbf{n}^{\mathrm{I}}(\mathbf{x}) = -\mathbf{n}^{\mathrm{II}}(\mathbf{x}) \quad \text{as } \mathbf{x} \in \Gamma^{c+a}. \tag{24}$$

Using (15), (17), and (20) through (23) to (18) and (19), we obtain

$$\begin{aligned} I_i(\mathbf{X}) = &\int_{\Gamma^{\mathrm{ex}}} \{u_{ij}^*(\mathbf{X}, \mathbf{x}) p_j(\mathbf{x}) - p_{ij}^*(\mathbf{X}, \mathbf{x}, \mathbf{n}) u_j(\mathbf{x})\} \, \mathrm{d}\Gamma(\mathbf{x}) \\ &+ \int_{\Gamma^c} \{u_{ij}^*(\mathbf{X}, \mathbf{x})(p_j^+(\mathbf{x}) + p_j^-(\mathbf{x})) \\ &- p_{ij}^*(\mathbf{X}, \mathbf{x}, \mathbf{n}^+) u_j^+(\mathbf{x}) - u_j^-(\mathbf{x}))\} \, \mathrm{d}\Gamma(\mathbf{x}), \end{aligned} \tag{25}$$

with

$$I_i(\mathbf{x}) = \begin{cases} u_i(\mathbf{X}) & \mathbf{X} \in \Omega, \\ c_{ij}(\mathbf{X}) u_j(\mathbf{X}) & \mathbf{X} \in \Gamma_{\mathrm{ex}}, \\ c_{ij}^+(\mathbf{X}) u_j^+(\mathbf{X}) + c_{ij}^-(\mathbf{X}) u_j^-(\mathbf{X}) & \mathbf{X} \in \Gamma^c, \\ 0, & \text{otherwise} \end{cases} \tag{26}$$

and

$$\begin{aligned} J_{ij}(\mathbf{X}) = &\int_{\Gamma^{\mathrm{ex}}} \{U_{ijk}^*(\mathbf{X}, \mathbf{x}) p_k(\mathbf{x}) - P_{ijk}^*(\mathbf{X}, \mathbf{x}, \mathbf{n}) u_k(\mathbf{x})\} \, \mathrm{d}\Gamma(\mathbf{x}) \\ &+ \int_{\Gamma^c} \{U_{ijk}^*(\mathbf{X}, \mathbf{x})(p_k^+(\mathbf{x}) + p_k^-(\mathbf{x})) \\ &- P_{ijk}^*(\mathbf{X}, \mathbf{x}, \mathbf{n}^+)(u_k^+(\mathbf{x}) - u_k^-(\mathbf{x}))\} \, \mathrm{d}\Gamma(\mathbf{x}), \end{aligned} \tag{27}$$

with

$$J_{ij}(\mathbf{X}) = \begin{cases} \sigma_{ij}(\mathbf{X}) & \mathbf{X} \in \Omega, \\ \sigma_{ij}(\mathbf{X})/2 & \mathbf{X} \in \Gamma_{\mathrm{ex}}, \\ (\sigma_{ij}^+(\mathbf{X}) + \sigma_{ij}^-(\mathbf{X}))/2 & \mathbf{X} \in \Gamma^c, \\ 0, & \text{otherwise,} \end{cases} \tag{28}$$

where the superscript $^+$ indicates one side of a crack, and the superscript $^-$ indicates the corresponding opposite side, instead of the superscripts I and II. The positive side of a crack may be chosen arbitrarily, for convenience. Equations (25) and (27) are so-called the *single-domain dual integral equations of a cracked structure*; in these equations the integrals are taken only along the regular external boundary and one side of a crack. The artificial boundaries that appeared due to the cutting process have been entirely eliminated from consideration. If a structure contains multiple cracks, the single-domain dual integral equations are given in the same forms as (25) and (27).

Equations (25) and (27) with a source point on the boundary, i.e., $\mathbf{X} \in \Gamma^{\mathrm{ex}+c}$, are of the most importance, in the formulation of boundary integral equations for elastostatic problems. In formulating a general boundary integral method using (27), it is advantageous to use tractions instead of stresses since the number of unknowns can be reduced. If (27) are 'multiplied' on both sides by the outward normal at the boundary point $\mathbf{X}$ as usual the traction version of (27) with $\mathbf{X} \in \Gamma^{\mathrm{ex}+c}$ can be obtained as

$$\begin{aligned} J_{ij}(\mathbf{X})n_j(\mathbf{X}) = & \int_{\Gamma^{\mathrm{ex}}} \{U^*_{ijk}(\mathbf{X}, \mathbf{x})p_k(\mathbf{x}) - P^*_{ijk}(\mathbf{X}, \mathbf{x}, \mathbf{n})u_k(\mathbf{x})\}n_j(\mathbf{X}) \\ & + \int_{\Gamma^c} \{U^*_{ijk}(\mathbf{X}, \mathbf{x})(p^+_k(\mathbf{x}) + p^-_k(\mathbf{x})) \\ & - P^*_{ijk}(\mathbf{X}, \mathbf{x}, \mathbf{n}^+)(u^+_k(\mathbf{x}) - u^-_k(\mathbf{x}))\}n_j(\mathbf{X})\, \mathrm{d}\Gamma(\mathbf{x}). \end{aligned} \tag{29}$$

Note that, as $\mathbf{X} \in \Gamma^c$, $n_j(\mathbf{X})$ is taken to be the outward normal of the positive crack side, i.e., $n_j(\mathbf{X}) = n^+_j(\mathbf{X})$.

Note that on the crack surfaces, the tractions are self-equilibrating and hence

$$n^+_j(\mathbf{X})(\sigma^+_{ij}(\mathbf{X}) + \sigma^-_{ij}(\mathbf{X}))/2 = p^+_i(\mathbf{X}) \quad \text{and} \quad p^+_i(\mathbf{x}) + p^-_i(\mathbf{x}) = 0. \tag{30}$$

Equations (30) can be used to simplify (25), (27) and (29) when applying to cracks. In this case, traction and displacement jump on a crack may be regarded as independent variables in a numerical formulation, with the least unknowns.

3. Cohesive zone model for a crack in elastostatics

The dual-boundary integral equations in (25) and (29) are valid for the whole body. However, while the external boundary conditions are given explicitly, special attention must be paid to the crack boundaries as they evolve, particularly if the fracture process zone near the crack tip is to be modeled. In this section, we consider the application of the boundary element method to the two dimensional problem of a crack with a line fracture process zone. This model of the fracture process zone is motivated by the fact that in some materials such as concrete, brittle polymers, fiber-reinforced composites, tough ceramics and some alloys, the crack surfaces are usually not separated completely behind the (fictitious) crack tip. There exists a relatively long extension of the crack – variously called the wake zone, the bridging zone, or cohesive zone – where tractions can be transferred across the crack line. The mechanisms responsible for the development of this kind of a process zone can be bridging of long-chain molecules in polymers, bridging of fibers or whiskers in composites, interlocking of grain boundaries in alloys and ceramics, and so on. For these materials, the cohesive zone model of a crack

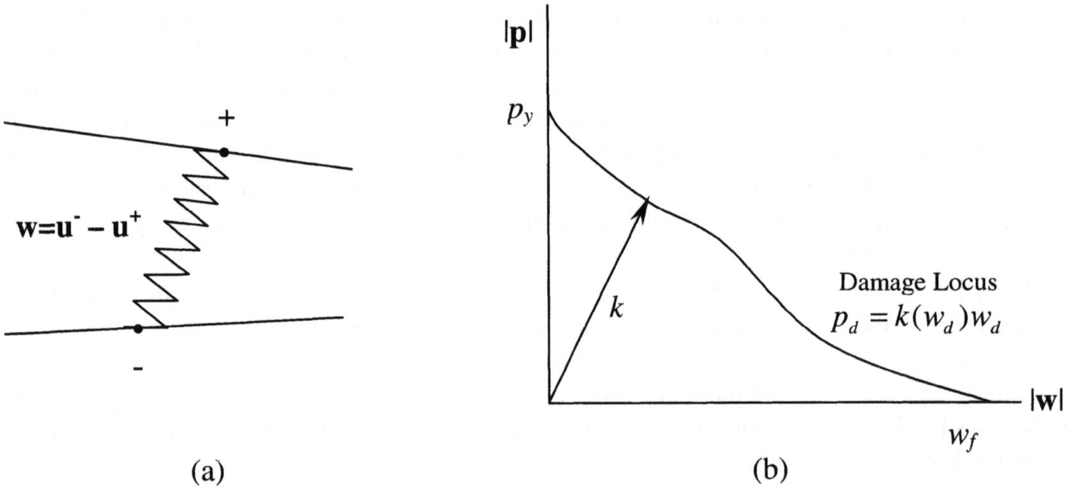

Figure 3. (a) Illustration of the two originally coincident points of a cohesive zone in the opening mode, connected by the line spring; (b) schematic diagram of the constitutive law of the line spring in terms of the traction and displacement jump. Note that instantaneous loading and unloading of the line spring is given by the slope k.

seems to be the appropriate model. The key assumption in this model is that material softening beyond the peak load is localized in a narrow layer behind a fictitious crack tip, whose volume is negligible and whose action is replaceable by cohesive forces. Typically, two types of constitutive laws are used in the literature for cohesive materials: one is characterized by a traction-displacement relationship, the other one by a material constitutive law defined in terms of stress and strain accompanied with a thickening law of the layer. In the latter, thickening of a cohesive layer is decoupled which may be worthy of studying separately in various materials considering rate-dependence and under dynamic loading. However, for quasistatic loading – the case considered predominantly in the literature – a law describing the traction-displacement relation is sufficient for modeling of the cohesive zone. We present below one such law for a cohesive crack represented by a line spring connecting two coincident crack points – this seems to be the most appropriate model for crazing-dominant brittle polymers and fiber-reinforced composites.

Consider a representation of the development of a crack shown in Figure 3(a). We suggest that two points, $\mathbf{x}^+$ and $\mathbf{x}^-$, originally coincident on opposite sides of a line, separate into two distinct points, connected by the cohesive zone material; continued straining increases the separation between these two points and eventually leads to cracking. The kinematics of this separation process are assumed to be described completely by the crack face separation, $\mathbf{w}$. Introducing the local normal and tangent directions at the cohesive zone tip, $\mathbf{w}$ can be resolved into the normal separation distance (or the cohesive zone opening displacement) component, $w_n = u_n^- - u_n^+$ and the tangential separation distance (or the sliding displacement) component $w_\tau = u_\tau^- - u_\tau^+$. In order to prevent the inter-penetration of the cohesive zone, $w_n \geqslant 0$; equality holds only in the case that there is contact between the top and bottom surfaces of the cohesive zone and in this case, we must also provide a description of the frictional resistance on the surface as well. We consider first the case of a locally opening mode crack with $w_n > 0$ and describe the force-separation law for the cohesive zone. The cohesive material may be modeled by a simple line spring that behaves according to the following

$$\mathbf{p} = k(w_d)\mathbf{w}, \tag{31}$$

or in component form

$$p_n = k(w_d)w_n, \qquad p_\tau = k(w_d)w_\tau, \tag{32}$$

where $\mathbf{p}$ is the traction vector with normal and tangential components, p_n and p_τ, respectively. w_d is the maximum separation distance between two originally coincident points on the crack over the entire loading history, and is used as a damage parameter. The stiffness of the cohesive zone material is denoted by $k(w_d)$ and is assumed to depend on the current state of damage. Note that $k(w_d)$ is a decreasing function of w_d, indicating the softening behavior of the material; this imposes an irreversibility of the damage process under unloading. Corresponding to each w_d, there exists a traction, p_d, and a damage locus derived from (31)

$$p_d = k(w_d)w_d. \tag{33}$$

The constitutive law described above is illustrated schematically in Figure 3(b). Note that the description of the cohesive zone material through the damage parameter and stiffness allows for irreversibility of damage. Upon unloading, the points on the cohesive zone unload linearly with a stiffness $k(w_d)$, whereas in most cohesive zone models, unloading occurs along the damage locus. Including the unloading effect has expanded the capability of the cohesive crack model in simulating realistic fracture behavior significantly. There are two critical states along the damage locus. The first one, at $w_d = 0$ and $p_d = p_y$, represents the maximum traction that can be sustained by the material before the cohesive zone begins to develop; beyond this critical level, separation processes begin and w_d increases. The point on the specimen that is at this state is usually called the fictitious crack tip or the cohesive zone tip. The second critical point on the damage locus occurs at $w_d = w_f$ and $p_d = 0$; this point represents the maximum displacement jump across the cohesive zone that can be sustained before cracking; beyond this level, the traction goes to zero and the two initially coincident points are now completely separated. The point on the specimen that is at this state is usually called the physical crack tip. Hence, (31) and (33) together, define the complete process of separation of a material point into a crack, as long as $w_n > 0$. A constitutive law in stress-strain for a cohesive material can be described in a similar manner (see Yang and Ravi-Chandar, 1996). The transition of that law to this law is apparent.

We now turn to the case when $w_n = 0$. Under arbitrary loading, contact of two crack surfaces could occur. In such cases, as shown in Figure 4a, we must provide an appropriate description for the development of friction along the cohesive zone, in addition to the line-spring model described above. We model the frictional interaction simply through a Coulomb type law. Thus, the tangential interaction of the crack surfaces in contact is given by

$$p_\tau = k(w_d)w_\tau - \operatorname{sgn}(w_\tau) f p_n, \tag{34}$$

where f is frictional coefficient. The equation for the determination of the normal component of the traction p_n is obtained by enforcing the contact condition $w_n = 0$. In (34), the first term represents the tangential traction contribution due to the line spring described above (nontrivially if it is not broken completely) and the second term represents the Coulomb friction component. In order to attain smooth transition of frictional force near $w_\tau = 0$, the frictional coefficient f may be assumed to be

$$f = \begin{cases} f_0 \dfrac{|w_\tau|}{w_{\tau d}}, & |w_\tau| \leqslant w_{\tau d}, \\ f_0, & \text{otherwise}. \end{cases} \tag{35}$$

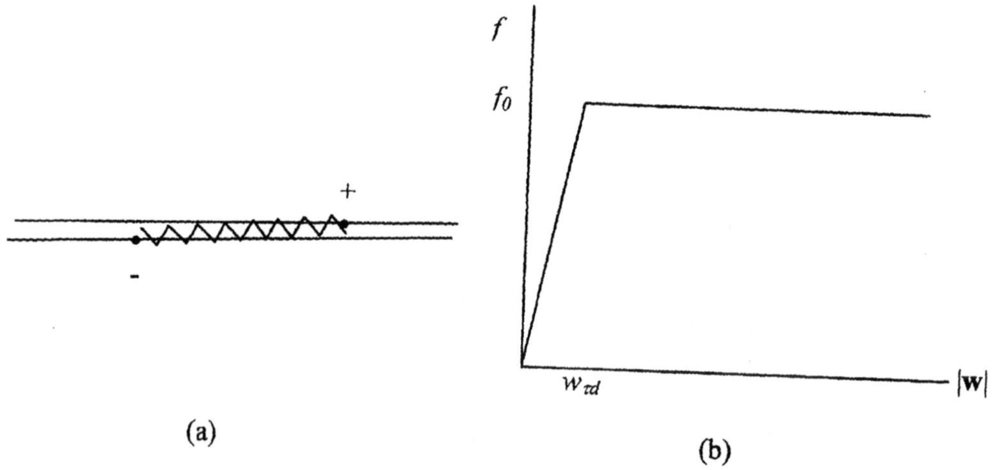

Figure 4. (a) Illustration of the two originally coincident points of a cohesive zone in the contact/sliding mode. The tangential interaction between the two points is modeled using Coulomb frictional force and the cohesive force if the cohesive line spring is not already broken; (b) sketch of the variation of the frictional coefficient of the contact surface.

where f_0 and $w_{\tau d}$ are both non-negative material constants. Function f is plotted in Figure 4(b). This completes the description of the cohesive zone material behavior.

There still remains the issue of deciding on the appropriate incorporation of this model into an elastostatic crack problem. One major question that arises is the following: *What is the criterion that can be used to grow the cohesive zone from a stress contentrator?* The incorporation of the cohesive zone model eliminates energy release criterion from consideration since the energy release rate will always be the same regardless of the direction of crack extension – it is simply the area under the damage locus, assuming a fully developed cohesive zone. The most plausible criteria, particularly for the line-spring nature of the cohesive zone mode, are stress based criteria such as the maximum principal stress criterion or the maximum tangential stress criterion at the fictitious crack tip. Assuming that one of these criteria would be appropriate, a second major question arises: *What should be the step size in extending the crack along this direction?* In finite element formulations, such as those of Xu and Needleman, (1994) and Ortiz (1996), cohesive zones are forced to develop along element boundaries and the extension is over the side of one entire element. The approach in these models is to make the element size so small that the overall crack growth behavior is captured adequately. In other words, the macroscopic crack path is suggested to be independent of the length scale of the discretization when the latter is sufficiently small compared to characteristic structural length scale. In the present work, we choose the maximum principal stress criterion for determining the crack increment. If the maximum principal stress at a fictitious crack tip reaches the critical value p_y, this tip is ready to run under further loading. The direction in which the tip advances is perpendicular to the direction of the maximum principal stress at that point, and the extension is such that the maximum principal stress at the new tip position is kept at the critical value p_y, during continued loading. However, note that, structural instability may occur while a crack is advancing; i.e., extension of a crack may enhance the stress state at a crack tip rather than release it. In this case, the crack may run fast and inertia effects may have to be included.

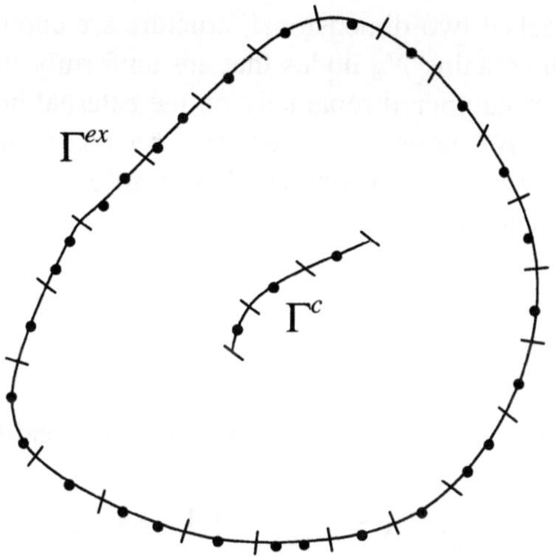

Figure 5. The discretization of the boundary into elements is shown in this figure. Each element contains one or more nodes distributed uniformly within the element. The nodes are internal to the element, indicating discontinuous elements.

4. Iterative boundary element method of successive over-relaxation

The behavior of the cohesive zone materials and the criterion for crack advance described in the last section require that an incremental loading procedure be used when solving boundary value problems. Moreover, an iteration process is indispensable in each loading step in general due to the fact that the constitutive law of the cohesive zone material, given in (31)–(35), involves irreversible damaging process, and is essentially history-dependent. Furthermore, the stiffness of the model line spring of the cohesive zone material $k(w_d)$, is in general a part of the solution, which may or may not be dependent on the current displacement discontinuity; in other words, there is no simple relationship between the traction and displacement discontinuity on a crack which may be used to achieve a linear system of equations to solve the problems numerically. Two strategies of iteration are possible in general: in the first, a linear system of equations of the discretized problem is formulated using the stiffness of the cohesive zone material obtained in the previous iteration step, and is solved by a typical solver either direct or iterative. The stiffness of the cohesive zone material is then modified based on this solution, and the process is repeated until the solution of the desired accuracy is achieved. In the second strategy, a nonlinear system of equations of the discretized problem is formulated using the stiffness of the cohesive zone material as unknowns, and is solved iteratively. By the first strategy, the iterative procedure is very clear and the convergence of the iteration process is expected. However, it is apparently time-consuming even with a very efficient solver of a linear system of equations; the linear system of equations is solved fully many times until a solution is achieved. On the other hand, the total time for a solution by the second strategy, if a good iterative procedure is found, would be in the same order as that for one step of the iteration by the first strategy (solving the linear system of equations once). In the present work, we adopt the second strategy and formulate an iterative method of successive over-relaxation for the present nonlinear problem of a cohesive crack. The solution procedure of this iterative method will be described below, following discretization of the boundaries and the boundary integral equations.

The boundaries of a cracked two-dimensional structure are approximated by straight elements, Γ_{el}s, of which each contains N_{el} nodes that are uniformly distributed in it, as shown in Figure 5. These nodes are numbered separately on the external boundary and on the crack locus. Assume that we have N^{ex} nodes on the external boundary and N^c nodes on the crack locus. A field quantity, $\mathbf{q}(\mathbf{x})$ can be approximated over an element Γ_{el} by interpolating the nodal values $\mathbf{q}^n$ in this element as

$$\mathbf{q}(\mathbf{x}) = \sum_{n=1}^{N_{el}} \phi^n(\mathbf{x})\mathbf{q}^n, \tag{36}$$

where the interpolation function, $\phi^n(\mathbf{x})$ satisfies the following conditions

$$\phi^n(\mathbf{x}^m) = \begin{cases} 1 & m = n, \\ 0 & m \neq n, \end{cases} \quad \text{and} \quad \sum_{n=1}^{N_{el}} \phi^n(\mathbf{x}) = \begin{cases} 1 & \mathbf{x} \in \Gamma_{el}, \\ 0 & \mathbf{x} \notin \Gamma_{el} \end{cases} \tag{37}$$

The interpolation functions may be constant, linear or quadratic, depending on the required accuracy of the representation. Substitute (36) into (25) and (29) with $\mathbf{X} \in \Gamma^{ex+c}$, the discretized forms of the dual boundary integral equations are obtained. Thus, one obtains

$$\sum_{n=1}^{N^{ex}} (g_{\alpha\beta}^{mn} p_\beta^n - h_{\alpha\beta}^{mn} u_\beta^n) + \sum_{n=N^{ex}+1}^{N^{ex}+N^c} h_{\alpha\beta}^{mn} w_\beta^n - I_\alpha(\mathbf{X}^m) = 0,$$
$$m = 1, 2, \ldots, N^{ex} + N^c \tag{38}$$

and

$$\sum_{n=1}^{N^{ex}} (G_{\alpha\beta}^{mn} p_\beta^n - H_{\alpha\beta}^{mn} u_\beta^n) + \sum_{n=N^{ex}+1}^{N^{ex}+N^c} H_{\alpha\beta}^{mn} w_\beta^n - J_{\alpha\beta}(\mathbf{X}^m) n_\beta(\mathbf{X}^m) = 0,$$
$$m = 1, 2, \ldots, N^{ex} + N^c. \tag{39}$$

Equations (38) and (39) each represent $2(N^{ex} + N^c)$ equations that are the discretized version of (25) and (29). In these equations, $g_{\alpha\beta}^{mn}$, $h_{\alpha\beta}^{mn}$, $G_{\alpha\beta}^{mn}$, and $H_{\alpha\beta}^{mn}$ are given explicitly by

$$g_{\alpha\beta}^{mn} = \int_{\Gamma^n} u_{\alpha\beta}^*(\mathbf{X}^m, \mathbf{x})\phi^n(\mathbf{x})\, \mathrm{d}\Gamma(\mathbf{x}), \tag{40}$$

$$h_{\alpha\beta}^{mn} = \int_{\Gamma^n} p_{\alpha\beta}^*(\mathbf{X}^m, \mathbf{x}, \mathbf{n})\phi^n(\mathbf{x})\, \mathrm{d}\Gamma(\mathbf{x}), \tag{41}$$

$$G_{\alpha\beta}^{mn} = \int_{\Gamma^n} U_{\alpha\delta\beta}^*(\mathbf{X}^m, \mathbf{x})\phi^n(\mathbf{x}) n_\delta(\mathbf{X}^m)\, \mathrm{d}\Gamma(\mathbf{x}), \tag{42}$$

$$H_{\alpha\beta}^{mn} = \int_{\Gamma^n} P_{\alpha\delta\beta}^*(\mathbf{X}^m, \mathbf{x}, \mathbf{n})\phi^n(\mathbf{x}) n_\delta(\mathbf{X}^m)\, \mathrm{d}\Gamma(\mathbf{x}), \tag{43}$$

where Γ^n is the element where the nth node is located. Equations (40) through (43) may be evaluated numerically if $\mathbf{X}^m \notin \Gamma^n$, and analytically if $\mathbf{X}^m \in \Gamma^n$. Note that the condition for a self-equilibrating crack, Equation (30), has been used in the above equations. If **u** and **p**, are a trial set of displacements and tractions, Equations (38) and (39) would not be satisfied in general; instead, we obtain the following residuals

$$r_\alpha^m(\mathbf{u},\mathbf{p}) = \sum_{n=1}^{N^{\text{ex}}}(g_{\alpha\beta}^{mn}p_\beta^n - h_{\alpha\beta}^{mn}u_\beta^n) + \sum_{n=N^{\text{ex}}+1}^{N^{\text{ex}}+N^c} h_{\alpha\beta}^{mn}w_\beta^n - I_\alpha(\mathbf{X}^m),$$

$$m = 1, 2, \ldots, N^{\text{ex}} + N^c. \tag{44}$$

$$R_\alpha^m(\mathbf{u},\mathbf{p}) = \sum_{n=1}^{N^{\text{ex}}}(G_{\alpha\beta}^{mn}p_\beta^n - H_{\alpha\beta}^{mn}u_\beta^n) + \sum_{n=N^{\text{ex}}+1}^{N^{\text{ex}}+N^c} H_{\alpha\beta}^{mn}w_\beta^n - J_{\alpha\beta}(\mathbf{X}^m)n_\beta(\mathbf{X}^m),$$

$$m = 1, 2, \ldots, N^{\text{ex}} + N^c. \tag{45}$$

We call r_α^m the displacement residual and R_α^m the traction residual. If the residuals are close enough to zero, the trial set of **u** and **p** represents an approximate solution of the boundary value problem. At all regular boundary points, either the displacement or the traction vector components are prescribed; thus only one of (44) and (45) would be used in assembling the overall system of equations. For points that are in the cohesive zone, both the traction vector and the displacement discontinuity vector are unknown but related to each other through the constitutive description of the cohesive zone material provided in (31)–(35). From (45), for a point m that lies on the cohesive zone, the residual can be written as a relation between the displacement jump vector $\mathbf{w}(\mathbf{X}^m)$ and the traction vector $\mathbf{p}(\mathbf{X}^m)$. Note that points on the crack are simply special cases of the cohesive zone where the stiffness has decreased to zero and where $\mathbf{w}(\mathbf{X}^m)$ is unrestricted except that $w_n(\mathbf{X}^m) \geqslant 0$. Hence the complete set of all residuals at all nodal points can be calculated. Note that in order to incorporate the crack opening, crack sliding, and crack contact modes of deformation in a convenient way, field quantities in the discretized equations must be transformed into the local orthogonal coordinates in terms of the normal and tangential directions. We now turn to a description of an iterative solution scheme for solving (44) and (45) for **u** and **p**.

For a node m on the external boundary, either (44) or (45) could be used, depending on whether the imposed boundary condition at that point is a displacement or a traction condition. For the iterative solution scheme, in general, a field quantity $q_\alpha^{m,l+1}$ at the $(l+1)$th iteration is written in terms of its value at the lth iteration $q_\alpha^{m,l}$, and an increment that depends on the residuals as

$$q_\alpha^{m,l+1} = q_\alpha^{m,l} - \left.\frac{\overline{R}}{\partial\overline{R}/\partial q_\alpha^m}\right|^l, \tag{46}$$

where $\overline{R}$ is the appropriate residual for the field quantity q_α^m at the lth iteration. At the $(l+1)$th iteration step, the displacement component at this node $u_\alpha^{m,l+1}$, if not prescribed as a boundary condition, is calculated using the results at the lth iteration by

$$u_\alpha^{m,l+1} = u_\alpha^{m,l} + \omega r_\alpha^m(\mathbf{u},\mathbf{p})/(c_{\alpha\alpha}^m + h_{\alpha\alpha}^{mm}), \quad \text{(no sum on } \alpha), \tag{47}$$

where ω is an adjustable factor of relaxation. The displacement residual, r_α^m $(\mathbf{u},\mathbf{p})$, is computed using the nodal values at the $(l+1)$th iteration step if available, or at the lth iteration step. Similarly, a traction component $p_\alpha^{m,l+1}$, if not prescribed as a boundary condition, is calculated using the results at the lth iteration by

$$p_\alpha^{m,l+1} = p_\alpha^{m,l} + \omega R_\alpha^m(\mathbf{u},\mathbf{p})/(0.5 - G_{\alpha\alpha}^{mm}), \quad \text{(no sum on } \alpha), \tag{48}$$

where the traction residual R_α^m $(\mathbf{u},\mathbf{p})$, is dealt with in the same way as r_α^m $(\mathbf{u},\mathbf{p})$. For a node m which is on the cohesive zone, we calculate the displacement discontinuity $w_\alpha^{m,l+1}$, in opening mode by

$$w_\alpha^{m,l+1} = w_\alpha^{m,l} + \varpi R_\alpha^m(\mathbf{u},\mathbf{p})/(k^{m,l} - H_{\alpha\alpha}^{mm}), \quad \text{(no sum on } \alpha), \tag{49}$$

where ϖ is another adjustable factor of relaxation, and $k^{m,l}$ is calculated using (31) with $|\mathbf{w}^{m,l}|$. If the normal component of $\mathbf{w}$ at the $(l+1)$th iteration step $w_n^{m,l+1}$, is negative, it is set to be zero for no penetration of the crack surfaces. The tangential component of $\mathbf{w}$ at the $(l+1)$th iteration step $w_\tau^{m,l+1}$, is re-calculated by

$$w_\tau^{m,l+1} = w_\tau^{m,l} + \varpi R_\tau^m(\mathbf{u},\mathbf{p})/(k^{m,l} - f_0 p_n^{m,l+1}/\bar{w}_{\tau d}| - H_{\tau\tau}^{mm}), \quad \text{(no sum on } \tau), \tag{50}$$

where the frictional mode of the cohesive crack is activated; also $\bar{w}_{\tau d} = w_{\tau d}$ if $w_\tau^{m,l} \leqslant w_{\tau d}$; otherwise, $\bar{w}_{\tau d} = |w_\tau^{m,l}|$. Note that no special iterative process for mixed crack opening and contact is necessary.

We now describe the marching scheme used in the numerical solution; the solution procedure is implemented in three steps. Suppose that the solution at a previous loading step is already known. In order to solve the problem in the current loading step, first the cracked body is iterated to be in equilibrium with all the current cohesive nodes being held at the same displacement, using (47) and (48); in this step, the cohesive zone material law is not used. In the second step, the cohesive zone nodes are made free to displace according to the constitutive laws described above. The whole body is iterated again to be in equilibrium using (47) through (50). This holding and releasing process is found to produce quicker convergence for the iteration process than other schemes when a crack is advancing. The absolute difference of either displacement or traction at a boundary node between two next iteration steps is used to judge the convergence. The best relaxation factors of w and $\bar{w}$ can be obtained through some trial computations and these factors vary in general with the geometrical configuration of the body and loading conditions. Note that in the second step, the fictitious crack tip has not been allowed to move. After the equilibrium of the cracked body is obtained without crack advance, in the third step, stresses at a point of the fictitious crack tip are calculated using (27); note that a constant element is always used at the fictitious crack tip avoiding the difficulty of the integration due to the hypersingularity in the fundamental solution. If the maximum principal stress at this point is over the critical value, i.e. p_y, the fictitious crack tip is forced to advance a small increment in the direction perpendicular to the maximum principal stress. This is accomplished by increasing the length of the last cohesive tip element if it is smaller than a specific value and adding a new tip element otherwise. In this paper, we set the crack advance step to be equal to the cohesive zone tip element size; the ratio of the crack advance step to the crack element size could play a role in determining the local crack patterns in situations where the crack path is unstable, but this issue is not considered in this paper. Note that in general, a numerical stepwise scheme of crack advance may generate sharp corners between the approximating straight elements of a crack which may concentrate stresses; these stress concentrations are of a lower order than the crack tip concentration. Investigating the role of these corners is of less interest than that of examining the crack tip; thus, in the present paper, we neglect the weaker corner singular point while dealing with a crack. Note also that at the current loading level, the above procedure of crack advance is repeated, and the iterative solution over whole body is obtained until the maximum principal stress is not above p_y at any point in the body. This procedure achieves the solution in the current loading step. The load is then incremented and the procedure repeated to march both the loading and the crack extension. In the following, we shall apply the method formulated above to some problems of a cohesive crack, and demonstrate the capacity of this method.

5. Applications

We turn to a demonstration of the capability of the boundary element strategy for handling of the cohesive cracks described above. In order to explore different aspects of the constitutive model for the cohesive zone that has been assumed in this paper, we consider three problems. The first problem concerns a single-edge-crack in a rectangular specimen under uniform far field tensile loading. For small initial crack lengths the crack extension in this configuration is unstable under displacement control; however, deep cracks exhibit stable crack growth. The results of this problem are discussed in Section 5.1. The second problem considered is that of an angled internal crack, under compressive loading; in this case, the cohesive zone develops contact and shear. The resulting path of the crack is the familiar wing crack pattern and is described in Section 5.2. The final problem considered is a demonstration of the capability of the formulation to handle mixed-mode crack growth and crack interaction; using the same loading as in the first problem, but with an offset double-edge crack configuration, the trajectories of the two edge cracks are tracked. The results, described in Section 5.3, show that the model duplicates the commonly observed behavior of two approaching cracks, and that the crack patterns exhibit a sort of bifurcation with the offset.

The following conditions describe the details of the simulation that are common to all three problems. The domain of interest is a rectangular two-dimensional region of length l and height h, taken equal to l in these simulations. In these simulations, all length quantities are normalized by l; while this is not a natural length scale for the fracture problem, it is convenient and easily interpreted. If it is desired, a scaling to an intrinsic length scale, such as the critical crack opening displacement for the cohesive zone w_f, may be effected easily. The deformation of the specimen is assumed to be in plane-strain. An initial process zone of length 0.02 is assumed to exist at the tip of all cracks in all simulations described in this paper; the damage parameter w_d is assumed to increase linearly from zero at the fictitious crack tip to w_f at the physical crack tip. Note that this initial damage zone can be prescribed arbitrarily and is in general unknown in the physical problem. The boundaries are discretized into discontinuous elements with a constant or quadratic interpolation over the element. Specifically, the straight-line segments on the external boundary are divided into 10 elements with quadratic interpolation; the length of each element is 0.1 and the nodal spacing is 0.0333. The initial crack (outside of the cohesive zone) is also divided into quadratic elements with a nodal spacing 0.01. The initial process zone is divided into equally-spaced elements of size 0.01 with quadratic interpolation, except for the element at the fictitious crack tip which is taken to be a constant element. When a fictitious crack tip advances, one new constant element of size 0.01 is added as the corresponding new fictitious crack tip element. The old fictitious crack tip element that is now an interior element within the cohesive zone is then redefined

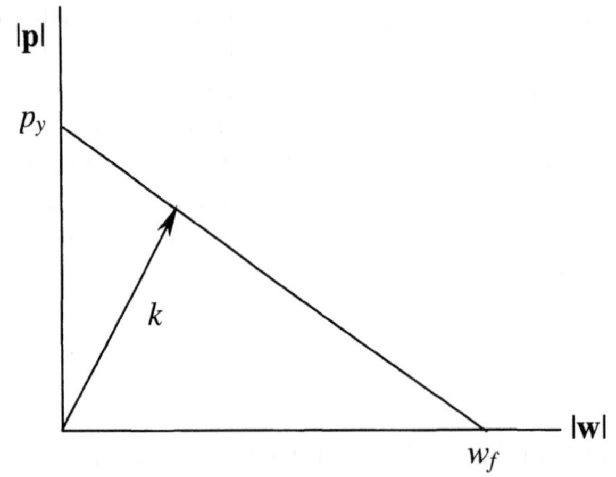

Figure 6. The two parameter constitutive law for the cohesive zone material used in the simulations.

as a quadratic element using three nodes inside it. The constitutive law of the cohesive crack is assumed to be represented by a straight line as shown in Figure 6, which is certainly the simplest one defined by the two parameters p_y and w_f; these parameters are taken to be $p_y = 0.01$ and $w_f = 0.001$. Note that all stress and traction quantities are normalized by μ. Poisson's ratio is taken to be 0.3. The unloaded specimen is initially in a stress free state. It is loaded incrementally in the direction perpendicular to the top and bottom boundaries under displacement control. The increment of the loading displacement is taken to be 1E-5 in tension and −1E-4 in compression. The relaxation factors were chosen initially by trial and error and were subsequently fixed at $\omega = 0.6$ and $\varpi = 1.4$.

In the following simulations, the boundary (including the cracks) is discretized into 150 to 350 nodes depending on the crack length resulting in a system of 300 to 700 equations (or degrees of freedom). In addition there are a few equations that provide the connection between tractions and crack opening displacements on the cohesive zone. If the stiffness of the cohesive zones is not modified by damage, for each loading step, fewer than 100 iteration steps are needed to achieve a solution to the system of equation, with an accuracy of 1E-8 in traction components and 1E-9 in displacement components. If the stiffness of the cohesive zones changes through damage, but still without propagation of the fictitious crack tips, slightly more than 100 iteration steps are required for each loading step. If a new cohesive element is added and accumulates damage, each loading step requires between 150 and 400 iteration steps. If the fictitious crack tip movement is sensitive to the load, quite a few new crack elements are added in one loading step; in this case, the required iteration steps may be as large as 1000, depending on the number of the new crack elements; fortunately, this does not happen often. We ran the code for the following problems on a Digital Alpha Workstation (200 4/100). Depending on the size of the discretized system, between 10 and 30 iteration steps can be performed in one second. Run times are as short as $\frac{1}{2}$ hours for small crack extensions; if long crack extensions are to be simulated the time required increases to the order of a few hours.

5.1. Single Edge Crack under Tension

A number of simulations were performed with a single-edge-crack in the rectangular geometry shown in Figure 7. The crack was considered to be along the line of symmetry and its initial

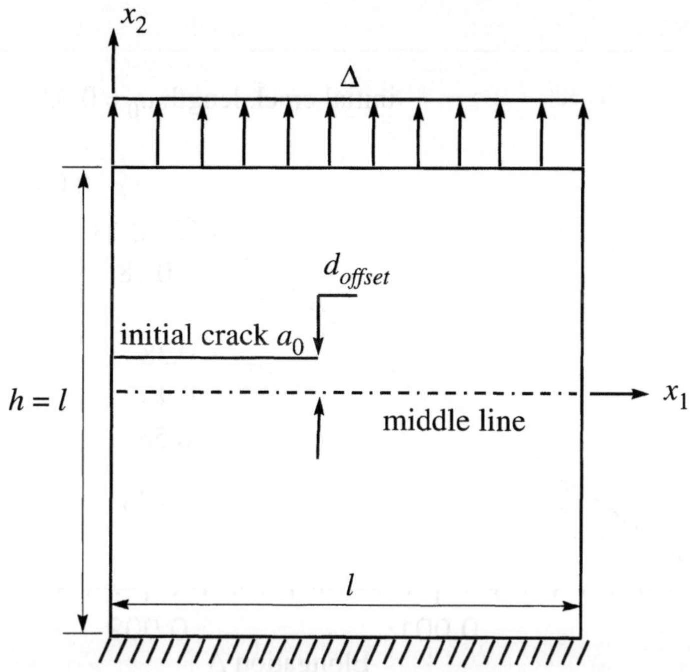

Figure 7. The configuration of a rectangular specimen with a single edge crack of initial length a_0, under displacement controlled elongation. This initial crack is parallel to the loading boundaries, and may be offset from the symmetry line in a distance.

length a_0, was varied from 0.03 to 0.78. The variation of the total load on the top boundary with the imposed elongation obtained from the simulations is shown in Figure 8. The main result, of course, is that in these simulations, simply by prescribing the applied loads and the cohesive law, initiation and growth of the crack appear naturally. These results are the mode-I analog of results that were obtained under mode III by Yang and Ravi-Chandar, (1998), using a finite difference scheme. A number of remarks regarding these simulations are listed below:

- During the loading process of a stable configuration, the fictitious crack tip starts to advance first. Note that the critical point of load at which the fictitious crack tip starts moving is of no significance since it is dependent on the initial state of the cohesive zone. With continued loading, the physical crack tip also starts to move. At this stage, the two crack tips advance at the same rate, keeping the size of the cohesive zone-defined as the distance between the fictitious and physical crack tips – a constant at about 0.1. This is considered to be the fully developed equilibrium cohesive zone size under this condition; of course, the size depends on the cohesive material model and the geometrical constraint imposed in the specimen.
- If the initial crack length a_0, is small, crack initiation is unstable even under displacement controlled loading. In the numerical procedure, this is manifested by the fact that the stress, calculated at a point ahead of the fictitious crack tip is larger than p_y even after the crack extension procedure has been applied over a length of many cohesive elements. This can also be inferred from the fact that the critical load for the onset of instability decreases with increasing initial crack length. Thus, crack extension and structural instability coincide in these cases. If the initial crack length is larger than about 0.48, crack extension is

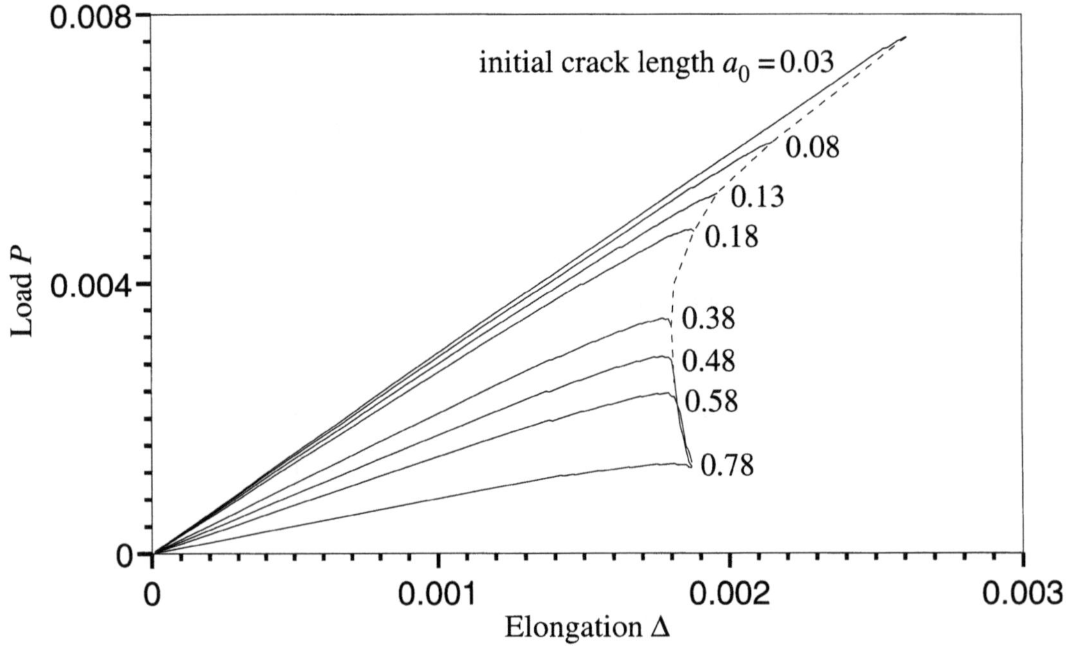

Figure 8. Load-elongation diagram of a set of simulations with a crack initially lying on $x_2 = 0$ and with a length a_0 from 0.03, to 0.78. An initial cohesive zone of size 0.02 is assumed in all the simulations. The crack growth is in the pure mode I. For $a_0 < 0.48$ approximately, crack extension is unstable under displacement control.

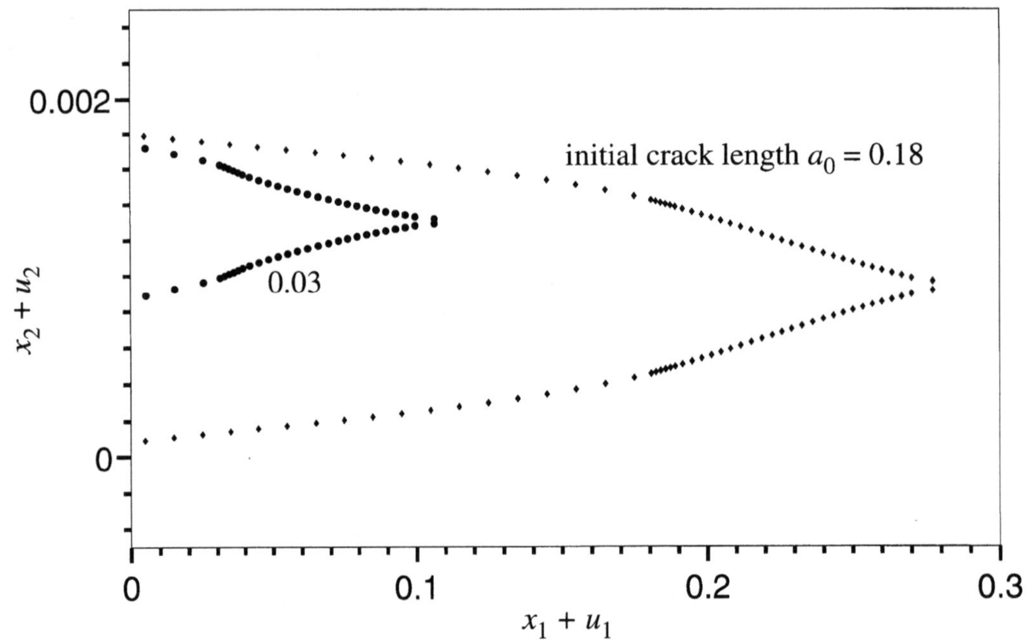

Figure 9. Crack opening profiles at the critical point of instability in two of the simulations for $a_0 = 0.03$ and 0.18.

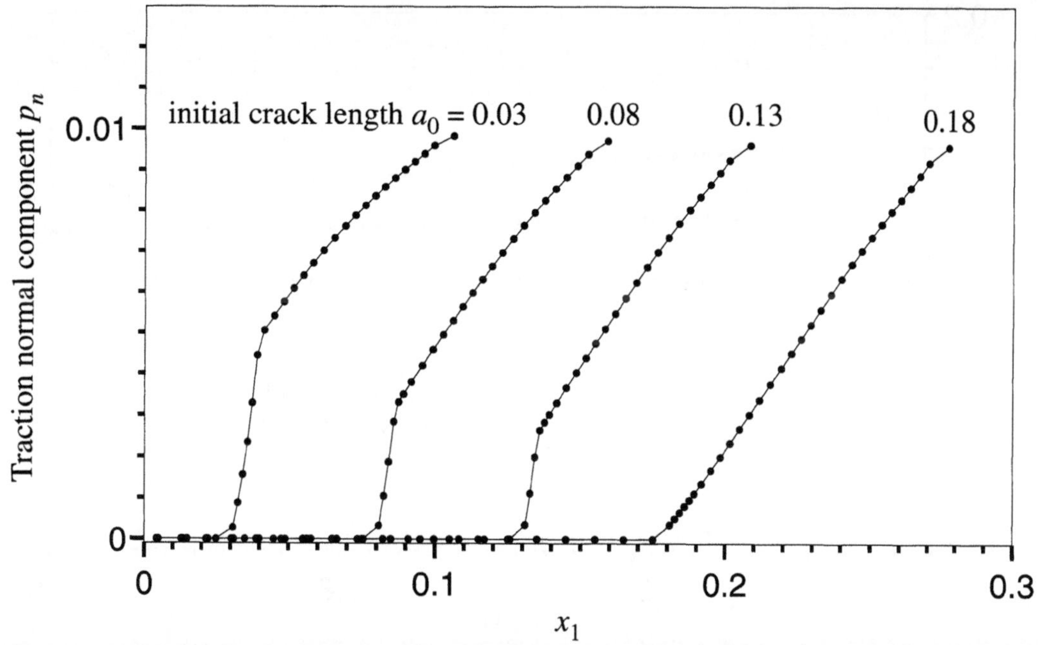

Figure 10. Traction normal component p_n along the cohesive zone at the critical point of instability in four of the simulations for $a_0 = 0.03$, 0.08, 0.13, and 0.18. The tangential component is zero indicating a pure opening mode for the cohesive zone. Note that, for very short initial crack lengths, the cohesive zones are not fully developed.

stable under displacement control. We examine later some issues related to crack paths in this range of stable crack growth. The behavior indicated by the sudden drop of load, for initial crack lengths less than 0.48 is usually referred to as a snap-back instability in structural mechanics. Note that the snap-back instability is also predicted by the stress intensity factor based linear elastic fracture mechanics and is simply a structural feature of this configuration; it is not influenced qualitatively by the fracture model.

- The evolution of the crack opening profiles and the cohesive tractions at the onset of unstable crack extension are shown in Figures 9 and 10 for a few initial crack lengths in the unstable region. Note that all these cracks had an initial cohesive zone of length 0.02 prior to load application. For long initial crack lengths, a steady-state cohesive zone of length 0.1 develops. However, for small initial crack lengths, at the onset of unstable crack extension, the cohesive zone is not completely developed. In other words, for short initial crack lengths, structural instability appears before development of a self-similar crack tip process zone. If the dissipation in the cohesive zone is computed, the result indicates that the dissipation is a function of crack length. Note that the data used for Figures 9 and 10 were obtained based on the numerical procedure of the loading displacement increment equal to 1E-5. A finer loading displacement increment may produce more developed cohesive zones representing their critical states instead of the ones plotted in Figures 9 and 10; however, this characteristic of incomplete development of the critical process zone at the onset of unstable crack initiation will not change. A similar behavior was identified in the mode III simulation by Yang and Ravi-Chandar (1998) and used to determine the range of applicability of a single parameter characterization of fracture.

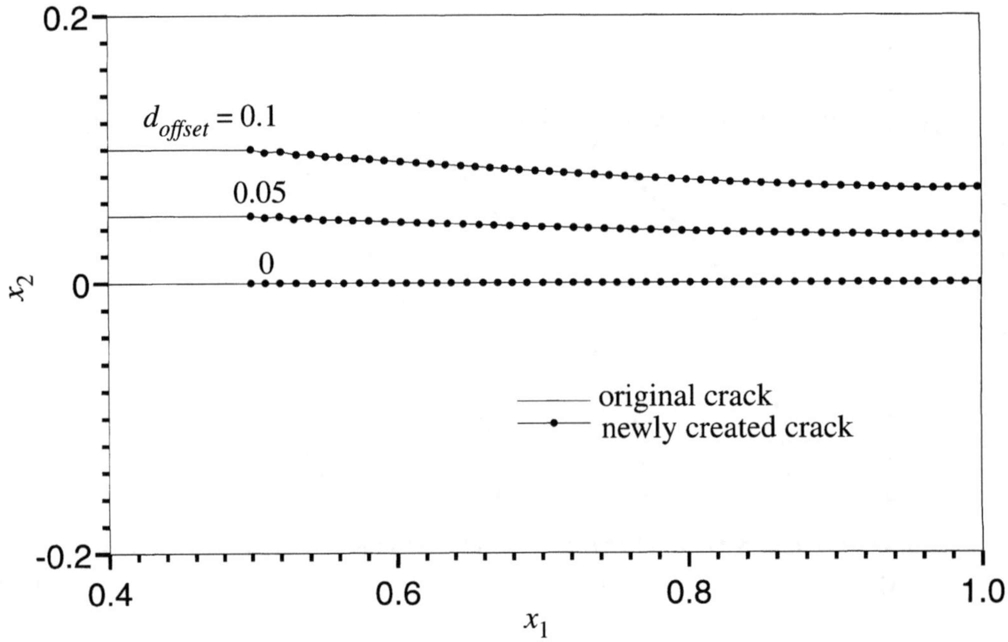

Figure 11. Crack trajectories in the simulations with different offsets of initial cracks. Nodes along the newly-created cracks are indicative of the new crack element ends.

In addition to the structural stability, the stability of the crack path is also of interest in crack problems. The single-edge-notched geometry used here is very stable to perturbations in the crack path. This is demonstrated by the results of the simulations shown in Figure 11. Here the crack paths obtained from three simulations are shown. The only difference between these simulations is that the edge crack is offset from the line of symmetry by d_{offset}. This offset, if nonzero, imposes a mixed-mode loading at the crack tip with the result that the crack path is no longer straight. The path stability of the crack is indicated by the fact that these cracks tend towards the line of symmetry. The main idea here is to demonstrate the capability of the present boundary element formulation to track arbitrary crack-path evolution. Further investigations into the influence of the loading on the crack-path stability are still under progress.

5.2. ANGLED CRACK UNDER COMPRESSION

The problem of cracks under compression has been of long-standing interest due to its importance in rocks and other brittle materials. Brace and Bombolakis (1963) made a crucial observation regarding the growth of angled cracks under compression: 'crack growth occurred by the extension of the initial crack along a curving path which gradually became parallel with the direction of compression'. Moreover, the extension of the crack under compression was found to be stable even under prescribed loads. A number of other researchers have examined this problem since these early observations (see Nemat-Nasser and Horii, 1982, Steif, 1984, Horii and Nemat-Nasser, 1986). The overall objective in these studies has been to determine the stress intensity factors at the wing-crack tips so that a failure criterion based on a critical stress intensity factor criterion may be applied to determine the onset of crack extension. The analytical complications associated with such modeling forces one to assume a straight crack

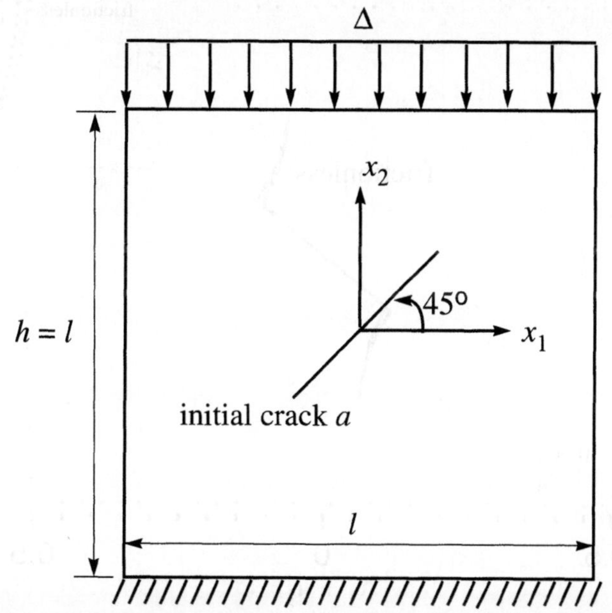

Figure 12. The loading configuration of a rectangular specimen with an internal crack inclined in an angle, subjected to displacement controlled compression.

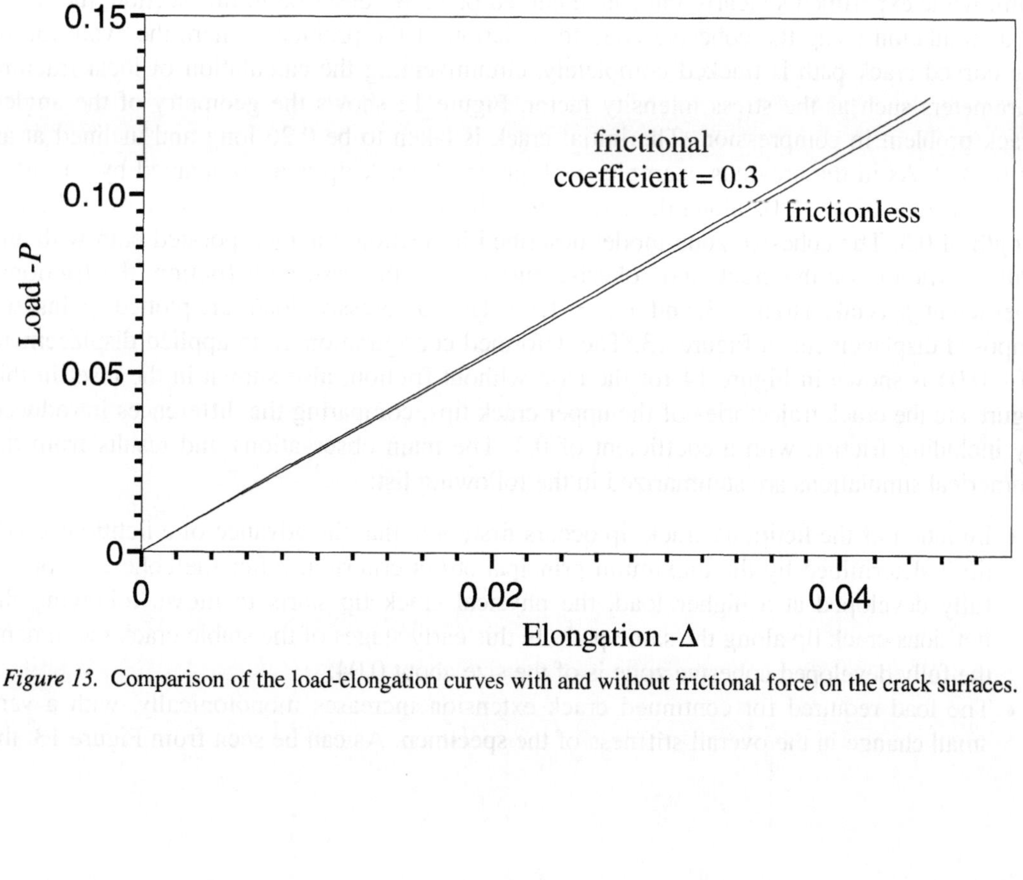

Figure 13. Comparison of the load-elongation curves with and without frictional force on the crack surfaces.

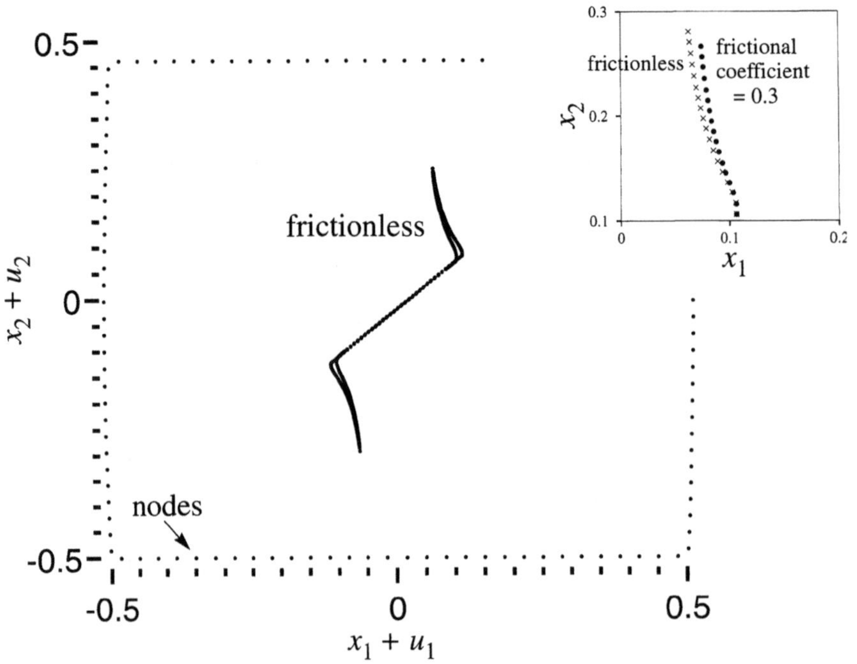

Figure 14. Deformed configuration of the specimen subjected to a compressive load of Δ equal to −0.04, without friction force on the crack surface. The wings are newly created cracks in opening mode. The final crack trajectories with and without frictional force on the crack surfaces are compared in the inset on the right and top corner, on the background of the original coordinates.

path, while experiments clearly indicate a curved path. We describe in this section, the result of a simulation using the cohesive zone formulation of the problem, where the evolution of the curved crack path is tracked completely, circumventing the calculation of local fracture parameters such as the stress intensity factor. Figure 12 shows the geometry of the angled crack problem in compression. The initial crack is taken to be 0.26 long and inclined at an angle 45°. As in the previous example, each physical crack tip is accompanied by an initial cohesive zone of size 0.02 along the direction of the initial crack, resulting in an overall crack length of 0.3. The cohesive zone model described in Section 3 is incorporated both with and without friction on the crack and cohesive surfaces. In the case with friction, the frictional coefficient f_0 is taken to be 0.3, and $w_{\tau d} = 1\text{E-}4$. The compressive loads are plotted against the imposed displacement in Figure 13. The deformed configuration at an applied displacement of −0.04 is shown in Figure 14 for the case without friction; also shown in the inset in this figure are the crack trajectories of the upper crack tips, comparing the differences introduced by including friction with a coefficient of 0.3. The main observations and results from the numerical simulations are summarized in the following list:

- Initiation of the fictitious crack tip occurs first; note that the advance of a fictitious crack tip is determined by the maximum principal stress criterion. After the cohesive zone is fully developed at a higher load, the physical crack tip starts to move, following the fictitious crack tip along the same path. In this early stages of the stable crack movement, the fully developed cohesive zone is of the size about 0.04.
- The load required for continued crack extension increases monotonically, with a very small change in the overall stiffness of the specimen. As can be seen from Figure 13, the

influence of friction on the global response is also small. Of course, the magnitude of the stiffness change is influenced by the dissipation assumed in the cohesive zone model for the crack. Note that the energy release rate for a stable crack in steady state – which is simply equal to $p_y w_f/2$ – was assumed to be 5E-6. At the imposed displacement of -0.04, the total dissipation due to the crack advance can be estimated to be three orders of magnitude smaller than the strain energy stored in this system. If the specimen is unloaded, only a very small loop of dissipation can be observed.

- Due to the continuous evolution of the stress field as the crack and cohesive zone extend, a curved path is followed. The influence of friction on the overall evolution of the crack path is also not very significant, as shown in Figure 14. The curved path bears a good resemblance to the path observed in experiments.
- After the crack path aligns itself parallel to the direction of compression, the crack does not extend further under increased global compression; the applied compression is transmitted through contact on the inclined initial cracks and the extended tips of the wing cracks appear not to be loaded any further. This can be understood simply by considering that the wing crack tips are parallel to the applied loading and hence do not generate a stress concentration at their tips. This observation is confirmed by the approximate solutions of Steif (1984) and Horii and Nemat-Nasser (1986b) which show the diminishing influence of the shearing of the original crack faces on the wing crack tip and the approach of the mode I stress intensity factor at the wing crack tip to zero as the wing crack length becomes long. For further failure of the block, other mechanisms, not included in this simulation, are responsible.
- After the cohesive zone is fully developed the cohesive zones at the tips of both cracks advance at the same rate. However, the size of the cohesive zone was found to decrease with continued loading; for instance, the cohesive zone size at the loading displacement of -0.04 is only about half of the full cohesive zone size 0.04 at the loading displacement of -0.015. This is explained by the effect of the component of the normal stress parallel to the crack (the T-stress). The large T-stress parallel to the growing wing crack suppresses the expansion of the cohesive zone.
- Note that in this problem, two crack and cohesive zone tips are involved, but the formulation of thc boundary element method with a cohesive zone is able to track both tips without any additional complications in the numerical scheme.

5.3. Interaction of cracks in the double edge cracked specimen

The last problem we consider in this paper is a double-edge-crack loaded in tension allowing crack interaction. This problem also provides an opportunity to examine cracks that exhibit structural instability. The geometry of the specimen is shown in Figure 15. The initial lengths of the two edge-cracks, a_{1o} and a_{2o}, are both set equal to 0.28. Each crack tip is again provided with an initial cohesive zone of length 0.02. In order to examine the approach of the two cracks towards each other, an offset d_{offset} is provided between the two edge cracks. The load-elongation curves for two cases of d_{offset} equal to 0.1 and 0.2 respectively are shown in Figure 16. The deformed configurations showing the crack paths determined from the simulations are presented in Figure 17 and 18. The main observations and results are described as follows:

- Initiation of the fictitious and physical crack tips is very similar to the cases described in Section 5.1. Note that the full geometry of the specimen is modeled and the anti-symmetry

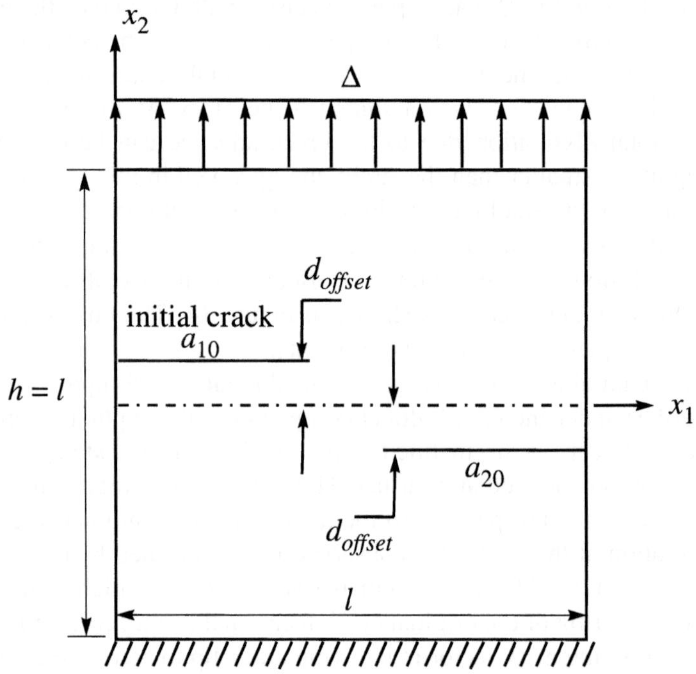

Figure 15. The loading configuration of a rectangular specimen with two edge cracks on the opposite sides, under displacement controlled elongation. The cracks are initially parallel to the loading boundaries, and may be offset from the symmetry line by an offset d_{offset}.

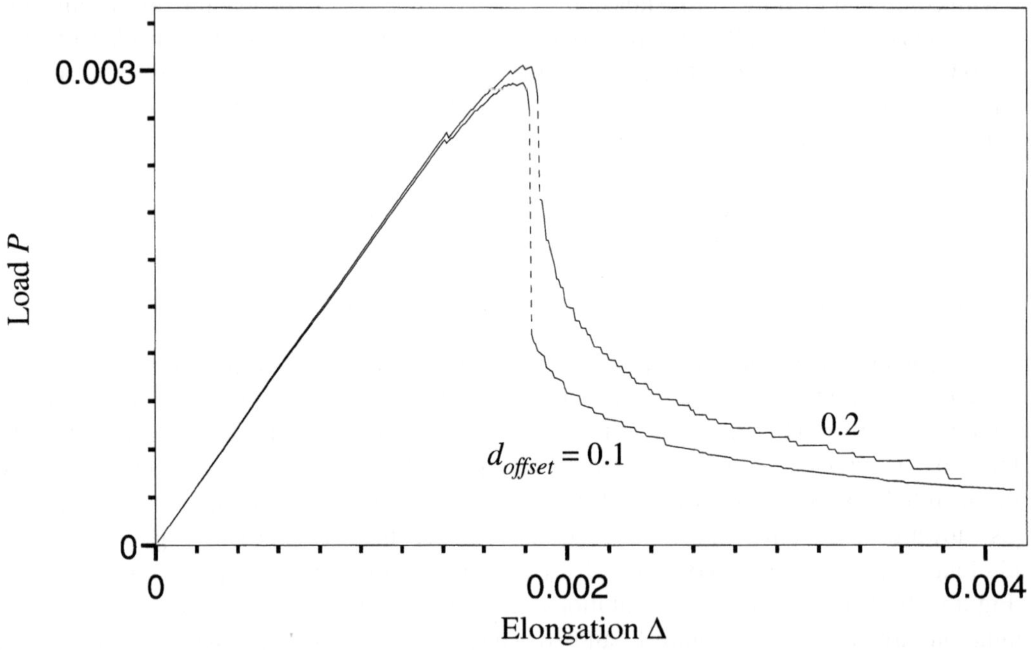

Figure 16. Load-elongation curves in the simulations of crack interaction with the offset d_{offset} as given above. Besides, the cracks $a_{10} = a_{20} = 0.28$ initially, and each crack tip was assumed to have an initial cohesive zone of size 0.02. The dashed lined indicate the loss of structural stability.

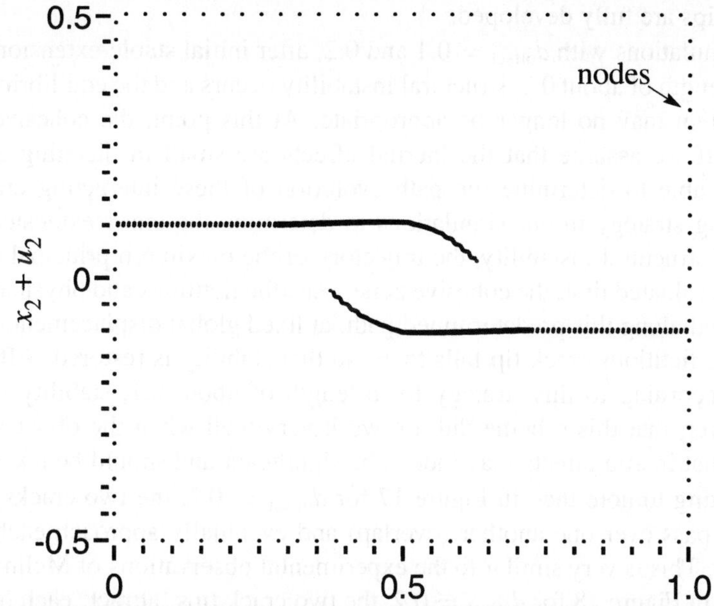

Figure 17. Deformed configuration of the specimen with two edge cracks with the offset $d_{\text{offset}} = 0.1$, corresponding to an extension $\Delta = 0.0038$.

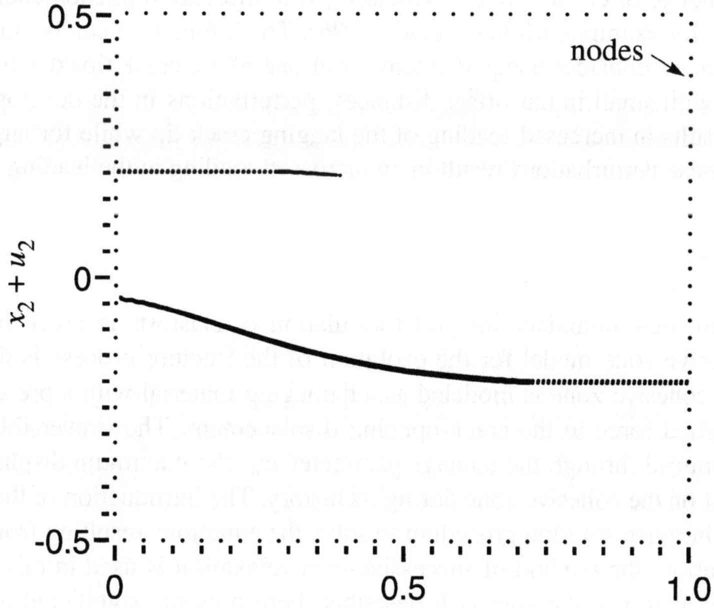

Figure 18. Deformed configuration of the specimen with two edge cracks with the offset $d_{\text{offset}} = 0.2$, corresponding to an extension $\Delta = 0.0038$.

of the offset cracks has not been imposed externally. For both cases, the cohesive ones at both crack tips are fully developed.

- For both simulations with $d_{\text{offset}} = 0.1$ and 0.2, after initial stable extension of both crack tips over a length of about 0.1, structural instability occurs and the equilibrium formulation of the problem may no longer be appropriate. At this point, the cohesive zone is fully developed. If we assume that the inertial effects are small in dictating the crack path, we may be able to determine the path evolution of these interacting cracks. We used the following strategy in our simulations to determine the crack extension behavior. At the onset of structural instability, the trajectory of the maximum principal stress from the crack tip is evaluated first; the cohesive zone – i.e., the fictitious and physical crack tips – is then extended along this predetermined path, at fixed global displacement, until the stress ahead of the fictitious crack tip falls to p_y so that stability is restored. After propagating the crack according to this strategy for a length of about 0.1, stability is restored. We note, however, that this scheme did not work very well when the offset was very small indicating that inertia effects may indeed be significant and should be taken into account.
- It is interesting to note that, in Figure 17 for $d_{\text{offset}} = 0.1$, the two cracks initially 'repel each other, pass over one another (overlap) and eventually approach each other along a curved path. This is very similar to the experimental observations of Melin (1983). On the other hand, in Figure 18 for $d_{\text{offset}} = 0.2$, the two crack tips 'attract' each other right from initiation. From a different point of view, the two crack tips approach the line of symmetry separately without much interaction in the early stages; we note that a single offset crack shows the tendency to approach the line of symmetry as demonstrated in Section 5.1. After initial extensions of the both cracks over a length of about 0.1 each, the left crack is completely shielded by the growth of the right-side crack and only one crack grows in this case. The change of crack patterns with d_{offset} is a bifurcation phenomenon of significant interest (see for example, Mulhaus et al., 1996). The bifurcation can be understood in the following terms: consider unequal extension of one of the crack tips due to perturbations. For cracks with small initial offset distances, perturbations in the development of either crack tip results in increased loading of the lagging crack tip while for large initial offset distances, these perturbations result in an increased loading at the leading crack tip.

6. Conclusions

A single-domain, dual-boundary integral formulation of elastostatic crack problems incorporating a cohesive zone model for the evolution of the fracture process is demonstrated in this paper. The cohesive zone is modeled as a damaging material with a prescribed behavior relating the applied force to the crack-opening displacement. The irreversible nature of the damage is introduced through the damage parameter w_d, the maximum displacement experienced by a point on the cohesive zone during its history. The introduction of the cohesive zone necessitates an iterative solution procedure to solve the equations resulting from the boundary integral formulation; the method of successive-over-relaxation is used in this study. In terms of numerical simulations, the approach described here presents significant advantages over grid-based finite element methods since the present formulation

(i) does not force development of the crack along element boundaries and thereby introduce mesh size and element geometry dependencies,
(ii) does not require fine nodal spacing except within the cohesive zone, and

(iii) does not require remeshing with crack extension.

Three example problems were considered to demonstrate the capability of this formulation to handle arbitrary in-plane crack problems, including mixed-mode problems, contact problems, and crack interaction problems.

Acknowledgments

The support of the Air Force Office of Scientific Research during the course of this work is gratefully acknowledged.

References

Aliabadi, M.H. (1997). Boundary element formulations in fracture mechanics. *Applied Mechanics Reviews* **50**, 83–96.

Ameen, Mohammed and Raghuprasad, B.K. (1994). A hybrid technique of modeling of cracks using displacement discontinuity and direct boundary element method. *International Journal of Fracture* **67**, 343–355.

Blandford, G.E., Ingraffea, A.R. and Liggett, J.A. (1981). Two-dimensional stress intensity factor computations using the boundary element method. *International Journal of Numerical Methods in Engineering* **17**, 387–404.

Brace, W.F. and Bonbolakis, E.G. (1963). A note on brittle crack growth in compression. *Journal of Geophysical Research* **68**, 3709–3713.

Brebbia, C.A., Telles, J.C.F. and Wrobel, L.C. (1984) *Boundary Element Techniques: Theory and Applications in Engineering*, Springer-Verlag.

Chen, W.H. and Chen, T.C. (1995). An efficient dual boundary element technique for a two-dimensional fracture problem with multiple cracks. *International Journal of Numerical Methods in Engineering* **38**, 1739–1756.

Crouch, S.L. (1976). Solution of plane elasticity problems by the displacement discontinuity method. *International Journal of Numerical Methods in Engineering* **10**, 301–342.

Cruse, T.A. (1996). BIE fracture mechanics analysis: 25 years of developments. *Computational Mechanics* **18**, 1–11.

Hartmann, F. (1980). Computing the C-matrix in non-smooth boundary points. In: *New Developments in Boundary Element Methods*. (Edited by Brebbia, C.A.), Butterworths, London, 367–379.

Hong, H. and Chen, J. (1988). Derivations of integral equations of elasticity. *Journal of Engineering Mechanics* ASCE, **114**, 1028–1044.

Horii, H. and Nemat-Nasser, S. (1986). Brittle failure in compression: Splitting, faulting and brittle-ductile transitions. *Philosophical Transactions of the Royal Society of London* **A319**, 337–374.

Love, A.E.H. (1944). *A Treatise on the Mathematical Theory of Elasticity*, Dover, New York.

Melin, S. (1983). Why do cracks avoid each other? *International Journal of Fracture* **23**, 37–45.

Mulhaus, H.-B., Chau, K.T. and Ord, A. (1996). Bifurcation of crack pattern in arrays of two-dimensional cracks. *International Journal of Fracture* **77**, 1–14.

Nemat-Nasser, S.N. and Horii, H. (1982). Compression-induced nonplanar crack extensionwith application to splitting, exfoliation and rockburst. *Journal of Geophysical Research* **87**, 6805–6821.

Ortiz, M. (1996). Computational micromechanics. *Computational Mechanics* **18**, 321–338.

Portela, A., Aliabadi, M.H. and Rooke, D.P. (1991). The dual boundary element method: effective implementation for crack problems. *International Journal of Numerical Methods in Engineering* **33**, 1269–1287.

Snyder, M.D. and Cruse, T.A. (1975). Boundary integral equation analysis of cracked anisotropic plates. *International Journal of Fracture* **11**, 315–342.

Steif, P.S. (1984). Crack extension under compressive loading. *Engineering Fracture Mechanics* **20**, 463–473.

Wen, P.H. (1996). Dynamic fracture mechanics: Displacement discontinuity method, Computational Mechanics Publications.

Xu, X.-P. and Needleman, A. (1994). Numerical simulations of fast crack growth in brittle solids. *Journal of Mechanics and Physics of Solids* **42**, 1397–1434.

Yang, B. and Ravi-Chandar, K. (1996). On the role of the process zone in dynamic fracture. *Journal of the Mechanics and Physics of Solids* **44**, 1955–1976.

Yang, B. and Ravi-Chandar, K. (1998). Anti-plane shear crack growth under quasistatic loading in a damaging materials. *International Journal of Solids and Structures* **35**, 3695–3715.

Young, A. (1996). A single-domain boundary element method for 3-d elastostatic crack analysis using continuous elements. *International Journal of Numerical Methods in Engineering* **39**, 1265–1293.

International Journal of Fracture **93:** 145–159, 1998.

Nonsteady crack and craze behavior in PMMA under cyclical loading: I. Experimental preliminaries

GUILLERMO C. PULOS* and WOLFGANG G. KNAUSS
Graduate Aeronautical Laboratories, California Institute of Technology, Pasadena, California, U.S.A.
e-mail: wgk@atlantis.caltech.edu

Received 21 November 1996; accepted in final form 15 July 1998

Abstract. This is the first of three papers devoted to the study of nonsteady crack propagation under cyclic loading in polymers, specifically PMMA. Drawing on experimental tools and measurement methods offering high spatial and real-time resolution at the subcycle level, we find that the change in craze behavior with load history is more complex than believed to date. At room temperature the change in fatigue cycle frequency on the order of slightly more than a decade gives rise to considerably different fracture surface morphologies, which are load history dependent. These observations are categorized and deductions for the variable cohesive force transmission by the craze are considered. In this first paper we outline the design of a special loading device which allows ultra-precise load or displacement control commensurate with the high resolution measurements of the crack tip material response. This development illustrates the precision of control for the experiments described in the subsequent papers. In addition, we delineate the software control and data acquisition methods to highlight the spatial resolution which provides the micron accuracy for measuring sub-cyclical crack propagation necessary for the later accounts.

Key words: Polymer fatigue; time dependent crack growth; polymer fracture; evolution of cohesive forces.

1. Introduction

Polymers have come to be widely used in engineering applications. Because they are often employed in designs requiring long term sustaining of loads it is important to understand their durability. From a materials or synthesis point of view it is of interest to relate 'performance' to the microstructure. One such topic of interest is the relation of their microstructural (craze) behavior to the more global aspects of failure. It is the intent here to contribute to this understanding by studying the load-history dependent evolution of the cohesive forces provided by the craze material at the tip of a propagating crack under cyclic loading (fatigue).

The traditional engineering approach to fatigue design with metals and in the presence of a crack is to relate the stress intensity factor to the average crack propagation rate (per cycle). With respect to polymers there exist two features that set these materials apart. One is the fact that because of their intrinsic time or rate sensitive characteristics the growth of a crack does not depend so much on the number of cycles but (also) on the frequency as the parameter representing the time history (Chang, 1983; Knauss, 1989; Gregory and Botsis, 1991). Furthermore, there have been reports (Elinck et al., 1971; Döll and Könczöl, 1990; Hertzberg and Manson, 1980) that the crack does not propagate steadily but in 'spurts', the time frame of each spurt encompassing possibly many cycles. It turns out, ultimately, that these spurts depend on the load level and time history, a fact that is readily missed when

* Currently at the Instituto de Investigaciones en Materiales-UNAM, Mexico.

studies are conducted at a single frequency. In the latter context the question thus arises as to how the craze participates in this nonsteady behavior: Is the nonsteady 'periodicity' associated with a total break down of the craze thus advancing the crack by the length of the craze, or is such periodicity associated with a break down within the craze so that advancement is only partial?

The global aim of this study is to gain better knowledge of the relevant material behavior at the crack tip under time varying loading through its detailed observations with the aid of a microscope and optical interferometry. Because the size scale of these observations is at the micron level conventional engineering test facilities are inadequate to guarantee sufficient control over the test conditions and special facilities must be designed and constructed for allowing displacement control to the same level of precision. In addition, there exists the perennial question of whether the observations made on the surface of a specimen truly reflect the behavior of the crack front throughout the thickness dimension of the specimen. This latter concern is particularly important when one deals with measurements intended to distinguish differences at the micron scale. Further, because many such measurements are likely to be required, a method must be devised which allows (at least semi) automatic data acquisition in real time in order to reduce the inordinate amount of detailed data.

The extent of the whole work is beyond the scope of a single paper. We thus divide its presentation into three accounts that summarize:

(1) the experimental requirements and equipment design,
(2) the nonsteady crack propagation and
(3) the effect of load history on the material behavior at the crack tip.

Specialized loading equipment for studying the behavior of the craze at the crack tip under quasi-static loading was built by Döll and coworkers some twenty years ago (Weidmann and Döll, 1976); later on a device, based on a 240 watt electromechanical shaker, became available for fatigue studies (Schirrer et al., 1981; Döll et al., 1982). This loading apparatus was capable of loads up to 30 N and frequencies to 500 Hz. Döll's and Schirrer's studies used very small compact tension specimens ($8 \times 10 \times 4\,\text{mm}^3$) and a flash lamp capable of being triggered at any moment during the loading cycle. In our present examinations, we were interested in a very stiff loading device and larger specimens. When interferometric methods are employed to determine the craze parameters, the size of the specimen is determined by the working distance of the microscope objectives. The availability of microscope objectives with larger working distances thus allowed us the use of larger specimens which resulted in smaller relative positioning errors. The use of a stiff loading device with larger loading capability makes for a system insensitive to changes in the compliance of the specimen as the crack propagates and thus would not induce nonsteady propagation behavior.

We start in this first paper with a discussion of how precisely controlled loading conditions can be assured for high resolution measurements. Because our measurement techniques hinge on optics it is our objective to provide displacement control in:

(a) arbitrarily prescribable time histories to
(b) within at least the wave length of light (micron scale).

The first requirement demands precision control which can only be achieved in the absence of vibrations and thus demands that the loading device be mounted on a vibration isolated optical table; this, in turn, places limitations on its physical size. Precision can be readily guaranteed through the use of thermal deformations (Liechti and Knauss, 1982) or through piezzo electric devices (Vendroux and Knauss, 1998). The former are limited in the speed of deformations while the latter are typically limited to very small loads and displacements. All three methods are capable of being computer controlled in real time for prescribing arbitrary load histories. Although hydraulic devices and associated linkages are difficult to control at the tolerances desired here, hydraulics provides the best overall answer for our study once special care in the design of the loading device was exercised so that the design ultimately settled on the choice of a computer controlled hydraulic loading device.

We delineate briefly the design of the system that provides displacement control in the micron range. Bearing in mind the need for microscope observation of the crack tip, loading must be accomplished such that the latter does not move appreciably during cycles of loading. Following a brief summary of the mechanical (servo-hydraulic) loading device in Section 2, the optical arrangements are discussed in Section 3. In Section 4, the collection and analysis of data as well as the specimens configuration are discussed, while the final section summarizes the overall experiment design, with a view towards its use in interaction with the optical interferometry measurements and resulting material behavior described in the subsequent papers.

2. Miniature hydraulic loading device

A facility to load small fracture specimens (with a typical dimension of 30 mm) was constructed with displacements being prescribed symmetric with respect to the crack line so as to allow continuous in-focus monitoring of the crack tip through a microscope. The schematic and a photograph of the loading device are shown in Figure 1. The loading facility is mounted on an optical breadboard to facilitate incorporating the optical components into the set-up. It consists of a single servovalve to control the position of two cylinders arranged hydraulically in series. A linearly variable differential transformer (LVDT)[1] is attached to each hydraulic cylinder to monitor its displacement. The range of the LVDTs is ± 1.3 mm and they provide an output[2] of approximately 5 mV/μm. A 1100 N load cell[3] attached to the left cylinder measures the load on the specimen and provides an output[4] of approximately 10 mV/N. The load cell resolution is pegged at 0.05 N. The most important figure of merit and the motivation for the design and construction of this loading device was its positional resolution so that either displacement or forces could be prescribed. The loading device has a positional resolution of $\pm 1\ \mu$m, a typical long term drift[5] of 0.5 μm and a typical rms. error[6] of 0.4 μm. The design parameters, servovalve and the arrangement of the hydraulic lines are not detailed here but may be found in reference (Pulos and Knauss, 1996).

[1] Lucas Schaevitz DC-E050 LVDT.

[2] The output derives from the signal conditioner connected to the LVDT.

[3] Sensotec 31/1432-05 250 lb. load cell.

[4] The output derives from the signal conditioner connected to the load cell.

[5] The deviation (as measured from the LVDTs) from the prescribed (preset) displacement amplitude (at the beginning of the test) over 5000 cycles at a frequency of 1 Hz.

[6] The rms. error after the prescribed displacement and the tracking error have been subtracted.

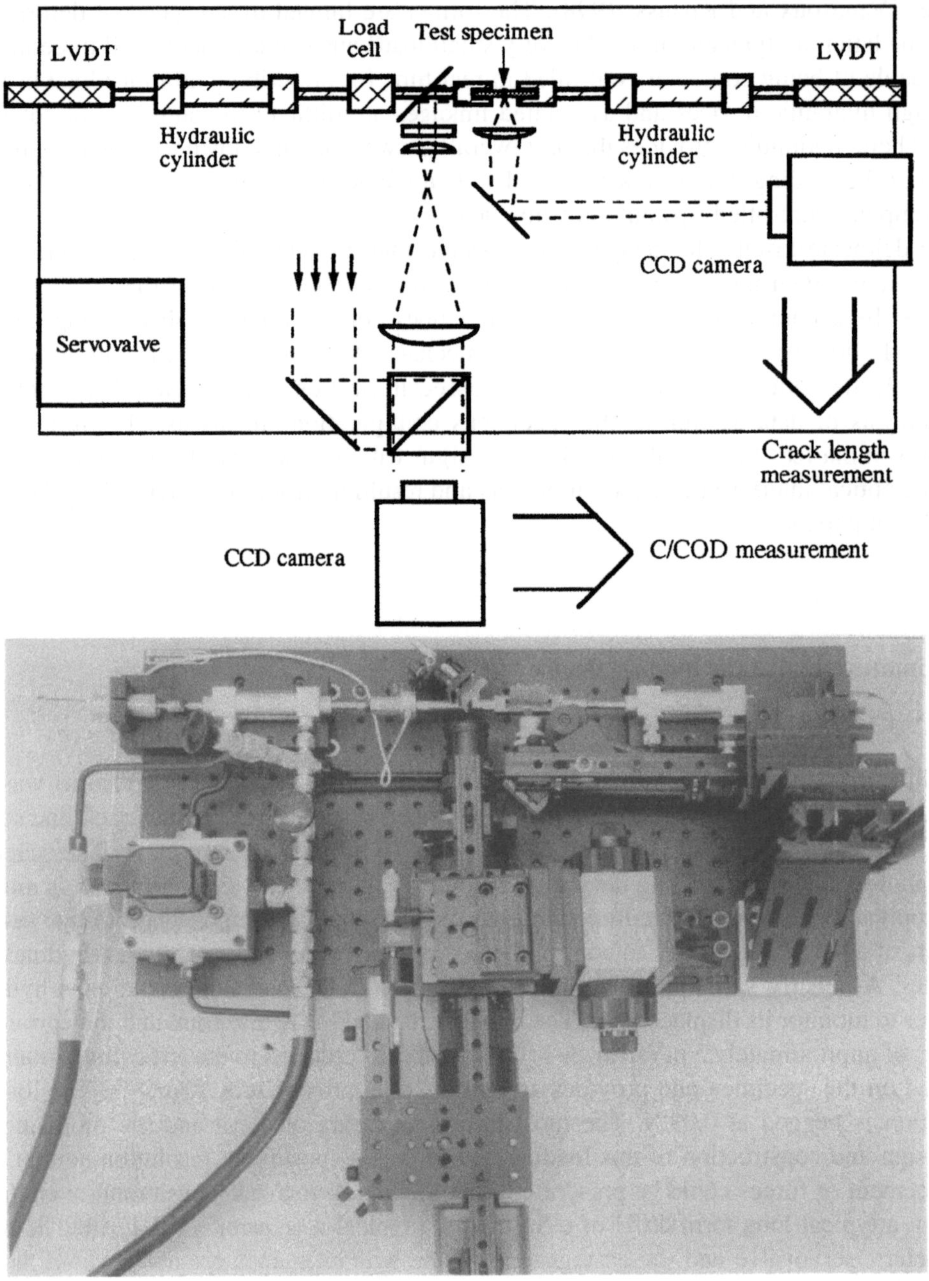

Figure 1. Schematic and photograph of the loading facility.

3. Optical arrangements

The crack length and the crack and craze opening displacements are measured simultaneously by two separate optical systems each ending in a charge-couple-device (CCD) camera. The crack length as monitored on the lateral specimen surface is determined by computer image tracking and processing through a microscope while the crack opening displacements are

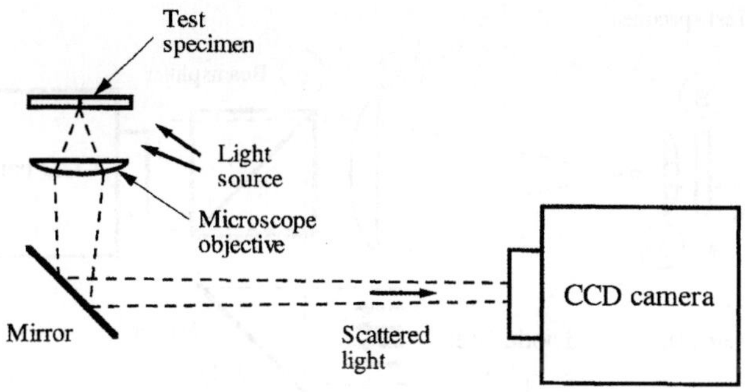

Figure 2. Schematic of the optical arrangement to record the crack length on the surface of the specimen.

collected interferometrically through a separate microscope with a view normal to the crack surface.

3.1. Crack length measurement

In order to achieve sufficiently high resolution for tracing differences in steady as opposed to possibly 'discontinuous' crack growth it is necessary to devise a means for the spatial resolution of crack tip location that is measured in units substantially smaller than the length of the craze. Thus (optical) microscope methods are indicated. At the same time the large number of data points required for this purpose raises the need for (semi)automatic means of monitoring the crack tip position.

Accordingly, the optical system for tracking the crack length consists of a CCD camera, a mirror and a microscope objective lens as shown schematically in Figure 2. The optical system is mounted on a mechanical, electrically controlled, traversing mechanism attached to the breadboard of the loading device to focus the optical system, as well as to maintain the crack tip within the field-of-view of the camera. By trading off between resolution and the size of the field-of-view, using a 5X microscope objective and a $\frac{1}{2}$-inch CCD sensor, we obtain a pixel dimension of 1.94 μm $\times$ 1.59 μm, where the smaller dimension corresponds to the direction of crack propagation. The resulting field-of-view (512 $\times$ 480 pixels) is approximately 1 mm by 0.75 mm. The specimen is illuminated by oblique light (dark-field configuration) and the computer obtains the crack length by determining the location of the crack tip relative to a small dimple or scratch on the surface of the specimen (reference mark; see Section 4.2). Using this arrangement under computer control it was possible to measure the crack length with a resolution of 1 micron during every loading cycle of a fatigue test.

3.2. Crack and craze opening displacement measurements

The arrangement to record the interferometric fringe pattern for reconstructing the craze/crack opening displacement profile consists of an afocal optical system shown schematically in Figure 3 which is formed by two lenses mounted back to back such that their focal points coincide.[7] In this system the front lens (A) (closest to the specimen) is a microscope objective with a large working distance and a focal length of approximately 52 mm. The back lens (B) is a telephoto lens with a focal length of 200 mm that is changed to 400 mm by a 2X tele-

[7] An afocal system has the advantage that the size of the image is independent of its position. The size of the image (magnification) is determined by the ratio of the focal lengths of the individual lenses.

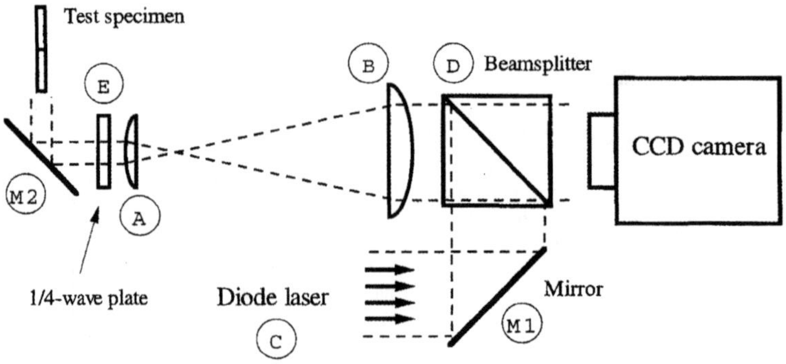

Figure 3. Schematic of the optical arrangement to record the fringe pattern of the craze and crack opening displacements.

converter, so that the crack and craze are viewed by the CCD camera at a magnification of 8X.

Illumination is provided by a 5 mW diode laser (C) collimated by a 10X Galilean beam expander; it enters the optical system through a mirror (M1) and a polarizing beam-splitter (D). The polarization of the beam is such that the beam splitter reflects (turns) the light and directs it to the specimen through the afocal system and a $\frac{1}{4}$-wave plate (E). A final mirror (M2) directs a 2.5 mm diameter beam through the specimen and illuminates a portion of the craze/crack front.

As the reflected light from the (internal) craze/crack interfaces passes twice through the $\frac{1}{4}$-wave plate its polarization has changed (rotated by 90°) so that as it exits the afocal system the beam-splitter transmits the light to the CCD camera. The '$\frac{1}{2}$-inch' CCD sensor, together with the afocal system, generates a field-of-view of 0.87 mm by 0.68 mm. By digitizing the TV frame into a 512×480 array each pixel measures 1.70 μm in the y-direction and 1.41 μm along the direction of crack propagation (x-direction). The details of the data collection during an experiment are discussed in the next section; in particular, Section 4.3 discusses the data collection from the CCD camera that records the opening displacement. With the aid of computer control this arrangement allowed tracking of the craze parameters during the fatigue test as well as during any particular loading cycle.

4. Data acquisition and analysis

A computer[8] controls the experiment by supplying the command signal to the servomechanism, while it simultaneously records and analyzes two different types of information: On the one hand it receives pictorial information from the CCD cameras; on the other hand, it accepts one-dimensional information from the signal conditioners (i.e., load and displacements). A schematic of the data collection is shown in Figure 4, which includes two sample images collected by these optical systems: The upper left corner represents a portion of the fringes of the opening displacement (cf. Section 4.3) while the upper right corner shows a portion of the image obtained from the CCD camera simultaneously viewing the lateral surface; data collected from that camera is used to determine the crack length record as discussed in Section 4.2.

[8] A PDP 11/73 was used for this study. However, while the programming details are dependent on the choice of computer, any computer (such as a 'PC') with a single-user operating system would be useful.

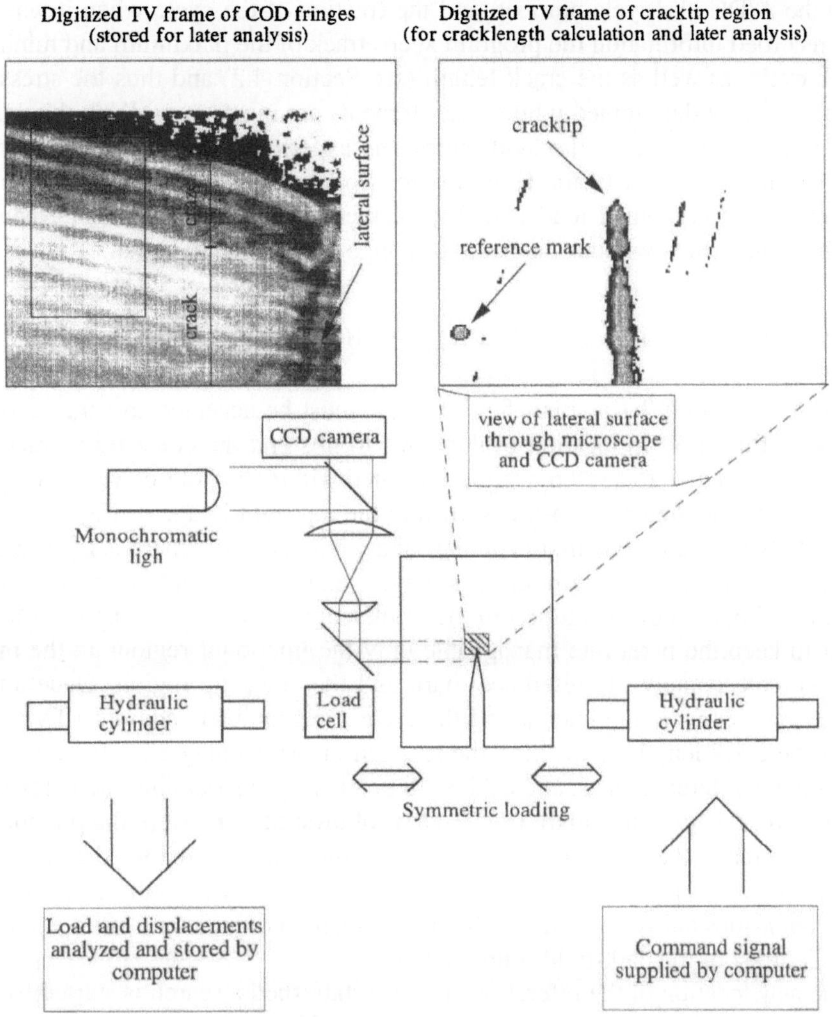

Figure 4. Schematic of data collection.

4.1. Data collection: one-dimensional information

The program controlling the experiment and collecting the data in 'real time' uses interrupt routines[9] to output the command signal through the digital-to-analog-to converter (D/AC) as well as to collect data from the load cell and the displacement of the cylinders through an analog-to-digital converter (A/DC). This function is executed in addition to the image acquisition and evaluation discussed above and below.

The sources for the interrupts come from 15 timers programmed separately to trigger at different times during a cycle. The load and displacement are digitized 60 times per cycle; those data, together with various other signals, combine to trigger the A/DC around 250 times during each cycle. After the data is digitized by the A/DC, it is stored in a buffer along with

[9] A short program or subroutine that can suspend the (main) program in response to a signal (an interrupt) leaving the internal registers unchanged so that the main program can resume its execution at the point where it was suspended without any errors or additional programming considerations.

the gain in the A/DC, the cycle number, and the fraction of the cycle when it was acquired. Using this recorded information the program keeps track of the maximum and minimum load during each cycle, as well as the crack length (see Section 4.2) and thus the stress intensity factor is continuously determined while measurements are in progress. With this information the computer is able to change the load during the experiment so as to follow a prescribed history for the stress intensity factor. However, for work with a constant stress intensity factor, a technique using a combined load and displacement feedback (Pulos and Knauss, 1993) produces the same results without the need to change the command signal.

4.2. Data acquisition for the crack length determination

As mentioned in Section 3.1 a small field-of-view must be accepted to achieve sufficiently high resolution for crack tip location evaluations; to this end the crack tip is tracked once[10] during each cycle relative to a reference mark located within the field-of-view. A typical view of the crack tip and a reference mark is shown in the top right image of Figure 4. Two types of measurements are made: The first is in real-time while the experiment is in progress and its ultimate purpose is to reduce the amount of storage[11]; the second, made after the experiment is completed, is to improve the resolution in the crack length record as described in Section 4.4.

In order to keep the pixel rate manageable only the important regions in the microscope view are recorded, namely the reference mark and the crack tip region. Once during each loading cycle, at the maximum load during the cycle, the computer acquires a TV video frame and obtains the crack length by tracking the location of two rectangles: a 16×32 pixel array centered around a reference mark and a 32×48 pixel array surrounding the crack tip.

The search for the reference mark proceeds as follows: Starting from the previous location of the reference mark, the computer searches in the surrounding area for the new location of the reference mark.[12] This is accomplished on a line-by-line basis as the computer looks for pixels with intensities above a given threshold. The search is simple and efficient, due, in part, to the high contrast of the dark-field illumination.

Once the new location of the reference mark is established a search is started for the crack tip location a few pixels behind its last known position. The computer searches for pixels with intensities above the threshold, and the scan line where the search ends is logged as the new crack length. The computer then stores the rectangular areas surrounding the reference mark and the crack tip as well as the relative distance between the rectangles.

Since the length of the crack is larger than the field-of-view of the camera which tracks the location of the crack tip, the crack length is determined by tracking reference marks within the field-of-view starting with a known position[13] up to the location of the tip. The calibration of the measurement is obtained by (previously) imaging a target with known dimensions and it is checked from the output[14] of the actuator that moves the optical system.

---

[10] The program was capable of obtaining 2 to 3 measurements per cycle at frequencies below 1 Hz and one measurement every two cycle for frequencies above 3 Hz. Most of the data was collected using one reading per cycle.

[11] Prior to the development of the combined feedback technique (Pulos and Knauss, 1993) the real-time crack length record was also used to control the stress intensity factor.

[12] Slight movements of the specimen relative to the optical system make this step necessary.

[13] In this case, the corner of the 12.7 mm machined notch (Figure 5).

[14] The actuator has a resolution of 0.1 μm.

4.3. Data acquisition of the crack and craze opening displacements

A second input to the frame-grabber is connected to the CCD camera that records the fringes from the crack and craze opening displacements (COD); thus, as the computer switches the input to the frame-grabber, which is normally viewing the lateral surface and recording the crack length, a TV frame of the COD can be recorded. A portion of such a frame is shown in the top left corner of Figure 4: The area at the top of the picture, which does not exhibit any fringes, is the unbroken bulk material ahead of the craze. The broad and brighter set of (left-to-right) interference fringes covering most of the photograph are caused by the crack opening while the first three (faint) fringes are associated with the craze opening displacement. The area where the fringes end (right side of the photograph) is the lateral surface of the specimen. It is on this surface where the optical system described above (in Sections 3.1 and 4.2) records the crack length. The region in the upper right of the photograph and marked by the rectangle is a 64×128 pixel array that is monitored by the computer. This area is always chosen to be close to the free surface[15] of the specimen so that the craze/crack opening displacements (COD) measurements can be correlated with the surface crack length measurements; the latter considerations are detailed in a subsequent paper of this series (Pulos and Knauss, 1998, II).

Due to the large number of tasks which the computer must perform at any time and because of the rather large scale simultaneous processing requirements for analyzing and storing visual data in real-time, the computer cannot acquire frames of the opening displacements continuously. Instead, once every four cycles and during the minimum load of the cycle the computer 'grabs a frame' and stores the 64×128 pixel byte array of the area that is being monitored. For studying the loading/unloading effect on the craze, 11 frames are acquired during a cycle and the area within the monitoring rectangle is stored for later analysis; the particular cycle when these operations take effect was programmed to occur at fixed intervals, typically every 100 cycles. Finally, a full frame at both maximum and minimum loads during a cycle is acquired and stored every 250 cycles.

4.4. Detailed analysis

After the measurements are completed the log files are transferred to a separate computer for analysis. First, the information on the log files is separated into individual files. The cyclic information (load and displacements) is analyzed to determine the extrema in each cycle; for each file the minima, maxima, amplitude and the phase lag relative to the command signal are calculated and stored.

Further and more refined analysis is performed on the images from the lateral surfaces (used in determining the crack length) beyond the in-process determinations. This post-experiment processing improves on the 'real-time' data to produce a higher resolution (of about 2 μm) in the crack length record since it is possible to fit functions to the intensity of the pixels that make up the reference mark and crack tip; determining positions from the fitted functions is, in essence, an averaging technique that reduces noise and thus improves the resolution of the measurements.

[15] In most other studies the monitored area is chosen near the center of the specimen. The dimensions of the specimens are such that the recorded opening displacement can be considered to be under plane strain conditions. In this study the measurements are within the three-dimensional zone associated with the intersection of the crack front with the free surface.

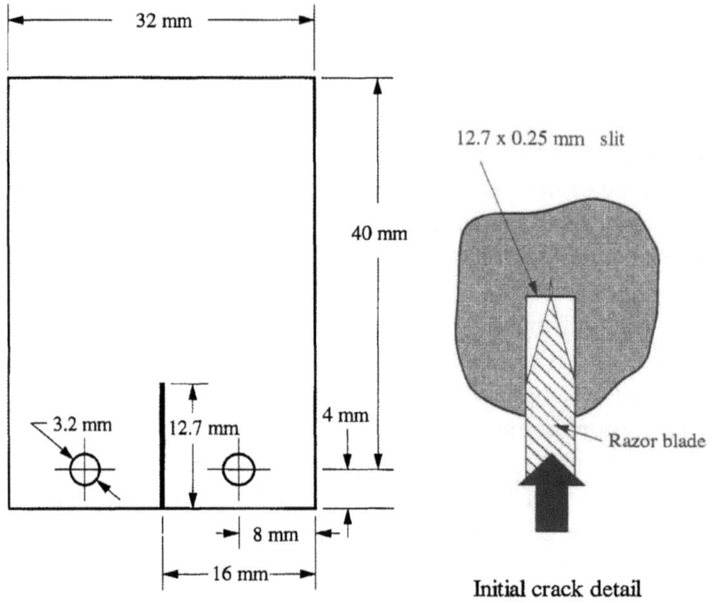

Figure 5. Specimen geometry and initial crack detail.

In addition, a visual inspection of the rectangle surrounding the crack tip allows a higher (dynamic) threshold to be used in determining the crack tip position. Although there exists high contrast between the crack and its background, the threshold for the crack tip has to be maintained low so that the computer (without the help from the operator) does not 'loose' the crack tip during the experiment. This reduction in the threshold causes the program to introduce noise in the crack length record as it (incorrectly) responds to pixels in the background with a high intensity as belonging to the crack. This type of noise is eliminated as the rectangular pixel array is visually inspected by the operator. Results from a typical crack length record are discussed later in the section.

4.5. Specimen detail and sample records

Finally, we outline specifics regarding the specimen configuration. Two different types of PMMA were used in this study. Most of the specimens were prepared from commercial-grade PMMA (Plexiglas G, The Rohm and Haas Company) while some were cast from medium molecular weight PMMA (The Aldrich Company; the typical weight-average molecular weight was 10^5). The pellets were formed into a plate in a mold (Washabaugh, 1990). To this end the pellets are heated under vacuum to 150°C and pressurized to 1.4 MPa for about 30 hours before slowly cooling, after the pressure is removed, at a rate of 1.7 degree/hr.

The geometry used for the experiments is shown in Figure 5. The thickness of the specimens was about 2.85 mm ($\frac{1}{8}$-inch nominal) for the commercial PMMA and 3.1 mm for the cast specimens. The stress intensity factor was calculated by way of the formula collected by Tada et al. (1973).

It should be pointed out that the geometry is not exactly that of a compact tension specimen; it was decided to use this geometry to allow longer crack propagation phases from a single specimen. As pointed out in the introduction, the width of the specimen (32 mm) is determined by the working distance of the optical system that monitors the craze; however, the length has

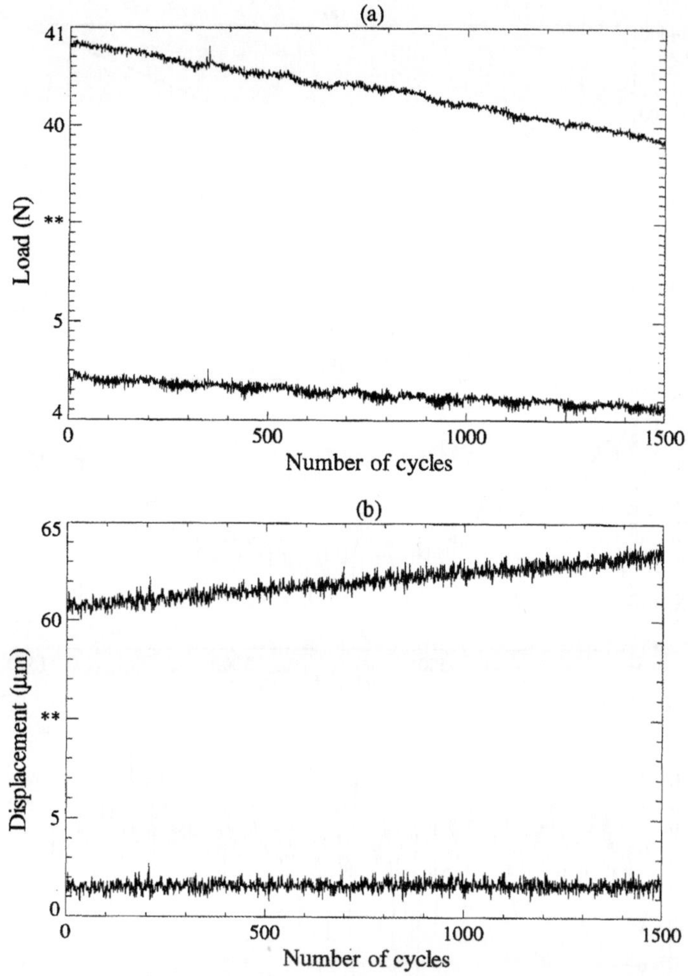

Figure 6. Load and load-point displacement records for a typical fatigue experiment.

no such constraints other than buckling if it is too large (and the specimen is not thick enough). The choice of the compact tension geometry would require a length of 26.7 mm and it was deemed too small to carry out several tests on a single specimen. As it turned out, the use of the combined feedback technique (Pulos and Knauss, 1993) to control the stress intensity factor also depends on the geometry and it may be that the optimum geometry for a constant stress intensity factor may not correspond to that of a compact tension specimen.

The top and bottom (narrow) sides of the specimens were polished to obtain a clear in-plane view through the specimen and to allow the measurement of the craze and crack opening displacements. The lateral surface of the specimens were coated with thin (15 nm) layers of Aluminum to provide a focusing plane for the crack length measurements. Finally, a razor blade, the thickness of which is the same as the width of the slit, was used to make a sharp notch (by impact) along the middle of the specimen (see initial crack detail in Figure 5) from where a fatigue crack is started; the crack is allowed to sharpen by propagating for about 2.3 mm under cyclic loading (1 Hz) to eliminate the effects of the razor blade operation. The loading was adjusted to obtain crack growth rates around 1 μm/cycle.

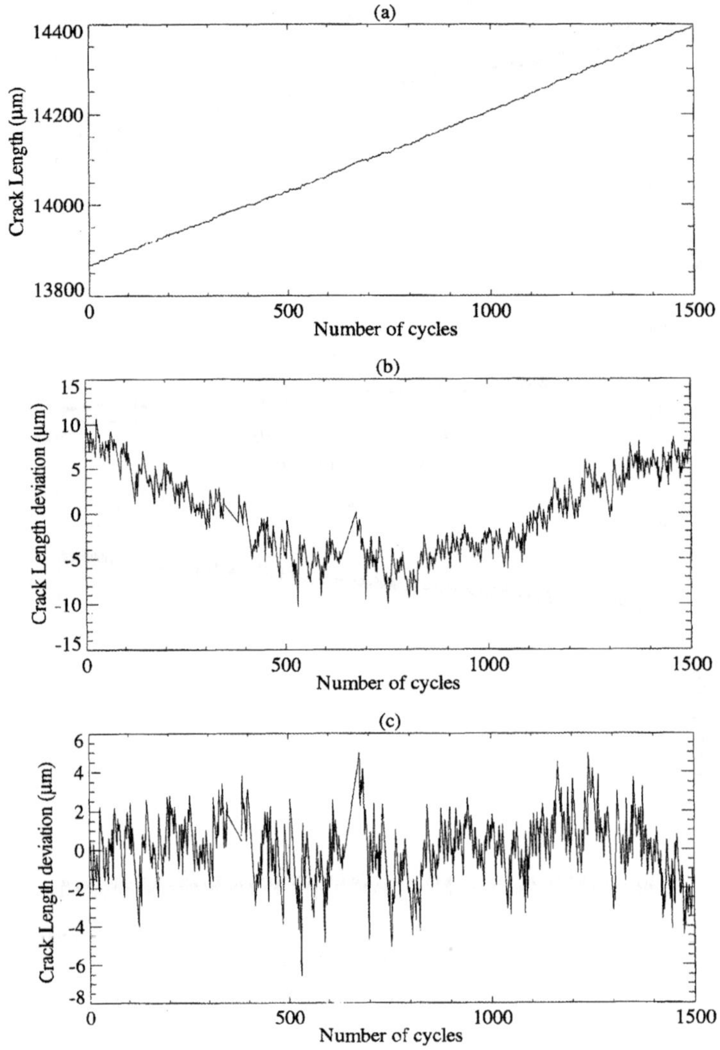

Figure 7. Crack length and crack length deviation records for a typical fatigue experiment; (a) crack length record; (b) crack length deviation record for a straight line fit to (a); (c) crack length deviation record for parabolic fit to (a).

Load[16], load-point displacement and crack length record for a typical fatigue experiment[17] (loading frequency of 1 Hz) are shown in Figures 6 and 7a. Both these load and displacement records are obtained from the raw data collected during the experiment which was sampled at 40 to 60 times the loading frequency. The record for the crack length contains one datum point per 2 cycles (see Section 4.2). It should be noted that as the crack propagated under (an almost) constant stress intensity, the loads decreased while the displacements simultaneously increased; this control was achieved by the combined feedback method (Pulos and Knauss,

[16] A change in the scale of the ordinates occurs at **; the lower scale corresponds to the minimum (bottom curve) while the upper one corresponds to the maximum (upper curve).

[17] The stress or stress intensity factor ratio (R) typically used was 0.1. However, for experiments under combined or load feedback the minimum load was chosen so as to prevent the servo control from becoming unstable.

1993). For the data shown the feedback consisted of 70 percent load and 30 percent displacement and the variation of the (maximum) stress intensity factor was no more that 0.15 percent which is 1/50 of the variation if the experiment had been carried out under a constant load or constant displacement.

To assess the resolution of the data shown in Figure 7a it is necessary to describe the data analytically. Since most of the data in these studies was collected at near-constant stress intensity factors, it can be described in terms of polynomials; the difference between the experimental data and its analytical representation of the crack length record is shown in Figure 7b and 7c.

While typical deviations (defined as the difference between experimental data and its analytical, polynomial description) in the amplitude of the load and displacement records can be characterized as noise with average absolute values of 0.05 N and 0.5 μm respectively, it is not clear whether the deviations in the crack length record can be characterized as noise. In this case, however, the crack does not propagate in a discontinuous (retarded) fashion and the apparent 'signal' in the deviation record comes from the choice of polynomial in describing the crack length record. The deviation in Figure 7b comes from fitting a straight line to the crack length record while that in Figure 7c uses a parabola to account for the slight acceleration in the crack growth. Thus the deviation in the crack length record is $\pm 2\,\mu$m (rms. of 1 micron); the deviation corresponds to noise and the crack is propagating 0.35 microns every cycle. An example illustrating retarded crack growth is described in (Pulos and Knauss, 1998, II and III).

4.6. Frequency dependence of Young's modulus

Data from compliance measurements obtained during fatigue experiments yielded the values for Young's modulus that are shown in Figure 8. The modulus is normalized by $E_0 = 3.4$ GPa, a typical value for the small strain modulus of PMMA; the increase in the modulus is about 8 percent per decade, the same as obtained by Lu and Knauss from relaxation experiments (Lu et al., 1997). Similar frequency dependence was observed by Hertzberg and coworkers (Hertzberg et al., 1979) whereas measurements from craze data and a Barenblatt/Dugdale analysis made by Döll, Könczöl and Schinker (Döll et al., 1983)[18] showed increases in modulus with frequency of about 6 percent per decade.

4.7. Thermal cracktip measurements

A third input to the frame-grabber was connected to a thermal camera[19] that measured the local temperature rise in the vicinity of the cracktip region during cyclical loading. However, for the loading and frequencies (0.1 to 3.3 Hz) used in the fatigue experiments, the temperature rise was below the resolution of the thermal camera (0.1°C).

5. Summary

A miniature loading device allowing the prescription of rather arbitrary load histories has been described which permits a positional resolution of 1 μm through a single servovalve

[18] In this case, the measurements come from crack tip data and the values are 50 percent smaller due to the higher strain state of the material surrounding the crack tip.

[19] Inframetrics model 600.

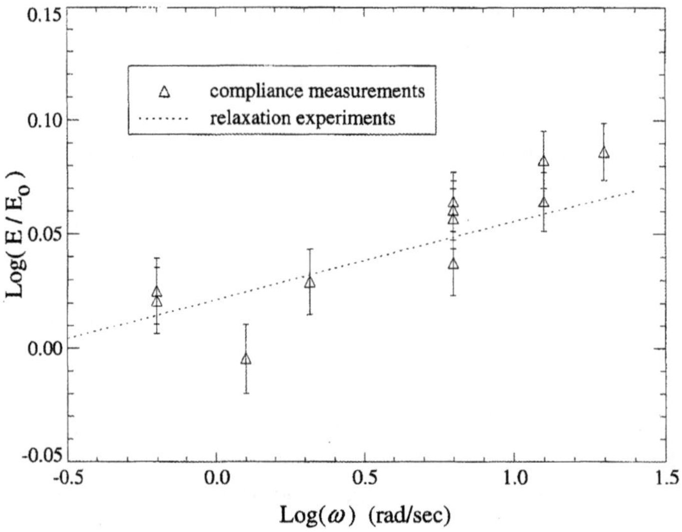

Figure 8. Frequency dependence of Young's modulus.

controlling two cylinders to maintain the crack tip of a specimen at closely the same location in space during repeat loading. Two optical systems simultaneously record the crack length and the craze and crack opening displacements. In anticipation of studies reported later (Pulos and Knauss, 1998, II; 1998, III) we have delineated an experimental arrangement whereby a computer controls the experiment and performs real-time measurements of the specimen load, load-point displacement, displacement of the crack/craze opening profile as well as crack length in an automated fashion; displacement control and crack tip position resolution is typically at the micron level. For normal cyclic data acquisition the computer is able to record data for about 5000 cycles; more modern platforms can exceed this number by several factors. The computer integrates the control of the loading facility with the measurements of one-dimensional data as well as two-dimensional (visual field) information thus creating a multi-media recording and control facility.

This facility has been employed to investigate the internal stress state of the craze in PMMA forming ahead of the crack tip under different time histories. The use of the loading device coupled to the computer allows keeping track of the craze as the material is loaded cyclically and as the crack propagates through the craze, or moves with it.

Acknowledgments

The authors wish to acknowledge various sources of support contributing to this work: ONR through grant #N00014-91-J-1427 with Dr. Peter Schmidt as the technical monitor; E.I. DuPont de Nemours Company through initial contact with Dr. Manuel Panar and the Newport Corporation for liberal equipment support.

References

Chang, S.K.Y. (1983). Crack propagation in viscoelastic materials under transient loading with application to adhesively bonded structures. *Aeronautical Engineer Thesis*, California Institute of Technology.

Döll, W. and Könczöl, L. (1990). Micromechanics of fracture under static and fatigue loading: Optical interferometry of crack tip craze zones. *Advances in Polymer Science: Crazing in Polymers* (Edited by H.H. Kausch) 91/92, Springer-Verlag, Berlin, 137–214.

Döll, W., Schinker, M.G. and Könczöl, L. (1982). Direct measurements of craze zones at crack tips in glassy thermoplastics during fatigue crack growth. *Deformation, Yield and Fracture of Polymers*, The Plastics and Rubber Institute, London, pp. 20.1–20.6.

Döll, W., Könczöl, L. and Schinker, G. (1983). Size and mechanical properties of craze zones at propagating crack tips in poly(methylmethacrylate) during fatigue loading. *Polymer* **24**, 1213–1219.

Elinck, J.P., Bauwens, J.C. and Homes, G. (1971). Fatigue crack growth in poly (vinyl chloride). *International Journal of Fracture Mechanics* **7**, 277–287.

Gregory, B.L. and Botsis, J. (1991). Experimental investigation of the effects of stress rate and stress level on fracture in polystyrene. *Journal of Materials Science* **26**, 1015–1026.

Hertzberg, R.W. et al. (1979) Comments on 'a model of fatigue crack growth in polymers'. *Journal of Materials Science* **14**, 1754–1758.

Hertzberg, R.W. and Manson, J.A. (1980). *Fatigue of Engineering Plastics*. 1st edn, Academic Press, New York.

Knauss, W.G. (1989). Time dependent fracture of polymers. *Advances in Fracture Research: Proceedings of the 7th International Conference on Fracture* 4, (Edited by K. Salama *et al.*), Houston, pp. 2683–2711.

Liechti, K.M. and Knauss, W.G. (1982). Crack propagation at material interfaces: I. Experimental technique to determine crack profiles. *Experimental Mechanics* **22**(7), 262–269.

Lu, H., Zhang, X. and Knauss, W.G. (1997). Uniaxial, shear and poisson relaxation and their conversion to bulk relaxation: Studies on poly(methyl methacrylate). *Journal of Polymer Science and Engineering* **37**(6), 1053–1064.

Pulos, G.C. and Knauss, W.G. (1993). Constant stress intensity factors through closed-loop control. *International Journal of Fracture* **63**, 101–112.

Pulos, G.C. and Knauss, W.G. (1996). Nonsteady crack and craze behavior in PMMA under cyclical loading: I. Experimental preliminaries. Report SM 96-9A, California Institute of Technology.

Pulos, G.C. and Knauss, W.G. (1999). Nonsteady crack and craze behavior in PMMA under cyclical loading: II. Effect of load history on growth rate and fracture morphology, in this issue.

Pulos, G.C. and Knauss, W.G. (1999). Nonsteady crack and craze behavior in PMMA under cyclical loading: III. Effect of load history on cohesive force distribution of the craze, in this issue.

Schirrer, R. et al. (1981). Interferenzoptische Untersuchungen zur Bruchausbreitung unter Wechsellast in PMMA. *Colloid and Polymer Science* **259**, 812–817.

Tada, H., Paris, P.C. and Irwin, G.R. (1973). *The Stress Analysis of Cracks Handbook*, Del Research Corporation, Hellertown.

Vendroux, G. and Knauss, W.G. (1998). Submicron deformation field measurements: Part 1. Developing a digital scanning tunneling microscope. *Experimental Mechanics* **38**(1), 18–23.

Washabaugh, P.D. (1990). *An Experimental Investigation of Mode-I Crack Tip Deformation*, Doctoral Thesis, California Institute of Technology, Pasadena.

Weidmann, G.W. and Döll, W. (1976). Interferenzoptische Vermessung der Craze-Zone vor der Rißspitze in PMMA Unterschiedlichen Molekulargewichtes. *Colloid and Polymer Science* **254**, 205–214.

International Journal of Fracture **93:** 161–185, 1998.

Nonsteady crack and craze behavior in PMMA under cyclical loading: II. Effect of load history on growth rate and fracture morphology

GUILLERMO C. PULOS* and WOLFGANG G. KNAUSS
Graduate Aeronautical Laboratories California Institute of Technology Pasadena, California, U.S.A.
e-mail: wgk@atlantis.caltech.edu

Received 21 November 1996; accepted in revised form 15 July 1998

Abstract. This is the second of three papers devoted to time or frequency effects on crack propagation in PMMA under cyclic loading with high resolution measurements of crack and craze growth in the micron range and with subcycle temporal resolution. It is demonstrated that the breakdown rate of the craze leads to quasi-periodic spurts of propagation and growth retardation in dependence on both the magnitude of the stress intensity factor and on the imposed load frequency. The surface morphology is governed more by the time required to build and deteriorate the craze (process time) than by the amplitude of the stress intensity factor.

Three different growth ranges are distinguished: At the highest load levels ($K \approx 1\,\mathrm{MPa\,m}^{1/2}$) multiple crazes occur at various distances from the major crack plane which coalesce and give rise to an incoherent surface structure not characterizable in terms of a single craze, although the front of the craze/crack is still represented by a smooth curve; crack advance occurs over domains that are larger than a typical craze length. At lower stress levels ($K \approx 0.6\,\mathrm{MPa\,m}^{1/2}$) the craze breaks at different levels within the craze structure, the spatial breakdown of the craze being limited to dimensions smaller than the craze length but with a correlation or memory effect that generates terraces (on the fracture surface) on a scale larger than the craze length. In a transition between this stage and the lowest stress level considered, both features may exist simultaneously on a fracture surface, one increasing or decreasing relative to the other. Increase/decrease of one phenomenon relative to the other does not occur at a fixed stress intensity range but depends on the past history. When a change in the loading produces a change in the fracture mode, it is possible to associate, during the transition, decreasing crack growth rates with increasing stress intensity factors. At the lowest loading ($K \approx 0.35\,\mathrm{MPa\ m}^{1/2}$) a craze breaks essentially through its center leaving a submicroscopically smooth surface; the associated crack growth and craze break down occurs with more or less periodic increments that are each smaller than the full 'equilibrium' craze length.

Key words: Polymer fatigue, time dependent crack growth, polymer fracture, evolution of cohesive forces.

1. Introduction

Failure of thermoplastic materials at the tip of propagating cracks involves the development of a craze ahead of the crack tip (Hertzberg and Manson, 1980). Because craze formation and propagation involves a transition from the bulk polymer state through microvoid formation into fracture, it plays a key role in controlling the fracture process. Usually, this zone is referred to in standard fracture mechanics parlance as the process zone, in which the material reaches a stress level which allows the molecular chains to become oriented into fibrils parallel to the (crack opening) tension stress.

The size of the craze zone or craze length is determined by the two different processes of the (rate of) fibrillation at the craze tip and the (rate of) failure of the fibrils at the base or

* Currently at the Instituto de Investigaciones en Materiales-UNAM, Mexico

trailing end of the craze (at the crack tip proper). Both these processes determine the length of the craze and determine the mode of propagation. Specifically, if both processes occur in phase and at the same rate the propagation is the same on a cycle per cycle basis; this type of propagation mode has been reported as 'normal' or 'continuous[1]' (NCG) and constitutes a self-similar translation of the crack and craze tip with each cycle. It occurs without change in the craze length (Döll and Könczöl, 1990).

A second mode of crack and craze advance is possible when both processes are out of phase with respect to each other. In the extreme case the crack remains stationary for a number of cycles while fibrillation enlarges the craze: during one particular cycle, the crack breaks through the craze or through a part of it. This mode has also been observed (Elinck, et al., 1971), (Skibo, et al., 1977) and is referred to as 'retarded' (RCG) or 'discontinuous.' Implicit in the notion of retarded crack propagation is the idea that the length scales involved (jump distances) are equal to or smaller than the length of the craze. The retarded mode of crack growth under fatigue loading was first observed by Elinck et al. (1971) who, by measuring the crack length as a function of time and taking the load cycle frequency into account, determined that the markings on the fracture surface of a PVC specimen corresponded to crack growth at approximately every 370 cycles.

Such diverse observations of 'nonunique' propagation prompt the question of how a crack/craze configuration propagates under varying loading conditions, including frequency changes. A further motivation in this quest was the observation that under truly dynamic crack propagation conditions (crack speeds on the order of hundreds of meters per second) the fracture surfaces in PMMA exhibited a periodic structure, which can be correlated with timewise periodically emitted surface waves (Washabaugh and Knauss, 1993). What is particularly surprising, and not necessarily explained, is that this spatial periodicity in dynamic crack propagation experiments correlates well with the periodicity observed under cyclic loading (Hertzberg et al., 1978) even though the time scales as well as load histories are widely different.

Anticipating later discussions, we note that nonsteady crack growth behavior exists in which the crack never comes to a complete stop before jumping. Rather it propagated with periods of acceleration and deceleration. Thus, the growth behavior was intermediate to the 'continuous' and the 'retarded or discontinuous' growth cases. Central to this behavior is the issue of how to detect and quantify the nonsteady crack growth rate, as well as how to quantify the (fluctuating) periodicity of the accelerating/decelerating stages. An answer to the previous questions hinges on obtaining crack length records with a large number of data points and a sufficiently high resolution so as to detect crack propagation in units significantly smaller than the length of the process zone (craze). In our case, the crack length records contain anywhere from 1500 to 5000 data points with a resolution of about 2 μm which typically amounts to 1/20 of the craze length (craze lengths were between 35 and 45 μm).

The following describes observations on details of crack tip motion in PMMA and high precision measurements made on the surface of PMMA compact tension specimens[2] through

[1] In his second review article (Döll and Könczöl, 1990) Döll argues that all crack/craze propagation is discontinuous in nature since it only occurs during part of the loading cycle. Thus, he denotes the continuous mode of propagation as normal (fatigue) crack growth and the discontinuous mode as retarded crack growth.

[2] The specimens are 32 mm wide, 40 mm long and 3 mm thick. The dimensions do not correspond to the standard specimen as it was decided that longer specimens would allow more tests to be carried out on a single specimen. The material used was commercial PMMA from Rohm and Haas (Plexiglas G) and medium molecular weight PMMA from Aldrich.

an optical system. Specifically, the objective is to observe in real-time[3] the crack growth on a cycle per cycle basis. The experimental set-up, including the optical arrangement, illumination and data collection, is discussed in reference (Pulos and Knauss, 1998, I). The procedure used for determining the nonsteady crack growth is described in the next section. Section three discusses the crack growth records in relation to the fracture surface morphology resulting from three different levels and histories of loading, while section four summarizes the conclusions.

2. Crack growth analysis

It is customary in crack propagation studies under fatigue loading to record the crack propagation rate as a function of a stress intensity variation (ΔK) under the assumption that the crack propagates with a constant rate, and one assigns then an average propagation rate to such a measurement. Here interest centers on the *variation* in crack propagation from such an average in order to assess the (possible) lack of steadiness in the propagation behavior. To characterize the nonsteady crack growth behavior two different approaches to the data analysis are considered. The first one examines the growth rate by differentiation and smoothing of the crack length record. This procedure introduced unacceptable subjective arguments into the physical record. In a second, improved way to analyze the unsteadiness in the growth rate, the difference in crack tip position between experimental and averaged data is obtained and Fourier transforms are used to quantify the periodicity of the unsteady growth.

Figure 1 shows a full field digitized TV frame of the lateral surface for which the crack length was monitored[4] at a frequency of 0.5 Hz along with the stress intensity factor as calculated from the load and crack length records[5]. Rectangles surrounding the crack tip (32 × 48 pixels) and the reference mark (16 × 32 pixels) as discussed in Pulos and Knauss (1998, I) are also indicated. This particular record extended over approximately 600 μm. Since this figure does not involve the issue of retarded crack propagation, the rate of crack growth was obtained by differentiating the crack length record after some modest data smoothing and then plotting it against the crack length.

To examine the effect of transient loading, consider first a three segment sequence of loading: After starting with a loading (0.75 MPa m$^{1/2}$ and amplitude of 0.77 MPa m$^{1/2}$) such that the crack grew at an average speed of 1 μm/cycle for 125 μm, the stress intensity factor was increased to 0.88 MPa m$^{1/2}$(amplitude of 0.70 MPa m$^{1/2}$) so that the crack began to propagate with an average speed of 2 μm/cycle. After propagating for 200 μm the loading was reduced again (maximum K_I of 0.75 MPa m$^{1/2}$ and amplitude of 0.50 MPa m$^{1/2}$) with the attendant average growth rate dropping to 0.5 μm/cycle.

Two observations impact the method of crack growth analysis. Referring to Figure 1 observe that[6] the advancing crack is accompanied by a dark band of varying thickness. This band

[3] In this context real-time refers to the time while the experiment is running. Observations are made during the fracture process with a sub-cycle time resolution.

[4] As was mentioned in (Pulos and Knauss, 1998, I) the recording system uses dark-field illumination (background is dark). However, for display purposes the picture has been digitally enhanced by the process of thresholding (Gonzalez and Wintz, 1987) where every pixel with an intensity below 105 had its intensity changed to 255 which prints as white.

[5] The solid line represents the maximum and the dashed line the amplitude.

[6] The crack has not propagated in a straight line but deviated by 150 μm when its length was 11000 μm. The angle of deviation is approximately 2.5°. However, this angle is small enough and this situation did not occur often enough that it presented a systematic problem in the analysis process.

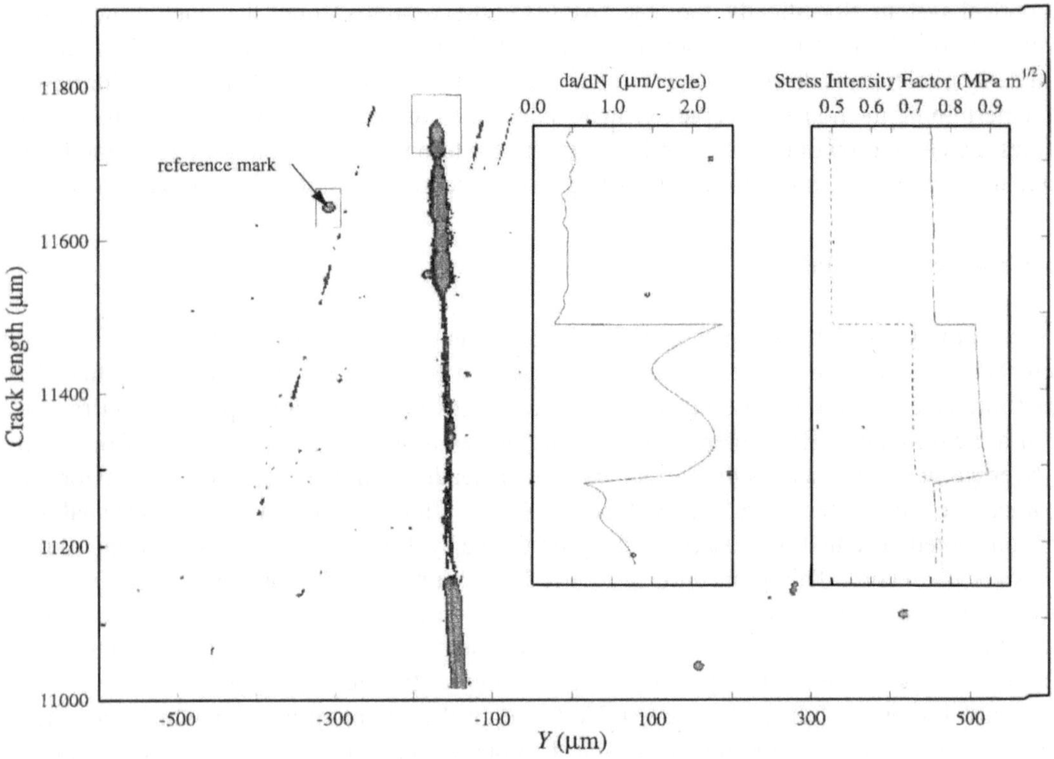

Figure 1. Digitized TV frame of the crack tip region.

is the result of surface deformations through a 'dimpling' in the wake of the crack tip. Within the time frame of the experiment this surface deformation is permanent (it lasts for weeks) and similar to 'plastic deformation' in metal specimens. Note that the width of this band changes in coordination with the loading level, being wider for slow crack tip motion (lower stress intensity). This 'band-dimpling' is thus a time dependent creep effect, and a greater width is associated with longer time exposure to the crack tip stresses.

While this phenomenon could impact the unique interpretation of crack advance rates when the front of the 'band-dimple' is used as a crack tip position indicator, there are more serious concerns with establishing growth rates by straightforwardly differentiating the record of crack tip position versus time. Numerical differentiation accentuates data noise. To reduce such noise data smoothing was attempted by computational filtering based on least square fitting an odd number of points and passing a best-fit low order polynomial through the data (typically a parabola over seven points).[7] In such an approach one must determine the 'quality' of the data subjectively by deciding on what constitutes signal content and what represents noise. This and the companion papers (Pulos and Knauss, 1998, I, III) are concerned with observing nonsteady crack growth, which may express itself in large-scale cyclical phenomena on the order of several hundred cycles. It is thus questionable whether such a subjective means of data analysis is permissible. For example, it is clear that the oscillations in the growth rate between 11300 μm and 11500 μm are indeed the result of material behavior, but it is difficult

[7] If the filtering program takes care of the end points in such a way that the new data set has the same number of points; the program can then run *ad-infinitum* and each new data set is smoother than the previous one.

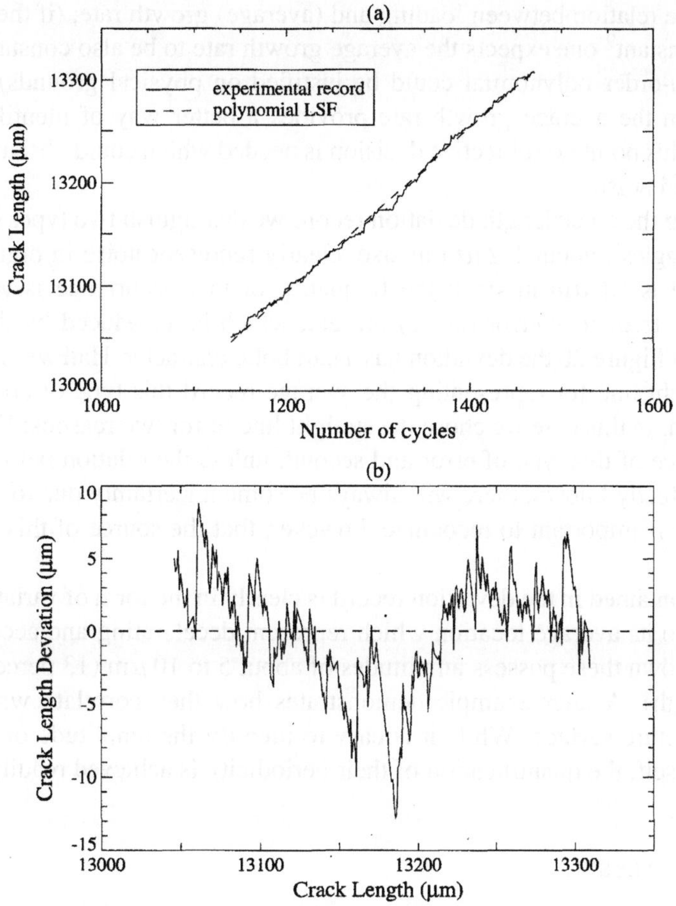

Figure 2. (a) Crack length and (b) crack length deviation records.

to attach much meaning to the (small) oscillations in the growth rate after 11600 μm. For this reason a second approach to crack propagation analysis was chosen as described in the next section.

2.1. CRACK LENGTH DEVIATION

Instead of fitting a low-order polynomial to a small portion of the crack length record for differentiation purposes it is preferable to fit a low-order polynomial (e.g. a straight line) to the whole record[8] and define the '*crack length deviation*' ('crack deviation' for short) as the difference between the 'fitted' location of the crack tip and the measured value at any instant of time. Thus if the crack accelerates and decelerates *as it progresses* the deviation should be a cyclical kind of record with time-averaged zero value. Figure 2a exemplifies an original data record as well as the fitted function; a plot of the deviation against crack length is shown in Figure 2b.

This procedure offers several advantages over the first method. Advantages are that (a) the choice of the fitting function can be justified since the loading is prescribed and knowledge is

[8] In records where the loading is changed during the experiment the record is subdivided into portions and each portion is treated as a separate experiment.

available about the relation between loading and (average) growth rate; (if the stress intensity factor remains constant[9] one expects the average growth rate to be also constant so that fitting a first (or second)-order polynomial could be justified on physical grounds); (b) separating the deviation from the average growth rate provides a better way of identifying nonsteady growth; (c) up to this point no subjective decision is needed which could obscure the difference between signal and noise.

For interpreting the crack length deviation record we distinguish two types of noise and one signal: Small 'wiggles', about 1-2 μm in size, clearly represent noise in digitizing the video signal (each pixel is 1.6 μm in size); the frequency of this occurrence is very high. There is also a very low frequency error (noise) present, which is introduced by the choice of the fitting function: In Figure 2b the deviation has a parabolic character. Had we chosen a parabola instead of a straight line for representing the average record this type of error (noise)would have been reduced. In this case we chose the straight line fit for two reasons: First, it illustrates clearly the presence of this type of error and second, unless the relation between loading and growth rate is perfectly known, there will always be some uncertainty due to the choice of the fitting function. It is important to recognize, however, that the source of this noise or error is then identifiable.

The 'signal' contained in the deviation record is clearly in the form of variations in crack tip position relative to its average location which represent decelerating and accelerating stages. In the example shown these possess amplitudes of about 5 to 10 μm (13 percent to 25 percent of the craze length). A later example demonstrates how they correlate with the markings present in the fracture surface. While it is easy to identify the amplitude of these variations from the record itself, the quantification of their periodicity is achieved readily with the aid of Fourier analysis.

2.2. FOURIER ANALYSIS

Two types of information can be obtained from the frequency information contained in the deviation record: If the abscissa records the instantaneous crack length, the frequency content signifies how far apart these stages occurred spatially; if the abscissa records numbers of cycles, the Fourier analysis yields how often acceleration/deceleration stages occurred.

The Fourier transform (magnitude) of the crack length deviation record (cf. Figure 2b) (Press, et al., 1986) is plotted against the *reciprocal* frequency in Figure 3a. The abscissa in this case is then the spatial period in microns since the abscissa of Figure 2b was the crack length. The plot of Figure 3b shows the transform against the number of cycles and the periodicity is now quantified in terms of the number of cycles between deceleration/acceleration stages. Barring identification of the harmonics of low frequency errors (noise) the peaks in Figure 3 indicate acceleration/deceleration spans of 17 to 53 μm and 20 to 65 cycles, respectively.

While it may appear on first sight that one has traded the uncertainty occasioned by the subjective smoothing (first approach) for an uncertainty in high and low frequency interpretation in the transformed records, it is now possible to discern between the harmonics of the low frequency noise and the periodicity of the deceleration/acceleration stages. For example, it will be possible in Section 3.2 to identify the harmonics of the low frequency noise and thereby deduce the periodicity associated with discontinuous or nonsteady crack growth as correlated with small scale features on the fracture surface.

[9] A procedure in which a combination of load and displacement produce extremely small variations in the stress intensity factor for a compact tension geometry is described by Pulos and Knauss (1993).

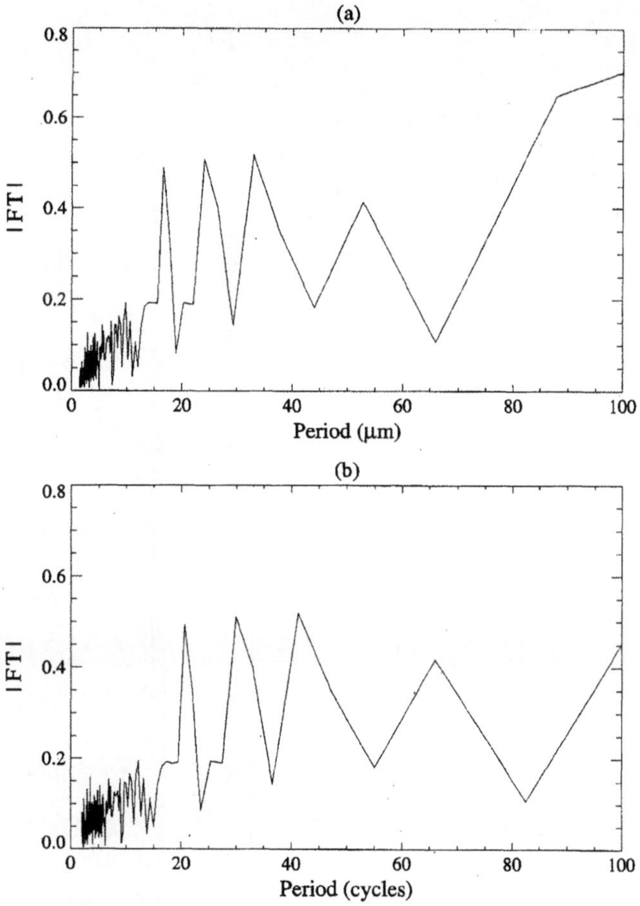

Figure 3. Magnitude of the Fourier transform of the length deviation record. Crack length deviation record was a function of the crack length (a) and a function of the number of cycles (b).

2.3. AVERAGE CRACK GROWTH

If a low order polynomial is fitted to the whole (crack length) record, the fitted function can be used (differentiated) to obtain a smooth representation of the average crack growth. In this way, macroscopic issues of crack growth versus stress intensity factors can be described using this smooth representation while microscopic issues of unsteady growth are detailed from the crack length deviation.

It is clear that the choice of the polynomial (order) will affect the apparent magnitude of the growth rate; however, the changes are small since low order polynomials are insensitive to the high frequency noise present in the (experimental) crack deviation record. By way of example, consider the crack length record of Figure 2a. The average growth rate of 0.96 μm/cycle (LSF of a straight line) will be reduced to 0.89 μm/cycle as the crack deviation changes from $+4$ μm (crack length of 13050 μm) to about -4 μm when the crack length reaches 13150 μm. Differentiating the experimental record gives growth rates that vary from about 0.01 μm/cycle to 1.75 μm/cycle and depend strongly on the amount and type of data smoothing.

In what follows, crack deviations are used to characterize unsteady growth, the magnitude of the Fourier Transform of the deviations are used to characterize the temporal or spatial peri-

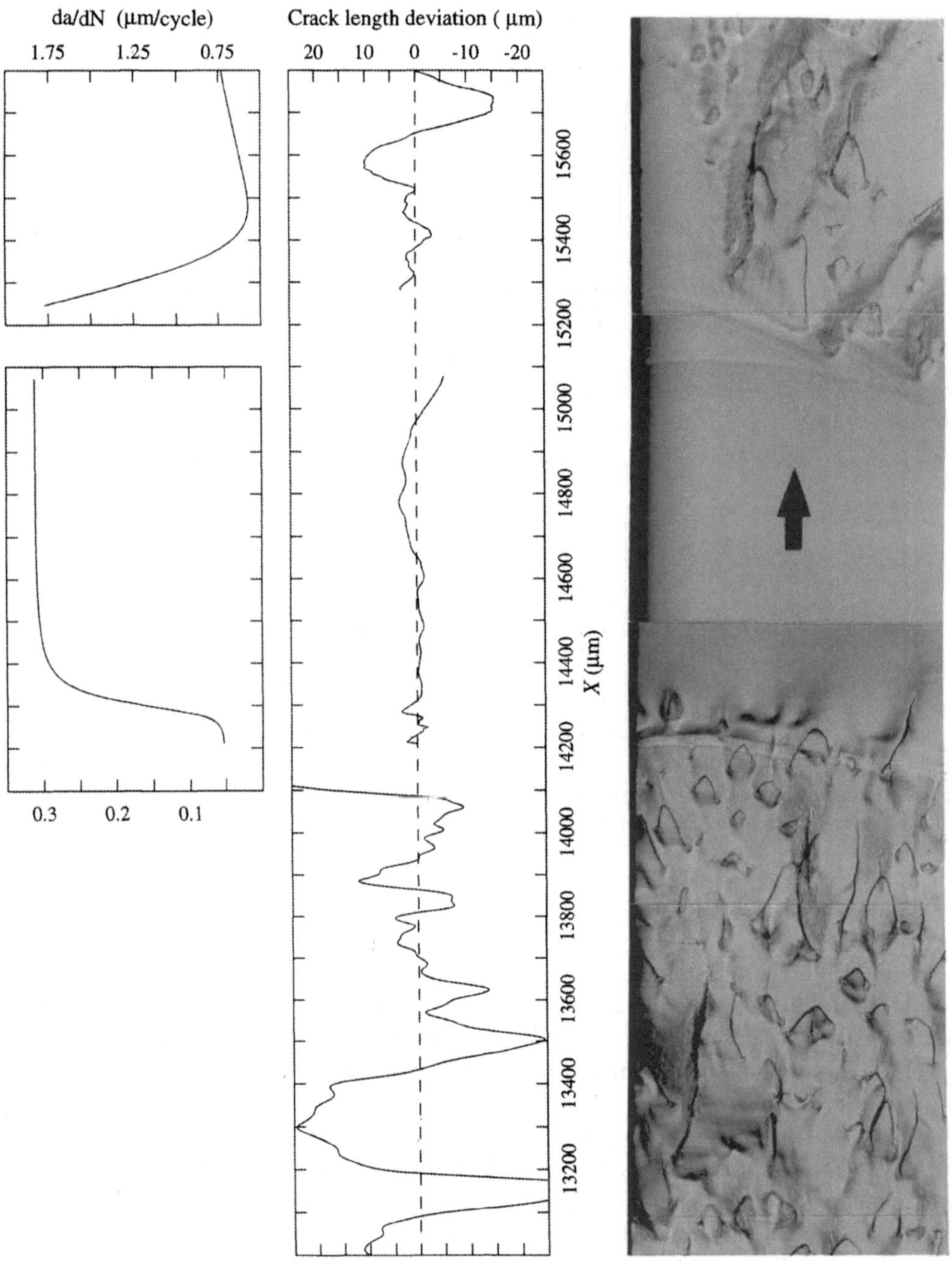

Figure 4. Average growth rate, crack length deviation and fracture surface morphology at high (0.97, 1.05 MPa $m^{1/2}$) and lower (0.8 MPa $m^{1/2}$) stress intensity factors (crack propagation in the direction of arrow).

odicity of the unsteady growth and the derivative of the fitted polynomial is used to represent the average crack growth rate.

3. Fracture surface morphology

We examine crack propagation for three different load (K) levels, each of which develops a different surface morphology depending on the time history (imposed frequency).[10] (1) At the highest load levels ($K \geqslant 0.9\,\mathrm{MPa\,m^{1/2}}$) multiple crazes form at small but variable distances from the crack plane; as the major crack front propagates, it connects to different crazes, leaving an incoherent and rough surface structure. (2) At somewhat lower loads ($K \leqslant 0.8\,\mathrm{MPa\,m^{1/2}}$) a single craze forms at the crack tip, which breaks according to time dependent constraints. If sufficient time is available for the craze to process the material, (low imposed frequencies) the crack breaks through the middle of the craze. However, as the process time is reduced, an incoherent instability forms in the form of wedge-shaped regions develop on the fracture surfaces that contain different craze failure modes. (3) Finally, for still lower loads ($K \leqslant 0.65\,\mathrm{MPa\,m^{1/2}}$) and longer process times (lower applied frequencies) the crack/craze system propagates in either a normal or a retarded mode with fracture consistently through the middle of the craze.

For the last case we consider the detailed growth of the crack as it propagates through part of the craze in stages of deceleration/acceleration. Since the craze cohesive stresses and deformations are most clearly identifiable in this loading regime this information becomes the basis for studying the cohesive stresses and craze opening displacements during multicyclical crack propagation in the companion paper (Pulos and Knauss, 1998, III).

However, before considering the details of crack growth under variable loading, it is important to emphasize two differences visible through optical and SEM records. Both bulk and crazed PMMA posses optical properties that allow light penetration into the broken craze along the fracture surface; optical images thus represent a record of the broken craze depth rather than a mere trace of the fracture topology. As a consequence the two sides of a broken craze as revealed on the two fracture surfaces may not be identical if the craze did not break through its middle. This fact is accentuated by different colors displayed over the fracture surface (Berry, 1961). On the other hand, SEM records require deposition of a thin conductive film (about 25 nanometer in thickness) so that the image represents a true image of the surface topology. Smoothness/roughness features observed on the surface must thus be moderated by the observation tool used to evaluate the record.

3.1. Variable stress intensity factor (high stress intensity)

For this sequence the amplitude of the stress intensity factor varies but is chosen large enough so that multiple crazes form ahead of or near the crack on different planes, with the main crack connecting to these different crazes (Döll, 1989) with $K = 0.97$ and $\nu = 1.33$ Hz. The fracture surface morphology (optical microscopy) is shown in Figure 4. Along with this morphology we plot the properly scaled average growth rate and crack length deviation record.[11] Thus,

[10] For all the cases presented here, the specimens were precracked by driving a razor blade into the specimen (crack length of approximately 9 mm); to erase any history effects, the crack was then propagated by fatigue (1 Hz, K 0.4 $\mathrm{MPa\,m^{1/2}}$) and as the crack length increased to 11 mm the frequency and stress intensity factors were adjusted to coincide with those used in the test.

[11] See Section2.1. The high frequency noise in the deviation record has been filtered out for display purposes.

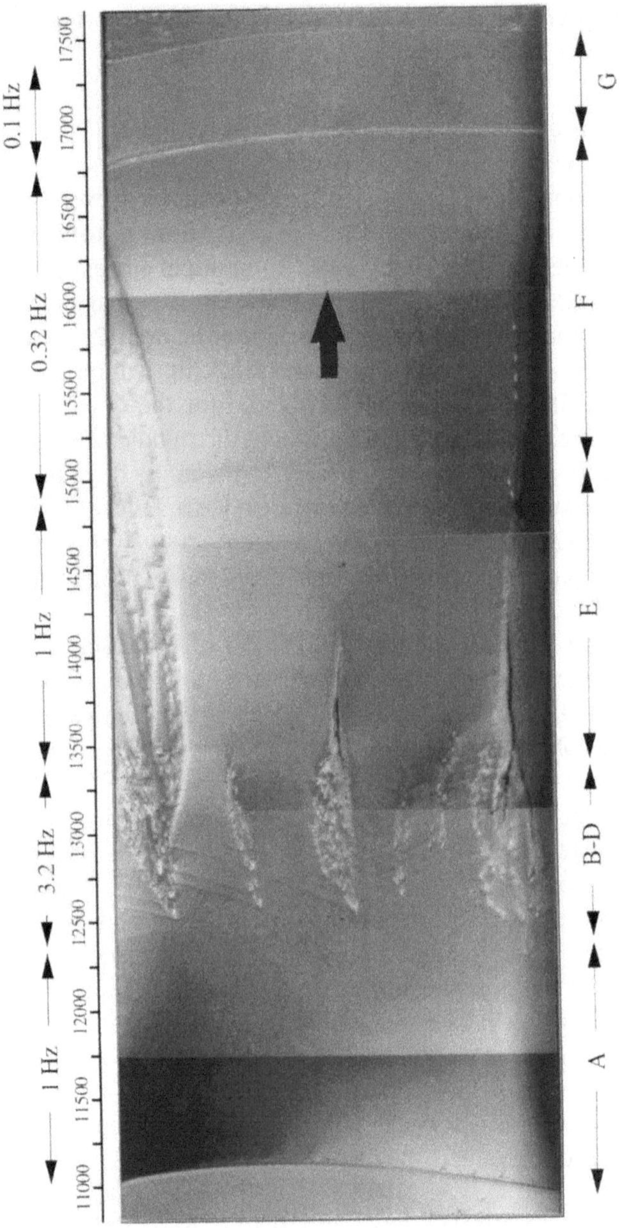

Figure 5. Fracture surface morphology at a constant stress intensity factor (0.61 MPa $m^{1/2}$) and test frequences of 1 Hz, 3.2 Hz, 1 Hz, 0.32 Hz and 0.1 Hz (scale in microns; crack propagation in the direction of arrow).

the scale denoted by X (in microns) serves for the growth and deviation records and also for the photograph. The crack length was measured on the edge or side surfaces located at the left side of the surface photograph. For comparison purposes we identify a measure of the craze size by noting that in later measurements when the fracture surface is very smooth and a craze length can be measured interferometrically (see (Pulos and Knauss, 1998, III)) it is determined to be 35 to 45 μm for a stress intensity that is only 16 percent lower than that used for Figure 4. Discounting large changes in size as a result of rate effects we estimate thus via a Barenblatt/Dugdale analysis that locally the length of the craze is on the order of 50 to 60 μm.

With respect to (the bottom of) Figure 4 we note first that the crack length deviation (from the average) is on the order of 30 to 50 μm, which is about the same as the (local) craze size. One is inclined to deduce thus, that the crack propagates locally in increments that are somewhat smaller than the craze size. However, the period or distance over which the associated nonsteady deviations occur are much larger, namely on the order of four to five times that dimension: This is evident by examining the peak-to-peak distances of the (spatially plotted) deviation record which yields typical acceleration/deceleration stages over 100 to 200 μm. It is further clear by comparing the deviation record with the lengths of the features on the fracture surface that the acceleration/deceleration stages correlate with these larger features.

As the crack length reaches the 14200 μm marker the loading is changed from the previous level of 0.97 MPa $m^{1/2}$ to 0.8 MPa $m^{1/2}$ (corresponding to a drop in the amplitude from 0.8 to 0.51 MPa $m^{1/2}$) at a frequency of 0.5 Hz. The thin band across the surface represents the stationary crack front while the crack was arrested (during load change) and connotes the start of the new crack propagation. It is interesting to note that even though this band occurs at the end of a noncoherent[12] fracture surface, it represents a rather smooth and continuous band across the width of the specimen. The noncoherent appearing features on the fracture surface (bottom of Figure 4) are thus, nevertheless, the result of a rather continuous craze/crack front, with the major spatial variation being in the direction above and below the crack plane. As the crack begins to propagate at a slower rate and through the middle of a single craze, the fracture surface becomes smooth and, correspondingly, the amplitude of the deviations decreases considerably. When the crack attained a length of 15200 μm, the stress intensity factor was again raised, (to 1.05 MPa $m^{1/2}$, $\nu = 0.5$ Hz) so that the crack reverts essentially to the previous propagation mode through different planes and different crazes (cf. bottom of Figure 4) and the corresponding effect can be detected in the deviation record.

Plotted also in Figure 4 are the average growth rates for the cases where the crack propagated from 14200 to 15100 μm and from 15200 to 15800 μm. In both cases the experiment is under load control and hence, the stress intensity factor is increasing as the crack propagates; however, the increases are only 4.2 percent for the first case and 2.8 percent for the second since the changes in crack length are small compared to the length of the specimen (40 mm). As the crack passes the 14200 μm marker the growth rate is about 0.05 μm/cycle and it starts to increase after the crack propagates about 100 μm; the change in growth rate (by a factor of 5 to 0.31 μm/cycle) occurs with no change in the loading.[13] It is clear from the (change in) appearance of the fracture surface that the change in growth rate comes from a change in the mode of crack propagation; the crack, which was previously propagating through multiple crazes, is now propagating through the middle of a single craze. Although lowering the stress intensity factor changes the mode of propagation, there is a history effect present which lasts from 2 to 3 craze lengths.

Although it is difficult to notice from looking at a narrow segment of the fracture surface, the crack front, which was leading at the 14200 μm marker, gets further ahead as the crack

[12] Because the markings on the fracture surface do not form single features throughout the thickness of the specimen these types of fractures are termed noncoherent. As we illustrate later on, and as has been reported in the literature, when the crack/craze system propagates in a retarded mode, the features (markings) on the fracture surface run through the thickness of the specimen. This type of behavior will be referred to as coherent fracture.

[13] 100 μm of crack propagation (0.25 percent of the length of the specimen) amount to about 0.5 percent increase in stress intensity factor in a load control experiment.

reaches the 15100 μm marker. The crack front (at the left edge of the photograph) increases its lead since, under load control, the stress intensity factor is larger for longer crack lengths; thus, if the crack front gets ahead, it has the tendency to move further ahead as it experiences larger stresses.[14].

An increase in the stress intensity factor as the crack reaches the 15200 μm marker changes the growth rate from 0.31 μm/cycle to about 2.5 μm/cycle; however, a change in the mode of crack propagation (single to multiple crazes) quickly reduces the growth rate to about 0.6 μm/cycle as the crack propagates for about 250 μm. The reduction in the growth rate occurs even though the stress intensity factor is slowly increasing (at about 0.5 percent every 100 μm); after the 15500 μm marker the average growth rate increases at about 8 percent every 100 μm, reaching 0.73 μm/cycle at the end of the record. Using a value of 8 (Hertzberg, et al., 1979) for the exponent in a Paris-type relation between average growth rate and ΔK, a 0.5 percent increase (per 100 μm) in the stress intensity factor would produce a 4 percent increase (per 100 μm) in the growth rate. The fact that the measured increase in growth rate is higher than the expected one can be (qualitatively) accounted for from the fracture surface since (1) the measurement was made on the leading edge and (2) the crack front near the edge is propagating through a single craze while, towards the center of the specimen, it begins to propagate through multiple crazes. The fact that the growth rate (at the beginning of the test) decreases while the stress intensity factor increases may be accounted for by postulating that crack propagation through multiple crazes is a tougher mechanism (rather than propagating through a single craze) and by realizing that the increase in loading (from one test to the other) produced a change in the mode of propagation.

3.2. Effect of Changing the Applied Frequency (Medium Stress Intensity)

Having demonstrated that a mere change of 15 percent in the loading produces very significant differences in the crack propagation process as evidenced through the features on the fracture surface and the growth rate records, we examine next the effect of frequency. The combined feedback technique (Pulos and Knauss, 1993) is used to maintain the maximum stress intensity factor at 0.61 MPa $m^{1/2}$ (with a variation below 0.75 percent[15]) and an R-ratio of 0.1 as the crack propagates for about 15% of the length of the specimen. A five-section sequence at four different frequencies is employed according to the table below:

Table 1.

to $X = 12450\ \mu$m	$\nu = 1$ Hz,
$12400\ \mu\text{m} < X < 13400\ \mu$m	$\nu = 3.2$ Hz,
$13400\ \mu\text{m} < X < 14900\ \mu$m	$\nu = 1$ Hz,
$14900\ \mu\text{m} < X < 16800\ \mu$m	$\nu = 0.32$ Hz,
from 16800 μm on	$\nu = 0.1$ Hz.

[14] The crack front could also move ahead if there is mode III (out of plane) loading; however, we found other cases in which the crack front was symmetrical and moved by the same amount all along the front. In the next section, we discuss a case where an incoherent type of propagation changes the shape of the crack front.

[15] A factor contributing to the variation is the slight change in Young's modulus that PMMA exhibits as it is tested at different frequencies.

A photograph of one of the fracture surfaces covering 88 percent of the specimen thickness is shown in Figure 5. The most striking features are the six wedge shaped markings (incoherent fracture) that appear at a crack length of about 12400 μm as the frequency is changed from 1 Hz to 3.2 Hz. For these loading cases the incoherent surface structures have dimensions along the direction of crack propagation that are larger than the craze length. For the subsequent discussion we focus mostly on the largest of these 'wedges' that formed near the left of the image in Figure 5. A higher magnification view of both fracture surfaces as viewed in color through optical microscopy is shown in Figures 6 and 8.

Starting at the crack length of 11000 μm the crack length increases by 1400 μm under a frequency of 1 Hz. The fracture surface (bottom of Figure 6) has a smooth appearance and the hue in the color photographs is very uniform but depends, of course, on the angle at which the fracture surface is viewed[16] as observed by Berry (Hertzberg et al., 1979) when he pointed out that the colors were produced by light interference occasioned by a layer of oriented fibrils (i.e., the broken craze). Because both contiguous surfaces show the same colors and because they are very uniform over areas much larger than the craze, one concludes that the thickness of both layers is approximately the same and that the crack is breaking through the middle of the craze as it propagates.

The crack and craze front at the end of the first (1 Hz) part of the sequence are shown at the bottom of Figure 6. At that point the frequency is changed to 3.2 Hz for a further crack length increase of 1000 μm.[17] As soon as this higher-frequency is imposed isolated wedge-shaped regions develop from discrete points across the specimen as shown in Figure 5. Close examination of these regions reveals complementary features inside the wedge regions. Locations where lines and features are sharply defined on one surface[18] correspond to places on the opposite surfaces that have a soft or out-of-focus appearance[19] (e.g., compare the places labeled AA' and BB' in Figure 6) indicating complementary 'hills' and 'valleys': Inside these regions the crack breaks the fibrils at different heights within the craze (meandering crack (Hull, 1973), see Döll (Berry, 1961); in the extreme case the crack breaks the craze at the craze-bulk interface leading to a 'mackerel' pattern, as first observed by Murray and Hull in Polystyrene (Murray and Hull, 1969; Doyle 1982) and by Doyle in PMMA (Lu et al., 1982)); these regions co-exist with regions of stable crack growth through the middle of the craze.

Though some of the wedges widen, some begin to contract already during the 3.2 Hz phase, the widest one remains nearly constant in the subsequent 1 Hz phase. The material outside the wedge shaped region has the same appearance (and hue) on both surfaces which indicates that outside these regions the craze is breaking at the middle as was true for the 1 Hz loading. There are, clearly, two mechanisms operating literally side by side: One fracture path develops within the wedge starting from a disturbance and grows into the wedge feature at the expense of the mid-craze fracture process. This process has the nature of an instability which, generally, grows until all the wedges interface and cover the surface completely.[20] The reason(s) for this wedge evolution are not clearly established, but apparently a significant structural disturbance

16 In this case, changing the magnification of the microscope objective changes the angle of the light path and of the hue since each objective has a different numerical aperture.

17 At three different crack lengths (12750, 13175 and 13275 μm) the test was interrupted for a brief amount of time; the arrested crack front left a marking on the fracture surface.

18 Well-defined color fringes appear next to the lines.

19 No fringes are present in this area and the colors are different.

20 Of the three largest wedge regions (left middle and right) the two at the edges continue throughout the second 1 Hz test portion, during which the left one decreases in width from 430 to 375 μm.

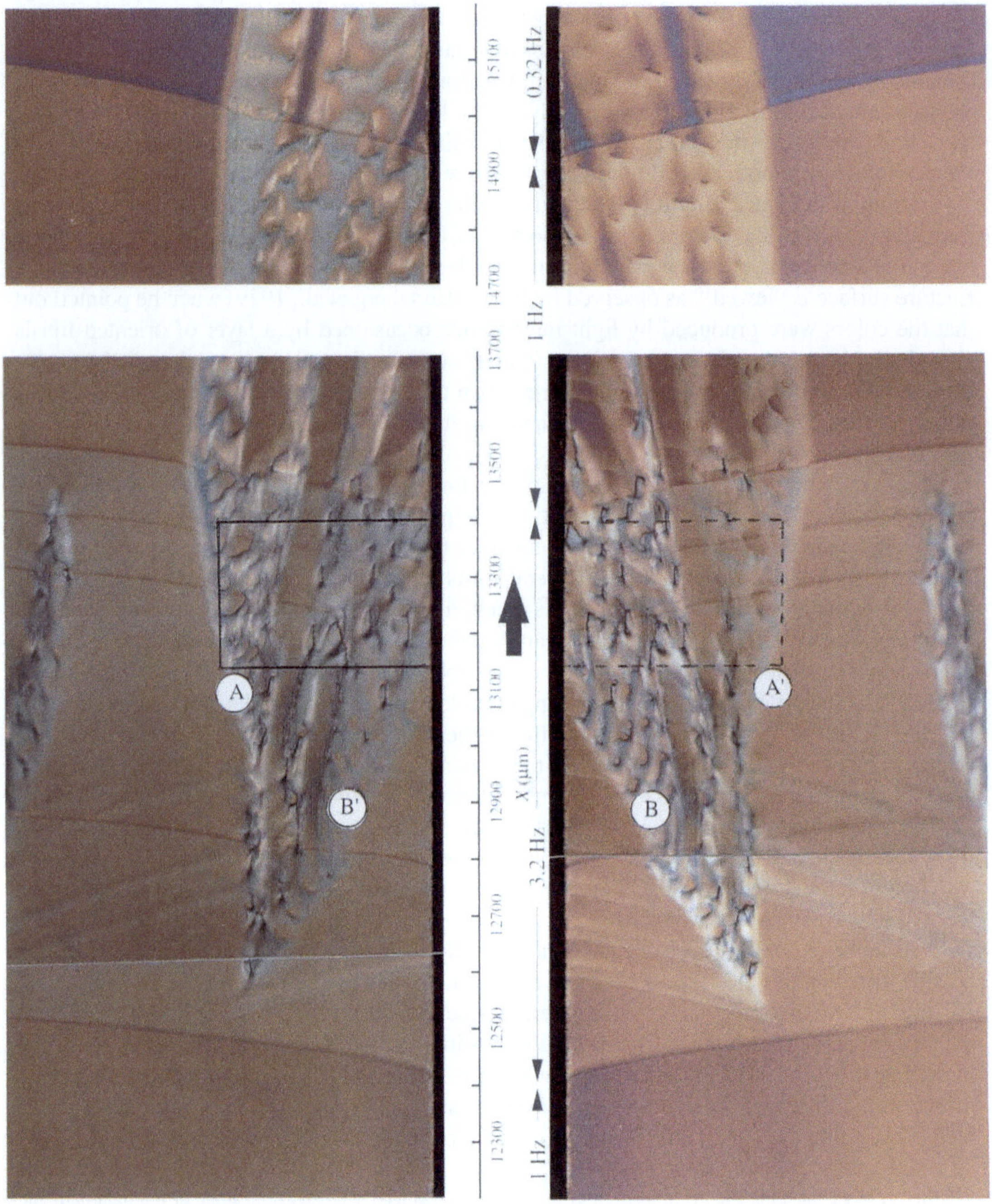

Figure 6. Fracture surface morphology at a constant stress intensity factor (0.61 MPa $m^{1/2}$) and test frequencies of 1 Hz, 32 Hz, 1 Hz and 0.32 Hz (crack propagation in the direction of arrow).

Table 2. Comparison of propagation rates and process time with imposed frequency. Letters refer to positions in Figure 5.

Location	Frequency Hz	$\bar{a}$ μm/cycle	$\dot{a}$ μm/sec	Process time 35 μm/adot (sec)
B-D	3.2	0.24–0.28	0.76–0.88	40–46
A, E	1	0.32–0.38	0.32–0.38	90–110
F	0.32	0.5–0.72	0.17–0.24	146–206
G	0.1	0.83	0.083	422

as well as the deformation frequency or associated craze process times have a controlling influence. To see this, it is instructive to determine the time it takes a material point to travel the length of the craze zone (on average) as illustrated in Table 2, where both the advances per cycle and the time-rate of growth are compared with the frequency (at constant stress intensity). While the advance per cycle ($\bar{a}$) changes moderately as a function of frequency, it is clear that the rate on a per second basis or the process time give a more distinct correlation with the frequency, since at a constant stress intensity the craze is approximately constant. Thus the time for processing the craze break-down has been reduced from about 100 seconds (1 Hz) to about 40 seconds (3.2 Hz) for the wedges of multi-level craze fracture to form.

This process appears to be reversible, possibly with some 'hysteresis', as illustrated in greater detail in Figure 6 for the most pronounced feature: After the frequency is changed back to 1 Hz (after about 1000 μm of spreading and at a crack length of 13400 μm) the wedge evolution terminates and is slightly reversed (cf. Figure 5) so that the width of the formerly wedge-shaped regions remains now almost constant[21], although in the previous 1 Hz test segment the crack/craze system did not even develop these regions.[22] These conditions did not change for the duration of this test segment which produced an additional crack extension of 1500 μm (shown at the top of Figure 6). To achieve a reduction in the extent of these regions similar to their spreading the frequency must be reduced further to about 0.32 Hz (increase in the process time to about 170 seconds) as evidenced in Figure 8. Thus the intra-craze fracture mode instability, specifically its growth and diminution all fall within the relatively narrow band of one decade of the applied frequency.

Figure 7 is an SEM detail[23] from Figure 6. Figure 7a corresponds to the rectangle in the left (optical) micrograph of Figure 7.[24] The white lines in Figure 7a, oriented in the direction of crack propagation, correspond to the dark lines in Figure 6 from which the color fringes originate. The 12 or so features (lines) in Figure 7a have lengths between 20 and 100 μm[25], which correspond to 60 to 300 percent of the craze length. The lines represent ridges that separate terraces of different heights which give rise to the color fringes in the optical micrographs;

[21] A slight decrease in width from 430 to 375 mm does occur. Other features in the middle of the specimen show similar behavior.

[22] Out of the 6 wedge-shaped regions that formed through the thickness of the specimen, 1 wedge was formed in the last 200 mm of the first 1 Hz test; however, its' width remained at 50 μm until the 3.2 Hz test, where it quickly increased to about 400 μm.

[23] Photographs taken with an Scanning Electron Microscope.

[24] The markings are more easily identified from the mirrored (right) micrograph of Figure 7 (dashed rectangle).

[25] The length of the line at the bottom edge of the photograph is actually about 100 μm long, as measured from the optical micrograph.

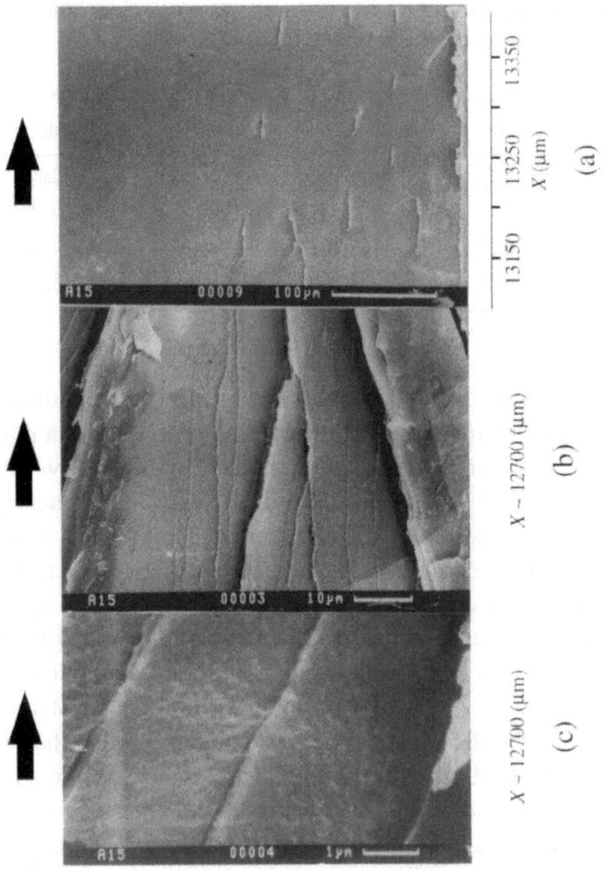

Figure 7. SEM photographs of wedge-shaped regions (crack propagation in the direction of arrow).

the higher resolution SEM photographs of Figure 7b and 7c indicate this terrace structure. Photograph 7c corresponds to a line, similar to those in 7a, that were formed when the crack length was about $12,700\,\mu$m; 7c is a detail of the left edge in Figure 7b: One concludes from these observations that in this region the craze does not rupture through its middle but at different height levels. Thus, as the craze breaks, regions on one fracture surface with longer fibrils complement regions of short or missing fibrils on the other surface.

As noted before, reduction in the width of the region fostering craze fracture in multiple planes occurred after the frequency was reduced to 0.32 Hz, increasing again the region of the normal and/or retarded mode (fracture through the middle of the craze) and in which the fracture surface is covered by bands which are shorter than the craze length. It is interesting to note that when these regions disappear (in a point, $X = 16400\,\mu$m in Figure 8) the fracture occurs through the retarded mode in which the fracture surface shows coherent or through-the-thickness bands spaced $28\,\mu$m apart. It is thus clear, that in this fracture process under a single loading history, fracture modes are developed that are (a) much larger than the craze zone and (b) are dependent on the initial conditions existing at the onset of the loading (on the past history).

A subsequent reduction in frequency to 0.1 Hz produces the same type of retarded crack growth with similar band spacing (about $26\,\mu$m in this case). Before analyzing this segment

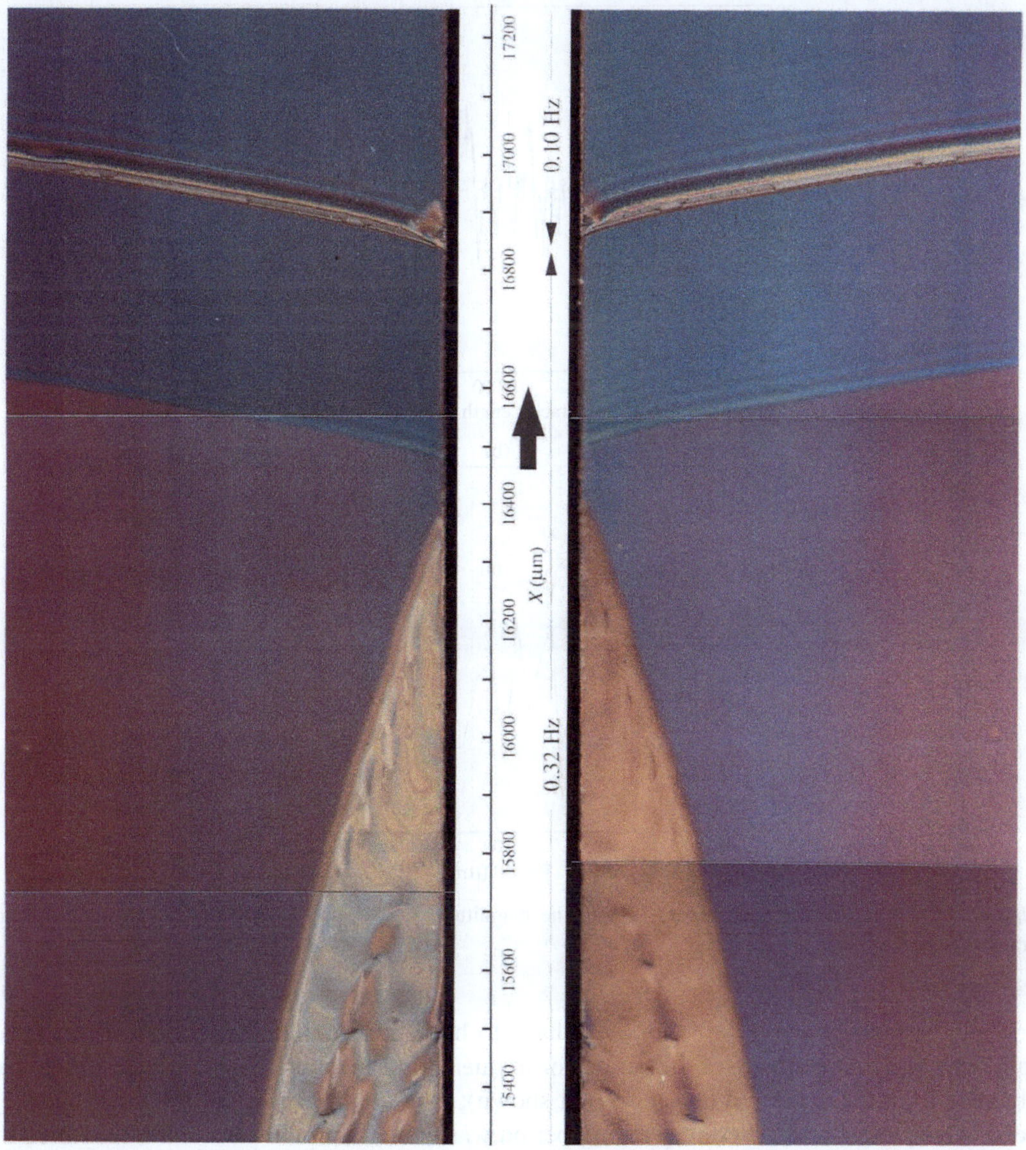

Figure 8. Fracture surface morphology at a constant stress intensity factor (0.61 MPa $m^{1/2}$) and test frequences of 0.32 Hz and 0.10 Hz (crack propagation in the direction of arrow).

of the test sequence in detail (in Section 3.3), we include here a Fourier analysis of data from that portion of the experiment during which the crack formed the wedge-shaped region to the time when the latter covered the whole thickness of the specimen.

The crack length deviation of an experiment in which features in the wedge-shaped regions have covered the whole width of the specimen is shown in Figure 9a which indicates that the amplitudes of the deviations are on the order of 10 to 20 μm. The main component of the low frequency noise introduced by the choice of fitted function (other than its possible harmonics) has been filtered by a low-pass filter in the frequency domain. The amplitude of the Fourier transform plotted against the applied period is shown in Figure 9b, in which

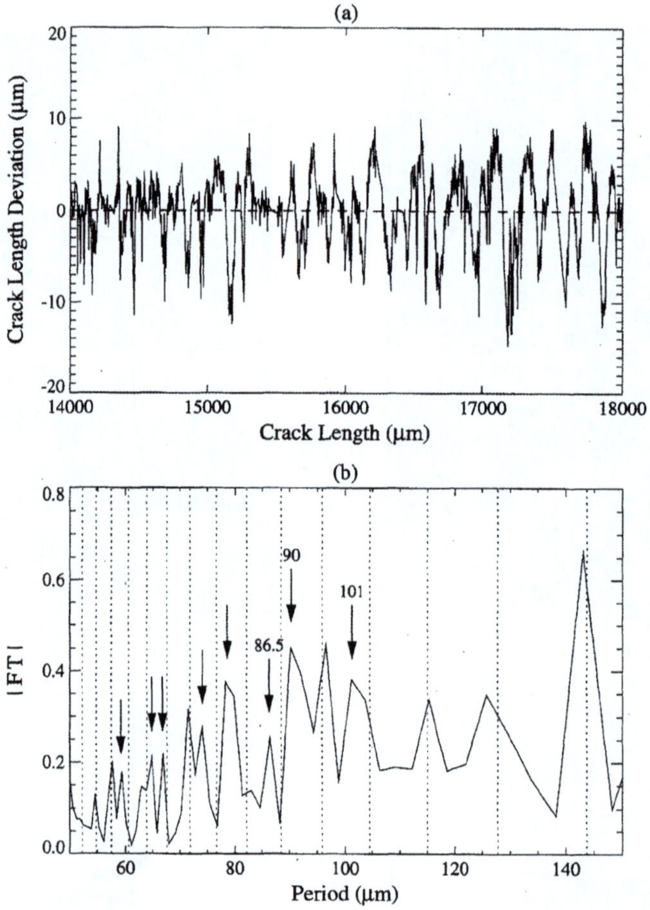

Figure 9. Crack length deviation record (a) and the magnitude of its Fourier transform (b) of an Incoherent Fracture (wedge markings) experiment.

possible harmonics of the low frequency have to be identified. This is accomplished in the following way: An estimate of the period (or frequency) of the noise introduced by the fit was obtained from the unfiltered deviation (not shown); then the harmonics of the low frequency noise were plotted and the corresponding period was varied until the harmonics, identified by the dashed lines, covered most of the peaks.[26] The remaining information has periods from about 60 to 100 μm which is very consistent with the features observed in the wedge-shaped regions.

We point out that subjective judgment was used in identifying the period of the low frequency noise (in this case 1150 μm) based on *a priori* knowledge of the length scales present on the fracture surface. In addition it was desirable to eliminate the higher noise harmonics so as to eliminate most of the peaks, including those at about 300 and 600 μm since these were clearly the second and first harmonics of the artificially introduced noise.

As a final note on the effect which the incoherent fracture features (wedges) have on the growth rate, it should be pointed out that incoherent crack growth is associated with reduced

[26] Thus the peaks not identified by arrows (which fall on dashed lines) are considered harmonics of the low frequency noise (e.g. peaks at 143 μm, 126 μm, 115 μm, etc.).

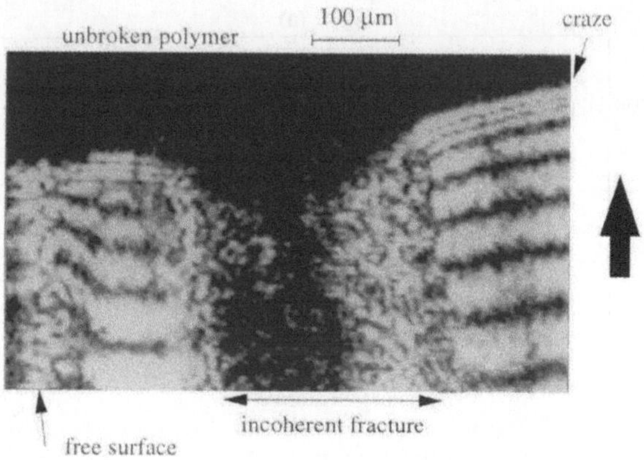

Figure 10. Interferometric fringes of the craze and crack opening displacement showing the slow down of the craze/crack front through an incoherent feature (crack propagation in the direction of arrow.

crack propagation rates when compared with coherent growth. Figure 10 illustrates an interferometric rendition of the craze/crack front (propagation towards top of figure) with a band of incoherent growth as indicated. It is clear that in the latter region the crack front lags behind that for the coherent growth. Moreover, one observes from Figure 5 that the crack front, which was leading on the left edge at $X = 11000\,\mu$m lags behind after the test ends ($X = 16800\,\mu$m) due to the formation of the (largest) wedge at the left edge. However, the difference in the velocity in this case is only on the order of 5 percent.

3.3. Constant stress intensity factor (low stress intensity)

We return, also for later reference (Pulos and Knauss, 1998, III), to the craze fracture under the low frequency history (0.1 Hz) indicated in the last section. Under these conditions the coherence (or through-the-thickness) instability leaves band markings on the fracture surface; the width of these marking are less than or equal to the length of the craze. These cases of retarded crack growth (RCG) are well documented in the literature (Hertzber and Manson, 1980; Döll and Könczöl, 1990). When the stress intensity factor and its time-rate are low enough to produce a single craze in which the fibrils brake (or disentangle) through the craze midplane, it is experimentally possible to record both the crack length on the surface of the specimen simultaneously with the (internal) crack/craze opening displacement. Since in this loading regime the crack/craze system propagates in a retarded manner these records are suitable to discuss the behavior of the crack/craze as it passes through (one of) the deceleration/acceleration periods.[27]; to this end consider the growth in terms of numbers of cycles instead of crack length advance.

The crack length record for a record sequence over 1500 cycles[28] is shown in Figure 11a which shows several occurrences of crack decelerations and accelerations or jumps. The jumps are best illustrated in the corresponding crack length deviation record of Figure 11b: As the crack decelerates the deviation decreases and as it accelerates or jumps the deviation increases. All of the jumps are about 5 μm in length excepting that just prior to cycle 1200

[27] The craze opening displacement as well as the stress distributions for this experiment are discussed in Pulos and Knauss (1998, III).

[28] The test frequency was 0.1 Hz, the maximum stress intensity factor was 0.46 MPa m$^{1/2}$ and the stress intensity factor amplitude was 0.35 MPa m$^{1/2}$.

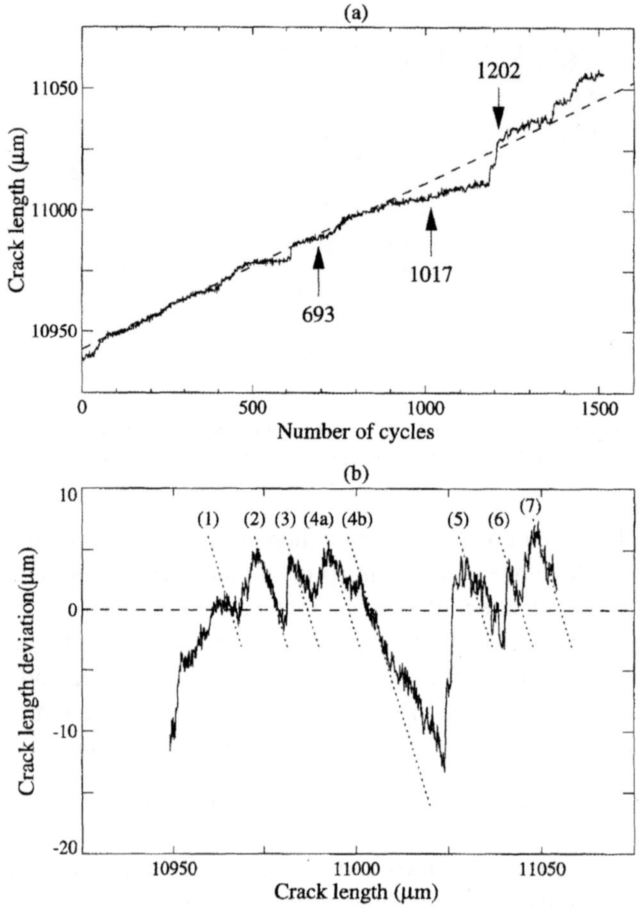

Figure 11. Crack length (a) and crack length deviation (b) records of an experiment where the crack jumped by 18 μm (RCG; Coherent fracture).

when the crack length increases by about 18 μm. The craze and crack tips move simultaneously during these deceleration/acceleration stages, but not in-phase. To demonstrate this note that at cycle 693 the craze length was measured[29] as 42 μm and decreased to 34 μm at cycle 1017; during the same time the crack length increased by 16 μm so that the craze tip advanced by only 8 μm. In contrast, between cycles 1017 and 1202 the crack tip passes through a deceleration/acceleration period in which the crack propagates about 6 μm through the deceleration period and 18 μm through the acceleration stage while the craze length only increased/decreased by about 5 μm. This particular growth history is examined in terms of craze profile and craze stress in (Pulos and Knauss, 1998, III).

To characterize the crack propagation just described in terms of normal/retarded[30] crack growth it is instructive to study the crack length deviation record (Figure 11b) in more detail.

[29] The craze lengths are obtained from the interferometric measurements of the craze/crack opening displacements as detailed in (Pulos and Knauss, 1998, III).

[30] Normal crack growth is a steady growth in which the crack length increases the same amount each cycle while the craze length does not change. In the retarded mode the crack remains stationary for a number of cycles while the craze lengthens and then, during one particular cycle, the crack breaks through the craze or part of it. For a discussion see (Hertzber and Manson, 1980; Döll and Könczöl, 1990).

There are seven distinct deceleration/acceleration stages which are labeled 1 through 7; at the beginning of each deceleration stage a line with a slope of -1 is plotted. A slope of -1 corresponds to the situation when the crack tip (trailing end of craze) stops completely: This is the case for the second, sixth and seventh deceleration stage. On the other hand, a vertical line (slope of + infinity) would correspond to a crack jump: Such is the case for the second acceleration stage. Thus, for the second stage, a 7 μm 'crack stop' has about an 8 μm period or spacing, denoting a retarded type of growth. However, for the stage 4b, the accelerating stage or 'jump' of 18 μm has a spacing or period of 27 μm; it is clear why the spacing is larger than the 'jump': the average slope of the deceleration stage does not quite reach -1, meaning that the crack does not completely stop, and the slope for the acceleration stage is not quite infinite, it takes about 30 cycles for the 'jump' to occur. In terms of the growth rate (see Figure 10a), the average speed of 0.07 μm/cycle is reduced to about 0.03 μm/cycle for about 270 cycles before propagating at an accelerated rate of 0.5 μm/cycle for about 30 cycles.

It is of interest to examine why a propagation event over distances smaller than the length of the craze can generate features on the fracture surface which are larger than the length of the craze: If the deceleration/acceleration stages or retarded growth occur on the tip of a moving crack, the spacing or periodicity of the surface markings can be larger than the craze length while the jumping or 'breaking-through-the-craze' is still confined to a portion of the craze.

Finally, to elucidate the relation of nonsteady crack growth to surface morphology, the crack growth record of Figure 11a has been scaled (and has had its axis inverted) so as to match an optical micrograph of the fracture surface; it is shown in Figure 12.

There are four (faint) markings in the fracture surface similar to those at the top of Figure 8 (constant stress intensity factor and varying frequency tests described in Section 3.2). The first marking has the higher contrast since it lies within the position of the craze prior to the start of the experiment and thus it may be associated with the end of the previous experiment (which was carried under similar conditions).

The second, third and fourth markings seem to be associated with the beginning of an acceleration period; by looking at the crack length record the impression prevails that the crack/craze system tended to propagate with decelerating and 'jumping' stages that had periods of a few microns, and when the crack/craze system failed to decelerate between cycles 800 to 900 (crack length of 11000 μm) and continued to propagate, the failure led to the large jump prior to cycle 1200. The behavior of the crack/craze system in this particular case is somewhat similar to the behavior observed by Lu, Qian and Brown (1991) on the retarded crack growth in Polyethylene pipes.[31]

4. Summary and Conclusions

The data suggests that under the prevailing loading conditions the crack propagates most of the time with measurable acceleration and deceleration phases, rather than exhibiting discrete and finite increments in length within the time frame of a cycle or less. Although there are instances when the crack appears to jump by a fraction of the craze length, the crack typically accelerates and decelerates temporarily rather than jumps. The deviations in the crack length record tracked on the side surface(s) of the specimen correlate well with the marking on the fracture surface and thus, the surface measurements reflect the process evolution inside the specimen very well. Even under constant stress intensity the overall growth of a crack is thus

[31] In that case the period of the oscillations is on the order of 1 mm.

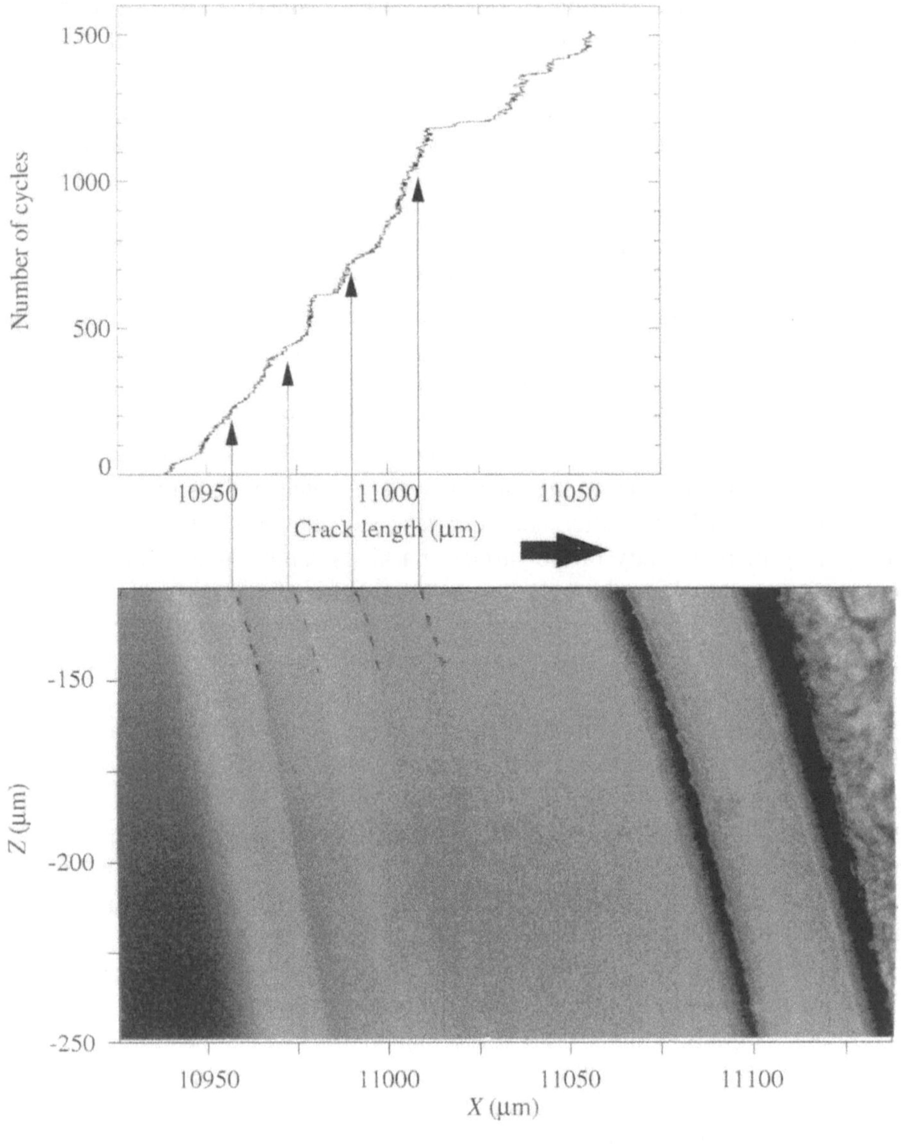

Figure 12. Crack length record (a) and fracture surface morphology (b) of an experiment where the crack jumped by 189 μm (RCG; Coherent fracture; crack propagation in the direction of thick arrow).

not a matter of distinct increments per cycle, but larger scale dimensions dominate. Typical propagation periods are on the order of 20–30 microns when the craze fractures through its middle.

High stress intensity failures are associated with multiple craze formations at the tip of a crack, which result in multiple sources of failure initiation to generate a rough surface with step-discontinuities. Sufficiently low stress intensity factors resulting in single crack tip crazes produce very smooth fractures surfaces, that lead to clearly observable accelerating and decelerating phases in the crack growth process. Inspite of the total smoothness of the fracture surface at the (sub-)micron size scale, the optically observed length scales associated with these deceleration/acceleration phases are typically on the order of two to four times the length of the (single) craze. This indicates that the crack propagation process depends on a rate dependent interaction between the bulk material surrounding the craze and the rate dependent response of the craze material and/or the changes occurring within it.

The craze with dimensions on the order of 30 to 40 μm is thus able to generate length scales (periodicity and/or length of surface markings) on the fracture surface that are larger than its own length, whether a single craze leads the crack tip or when the crack front is associated with multiple crazes. In the present case these length scales range between 50 to 250 μm. One possible phenomenon responsible for these length scales derives from the coupling between the fibrillation at the tip of the craze and the disentanglement or rupture of the craze fibrils at the trailing end of the craze (crack tip): Schirrer (1990) discusses such possible coupling in his description of the 'oscillating growth' observed in PVC. He argues that if the craze stress distribution is nonlinear rather than constant or corresponding to a simple proportionality along the craze axis, the feedback loop between these two processes may become unstable so that the growth would exhibit oscillations. He conjectures on molecular rather than continuum mechanical grounds that high molecular weight PMMA would not exhibit such oscillating propagation. It turns out, however, that the material in this study is indeed medium molecular weight material and the traction distribution supplied by the craze[32] is clearly nonlinear and of the type he considers.

While Schirrer's observations thus seems to fit the observations made here, the details of the 'feed back loop' performance are not clear and will, most likely, not be understood until extensive computational models are studied. The basis for this oscillatory behavior of the crack propagation rests with the details of the stress or strain history to which the craze material is subjected. Further investigation into these 'instability' phenomena must be coupled, however, with an improved constitutive formulation for the bulk and craze materials, and in this way the present work contributes to the advance of knowledge for this further phase of crack/craze propagation understanding. The next paper in this series (Pulos and Knauss, 1998, III) provides a contribution in that direction.

It should be clear, however, that while, in the light of the retarded crack growth model, one may find it difficult to explain how the craze can generate markings or length scales that are larger than its own length, by superimposing this oscillating growth on the tip of a moving crack, the (spatial) periodicity of events can be larger than its own amplitude (a portion of the craze length). Obviously, history effects in developing the craze and its associated cohesive forces play a significant role here, and this is here most clearly observable at the low stress intensity levels. In addition, the larger structures of multiple craze initiation sites (off the crack plane), coupled with these time or rate dependent effects control the evolution of fracture markings that are larger than the craze length itself.

Correlations between the average growth rate, crack length deviations and the fracture surface suggest that:

(1) high resolution measurements made on the (outer) edges of the specimen reflect the crack propagation occurring inside the specimen; however, care must be exercised in interpreting the results since the measurement is an integrated effect of the crack front through the thickness of the specimen.

---

32 We shall see that this is the case in (Pulos and Knauss, 1998, III).

(2) It may not be possible to find a relationship between stress intensity factor and growth rate if different propagation modes operate simultaneously and these modes are stress, time and history dependent.

Acknowledgments

The authors wish to acknowledge various sources of support contributing to this work: ONR through grant # N00014-91-J-1427 with Dr. Peter Schmidt as the technical monitor; E.I. DuPont de Nemours Company through initial contact with Dr. Manuel Panar as well as the Newport Corporation. Mr. E. Caballero prepared the material for Figures 5, 7 and 12.

References

Berry, J.P. (1961). Fracture processes in polymeric material. 1. The surface energy of poly(methylmethacrylate), *Journal of Polymer Science* **50**, 107–115.

Boyle, M.J. (1982). Fracture of crazes by the propagation of interfacial stress waves, *Journal of Materials Science* **17**, 204–210.

Döll, W. (1989). Fractography and Failure Mechanisms of Amorphous Thermoplastics, in *Fractography and Failure Mechanisms of Polymers and Composites* (Edited by A.C. Roulin-Moloney), pp. 387–436, Elsevier Science Publishers LTD, New York.

Döll, W. and Könczöl, L. (1990). Micromechanics of Fracture under Static and Fatigue Loading: Optical Interferometry of Crack Tip Craze Zones, in *Advances in Polymer Science: Crazing in Polymers*, (Edited by H.H. Kausch), Volume 91/92, pp. 137–214, Springer-Verlag, Berlin.

Doyle, M.J. (1982). Fracture of crazes by the propagation of interfacial stress waves. *Journal of Materials Science* **17**, 204–210.

Elinck, J.P., Bauwens, J.C. and Homes, G., (1971). Fatigue crack growth in poly (vinyl chloride), *International Journal of Fracture Mechanics* **7**, 277–287.

Gonzalez, R.C. and Wintz, P. (1987). *Digital Image Processing*, 2nd Edition, Addison-Wesley, Reading.

Hertzberg, R.W., et al. (1979). Comments on 'A model of fatigue crack growth in polymers' *Journal of Materials Science* **14**, 1754–1758.

Hertzberg, R.W. and Manson, J.A. (1980). *Fatigue of Engineering Plastics*, 1st Edition, Academic Press, New York.

Hertzberg, R.W., Skibo, M.D. and Manson, J.A. (1978). Fatigue Crack Propagation in Polyacetal *Journal of Materials Science* **13**, 1038–1044.

Hull, D. (1973). The microstructure and properties of crazes, in *Deformation and Fracture of High Polymers* (Edited by H.H. Kausch, J.A. Hassell, and R.I. Jaffe), pp. 171–189, Plenum Press, New York.

Könczöl, L., Schinker, M.G. and Döll, W. (1983). Interference optical measurements during continuous and discontinuous crack propagation in fatigue loaded polymers, in *Fatigue in Polymers* pp. 4.1–4.10, (London).

Lu, X., Qian, R. and Brown, N. (1991). Discontinuous crack growth in polyethylene under a constant load, *Journal of Materials Science*, **26**(4), 917–924.

Murray, J. and Hull, D. (1969). Nucleation and propagation of cracks in polystyrene, *Polymer* **10**, 451–465.

Press, W.H. et al. (1986). *Numerical Recipes: The Art of Scientific Computing*, Cambridge University Press, Cambridge.

Pulos, G.C. and Knauss, W.G. (1993). Constant stress intensity factors through closed-loop control. *International Journal of Fracture* **63**, 101–112.

Pulos, G.C. and Knauss, W.G. (1993). Nonsteady crack and craze behavior in PMMA under cyclical loading: I. Experimental preliminaries, in this issue.

Pulos, G.C. and Knauss, W.G. (1993). Constant stress intensity factors through close-loop control, *International Journal of Fracture* **63**, 101–112.

Pulos, G.C. and Knauss, W.G. (1999). Nonsteady crack and craze behavior in PMMA under cyclical loading: III. Effect of load history on cohesive force distribution of the craze, in this issue.

Schirrer, R. (1990). Optical interferometry: Running crack-tip morphologies and craze material properties, in *Advances in Polymer Science: Crazing in Polymers* (Edited by H.H. Kausch), Volume 91/92, pp. 215–261, Springer-Verlag, Berlin.

Washabaugh, P.D. and Knauss, W.G. (1993). Nonsteady, periodic behavior in the dynamic fracture of PMMA. *International Journal of Fracture* **59**(2), 189–197.

International Journal of Fracture **93:** 187–207, 1998.

Nonsteady crack and craze behavior in PMMA under cyclical loading: III. Effect of load history on cohesive force distribution on the craze

GUILLERMO C. PULOS* and WOLFGANG G. KNAUSS
Graduate Aeronautical Laboratories, California Institute of Technology, Pasadena, California, U.S.A.
e-mail: wgk@atlantis.caltech.edu

Received 21 November 1996; accepted in revised form 15 July 1998

Abstract. Measurements of crack opening and craze profiles are made under a range of loading histories including cyclical deformations that lead to nonsteady crack propagation histories. Of particular interest is the comparison of the distribution of traction transmission of a newly formed craze relative to a cyclically stressed one as it approaches the slow-down phase. Real time, interferometric measurements provide precise and multiple craze profiles during individual cycles. Cyclic deformations reduce the stiffness of a craze in its center resulting in a stress drop as part of the craze strength evolution; also, its thickness changes nonuniformly during the acceleration/retardation phases of the advancing craze/crack. The implications are, that for quasi-statically formed crazes the craze material can be reasonably well characterized by a stress-strain relation, while that is no longer readily true for a cyclically deforming crack, since rate dependent characteristics of the craze and bulk material intervene. Cracks unloaded as part of cyclical deformation histories exhibit crack closure (compression) near the trailing end of the craze.

Key words: Polymer fatigue, time dependent crack growth, polymer fracture, evolution of cohesive forces.

1. Introduction

Ever since Berry's (1961) early studies the role of crazes in polymer fracture has been investigated as a key element to polymer strength. Gradual improvement in understanding the role and behavior of crazes (Hsiao and Sauer, 1950); Kambour, 1964; Lauterwasser and Kramer, 1979) have made significant contributions to characterizing their role in material behavior in general, and in fracture in particular. Döll and his colleagues (Weidmann and Döll, 1976, 1978, 1980, 1982, 1983, Schinker et al., 1984), as well as Schirrer (1990) and Hertzberg and Manson (1980) have paid special attention to its 'break-down' role in fatigue behavior as summarized in a comprehensive review article: A systematic study of craze behavior under cyclic loading was contributed by Döll who concentrated on examining the effect of molecular weight on fatigue response of PMMA in terms of craze behavior (Weidmann and Döll, 1976). His review articles (Döll, 1983; Döll and Könczöl, 1990) still stand as an excellent summary of the state of the art on this particular topic.

Following leads from fatigue fracture of metals it has become customary to examine fatigue fracture (crack propagation) as if this were a steady state process in the sense that a crack propagates a given length for every cycle. In this context, crack-tip craze formation under fatigue crack propagation was investigated by Lang et al., (1984); while frequency had a

* Currently at the Instituto de Investigaciones en Materiales-UNAM,Mexico

pronounced effect on craze dimensions, it was found that the R-ratio did not affect craze parameters and that craze formation was governed by the maximum stress intensity factor. Moreover, there remain questions as to whether fatigue crack propagation in polymers is best 'correlated' on a 'per cycle' basis or relative to the time (frequency) at elevated load. Our own studies (Chang, 1983) with elastomers showed clear preference for the latter (time at load) basis, which was corroborated for structural polymers by Gregory and Botsis (1991). While this questions may be of secondary value in the evaluation of certain polymers for engineering purposes, the same cannot be said when basic understanding for materials engineering is important. In this context improved comprehension of any potential evolution of time dependent craze behavior, in particular any instability, is required, and this elucidation is the aim of the present study.

Motivated by this background the deformation of craze material at the tip of a crack is studied while the whole specimen is subjected to cyclical loading providing mode-I deformations. Interferometric measurements are used to determine real time craze profiles and crack openings with maximal resolution. These measurements are performed for ten time intervals while the crack tip undergoes deformation through a complete load cycle. Resulting displacements are then used to compute associated stresses and reaction-tractions on the craze/crack boundary. In these computations the idea of a primordial craze thickness is important. In the simplest notion this dimension specifies the 'gage length' of the deforming craze material so that an extensional strain distribution can be defined along the length of the craze. It will be seen that this concept is only partially successful, namely when the craze is newly formed; however, when the craze has been worked cyclically the definition of this thickness becomes less clear.

Studies of this type require that one be able to observe the crack with surface measurements as are made conventionally, and through observations that illuminate the crack front profile through the thickness of the specimen. While the experimental arrangement to make the surface measurements have been discussed elsewhere (Pulos and Knauss, 1998, I) this work delineates the observations on craze behavior as well as the analysis required for the evaluation of the optical data recorded through the thickness of the specimen. In this way one can reconstruct the craze/crack profiles and deduce the cohesive forces at the crack tip during individual cycles. Prominent in the notion of the primordial thickness (primordial craze) is the concept or belief of a distinct difference in the index of refraction appropriate to the craze material and the bulk solid. These concepts are discussed in Section 2. To deduce the cohesive forces supplied by the craze across the crack front one draws on (linearly) elastic stress analysis delineated as in Section 3. With these tools available, the results for monotonic, slow loading and unloading are presented in Section 4.1, while the fatigue counterpart (during a cycle) is documented in Section 4.2. The paper concludes with an overall evaluation of the observations in Section 5.

2. Craze profile measurements

Important in the determination of the craze tractions is the concept of the 'primordial (craze) thickness'. Crazing is the transformation of continuous and maximal density bulk material into a fibrillated and void-rich material of notably lower density, usually on the order of one half (Lauterwasser and Kramer, 1979). Thus a strip of bulk material with thickness τ is transformed into a craze thickness of $2w$. The thickness of the bulk material (which fibrillates into the

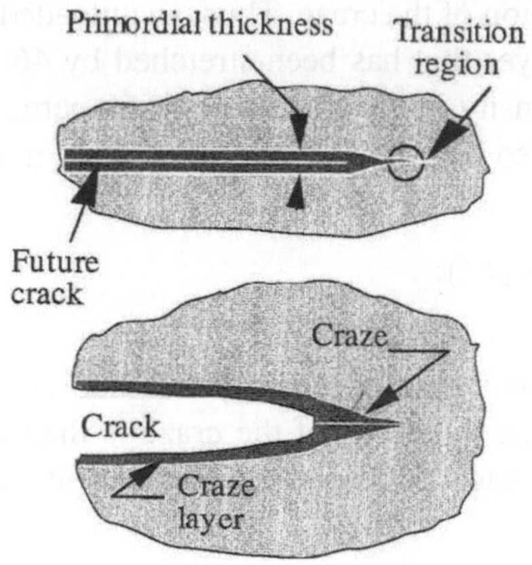

Figure 1. Primordial thickness defined.

craze) is called the 'primordial thickness'; a schematic representation is shown in Figure 1. According to the Lorentz-Lorentz equation the index of refraction n is a function of the density ρ *via*

$$\frac{n^2-1}{n^2+2}\frac{1}{\rho} = \text{const.}, \tag{1}$$

so that the bulk and craze material possess different indices of refraction; this allows total light reflection at the bulk-craze interface for interferometric purposes and thus the determination of the craze thickness. If one assumes that the craze extends without changing its in-plane dimensions then the density is inversely proportional to the stretch or thickness of the craze. Therefore, for crazes of width $2w(x)$ and constant index of refraction the Lorentz-Lorenz equation becomes

$$\frac{n^2-1}{n^2+2}2w(x) = \left(\frac{n^2-1}{n^2+2}2w(x)\right)_{\mathrm{REF}}, \tag{2}$$

where the reference 'state' is one of known index of refraction and thickness. In this section we examine briefly the consequences of Equation (2) and how it can be modified to accommodate an index of refraction that varies along the craze. A detailed discussion goes beyond the scope of this paper, and the reader is referred to additional discussion (Könczöl et al., 1990; Pulos, 1993).

Denoting the extension ratio representing the stretching of the primordial thickness to the current elastic–'plastic' boundary by Λ, one has

$$2w(x) = \Lambda\tau(x). \tag{3}$$

This equation may be applied, for example, to a deformed craze (where Λ may be a function of x) as well as to an unloaded craze. Using the primordial thickness (with index of refraction n_B[1]) as the reference state, substitution of (3) into (2) yields

$$\Lambda = \frac{n_{\mathrm{B}}^2-1}{n_{\mathrm{B}}^2+2}\frac{n^2+2}{n^2-1}, \tag{4}$$

[1] The index of refraction for the bulk polymer in PMMA is 1.49 (Kambour, 1964).

where n is the index of refraction of the craze. Thus, an unloaded craze with index of refraction $n_0 = 1.32$[2] is a fibrillated layer that has been stretched by 46 percent ($\Lambda = 1.46$) from the bulk; correspondingly, the density in the craze layer is 69 percent of that of the bulk polymer.

The unloaded craze can also be used as a reference state; in this case (2) yields

$$\frac{n^2-1}{n^2+2}2w(x) = \frac{n_0^2-1}{n_0^2+2}2w_0(x), \tag{5}$$

where $2w_0$ is the thickness of the unloaded craze. In addition, the index of refraction may be assumed to be constant; the thickness of the craze is then proportional to the number of extrema m of its interferometric fringe pattern and the index of refraction in the craze is obtained from

$$n^3 - \left(\frac{n_0^2+2}{n_0^2-1}\right)\left(\frac{m}{m_0}\right)(n^2-1) + 2n = 0 \tag{6}$$

with m_0 being the number of extrema of the unloaded craze. Under the assumption of a constant index of refraction throughout the craze, one can thus obtain the index of refraction of the craze, its extension ratio and reconstruct the elastic–plastic boundary for each fringe pattern; in addition, the primordial thickness can be obtained from (3).

On the other hand, if the index of refraction is not constant throughout the craze, one must assume that the primordial thickness $\tau(x)$ is known. To reconstruct the elastic–plastic boundary in this case for each fringe pattern[3], one proceeds as follows: Let the index of refraction be piece-wise constant along the craze so that

$$2w(x_i) = 2w(x_{i-1}) + \frac{\lambda}{4}\frac{1}{n(x_i)}, \tag{7}$$

where $n(x_i)$ is the (constant) index of refraction between x_i and x_{i-1} and combining (3) and (4) one writes a cubic equation for $n(x_i)$:

$$n(x_i)^3\left(\frac{2w(x_{i-1})}{\tau(x_i)} - \frac{n_B^2-1}{n_B^2+2}\right) + (n(x_i)^2-1)\frac{\lambda}{4}\frac{1}{\tau(x_i)}$$
$$-n(x_i)\left(\frac{2w(x_{i-1})}{\tau(x_i)} + \frac{n_B^2-1}{n_B^2+2}\right) = 0. \tag{8}$$

Using (7) and (8) one can reconstruct the opening profile starting at the craze tip ($2w(x_0) = 0$) continuing toward the crack tip where the x_i's are the location of the fringe extrema (for further detail see (Pulos, 1993)).

The above procedure requires the knowledge of the primordial thickness. One way to obtain this is to assume that at least one profile in the set corresponds to a constant index of refraction throughout the craze; then, (6) and (4) are used to obtain the index of refraction and the extension ratio while the primordial thickness is obtained from (3). For later reference it should be noted that in fatigue experiments it is common to choose the lower load limit in

[2] As it turns out Kambour later measured the index of refraction of the broken craze layer (Kambour, 1964) and found it to be equal to 1.32, the same as the unloaded craze.

[3] Using monochomatic light of wavelength λ.

the cycle low enough to assume that the index of refraction in the craze is constant and equal to that of the 'unloaded' craze (Könczöl et al., 1990). This practice is not necessarily valid if significant (nonlinearly) viscoelastic phenomena play a role.

3. Calculation of stresses

Once the boundary between the crazed and the bulk material has been constructed and the primordial thickness is known, a stress and deformation analysis yields the forces across the craze assuming that the deformations of this boundary result from an elastically deforming bulk medium consistent with a net opening displacement profile $v(x)$ of the craze boundary

$$2v(x) \equiv 2w(x) - \tau(x) = \text{fnc}(\sigma_{\text{CRAZE}}(x)). \tag{9}$$

The calculation of stresses follows the analysis of Ungsuwarungsri and Knauss (1988) with minor modifications: The coordinate system is located at the crack tip, the stress distribution $P(x)$ is piecewise continuous and the craze length s is considered to be small with respect to the crack length $a = c - s$.[4] In this case

$$2v(x) = \frac{2}{\tilde{E}}\left[\sigma_\infty\sqrt{2sc}\sqrt{1-x/s} - \frac{1}{\pi}\int_0^s P(T)\ln\left|\frac{1+\sqrt{\frac{1-x/s}{1-T/S}}}{1-\sqrt{\frac{1-x/s}{1-T/S}}}\right|\mathrm{d}T\right], \tag{10}$$

where σ_∞ is the far field stress and $\tilde{E}$ is the reduced modulus ($\tilde{E} = E$ the Young's Modulus for plane stress and for plane strain, with ν being Poisson's Ratio). The stress intensity factor becomes

$$K_\text{I} = \sigma_\infty\sqrt{c\pi} - \sqrt{2/\pi s}\int_0^s \frac{P(T)}{\sqrt{1-T/s}}\,\mathrm{d}T. \tag{11}$$

Defining the (Dugdale) stress σ_D as $\sigma_\text{D} = \pi\sqrt{c/8s}\sigma_\infty = \sqrt{\pi/8s}K_\text{I}$ and the nondimensional craze stress $Q(T)$ as $Q(T) = \frac{P(T)}{J_\text{D}}$ the (local) stress singularity is eliminated if

$$\int_0^s \frac{Q(T)}{\sqrt{1-T/s}}\,\mathrm{d}T = 2s. \tag{12}$$

Finally, defining $2v_\text{D} = \frac{K_\text{I}^2}{\tilde{E}\sigma_\text{D}} = \frac{8s}{\pi\tilde{E}}\sigma_\text{D}$ (10) becomes

$$2v(x) = 2v_\text{D}\left[2\sqrt{1-x/s} - \frac{1}{2s}\int_0^s Q(T)\ln\left|\frac{1+\sqrt{\frac{1-x/s}{1-T/s}}}{1-\sqrt{\frac{1-x/s}{1-T/s}}}\right|\mathrm{d}T\right]. \tag{13}$$

To deduce a stress distribution along the craze from the measured displacements one writes Equation 13 for a particular stress distribution in terms of a sequence of magnitude parameters and then performs a least-squares fit (LSF) to find their value. In this study we divide the craze into N (equal) segments and allow the stresses within each segment to vary linearly.[5]

[4] c is the location of the craze tip.

[5] In 1985 Imai and Ward used a bi-linear stress distribution across the craze in their fatigue study in PMMA (Imai and Ward, 1985).

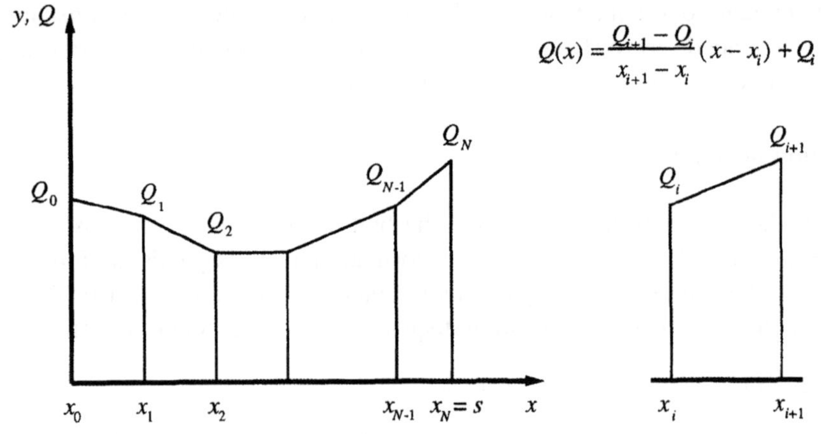

Figure 2. Discretization of the stress along the craze.

A trapezoidal stress distribution is shown in Figure 2. The integrals in (12) and (13) can be evaluated explicitly when $Q(T)$ is constant or proportional to T. The details for obtaining a final expression of the displacement in terms of the parameters that characterize the stress distribution are carried out in reference (Pulos, 1993). The number of segments N, as well as their location, for characterizing the stress distribution along the craze is chosen independently of the number and position of the extrema along the craze portion of the fringe pattern. Data from both the craze opening displacement as well as the crack opening displacement is used in the procedure since both opening displacements depend on the craze stress distribution.

One would expect from Saint-Venant's principle that the crack opening displacement will be independent of the details in the craze stress distributions as the point of observation is moved far away from the craze. However, the crack opening displacement is sensitive to the craze stresses near the crack tip. Thus, it is advantageous to use also the crack opening displacement near the craze for determining the stress distribution along the craze. In this study crack opening displacements within 3 craze lengths of the crack tip are used in the LSF procedure.

It is possible to obtain an expression which gives the stress along the craze directly in terms of the craze displacements; Wang and Kramer developed such a method based on a distributed dislocation analysis (Wang and Kramer, 1982a) and used it to study CO_2-induced crazes in polystyrene (Wang and Kramer, 1982b). The expression for the stress involves an integral of displacement gradients over the craze region only; while it is advantageous to obtain stresses without using a LSF procedure, ignoring the displacements from the crack tip region reduces the experimental data available for the calculation. For example, in an environmental craze study (Schirrer, 1988) which ignored the displacements outside of the craze zone, Schirrer pointed out the need for interpolation in light of the small number of (experimental) points that define the craze profile.

It is also possible to obtain stresses using the Finite Element Method (FEM). Bevan (Brevan, 1981, 1982, 1983) uses a 2-dimensional analysis wherein the applied (far field) stress and the experimental displacement of the craze profile are used as inputs to the FEM program. It appears that only the displacement from the craze region is used as a boundary condition in the analysis.

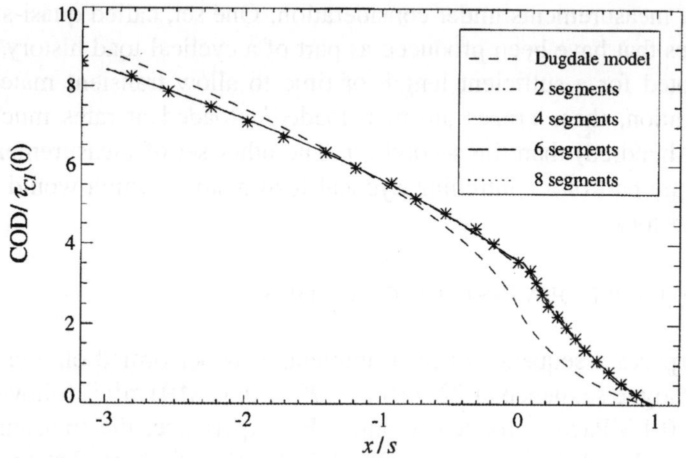

Figure 3. Profiles for different number of segments in the craze stress distribution as the craze is unloaded to $K_I = 0.339\ \text{MPa}\,\text{m}^{1/2}$.

Here the least squares fit is performed with respect to the stress values bounding the intervals. Starting with the Dugdale model (one segment with constant stress) one proceeds with $N = 1$ up to $N = 10$. An example of this procedure is illustrated in Figure 3 where the experimental and calculated crack/craze opening displacements (COD[6]) are normalized by the primordial thickness at the base of the craze.[7] The experimental data are the asterisks while the calculated data are represented by the different lines. The horizontal axis represents the distance (x) normalized by the craze length s; the craze lies between $x/s = 0$ and $x/s = 1$. The crack tip is located at zero and the crack extends over $x/s \leqslant 0$. Excepting the Dugdale model it is difficult to distinguish among the displacements represented by the stress distribution with different numbers of segments. The final choice is based on the smallest number of segment with the best or satisfactory fit to the measured displacements.

It should be pointed out that the stress distribution that corresponds to the profile shown in Figure 3 exhibits compression at the base of the craze (Figure 9a, discussed in Section 4.1); one of the effects of the compression is the jump in slope of the opening displacement that occurs at the crack tip ($x/s = 0$). The compression is also the reason why the Dugdale profile, which results from uniform tension of the craze, is so different from the other profiles.

4. Results and discussion

This section examines profiles and craze stress distributions obtained in near-compact tension specimens with equipment and procedures as described in (Pulos and Knauss, 1998, I).[8] There

[6] We use the term COD to denote the craze and crack opening displacement to avoid confusion between $2v(x)$ which is the opening displacement and $2w(x)$ which denotes the position of the elastic–plastic boundary. Thus, we have $\text{COD} = 2v(x) = 2w(x) - \tau(x)$.

[7] The primordial thickness at the base of the craze $\tau_{CI}(0) = 0.435\ \mu\text{m}$ is a convenient number for nondimensionalizing the COD. The nondimensional COD at the crack tip gives the average extension ratio of the craze.

[8] The specimens are 32 mm wide, 40 mm long and 3 mm thick. The dimensions do not correspond to the standard specimen as it was decided that longer specimens would allow more tests to be carried out on a single specimen. The material used was commercial PMMA from Rohm and Haas (Plexiglas G).

are two groups of measurements under consideration: One set, called quasi-static conditions, results from crazes that have been produced as part of a cyclical load history, which however has been interrupted for a sufficient length of time to allow transient material response to disappear; in addition, these crazes are then loaded/unloaded at rates much smaller (by a factor of ten to a hundred) than the second set. The other set of measurements is conducted on crazes which are part of a continuing cyclical load history, which would be typical of an ongoing fatigue history.

4.1. THE CRAZE UNDER QUASI-STATIC CONDITIONS

For reference purposes a sequence of measurements was performed on a craze that resulted from a previous fatigue sequence at 2 Hz ($0.1 < K < 0.35\,\mathrm{MPa\,m^{1/2}}$) followed by a constant loading at $K = 0.1\,\mathrm{MPa\,m^{1/2}}$ for ten minutes. By experience, the transient characteristics of the craze material had disappeared during this 'rest' period, so that the craze could be considered a 'settled' but not totally unloaded craze[9]. Two loading-unloading sequences were then imposed: The first with a maximum of $K = 0.3\,\mathrm{MPa\,m^{1/2}}$ (less than the previous maximal stress intensity) and extending over a period of 2 to 4 minutes, during which four craze profiles were recorded, two during loading and two during unloading. In the second case the craze was loaded to $K = 0.5\,\mathrm{MPa\,m^{1/2}}$ (50 percent higher than the maximum of the fatigue loading) and loaded/unloaded over a time of about two minutes.

A slow loading-unloading cycle with reduced stress intensity: Figure 4a[10] shows a comparison of the measured and computed[11] craze profiles for the first sequence. In this case a primordial craze thickness can be determined reasonably well which varies along the craze length, being thinnest at the craze tip and thickest at the crack tip as shown in Figure 5. One can then define a strain (stretch) of the craze material as a function of the position along the craze, as shown in Figure 4b (craze opening contour divided by the thickness of the primordial craze).

It is clear from Figure 4b that the extension ratios in the region close to the craze tip are higher on loading and the beginning of unloading than in the trailing end of the craze, while tending to small values at the craze tip proper. Moreover, it is clear that although there are history dependent effects operating on the craze material, at the present deformation rate ($\nu = 0.01$ Hz) the craze has a surprisingly large component of recoverable response.

Figure 6 presents the cohesive force distribution for these four cases. It is apparent that (a) the maximum stresses occur at the front of the craze where the material has undergone minimum 'working' and (b) the larger variation in deformation occurs near the middle of the craze but towards the leading end; (c) upon loading, the fibrils in the trailing half of the craze are stretched nonuniformly with the portion nearer to the craze middle providing some slack in the generation of a cohesive traction. One may interpret this behavior at the trailing end of the craze as the result of more pronounced fibril stretching by molecular orientation

---

[9] Upon measuring the stiffness of a PMMA craze (Schirrer, 1987), Schirrer detected an increase in the stiffness of the craze when it is left unloaded for 400 s. In this sense, the craze has not settled as re-entanglements may be increasing the 'knot' density; however, in this context, we refer to a settled craze as one where the elapsed time allows a change in the craze such that a primordial thickness can be defined and the craze parameters can be obtained.

[10] In all figures the craze tip is located at $x/s = 1$ and the trailing end (crack tip) at $x/s = 0$.

[11] For computations elastic analysis is employed throughout.

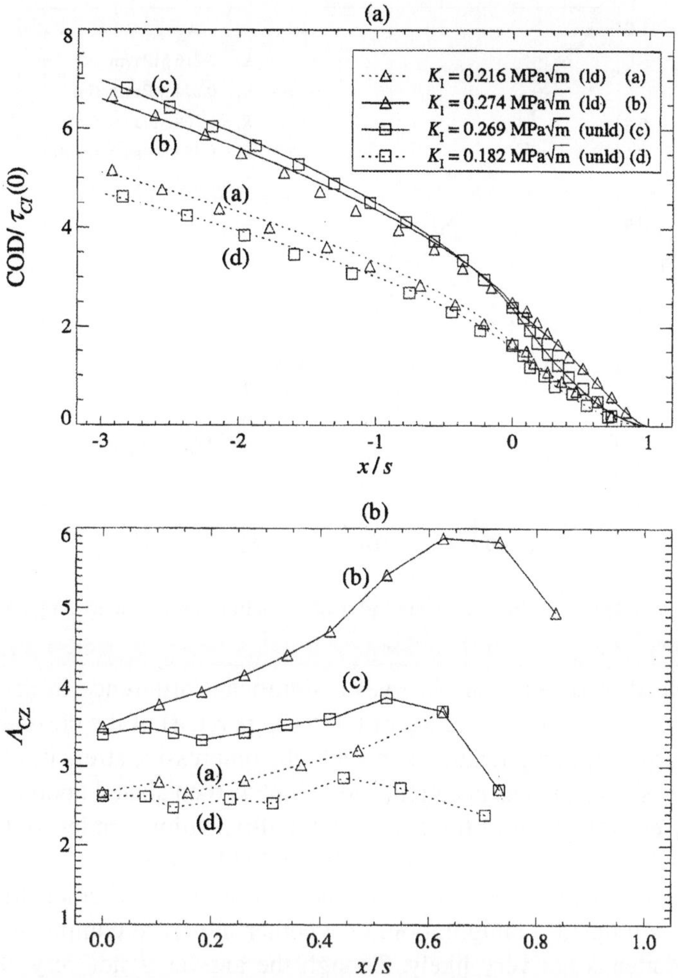

Figure 4. Measured and computed crack- and craze-opening profiles for a loading and unloading sequence in quasi-static (slow) deformations (a) and fibril stretch along craze (b); note that the data absence for $0.8 < x/s < 1$ derives from the absence of interference fringe data in the craze net to the tip. Letters on curves in part (a) and (b) correspond to plots in Figure 6 ($x/s = 1$ is the craze tip).

than in the region closer to the craze tip. It will be seen in the later examination of the traction distributions during a complete load cycle (as part of a continuing load history) that the 'slack' region is somewhat characteristic of a craze that has been 'worked' for some time.

Unloading from an overload–'craze closure' and compression: The second set of quasi-static (unloading) measurements (overload) renders the craze and crack opening profiles in Figure 7a along with the numerically modeled counterparts, along with the primordial craze thickness depicted in Figure 8. The tight definition of the latter for all unloading stages allows a similarly concise computation of the stretch ratios shown in Figure 7b. In contrast to the corresponding data for the previous (lower maximal load) quasistatic history, the stretch ratios are now nearly invariant along the craze for all stages of unloading. The letters on the curves identify again the unloading stages in correspondence with Figure 9, which shows the corresponding stress across the craze. In comparing these stress distributions with those in

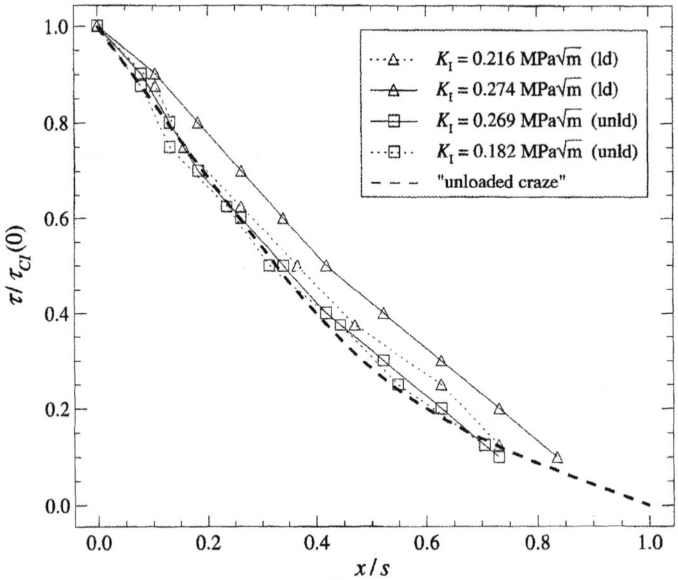

Figure 5. Primordial thicknesses using a constant index of refraction for a static loading and unloading case.

Figure 6 (no overload involved) one observes a significant difference in the fact that now a distinct compressive stress occurs over a part (20 to 25 percent) of the craze along the trailing portion. The, perhaps, surprising feature is that a high compressive stress has developed by the time half of the stress intensity has been reached, which is then abated upon further unloading. Compressive stresses at the base of the craze, with a distribution similar to that of Figure 9d, were found by Bevan et al. (1986) for an unloaded PMMA craze.

It is permissible to speculate here for discussion purposes on whether the craze 'buckles' or deforms into a 'crumpled' configuration, or whether the craze compresses uniformly like a continuum; the latter is not very likely. Though the answer is not very clear, the loading portions of Figure 6 provide additional input. It seems certain that upon (re)loading the compressed portion of the craze sustains higher stresses than the middle of the craze which would indicate at least that the trailing portion exhibits a higher degree of hysteresis than the rest of the craze. So much appears to emerge clearly, that the loading and unloading response of craze material does not represent trivial behavior.

4.2. THE CYCLED CRAZE

Real time measurements of the craze response are next considered for a complete cycle, which is one out of a large number applied continuously. In order to place these measurements into proper perspective relative to other work, we note that typically previous investigations have made observations at the maximum and/or the minimum of a cyclical load. Since the crack/craze does not propagate much during a cycle if at all (under the load amplitudes considered for studies of this type) one expects that steady state conditions prevail with respect to cycles. However, steady state conditions do not prevail if one follows the details of the deformations *during* a cycle; rather, one now addresses the truly transient behavior normally encountered in cyclical loading.

Again two variations are included which differ with respect to the naturally occurring acceleration/deceleration phases, but for neither of which the imposed loading was interrupted.

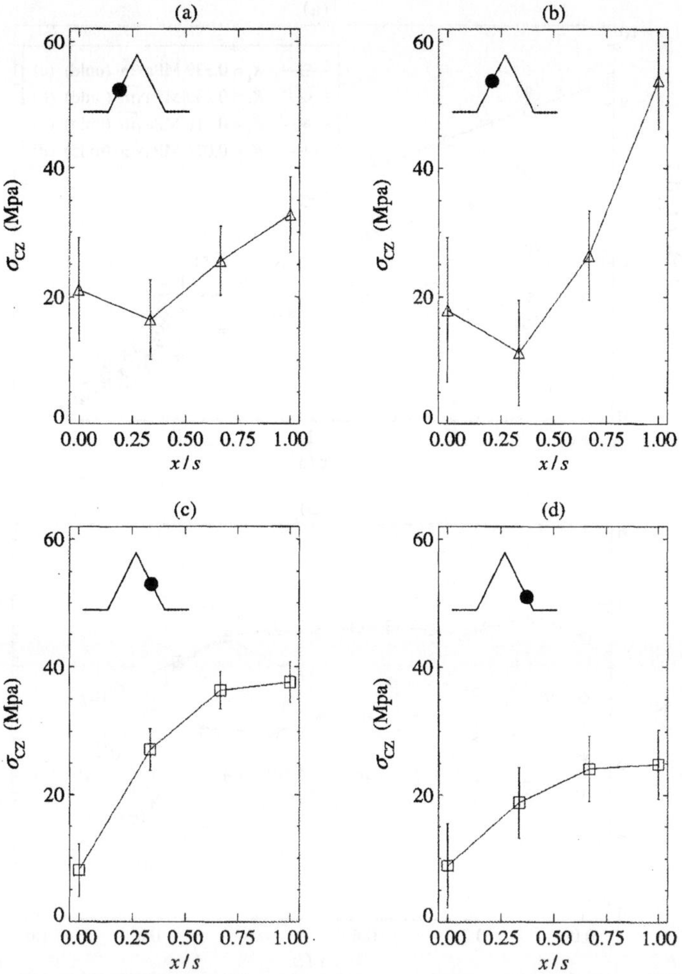

Figure 6. Cohesive force distribution for loading and unloading sequence in slow deformations. Letters on figures correspond to curves in Figure 4. ($x/s = 1$ is the craze tip).

The first set corresponds to a cycle embedded within others when no particular event occurred in the recent past or near future (it occurred at cycle # 693 in Figure 11a of Pulos and Knauss (1998, II)). The second were two cycles preceding and immediately following a major acceleration/deceleration event (cycles 1017 and 1202 in Figure 11a of Pulos and Knauss (1998, II)) respectively; the large deviation event with an 18 μm advance for the craze of 35 μm length). While the first case corresponds thus to a typical fatigue cycle, the latter represent a precursor of a pronounced propagation event and a relatively newly configured craze in which the craze properties should be quite different. In terms of load magnitude this loading falls into the category of 'low loads' for which very smooth fracture surfaces result. In the companion presentation (Pulos and Knauss, 1998, II) this type of crack propagation behavior has been classified as 'coherent'. In order to allow sufficient time to make eleven measurements bounding ten equal time increments the frequency needed to be held at $\nu = 0.1$ Hz.

Craze neither preceded nor followed by a special propagation event: The most distinctive difference from the earlier, quasistatic conditions is illustrated in Figure 10 where the prim-

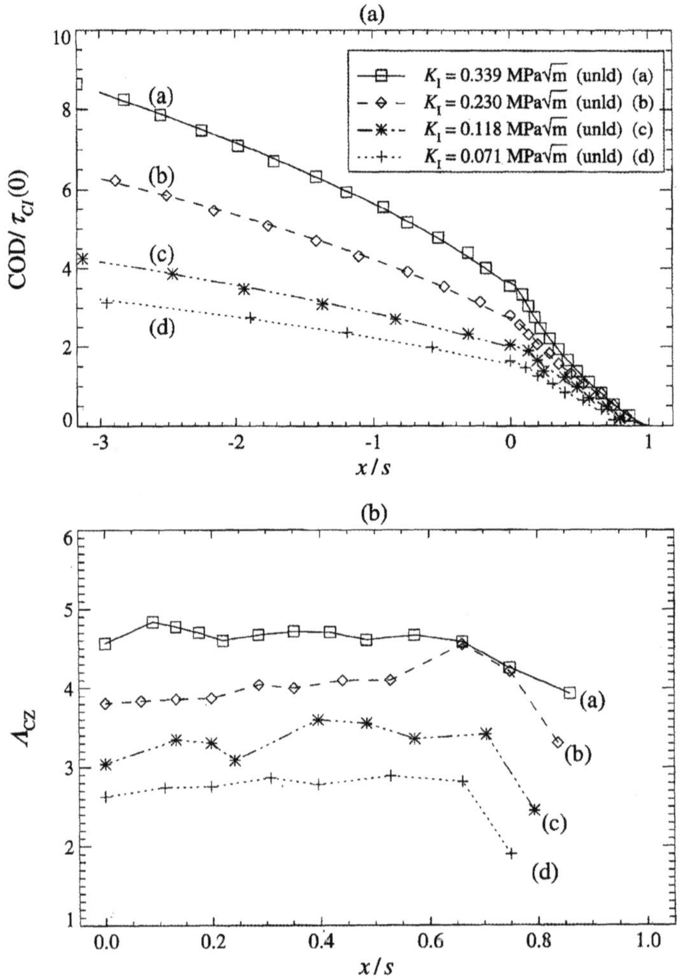

Figure 7. Measured and computed profiles (a) and extension ratios (b) for a static unloading case (letters identifying profiles and extension ratios correspond to stress distribution in Figure 9).

ordial craze thickness is shown for each portion of the cycle. Clearly, the traces for different increments vary too consistently to be attributable to statistical error. While the idea of the primordial thickness is to represent an (invariable, unloaded) craze configuration the data indicates a quantity that varies throughout a cycle: At the peak of the cycle the nearly linear profile towards the upper right in the figure results, while for lower portions of the load cycle the curves to the lower left are more appropriate[12]. Alternately, one could interpret the data by allowing that in the right half of the craze the primordial thickness vanishes during the nearly unloaded portion of the cycle. A direct consequence of this variation of the (apparent) primordial craze thickness is that a stretch ratio for the craze material cannot be defined uniquely by the methods used so far, and a 'stress-strain' behavior for the craze cannot thus be determined with confidence, either.

---

12 There are vestiges of this spread of the primordial thickness variation in the data for the quasi-static loading-unloading case represented in Figures 2 and 3, though that data is very much closer to that in Figure 4 than for the present case.

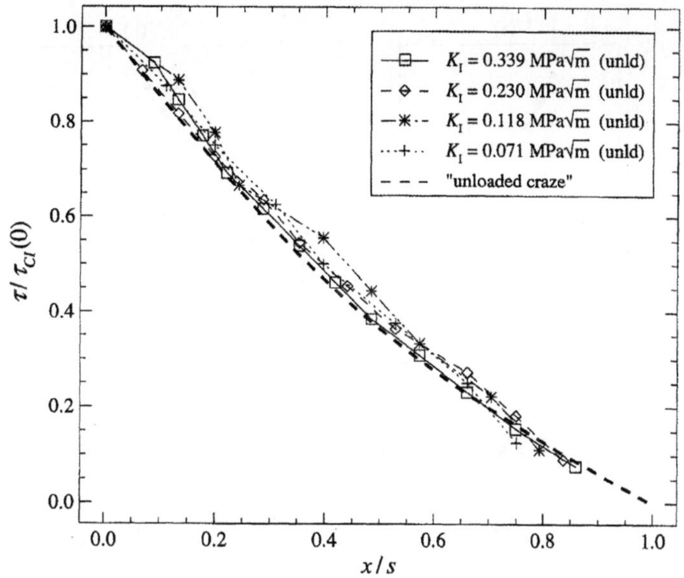

Figure 8. Primordial thickness at each measurement interval during a cycle (constant index of refraction).

The reason for this behavior, which is inconsistent with the earlier results as well as those of other investigators (who did not address the details of deformations during a single cycle), are not clear. Certainly, the assumptions underlying the deduction of the primordial thickness from optical measurements need to be questioned, since it seems unreasonable that the front half of the craze is optically similar to the bulk material (undamaged) during the low load portion of the cycle while having optical properties commensurate with the craze during the high load portion. For example, we have had to assume that the craze properties are invariant in the thickness direction so that the only reflection for interferometric purposes occurs at the bulk-to-craze boundary. On the other hand, if during part of a cycle the craze fibrils are 'buckled' or 'crumpled' a secondary structure of different density may form along the length of the craze as a result of the compaction at the low part of the cycle. On the other hand, there is no certainty relative to the time dependent effects by which the deformation occurs in the craze during a cyclic deformation and how these deformations produce the changes in the density and the index of refraction. Compared with the results in Section 4.1 only Figure 5 shows vestiges of this behavior, while the companion case of that section seems to agree well with all that is known about quasistatic deformations and associated craze configurations in the literature (Döll and Könczöl, 1990).

One possible source of this behavior may be associated with the optical properties of the craze. This possibility has been explored, but space in this presentation is insufficient to delineate details. The seemingly best choice resulted from accepting a primordial thickness represented by one half of the linear distribution in Figure 10 (maximal cycle load amplitude) and subsequent evaluations are based on that choice.

After much experimentation with potential average representation of the primordial craze for the behavior under steady state cyclic deformations it is possible to choose a single primordial craze configuration that does not conflict with the measured deformed craze profiles; this choice, depicted in Figure 11 (thick straight line denoted 'linear/2') along with selected curves

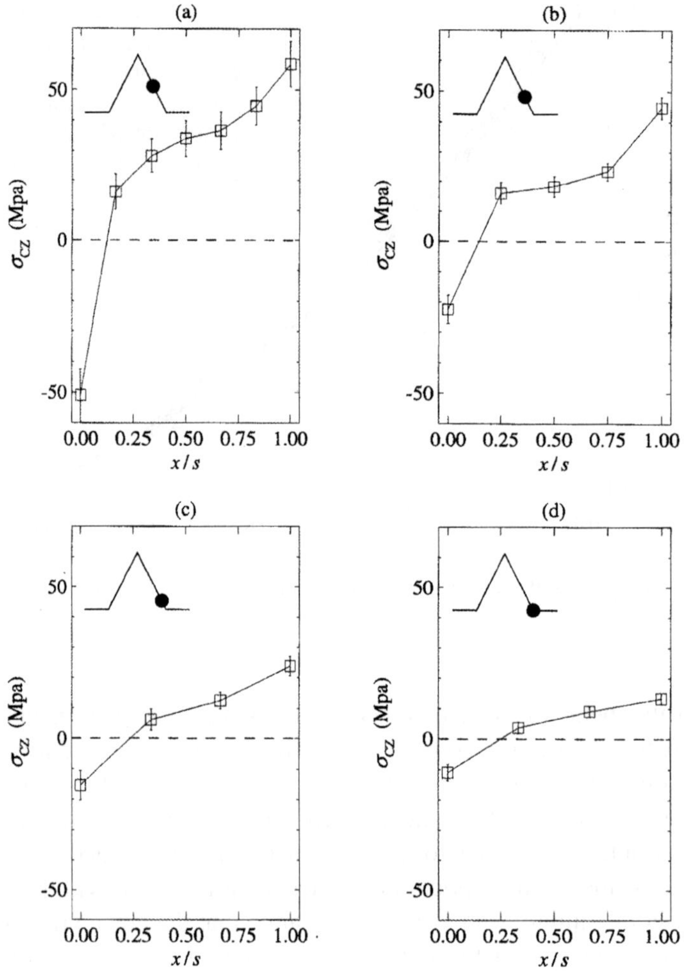

Figure 9. Cohesive force distribution in craze upon unloading from 'overload', showing compression at trailing end of craze. Letters on figures correspond to curves in Figure 7. ($x/s = 1$ is the craze tip).

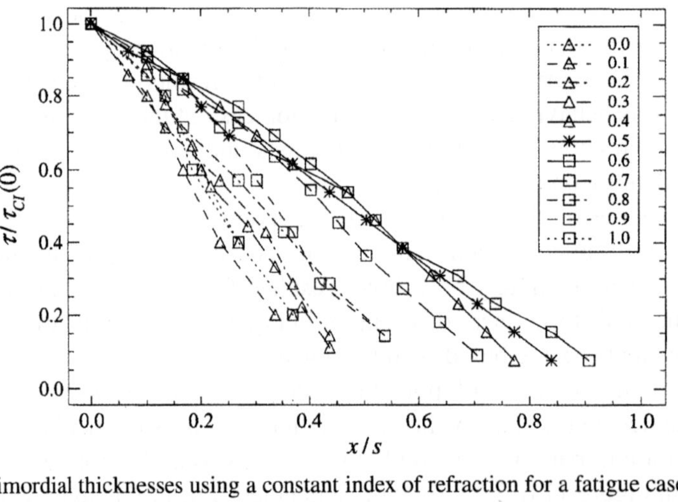

Figure 10. Primordial thicknesses using a constant index of refraction for a fatigue case (cycle 693).

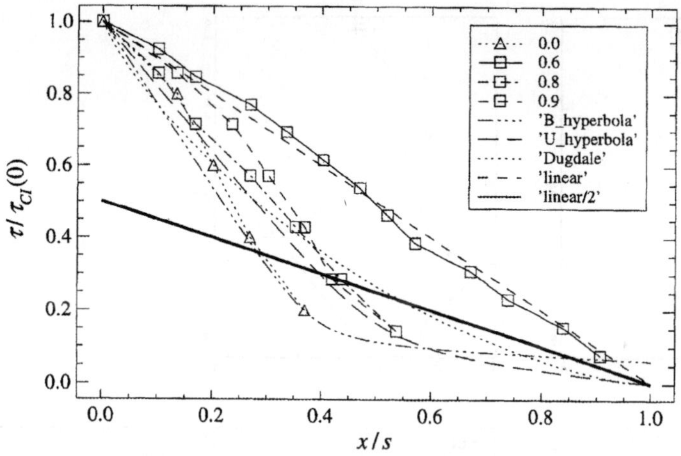

Figure 11. Definition of primordial thickness.

from Figure 10, is made for this and the following situations for reasons of consistency.[13] The heavy line in Figure 11 was thus used for estimating the stresses across the craze. While these stresses are subject to the uncertainties of the primordial craze definition, the points discussed for the craze tractions are reasonable[14] Once the primordial thickness is defined, all of the data for a single cycle are made to coincide with a single primordial thickness by allowing the index of refraction to vary along the craze.[15] Figure 12 shows the tractions for this sequence of loading segments. Two features are apparent: (a) the central to left portion of the craze develops a traction minimum during the highest portions of the cycle; (b) near the end of the cycle the trailing end of the craze undergoes mild compression. It appears thus that at least part of the craze breakdown involves tension/compression deformations of the crack-near portion of the craze.

Craze behavior during a deceleration and following an acceleration phase: In this section we examine the craze response during cycles 1017 and 1202, as well as the stresses during cycle 1202.

The primordial thicknesses for cycle 1017 during the deceleration phase of a deviation are very similar to those of cycle 693 except that the spread for the latter is slightly larger than that of cycle 693 (not shown), so that the analysis of cycle 1017 presents the same problem encountered before. However, after the crack passes through a deviation phase the primordial thicknesses are closer together as shown in Figure 13 and, except for the primordial thicknesses at the bottom part of the cycle (denoted by 0.0 and 1.0), they are amenable to defining a unique primordial thickness for the cycle. We note as part of the overall puzzle of the primordial craze definition that the primordial thicknesses for the low part of the cycle have now moved to the top right position in the diagram whereas, the high portion of the cycle is represented in the lower left of the figure, opposite to what the pre-deviation cycles 693 and 1017 rendered.

[13] There are constraints on how one chooses the primordial craze to avoid unreasonable negative displacements.

[14] Consideration of some of the other models for the primordial thickness yielded qualitatively similar stress distributions.

[15] See reference (Könczöl et al., 1990) for this type of calculation.

29

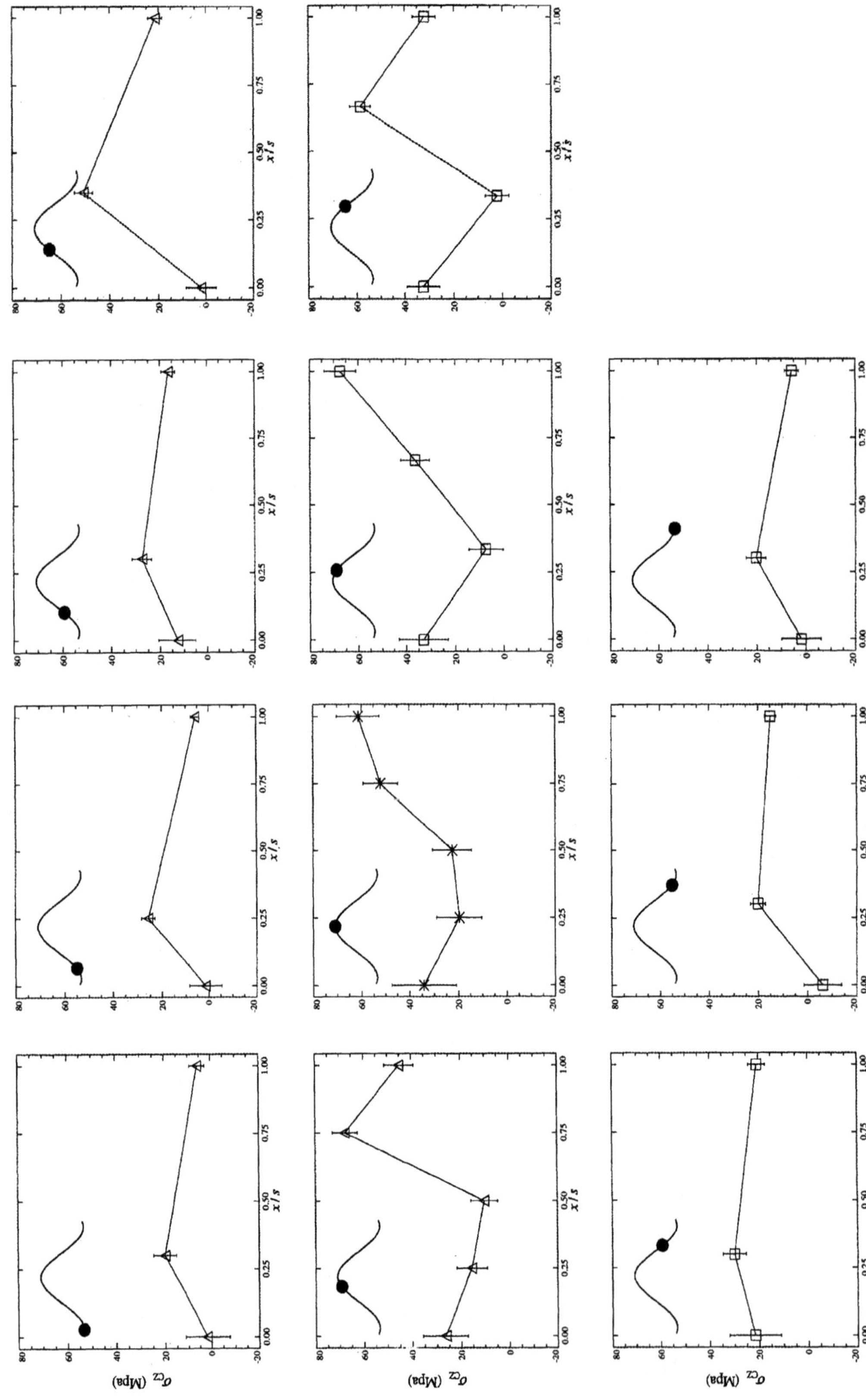

Figure 12. Stress distribution along craze for different load increments (cycle 693; before acceleration stage).

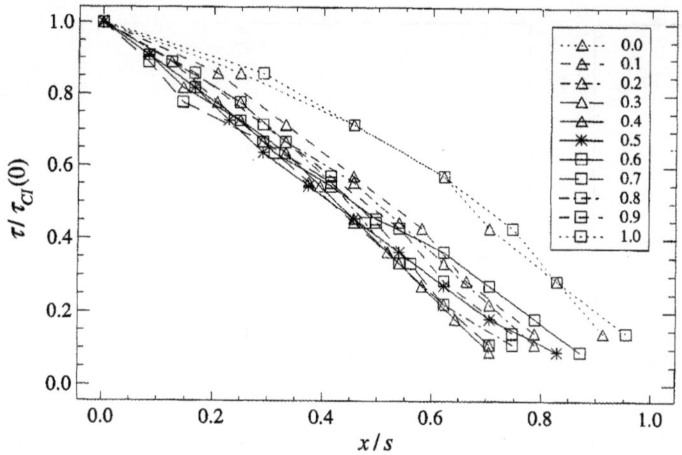

Figure 13. Primordial thicknesses using a constant index of refraction for cycle 1202.

The craze stresses during cycle 1017 are similar to those of cycle 693 and are not shown. The craze stresses during cycle 1202 are shown in Figure 14. The two main differences between these stress distributions for the 'new' craze and the previous ones are that (a) the 'sag' resulting from prolonged craze working in the trailing portion of the craze has disappeared so that there is little change in the distribution as the craze is loaded (or unloaded) and (b) that the distribution of stress is different at the bottom of the loading cycle: most of the craze is unloaded except for the tip (top left of Figure 14) whereas before the craze remained loaded (top left of Figure 12) at the bottom of the cycle.

Because it is not possible to measure a primordial craze in the present situations an extension ratio cannot be determined; a stress-stretch relation can thus also not be found. However, it is still possible to define a stress-displacement relation, for which a comparison between the pre- and post-deviation stages demonstrate a marked difference as illustrated in Figures 15 (cycle 693) and 16 (cycle 1202). Here the stress-displacement relations are given for all the increments of load during the two different cycles. (Note that the COD increases monotonically from the craze tip to the crack tip). While there exists substantial scatter, it is, nevertheless, clear that typical characteristics emerge; for the highest load levels these have been emphasized by the shaded bands or areas. The stress-displacement relation during cycle 1202 is almost linear with virtually no drop in the stress in or near the middle of the craze. Equally obvious is the fact that prior to the deviation process there are COD values spanned by craze material in excess of 1.5 μm; after the craze/crack has advanced by 18 μm this maximally stretched material is no longer present in the craze. However, the new stress for the range from 1 to 1.5 μm COD has increased over the previous value by about a factor of two, indicating a materially stiffened 'new' craze structure. This suggests that the fibrils have not sustained significant damage in the middle as a result of the fact that both crack and craze have propagated by a substantial amount (about 55 percent of the craze length) in about 50 cycles. Thus, the 'newly' formed craze material has had insufficient time to accumulate significant damage.

33

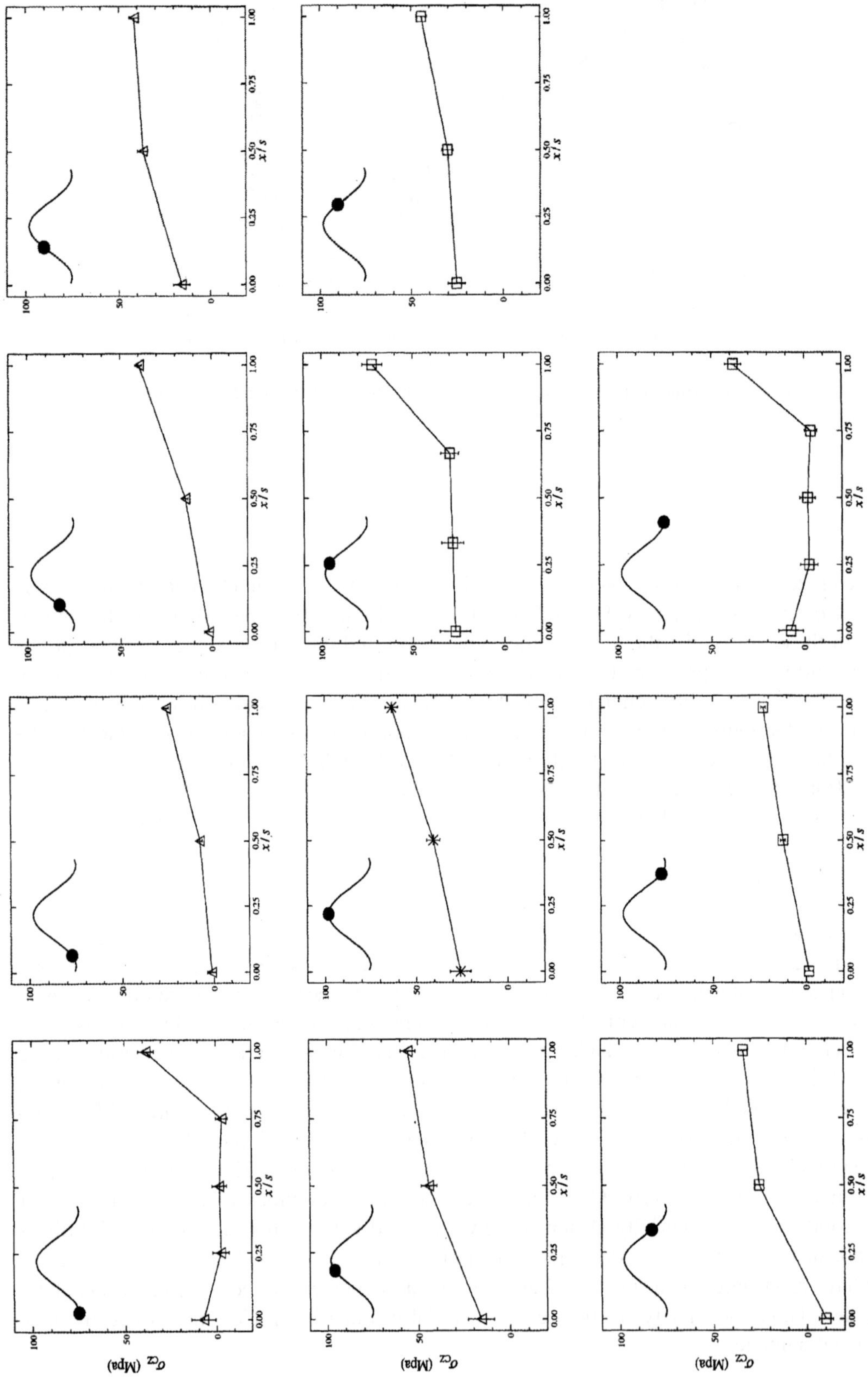

Figure 14. Stress distribution along craze for different load increments (cycle 1202; after acceleration stage).

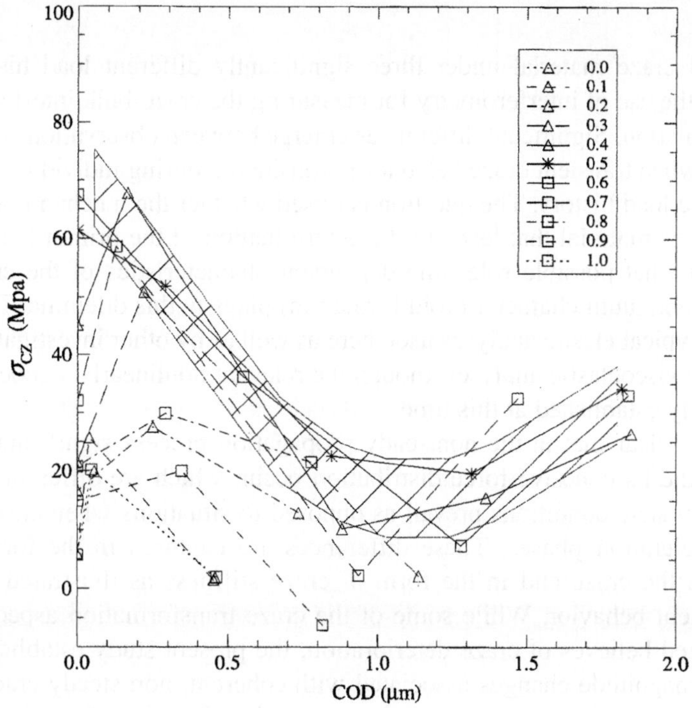

Figure 15. Stress-displacement relation during a fatigue cycle (693). The right portion of the COD scale corresponds to the crack tip.

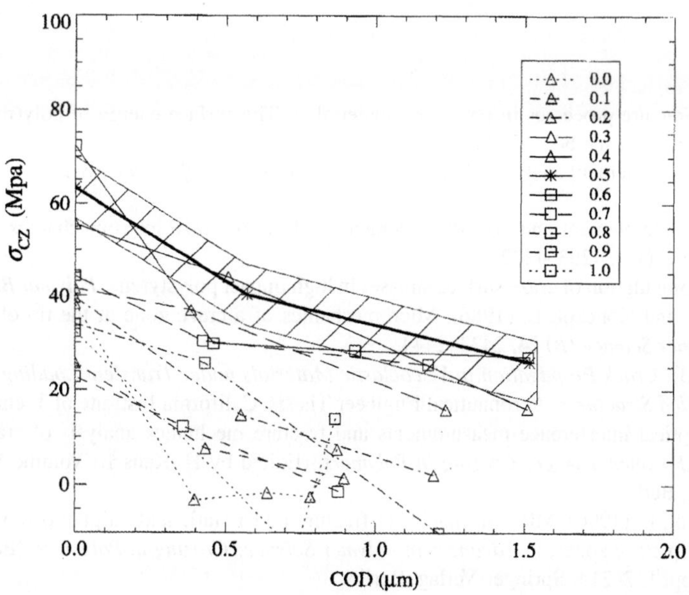

Figure 16. Stress-displacement relation during cycle 1202. The right portion of the COD scale corresponds to the crack tip.

5. Concluding remarks

The response of craze material under three significantly different load histories has been studied through the use of interferometry for measuring the craze-bulk interface deformations and traction estimation. Significant differences emerge between observations under quasistatic conditions, and when transient craze behavior is considered during individual cycles as part of an ongoing cyclic load history. The question is raised whether the mechano-optical characterization of the craze material that leads to the determination of the primordial craze thickness is adequate, and what possible role time dependent characteristics of the craze material as well as its non-continuum character (void formation) plays in this dilemma. The possibility is recognized that typical elastic analyses used here as well as by other investigators may have to be supplanted by viscoelastic analyses, though the relevant nonlinearly viscoelastic properties are not necessarily established at this time.

It has become clear that in the nonsteady propagation process significant changes in the craze structure and its cohesive force distribution occur, which are different in phases when essentially steady state conditions prevail as opposed to situations when crazes pass through deceleration/acceleration phases. These differences are captured in the form of stress distributions across the craze and in the form of craze stiffness as illustrated in terms of the stress-displacement behavior. While some of the craze transformation aspects confirm generally promulgated believes of craze deterioration, the present study establishes first bounds on the order of magnitude changes associated with coherent, non-steady crack growth under cyclical loading. In particular, the craze stresses tend to fall from the craze tip towards its trailing end (softening characteristic) with an intermittent increase (hardening) near the crack tip as the craze/crack prepares for a deviation event.

References

Berry, J.P. (1961). Fracture processes in polymeric material. 1. The surface energy of poly(methylmethacrylate). *Journal of Polymer Science* **50**, 107–115.

Bevan, L. (1981). Finite-element analysis of craze deformation. *Journal of Polymer Science: Polymer Physics Ed.* **19**, 1759–1768.

Bevan, L. (1982). Finite-element and boundary-element analysis of craze micromechanics. *Journal of Applied Polymer Science* **27**(11), 4263–4272.

Bevan, L. (1983). Computation of craze surface stresses in high impact polystyrene. *Polymer Bulletin* **10**, 187–190.

Bevan, L., Döll, W. and Könczöl, L. (1986). Micromechanics of a craze zone at the tip of a stationary crack. *Journal of Polymer Science (B)* **24**, 2433–2444.

Chang, S.K.Y. (1983). *Crack Propagation in Viscoelastic Materials under Transient Loading with Application to Adhesively Bonded Structures*. Aeronautical Engineer Thesis, California Institute of Technology.

Döll, W. (1983). Optical interference measurements and fracture mechanics analysis of crack tip craze zones, in *Advances In Polymer Science: Crazing in Polymers*. (Edited by H. Kausch) Volume 52/53, pp. 105–168, Springer-Verlag, Berlin.

Döll, W. and Könczöl, L. (1990). Micromechanics of fracture under static and fatigue loading: Optical interferometry of crack tip craze zones, in *Advances in Polymer Science: Crazing in Polymers*. (Edited H.H. Kausch), Volume 91/92, pp. 137–214, Springer-Verlag, Berlin.

Döll, W., Könczöl, L. and Schinker, G. (1983). Size and mechanical properties of craze zones at propagating crack tips in poly(methylmethacrylate) during fatigue loading. *Polymer* **24**, 1213–1219.

Döll, W., Seidelmann, U. and Könczöl, L. (1980). On the validity of the dugdale model for craze zones at crack tips in PMMA. *Journal of Materials Science* **15**, 2389–2394.

Döll, W., Schinker, M.G. and Könczöl, L. (1982). Direct measurements of craze zones at crack tips in glassy thermoplastics during fatigue crack growth, in *Deformation, Yield and Fracture of Polymers*, pp. 20.1–20.6, The Plastics and Rubber Institute, London.

Gregory, B.L. and Botsis, J. (1991). Experimental investigation of the effects of stress rate and stress level on fracture in polystyrene. *Journal of Materials Science* **26**, 1015–1026.

Hertzberg, R.W. and Manson, J.A. (1980). *Fatigue of Engineering Plastics*, 1st edition, Academic Press, New York.

Hsiao, C.C. and Sauer, J.A. (1950). On crazing of linear high polymers. *Journal of Applied Physics* **21**(11), 1071–1083.

Imai, Y. and Ward, I.M. (1985). A study of craze deformation in the fatigue fracture of polymethylmethacrylate. *Journal of Materials Science* **20**, 3842–3852.

Kambour, R.P., (1964). Refractive index and composition of poly(methylmethacrylate) fracture surface layers. *Journal of Polymer Science: Part A* **2**, 4165–4168.

Kambour, R.P. (1964). Structure and properties of crazes in polycarbonate and other glassy polymers. *Polymer* **5**, 143–155.

Könczöl, L., Döll, W., and Bevan, L. (1990). Mechanisms and micromechanics of fatigue crack propagation in glassy polymers. *Colloid and Polymer Science* **268**(9), 814–822.

Lang, R.W. et al. (1984). Craze development in poly(methyl methacrylate) during stable fatigue crack propagation. *Polymer Engineering Science* **24**, 833–842.

Lauterwasser, B.D. and Kramer, E.J. (1979). Microscopic mechanisms and mechanics of craze growth and fracture. *Philosophical Magazine A* **39**(4), 469–495.

Pulos, G.C. (1993). *Nonsteady Crack Propagation and Craze Behavior in PMMA*. Doctoral Thesis, California Institute of Technology, Pasadena.

Pulos, G.C. and Knauss, W.G. (1999). Nonsteady crack and craze behavior in PMMA under cyclical loading: I. Experimental preliminaries, in this issue.

Pulos, G.C. and Knauss, W.G. (1999). Nonsteady crack and craze behavior in PMMA under cyclical loading: II. Effect of load history on growth rate and fracture morphology, in this issue.

Schinker, M.G., Könczöl, L. and Döll, W. (1984). Micromechanics of fatigue crack propagation in glassy thermoplastics. *Colloid and Polymer Science* **262**, 230–235.

Schirrer, R. (1987). Interferometrical measurement of the craze stiffness and structure of the craze fibrils in PMMA. *Journal of Materials Science* **22**, 2289–2296.

Schirrer, R. (1988). Crazes in solvent vapours at very low pressure: craze stress and craze structure. *Polymer* **29**, 1615–1618.

Schirrer, R. (1990). Optical interferometry: Running crack-tip morphologies and craze material properties, in *Advances in Polymer Science: Crazing in Polymers*. (Edited by H.H. Kausch), Volume 91/92, pp. 215–261, Springer-Verlag, Berlin.

Ungsuwarungsri, T. and Knauss, W.G. (1988). A nonlinear analysis of an equilibrium craze: Part I. Problem formulation and solution. *Journal of Applied Mechanics* **55**(1), 44–51.

Wang, W.-c.V. and Kramer, E.J. (1982a). A distributed dislocation stress analysis for crazes and plastic zones at crack tips. *Journal of Materials Science* **17**, 2013–2026.

Wang, W.-c.V. and Kramer, E.J. (1982b). The micromechanics and microstructure of CO2 crazes in polystyrene. *Polymer* **23**, 1667–1674.

Weidmann, G.W. and Döll, W. (1976). Interferenzoptische Vermessung der Craze-Zone vor der Ri*β*spitze in PMMA Unterschiedlichen Molekulargewichtes. *Colloid and Polymer Science.* **254**, 205–214.

Weidmann, G.W. and Döll, W. (1978). Some results of optical interference measurements of critical displacements at the crack tip. *International Journal of Fracture* **14**, R189–R193.

International Journal of Fracture **93:** 209–226, 1998.

Crack growth behavior in an epoxy strip above and below T_g

KAZUO OGAWA[1], AKIHORO MISAWA[2] and MASAHISA TAKASHI[1]
[1]*Department of Mechanical Engineering, Aoyama Gakuin University, 6-16-1 Chitosedai, Setagaya-ku, Tokyo, 157-8572, Japan*
[2]*Department of Mechanical Engineering, Kanagawa Institute of Technology, 1030, Shimo-ogino Atsugi-shi, Kanagawa, 243-0292, Japan*

Received 15 December 1997; accepted in revised form 27 January 1999

Abstract. Crack growth behavior in a viscoelastic epoxy strip above and below the glass temperature T_g is extensively discussed by focusing on

(1) the threshold condition of crack growth;
(2) subsequent crack growth behavior; and
(3) the characteristics of fracture surface.

To realize good reproducibility of crack growth behavior that depends strongly on time and temperature, a new technique for introducing an initial crack simulating the mathematical model, i.e., a single plane crack with sharp tip, has been developed. Taking experimental evidence into account, an approach to the M_{ID} criterion, based on the displacement field around the crack tip for crack growth under monotonically increasing loading, is proposed. Two types of equations predicting crack growth rate are evaluated through the comparison with experimental results. As a measure of the critical condition of crack growth, an extended J-integral derived for linearly viscoelastic material, is discussed. The fracture surface is carefully investigated for morphological features of the submicroscopic structural change related with heterogeneous local deformation around the crack tip. Only stable crack growth of a single main crack is observed above T_g, while below T_g not only both stable and unstable crack growth, but sometimes branching to multi-cracks, is observed.

1. Introduction

It is well-known that the microscopic structure of polymeric materials, which consists of three-dimensional cross-links and entanglements of long molecular chains, is quite different from that of metallic materials. This type of structure can make not only their mechanical behavior dependent strongly on temperature and loading history, but also produce complicated fracture behavior. Thus, various difficulties still exist for the establishment of a unified understanding of fracture in this type of material.

The investigations of fracture and strength of polymeric materials are roughly classified into two categories. One is macroscopic treatment from the viewpoint of continuum mechanics (Knauss, 1970; 1974) and the other is microscopic study from a stand point of molecular morphology and polymer chemistry (Newman and Wolock, 1958). It seems, however, difficult and sometimes impossible to have discussions on the mechanisms and behavior of fracture combining the information obtained on different bases of sciences and with different scales of measurement. Thus, investigations based on semi-microscopic scale of measurement, e.g. μm$\sim$mm, are required in order to link together the results obtained in each category.

To make clear

(1) the threshold condition of crack growth,

(2) subsequent crack growth behavior, and
(3) the characteristics of fracture surface,

the authors have developed experimental techniques to obtain precise information on the fracture behavior of the material. Since the material used shows remarkable dependence on temperature and loading history, a new technique for introducing an initial crack simulating the mathematical model, i.e., a single plane natural crack with a sharp tip, which can realize good reproducibility of crack growth behavior, has been developed. In a theoretical aspect, the behavior of crack growth in a wide strip involving a semi-infinite crack is studied taking into account the pseudo-elastic treatment based on the extended correspondence principle proposed by Graham and Sabin (1973). A criterion for the crack growth threshold, i.e., M_{ID} criterion (Misawa et al., 1984), is discussed in terms of the displacement field around the crack tip. The authors investigate crack growth behavior in an epoxy strip above and below the glass temperature T_g. Two different criteria predicting crack growth rate, one proposed by Schapery (1975) and the other is the M_{ID}, are evaluated by comparing them with experimental results. As a measure of the critical condition of crack growth, an extended J-integral applicable to a viscoelastic strip having a crack (Misawa et al., 1990), is also discussed.

Studies on the morphological characteristics left on the fracture surface is another approach to identify some aspects of submicroscopic structural change related to heterogeneous local deformation in the vicinity of a crack tip. Regarding the characteristic features of the fracture surface above T_g, only wavy stream marks, of which the size and number are dependent on temperature, are observed over the whole area, while typical tongue marks covered with the similar wavy stream marks are seen below T_g. The characteristics of both cases are observed on the fracture surface formed during the period of stable crack growth. The shape and size of the tongue marks are also dependent on temperature below T_g.

2. Crack growth experiments

2.1. Preparations for crack growth experiments

2.1.1. *Initial crack and a moment constrained grip*

In crack growth experiments, the nature of the initial crack tip plays a definite role on subsequent crack growth behavior as well as on the condition for the onset of crack propagation, as pointed out in previous papers (Takashi et al., 1983; Ogawa et al., 1994). Inevitably, the initial crack front has to be very sharp and must be kept in the same plane during subsequent fracture processes in order to compare the results with those obtained by theoretical analyses. An initial crack with a sharp front is prepared by the following procedures:

(1) A razor cut of length 8 mm is naturally extended to 20 mm under a low constant displacement rate loading in a constant temperature chamber controlled to be slightly higher than T_g, in this case +8 K.
(2) The specimen is kept at a temperature in which the material shows rubbery behavior, for approximately 1 hour to remove residual stress and distortion.
(3) The prescribed temperature for the experiment has to be held at least 30 minutes in the case above T_g. With decreasing temperature, the duration of keeping the prescribed temperature is increased at least up to 60 minutes below T_g before the start of the crack growth experiment.

Figure 1. A typical example of fracture surface at initial crack front.

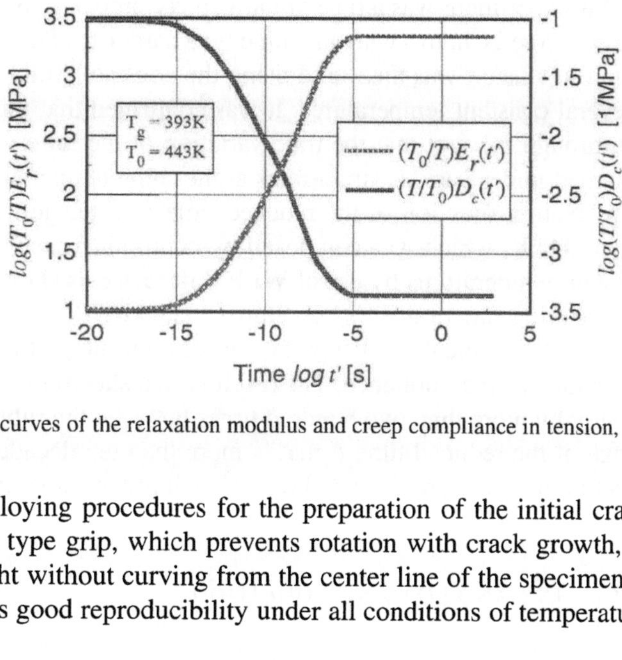

Figure 2. Master curves of the relaxation modulus and creep compliance in tension, $E_r(t')$ and $D_c(t')$.

Both upon employing procedures for the preparation of the initial crack and providing a moment constraint type grip, which prevents rotation with crack growth, it is found that the crack travels straight without curving from the center line of the specimen. Consequently, the crack growth shows good reproducibility under all conditions of temperature and loading rate tested.

2.1.2. *Measurement of crack growth*

In order to perform accurate and precise measurement of the increment of crack growth, a technique of applying slight cyclic tapping is adopted in this study. Slight cyclic impacts by tapping on the tensile rod of the testing machine easily form markings of the crack front on the fracture surface. The effect of the slight tapping on crack growth behavior itself is recognized not to be severe judging from the results of several tests under different intervals of tapping. Figure 1 shows a typical example of the fracture surface with crack front markings formed by cyclic tapping. Very precise and accurate measurement of crack length increment is performed by this technique since the number of markings shows one-to-one correspondence to that of tapping impact. Thus, very small increments in crack length of the order of one micrometer are easily measured.

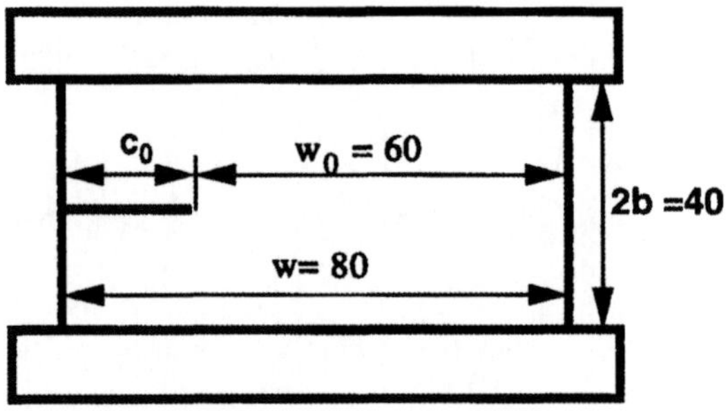

Thickness : 0.5 Unit : mm

Figure 3. Geometry of specimen.

2.2. Mechanical properties of the material

The material used in this experiment was a type of hard epoxy prepared by mixing Bisphenol-A type resin (Epikote 828, Shell Chem.) with an amine type hardener (Triethylene-Tetramine). The relaxation function in tension was measured using the constant strain rate test (Williams et al., 1963) under several constant temperatures. It was confirmed that this material exhibits linear viscoelasticity through the fact that the time variation of the ratio of stress σ to strain rate R, i.e., σ/R, obtained under various strain rates at the same temperature, coincides. The master curve of the relaxation function of the reduced time t' in tension $E_r(t')$ at the reference temperature $T_0 = 443$ K, which was composed by combining each relaxation function obtained at some constant temperatures by use of WLF shift factor, is shown in Figure 2. This master curve of the relaxation function $E_r(t')$ is approximated by Prony series and the creep compliance in tension $D_c(t')$ is calculated using the convolution integral relation between the relaxation function and the creep compliance. The result is also shown in Figure 2. The values of $E_r(t')$ and $D_c(t')$ vary by more than two hundred times between the rubbery and the glassy state over a wide range of the reduced time t' that is more than ten decades on a logarithmic scale. The glass transition temperature T_g is 393 K as shown in the same figure.

2.3. Specimen geometry and loading condition

Figure 3 shows the specimen geometry. The specimen is a strip with $w = 80$ mm in width, $2b = 40$ mm in height and $t = 0.5$ mm in thickness having an initial edge crack $c_0 = 20$ mm. The ratio of initial crack length to the other specimen dimensions is chosen to satisfy approximately the condition of the stress field caused by a semi-infinite crack in an infinite strip (Knauss, 1966). The length of the initial crack along the middle line of thickness is always longer than that at the side edge as shown in Figure 1. It is confirmed, however, that the small change in length of the initial crack does not have a slight significant influence on the crack growth behavior.

The constant rate of displacement loading, 8.33×10^{-6} m/s, is applied at the lower edge of the specimen by the grip fixed rigidly so as to not rotate with an increase of crack length. Twelve different constant temperatures are adopted to cover the whole range from the glassy to the rubbery state of the material.

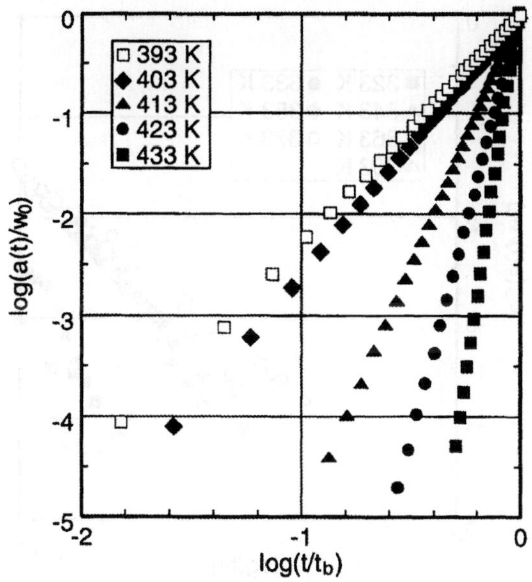

Figure 4. Several crack extension curves above T_g.

3. Experimental results and discussion

3.1. CRACK GROWTH BEHAVIOR

Using the moment constraint type grips, a constant rate of displacement loading of 8.33×10^{-6} m/s at several temperatures above and below T_g is applied to the lower edge of the specimen. The crack travels straight without deviation from the center of the specimen. The reproducibility of crack growth curve is excellent, being within ±2 percent deviation under the same experimental condition. A typical example of a fracture surface obtained above T_g in the neighborhood of the initial crack front is seen in Figure 1. The crack length measured along the middle line of thickness, $c_{\max}$, is always longer than that at the edge, $c_{\min}$, as seen in the same figure. Therefore, the increment of crack length $a(t) = c_{\max} - c_0$, where c_0 is the initial crack length, is obtained by measuring the interval of crack front markings along the middle line of thickness.

3.1.1. *Crack growth curve above T_g*

Figure 4 shows examples of crack growth curves obtained at different temperatures. The ordinate shows the logarithm of the ratio of crack growth $a(t)$ to specimen width w_0, which corresponds to the expected total length of crack path. Also, the abscissa shows the logarithm of the ratio of the time elapsed t, to the rupture time t_b, $\log(t/t_b)$. Each curve shows good linearity on a double logarithmic scale, from the first step of crack growth, on the order of magnitude of a micrometer, up to rupture. Since each curve passes through the point (0, 0), these curves are expressed as

$$\frac{a(t)}{w_0} = \left(\frac{t}{t_b}\right)^{\beta}, \tag{1}$$

where the exponent β is dependent on temperature. The lower the temperature, the smaller the value of β becomes. The temperature dependent properties of the material are reflected in

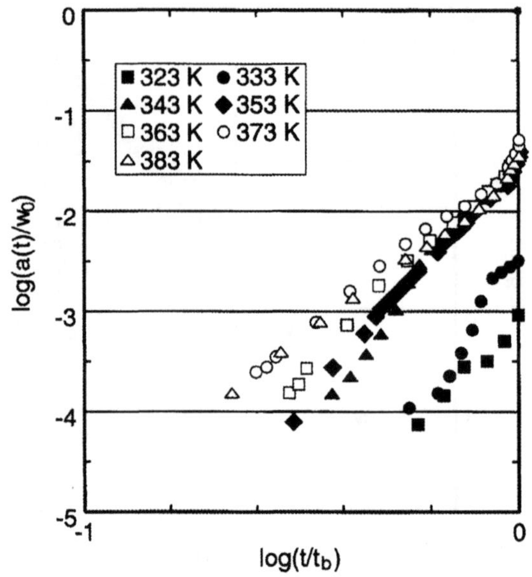

Figure 5. Several crack extension curves below T_g.

the variation of the slope of the curve. It is noteworthy that the crack growth is specified in the form of the βth power of time t and that only stable crack growth is observed up to rupture in this range of temperature.

3.1.2. *Crack growth curve below* T_g

Crack growth curves obtained at different temperatures below T_g are shown in Figure 5. Stable crack growth was observed in the early stage, then subsequent unstable crack growth follows. The linear relationship between $\log(a(t)/w_0)$ and $\log(t/t_b)$ is again observed from the early stage of crack growth to final rupture. Thus, the curve is expressed as

$$\frac{a(t)}{w_0} = \alpha \left(\frac{t}{t_b}\right)^{\beta}, \tag{2}$$

where α is the value of $a(t)/w_0$ at the time when unstable crack growth happens. The exponent β is also dependent on temperature. Lowering the temperature, the value of β shows somewhat complex variation, but it is not significant. The crack growth curve itself is shifted downward with temperature decrease. The duration of stable crack growth prior to the final unstable crack growth becomes shorter with decreasing temperature. Although the macroscopic mechanical properties such as the relaxation modulus are not so dependent on temperature below T_g, the crack growth behavior shows remarkable temperature dependence. In other words, crack growth mechanisms below T_g are not as simple as the source for macroscopic mechanical properties.

Slow and stable crack growth behavior prior to the final rupture is observed over the whole range of temperature, both above T_g and below T_g. From the start of loading to the time at which the crack grows unstably, no evidence of discontinuous change in crack growth behavior is observed. From that fact, the point of crack growth threshold may not be determined from these experimental data. In a later section of this paper, then, we have to assume the point of crack growth threshold in order to evaluate the fracture parameter, such as J'_{IC}.

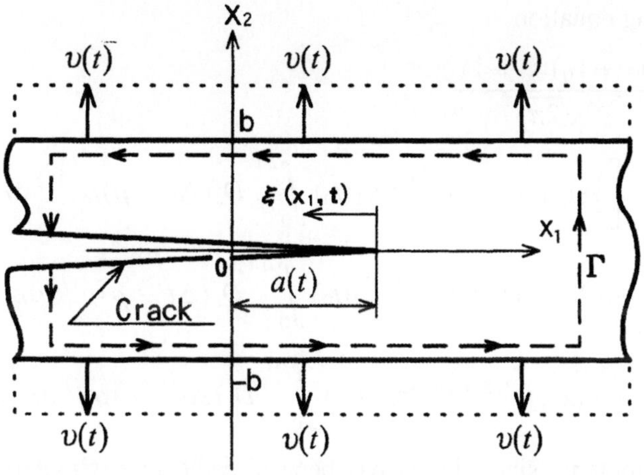

Figure 6. A strip specimen having a semi-infinite central crack and the path for $J'(t)$-integral.

3.1.3. *Crack growth rate*

Let us consider an infinite strip having a semi-infinite crack as shown in Figure 6. Here, we adopt a Cartesian coordinate system $o-x_1x_2$ with the origin at the tip of the crack. The strip is subjected to displacements $\pm v(t)$ in the direction of x_2 at the upper and lower edges $x_2 = \pm b$. Denoting the crack length (increment) at time t as $a(t)$, a new coordinate ξ is taken to the negative direction of x_1 from the tip of crack. Assuming that Poisson's ratio ν_0 is constant and denoting the time t_0 at which the crack tip has passed a point $(x_1, 0)$, the displacement u_2 in the direction of x_2 at a point ξ on the crack surface at time t is expressed as follows, using the extended correspondence rule (Graham and Sabin, 1973),

$$u_2(\xi, 0^+, t) = \frac{(1+\nu_0)(\kappa+1)}{\sqrt{2\pi}} \int_{t_0}^{t} D_c(t-\tau)\frac{\partial}{\partial \tau}[K_{\mathrm{I}}(\tau)\xi(x_1,\tau)^{1/2}]\,\mathrm{d}\tau,$$
$$0 < \xi(x_1, t) < a(t), \tag{3}$$

where K_{I} is the stress intensity factor. K_{I} and κ are given by

$$K_{\mathrm{I}}(t) = \frac{1}{\sqrt{(1-\nu_0^2)b}}\left[v(0)E_r(t) + \int_0^t E_r(t-\tau)\frac{\mathrm{d}v(\tau)}{\mathrm{d}\tau}\,\mathrm{d}\tau\right],$$

$$\kappa = \begin{cases} 3-4\nu_0, & \text{for plane strain,} \\ \dfrac{3-\nu_0}{1+\nu_0}, & \text{for plane stress.} \end{cases}$$

Adopting the following coordinate transformation after the technique proposed by Schapery (1975),

$$\rho = \tau - t_0, \qquad \Delta t = t - t_0,$$

also, paying attention to the relation

$$\xi(x_1, \tau) = a(\tau) - x_1 \approx (\tau - t_0)\dot{a}(t_0),$$

we have the following equation

$$u_2(\xi, 0^+, t) = \frac{(1+\nu_0)(\kappa+1)}{\sqrt{2\pi}}$$

$$\times \Bigg[\tfrac{1}{2}\xi(x_1, t)^{1/2} \Delta t^{-1/2} K_{\rm I}(t_0) \int_0^{\Delta t} D_c(\Delta t - \rho)\rho^{-1/2}\,{\rm d}\rho$$

$$+ \tfrac{3}{2}\xi(x_1, t)^{1/2} \Delta t^{-1/2} \dot{K}_{\rm I}(t_0) \int_0^{\Delta t} D_c(\Delta t - \rho)\rho^{1/2}\,{\rm d}\rho$$

$$+ \xi(x_1, t)^{1/2} \Delta t^{-1/2} \ddot{K}_{\rm I}(t_0) \int_0^{\Delta t} D_c(\Delta t - \rho)\rho^{3/2}\,{\rm d}\rho \Bigg]. \tag{4}$$

Equation (4) is computable since $D_c(t)$ has been already approximated by a Prony series. However, it can not be expressed by simple elementary functions. Then, assuming the form of

$$D_c(t) = Ct^n, \tag{5}$$

we define a displacement field function around the crack tip as

$$M'(t) = \lim_{\xi(x_1,t)\to 0^+} \frac{\sqrt{2\pi}\,u_2(\xi, 0^+, t)}{(\kappa+1)\xi(x_1, t)^{n+(1/2)}}.$$

Here, we assume that the $M_{\rm ID}$ criterion for crack growth in the viscoelastic material

$$M'(t) = M_{\rm ID} (= \text{const.})$$

is valid if $M'(t)$ were kept constant during the crack growth process.

Under the boundary conditions,

$$u_2(\xi, \pm b, t) = \begin{cases} 0, & t < 0 \\ \pm v_0 t, & t \geqslant 0 \end{cases},$$

the crack growth rate $\mathrm{d}a/\mathrm{d}t$ is predicted as

$$\frac{{\rm d}a(t)}{{\rm d}t} = \left[\frac{BC\sqrt{\pi}(1+\nu_0)\Gamma(n+1)v_0}{2\sqrt{(1-\nu_0^2)b}(n+\frac{1}{2})\Gamma(n+\frac{1}{2})M_{\rm ID}} \right]^{1/n} t^{m/n} \tag{6}$$

if the integral of the relaxation modulus $E_r(t)$ from time 0 to t can be approximated by

$$\int_0^t E_r(\tau)\,{\rm d}\tau = Bt^m.$$

Schapery (1975) has proposed a prediction of crack growth rate based on the COD criterion, as one example. For the case of the authors' experiment, his prediction will be expressed as

$$\frac{{\rm d}a(t)}{{\rm d}t} = \frac{\pi(v_0 B)^{2(1+(1/n))}}{2b(1-\nu_0^2)\sigma_m^2 I_1^2} \left[\frac{C\lambda_n}{8(1-\nu_0^2)b\Gamma} \right]^{1/n} t^{2m(1+(1/n))}, \tag{7}$$

where σ_m, I_1, λ_n and Γ are quantities dependent on the material. It is noteworthy that either prediction of crack growth rate is given by a power function of t. On the other hand, as mentioned earlier, the crack growth curve during the stable stage either above or below T_g is expressed by (1) or (2), respectively. Thus, the crack growth rate da/dt would be written as

$$\frac{da(t)}{dt} \propto t^{\beta-1}. \tag{8}$$

Since both (6) and (7) are, as shown, predicting the crack growth rate with a power function of t, the experimental data show good agreement with them qualitatively.

Table 1. Values of the exponents m and n at each temperature.

$T(K)$	m	n	$m+n$
423	1.0	0	1.0
413	1.0	0	1.0
403	0.806	0.157	0.963
393	0.645	0.347	0.992
383	0.776	0.268	1.04
373	1.0	0	1.0

Table 2. Comparison of the exponents in the power function of crack growth rate.

$T(K)$	m/n	$2m(1+1/n)$	$\beta-1$
423	–	–	7.41
413	–	–	4.03
403	5.13	11.9	1.58
393	1.86	5.01	1.31
383	2.90	7.34	2.50
373	–	–	2.61

Now, let us try to compare the exponents m/n in (6), $2m(1+1/n)$ in (7) and $\beta - 1$ in (8). The values of m and n are calculated from the master curves of $E_r(t)$ and $D_c(t)$ by using the WLF shift factor even for the temperatures below T_g. The results are shown in Table 1. Although the relation of $m + n = 1$ has to be realized from the theoretical viewpoint, we observe poor agreement at some temperatures. It should be pointed out that the accuracy of the Prony series approximation of $E_r(t)$ and $D_c(t)$ could be responsible for the deviation of the theoretical prediction from the experiment. Since there are only a few temperatures at which the value of n does not become zero, the comparison between the prediction and the experimental results is quite limited. In order to perform an accurate prediction by (6) or (7), it should be necessary to have a good approximation of $D_c(t)$ by (5) in the range of time at least between $t/10$ to t for (6) or $t/100$ to t for (7). Particularly at a higher temperature of 403K, an accurate prediction is impossible because a sufficient time range cannot be found for good approximation of $D_c(t)$. The values of the exponent, m/n in (6) and $2m(1 + 1/n)$ in (7) are compared with those of $\beta - 1$ in empirical formulae (1) and (2) in Table 2. As seen in the table, the values from (6) show fairly good agreement with the experiment, particularly at 383 and 393K. Although either (6) or (7) implies that the crack can grow instantaneously under the condition of $n \approx 0$, i.e., rubbery or glassy state, the experimental data do not support that prediction. In order to discuss the validity of (6) or (7), however, it will be necessary to perform not only a crack growth experiment for materials having long relaxation times, but also to provide more accurate measurement and precise approximations of $E_r(t)$ and $D_c(t)$.

3.2. RESISTANCE FOR CRACK GROWTH

3.2.1. *J′-integral*

Theoretical analysis of crack growth behavior in a viscoelastic body belonging to a mixed boundary value problem is very difficult by reason of the inapplicability of the well-known correspondence principle to the case that boundary itself varies with crack extension. Only the values of normal stress and normal displacement on the crack plane are readily found even if the extended correspondence principle (Graham and Sabin, 1973) can be applied. The authors have investigated whether or not the so-called J-integral is applicable, and concluded that it is not applicable to viscoelastic media. Therefore, the authors proposed a new concept of the J'-integral in one of their previous papers (Misawa et al., 1990), in which the strain energy density function in the J-integral was replaced by the free energy per unit volume ψ, as follows

$$J' = \int_{\Gamma} \psi \, \mathrm{d}x_2 - \sigma_{ij} n_j \frac{\partial u_i}{\partial x_1} \, \mathrm{d}s.$$

The free energy ψ is expressed as

$$\begin{aligned} \psi(t) &= \frac{1}{2} \int_{-\infty}^{t} \int_{-\infty}^{t} G_1(2t - \tau - \eta) \frac{\partial \varepsilon_{kl}(\tau)}{\partial \tau} \frac{\partial \varepsilon_{kl}(\eta)}{\partial \eta} \, \mathrm{d}\tau \, \mathrm{d}\eta \\ &\quad + \frac{1}{6} \int_{-\infty}^{t} \int_{-\infty}^{t} \{G_2(2t - \tau - \eta) - G_1(2t - \tau - \eta)\} \frac{\partial \varepsilon_{kk}}{\partial \tau} \frac{\partial \varepsilon_{ll}(\eta)}{\partial \eta} \, \mathrm{d}\tau \, \mathrm{d}\eta, \end{aligned}$$

where $G_1(t)$ and $G_2(t)$ are the relaxation functions in shear and isotropic compression, respectively. The proof of path-independence of the $J'(t)$-integral is easily shown by the same procedure as for the case of Rice's J-integral (1968) in which the stress components are given by

$$\sigma_{ij}(t) = \frac{\partial \psi(t)}{\partial \varepsilon_{ij}(t)}.$$

In the case that the specimen strip is loaded perpendicularly to the crack surfaces at the upper and the lower edges as shown in Figure 6, the boundary conditions are given by

$$u_2(\xi, \pm b, t) = \begin{cases} 0, & t < 0 \\ \pm v_0 t, & t \geqslant 0 \end{cases},$$

where v_0 is constant. Assuming that Poisson's ratio ν_0 is constant and integrating along a path Γ as shown in Figure 6, the J'-integral is expressed as

$$J'(t) = \frac{v_0^2}{(1 - \nu_0^2) b} \int_0^t \int_0^t E_r(2t - \tau - \eta) \, \mathrm{d}\tau \, \mathrm{d}\eta. \tag{9}$$

It is convenient for actual calculation to use the relaxation function in tension $E_r(t)$, represented by a Prony series approximation. The critical value of J', i.e., J'_{IC}, is calculated by assuming a certain time t^* for the crack growth threshold in (9).

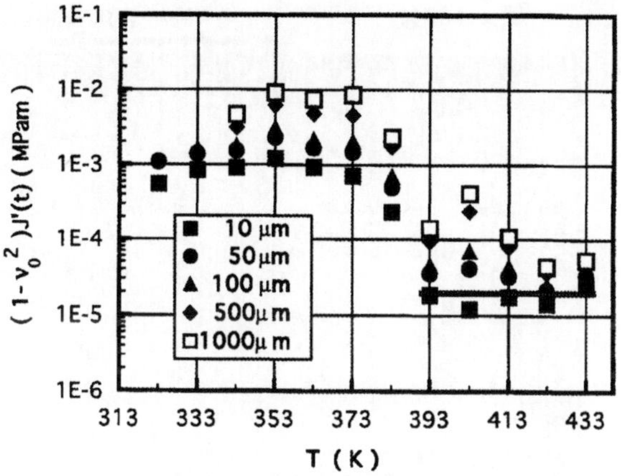

Figure 7. Temperature dependence of the values of $J'(t)$.

3.2.2. *An expedient J'_{IC} for crack growth*

As mentioned in Section 3.1.2, discontinuous phenomena both on the load-displacement curve and crack growth curve, from the start of loading up to the time of final rupture, or unstable crack growth, could not be observed, at least within the accuracy of measurement. Judging from the trend of crack growth curves shown in Figures 4 and 5, the empirical curves (1) and (2) could be extrapolated to a smaller increment of the crack length of some micrometers. Thus, the critical point for the crack growth threshold cannot be identified within the range of time or crack length measured, as a discontinuous point.

In this circumstance, it is necessary to develop an alternative method which can cope with the difficulty, as for example, a new factor or a technique more sensitive for minute discontinuous variations of a physical quantity at the instant of crack growth threshold than the current ones. When the increment of crack length reaches a specific minute value, for example 10 μm, if we can find a constant physical quantity independent of the experimental conditions, it could be very useful from the viewpoint of engineering. Although the quantity does not provide the critical condition for crack growth threshold, the value could be used as the expedient crack growth threshold. Taking the latter viewpoint, the authors paid attention to the J'-integral applicable to an growing crack in a viscoelastic strip. The value of J'-integral at the instant when the increment of crack length attains several steps of specified length within 10 to 1000 μm was calculated and examined. The time at which the crack growth reaches each specified value, considered an apparent critical time for macroscopic crack start in this study, is read out from the crack growth curves obtained under the condition of constant rate of displacement loading of 8.33×10^{-6} m/s.

(1) *Above T_g*

Strong temperature dependence of J' is seen when assuming a larger increment of crack growth for a macroscopic crack start, namely longer critical time, except in the case of the smallest crack length increment of 10 μm. The lower the temperature becomes, the greater value of J' is calculated, reflecting both the time and temperature dependence of mechanical properties and characteristic features of the crack growth curve. When we take the smallest critical crack length increment as 10 μm, which can be adopted as an apparent critical time with a reasonable experimental basis, the temperature dependence of J' vanishes, as shown in Figure 7; then the loading rate dependence of J' also vanishes. In other words, as the apparent

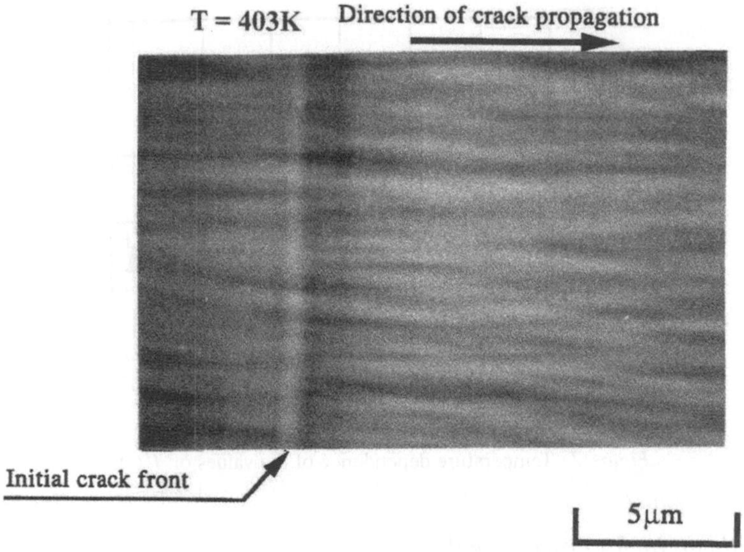

Figure 8. A typical example of fracture surface at initial crack front.

critical time for the macroscopic crack growth threshold becomes smaller, the value of the J'-integral decreases and its dependence on temperature also decreases until it becomes constant over a wide range of temperature. The material tested in this study may have its own critical crack growth resistance $J'_{\rm IC}$ independent of temperature. From a macroscopic point of view, this fact could be interpreted as a measure of an apparent critical time t^* for crack growth threshold which cannot be determined with ordinary tests on materials.

(2) *Below T_g*
When calculating J' below T_g (=393 K), it is assumed that the WLF equation is valid until 373 K, while $E_r(t)$ is constant below 363 K. Even below T_g, the temperature dependence of J' is seen between 323 K and 383 K, regardless of the crack length. The value of J'-integral increases gradually with increasing temperature until 353 K and decreases considerably until 383 K, although the modulus does not depend on temperature below 363 K. The value of J' calculated for the smallest crack length increment, 10 μm, on which J' is regarded as constant above T_g, shows temperature dependence; then it reaches the maximum value at 353 K. The values of J' from 323 to 383 K are very large as compared with the value above T_g. The reason will be discussed through observation of the fracture surface in Section 3.3.2.

3.3. Morphological characteristics of the fracture surface

3.3.1. *Characteristics of the fracture surface above T_g*

Fine wavy streamline marks as shown in Figure 1, which flow smoothly along the direction of crack growth, are observed over the whole area on the fracture surface. Reflecting the temperature and time dependent mobility of molecular chain and entanglement around a moving crack tip, the characteristic features of the fracture surface are mostly seen just ahead of the initial crack (Ogawa et al., 1994).

Since the characteristic region of the fracture surface is restricted in the vicinity of the tip of initial crack, a limited small area of 80 × 80 μm beyond 20 to 30 μm from the initial crack front is observed with a laser microscope and investigated with an image processing technique.

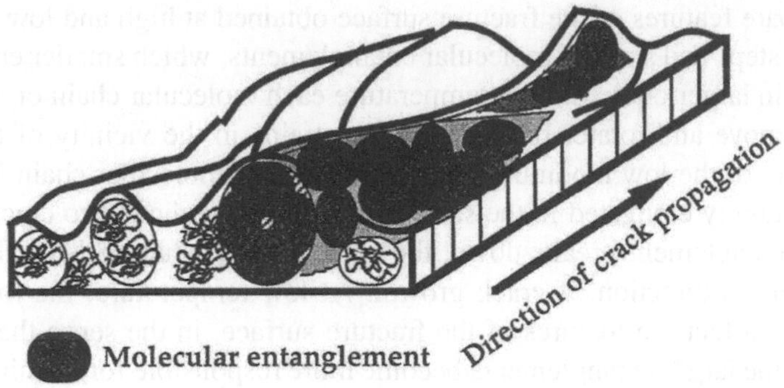

Figure 9. A schematic diagram of fracture surface indicating minute icicle marks.

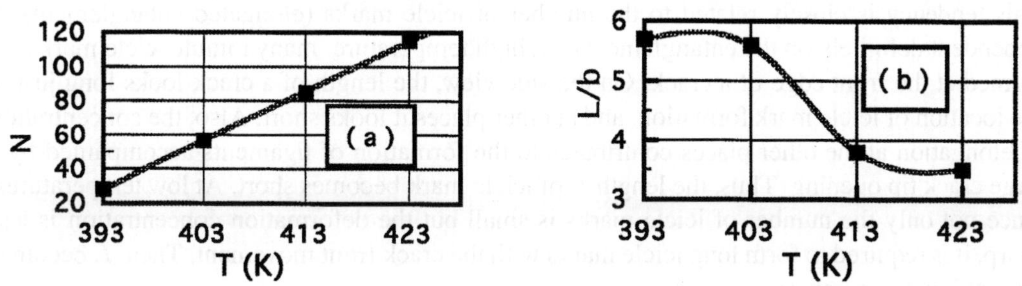

Figure 10. Temperature dependence of the characteristics of the minute icicle marks; (a) number; (b) aspect ratio.

Figure 8 shows a typical example of the fracture surface taken with a laser microscope. The fracture surface shows some microscopic structures which look like minute icicles lined up in the direction of crack growth. A streamline mark observed in low magnification corresponds to a series of the minute icicle marks at high magnification. Figure 9 shows a schematic diagram of the fracture surface including icicle marks. In order to evaluate the characteristics of this minute icicle mark quantitatively, the statistical parameter such as the number (N), the maximum perimeter (D), the maximum length (L), the maximum thickness (b), are extracted. Upon lowering the temperature, D, L and b increase while N decreases. Figure 10 shows temperature dependence of the number N of the minute icicle marks and the ratio of the maximum length to maximum thickness, L/b.

Let us propose a microscopic scale model for fracture at the crack tip, which could be related to the observed fracture surface. We assume that heterogeneity in polymeric materials consists of entanglement blocks of various sizes involving numerous molecular chains. The molecular density of the blocks could change broadly from sparse to dense. The larger block usually involves a lot of smaller ones. Also each block, regardless of its size, is connected by numerous molecular chains and/or links each other. So, the heterogeneity of polymeric material governing crack growth varies with the mobility of molecular chains and the gradient of the stress/strain distribution near the tip of the crack, depending on temperature and loading rate. Now, the following questions arise:

(a) Why does the number N of the icicle marks increase at high temperature?

In order to compare features of the fracture surface obtained at high and low temperature, we assume multiple steps and sizes of molecular entanglements, which smaller entanglements are mostly involved in larger ones. At high temperature each molecular chain or minute entanglement is easy to move and rotate; furthermore the strains in the vicinity of the crack tip are very high because of the low modulus. Thus, almost every molecular chain in the vicinity of crack tip can be highly elongated in the same direction perpendicular to crack surface. When some chain or entanglement breaks down, the effect of load sharing is easily transmitted to those lined up in the direction of crack growth. At low temperature, the mobility of larger entanglement can affect the features of the fracture surface, in the sense that upon lowering the temperature, the larger entanglements become more responsible for forming rough fracture surface.

(b) Why does the length of the icicle marks become short at high temperature?
This tendency is closely related to the number of icicle marks (elongated entanglement) N dependent definitely on the entanglements. At high temperature, many minute icicle marks are formed at the front edge of a crack. On the side view, the length of a crack looks long just at the location of icicle mark formation, and at other places it looks short. Also, the concentration of elongation at the other places contributes to the formation of ligaments accompanied by a large crack tip opening. Thus, the length L of icicle mark becomes short. At low temperatures, since not only the number of icicle marks is small but the deformation concentration is less sharp, it is required to form long icicle marks with the crack front movement. Then, L becomes long at low temperature.

(c) Why is the volume of single wavy mark ($L \times b$) so large in comparison to that of entanglements?
The size L and b are not a direct reflection of geometry of the entanglement, but of mechanical properties at the crack tip or of the mobility of the molecular chains. L has been mentioned in the previous paragraph. Regarding b, the minute icicle marks can be easily elongated in the direction of crack growth but not in the transverse direction. So, the dimension of b cannot depend strongly on temperature.

3.3.2. *Characteristics of the fracture surface below T_g*

Figure 11 shows examples of the fracture surface obtained at four different temperatures below T_g. The characteristic tongue mark is clearly observed. Therefore the variation of crack lengths with time was measured by the cyclic tapping method. At higher temperatures, that is, above T_g, the length of the tongue marks, $\Delta c (= c_{max} - c_{min})$, remains almost constant or increases slightly with time. But at lower temperatures it increases noticeably with crack length. The temperature dependence of Δc_{max}, which is the value of Δc when unstable fracture occurred, is shown in Figure 12. The value of Δc_{max} comes to a peak at 363 K, far below T_g. As shown in Figure 11, the shape and length of this type of tongue mark show remarkable temperature dependence even under the condition below T_g. With decreasing temperature from 383 to 363 K, the tongue becomes sharper and the curvature of the tongue tip increases gradually, but that tendency changes suddenly to a blunt head at 333 K, i.e., 60 K below T_g.

Next, let us consider the streamline marks indicating the path trace of crack growth. The fracture surface inside the tongue marks formed by stable crack growth shows clear evidence that the core of the crack develops in the middle of the specimen thickness and that the front of the crack moves towards both side edges symmetrically and perpendicularly to the sides

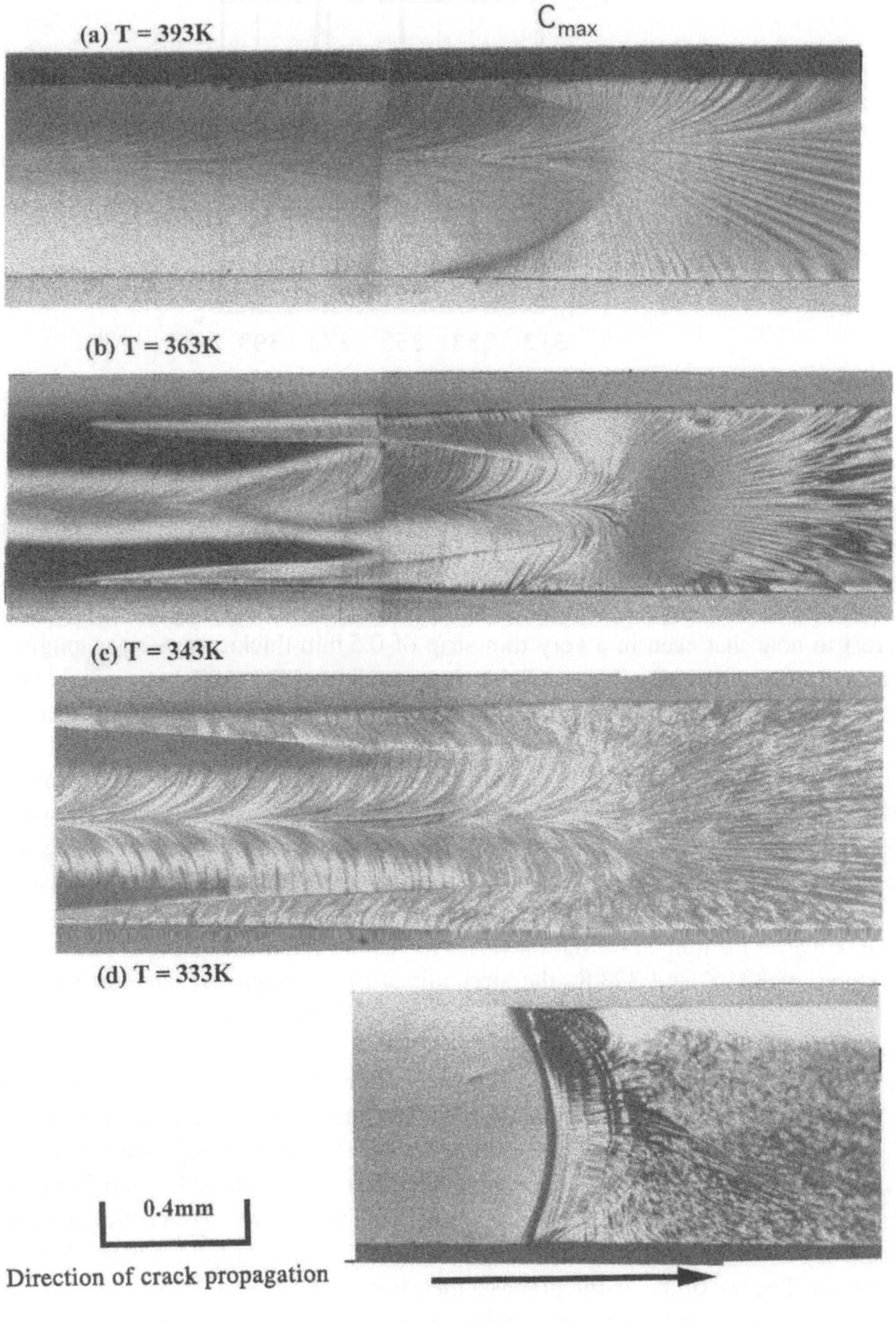

Figure 11. Typical examples of fracture surface below T_g.

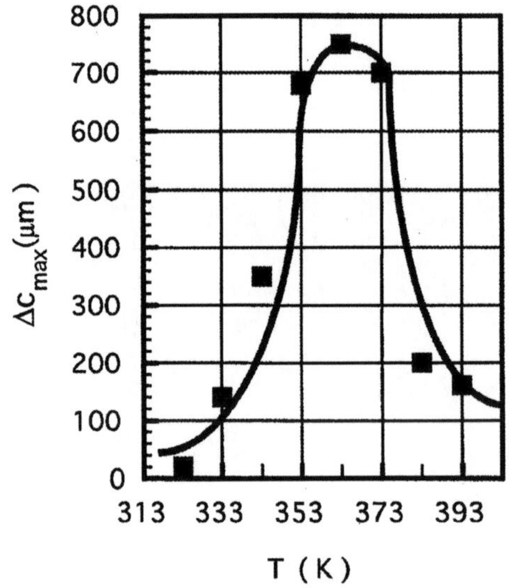

Figure 12. Temperature dependence of ΔC_{max}.

of the specimen. At the instant of the critical state for unstable crack growth, the crack bursts and extends in all directions, so that in certain portions, the crack propagates backwards. It is of interest to note that even in a very thin strip of 0.5 mm thickness, a clear tongue mark is formed on the fracture surface, not only by the complicated stress and strain field but also by the rapid change in stress or strain rate at the curved crack front. A possible explanation for the mechanism could be given as follows. Crack growth below T_g develops a plane strain state in the center portion of crack surface, but near side surfaces the plane stress state dominates. Thus the difference of triaxiality of stress state and the high deformability under the plane stress causes the larger delay of crack growth at the side surface. On the fracture surfaces obtained at 333 K and 323 K, the crack near the side surfaces can hardly extend up to unstable crack growth except for the middle portion. The fracture surface generated by rapid and unstable crack growth after the tongue formation varies its feature remarkably. Particularly at the lower temperatures of 333 K and 323 K, the streamlines of crack growth are not so clear, but a granular surface is observed over the whole surface reflecting an extremely high velocity of crack growth.

As mentioned before, the J'-integral calculated for the case of the smallest crack increment, 10 μm, has the maximum value at 353 K and shows temperature dependence between 323 K and 383 K, while the J'-integral does not show temperature dependence above T_g. The values of the J'-integral obtained at lower temperatures between 323 K and 373 K are much larger than those above T_g even in the case of the smallest increment of crack length of 10 μm. This tendency could be closely related to whether the direction of the crack front movement is simple or not. The use of the J'-integral as a measure for the critical condition of crack growth might be also possible at temperatures below 323 K, because any stable crack growth cannot be observed.

To explain the crack growth mechanism forming the characteristic fracture surface, the mechanical behavior between T_g and $T_g - 60$ K must be clarified and a new approach for the treatment of the complicated three dimensional stresses at crack front must be developed.

4. Concluding remarks

To investigate the crack growth mechanism and behavior of a viscoelastic epoxy strip above and below T_g, careful observations on the characteristic features of fracture surfaces and crack growth behavior were performed. The results obtained are briefly summarized as follows:

(1) Two equations predicting crack growth curve, one of which is based on the COD criterion and the other on the M_{ID} criterion, can explain the experimental results, except at the extremes of temperature in the rubbery and glassy states.
(2) The temperature dependence of the J'-integral is reduced with a decrease of the apparent critical time t^* or crack length increment. Then, the temperature independent critical crack growth resistance J'_{IC} appears to be a material constant for the crack growth threshold above T_g. The direct application of the J'-integral to stable crack growth between T_g and $T_g - 60\,\mathrm{K}$ is, however, difficult or impossible because of the very complicated crack tip configuration.
(3) Temperature dependent crack growth behavior was observed in the range of temperatures between T_g and $T_g - 60\,\mathrm{K}$. To explain this phenomenon, mechanical behavior including the time-temperature shift effect have to be investigated more precisely in conjunction with the material micro-structure and molecular mobility.
(4) Minute icicle marks lined up in the direction of crack growth are observed on the fracture surface obtained above T_g. The number and aspect ratios of these minute icicle marks depend on temperature.
(5) The tongue marks formed during stable crack growth before drastic unstable growth are observed on the fracture surface below T_g. The shape and size of the tongue marks depend on temperature.

Acknowledgment

The authors appreciate partial support of the Center for Science and Engineering Research Institute of Aoyama Gakuin University.

References

Graham, G.A.C. and Sabin, G.C.W. (1973). The correspondence principle of linear viscoelasticity for problems that involve time-dependent regions. *International Journal of Engineering Science* **11**, 123–140.

Knauss, W.G. (1966). Stresses in an infinite strip containing a semi-infinite crack. Transactions of ASME, Ser. E. *Journal of Applied Mechanics* **33**, 356–362.

Knauss, W.G. (1970). Delayed failure – The Griffith problem for linearly viscoelastic materials. *International Journal of Fracture Mechanics* **6**, 7–20.

Knauss, W.G. (1974). On the steady propagation of a crack in a viscoelastic sheet: Experiments and analysis. *Deformation and Fracture of High Polymers* (Edited by H.H. Kausch), Plenum Press, 501–541.

Misawa, A., Takashi, M. and Kunio, T. (1984). New criteria for onset and subsequent crack growth in a viscoelastic strip. *Proceedings of the 6th International Conference on Fracture, ICF-6, New Delhi* **4**, 2571–2578.

Misawa, A., Takashi, M. and Kunio, T. (1990). Application of the conservation law to threshold of crack growth in viscoelastic strip. *Proceedings of the 9th International Conference on Experimental Mechanics* **5**, 1786–1794.

Newman, S.B. and Wolock, K.I. (1958). Fracture phenomena and molecular weight in polymethyl methaacrylate. *Journal of Applied Physics* **29**, 49–52.

Ogawa, K., Misawa, A., Takashi, M. and Kunio, T. (1994). Temperature dependent morphology of fracture surface and crack growth mechanism in a viscoelastic epoxy resin. *Proceedings of the 10th International Conference on Experimental Mechanics* **2**, 1265–1270.

Rice, J.R. (1968). A path independent integral and the approximate analysis of strain concentration by notches and cracks. *Journal of Applied Mechanics* **35**, 379–386.

Schapery, R.A. (1975). A theory of crack initiation and growth in viscoelastic media II. Approximate methods of analysis. *International Journal of Fracture* **11**, 369–388.

Takashi, M., Ohtsuka, A., Misawa, A. and Kunio, T. (1983). Crack growth behavior and crack opening profile in viscoelastic strips. *Proceedings of the 26th Congress on Materials Research*, 247–252.

Williams, M.L., Kunio, T., Ferguson, C.W. and Arenz, R.J. (1963). GALCIT SM 63-31, California Institute of Technology.

International Journal of Fracture **93:** 227–246, 1998.

On the fracture of constrained layers

C.H. WANG
Aeronautical and Maritime Research Laboratory, Defence Science and Technology Organisation, Melbourne, Australia. e-mail: chun.wang@dsto.defence.gov.au

Received 11 November 1997; accepted in revised form 5 March 1998

Abstract. The problem of a crack embedded in a layer sandwiched between two elastic adherends is analysed accounting for the influence material property mismatch on the crack tip plastic deformation, which is contained in the layer. The cohesive crack model developed by Dugdale and Barrenblatt is adopted to model the strip yielding behaviour in a constrained layer. It is found that, due to the constraint imparted by elastic adherends with higher moduli, the near tip plastic deformation exhibits a sharp transition (plastic zone grows faster than the square of stress intensity factor) from small scale to large scale yielding. Because the region of singularity dominance for a crack embedded in a layer is generally much smaller than the layer thickness when the layer has a modulus much lower than the adherends, the prevailing failure mode of most bonded joints should be under large scale yielding conditions. A model based on energy balance is proposed to determine the fracture energy of bonded joints under such condition, taking into account of the plastic dissipation in the constrained layer. Comparison with experimental results demonstrates that the theory correctly predicts the dependence of fracture toughness on layer thickness as observed in experiments.

Key words: Layer, plasticity, fracture, constraint, bonded joints, strip yielding.

1. Introduction

Layered structures which are consisted of two or more dissimilar adherends joined together with thin, ductile layers have been employed in a wide range of applications. Examples include adhesively bonded joints, ceramics joined by metal foils, and a variety of composite materials. Of the many factors that are important to the mechanical performances of these structures, the integrity of the layer and/or the interface between the layer and the substrate is probably the most critical one. In this regard, fracture mechanics has been extensively used to quantify the resistance to cohesive or interfacial fracture, to identify mechanisms of failure, and to evaluate the residual strength and service life of sandwiched structures containing flaws.

Up to now in modelling the crack problems of layered materials, attention has been mainly focused on brittle layers (Hutchinson and Suo, 1992) having low fracture toughness, thus the extent of the failure process is confined to a region much smaller than the relevant geometric length scales, such as layer thickness. In other words, small scale yielding condition is often assumed to prevail near the crack tip, which ensures that there is exact similitude, and thus the critical value of fracture parameters can be considered as material properties. However, for materials used in adhesive bonding or metal foil bonding, the layer is often made of very ductile materials to enhance the damage tolerance of a bonded structure. For instance, the structural adhesives used in bonded joints are often toughened using additives such as rubber particles (Kinloch, 1987). Furthermore, due to the constraint imparted by the stiff adherends, for a crack embedded in a constrained layer, the region near the crack tip where the stress-strain distribution is dominated by the singularity associated with a crack can be much smaller than the layer thickness (Wang et al., 1978; Wang, 1997a; Becker et al., 1997). Therefore extensive plastic deformation may occur prior to the final fracture, with the plastic zone size at failure being much greater or at least comparable to the layer thickness (Dalgleish et al., 1988). In the case of metal foils (Varias et al., 1991), the fracture process has been found to be dictated by the nucleation and growth of voids at distances several foil thickness ahead of the crack tip, unlike for a crack in a bulk solid where voids nucleate and grow within distances comparable to the crack-tip opening displacement. The most important implication of this difference is perhaps that any of the fracture parameters alone may fail to provide a unique critical value for bonded structures. This is because the traditional framework of fracture mechanics relies on the assumption of small-scale yielding at the crack tip, and most fracture parameters, such as stress intensity factor or J-integral, characterise the asymptotic stress field near the crack tip region only. Therefore these parameters are unable to capture the essential features of fracture process occurring ahead of the crack tip at distance far greater than the crack tip opening displacement.

In the context of constrained layers, the small scale yielding condition required to ensure similitude, implies that the region within which plastic deformation occurs must be much smaller than any other geometric dimensions and, most importantly, the region over which these parameters dominate the stress field. This implies that the plastic zone must be much smaller than (i) the singularity dominance zone and (ii) the thickness of the layer. For example, in the case of prestressed thin films bonded on thick substrates, Wei and Hutchinson (1997) suggested the relevant condition of small scale yielding for thin film delamination is that the plastic zone size is small compared to film thickness. However, this similitude requirement has been largely overlooked and the small scale yielding condition has often been implicitly assumed in quantifying the fracture energy or toughness of adhesively bonded joints (Kinloch and Shaw, 1981; Chai, 1986; 1988; 1993; Hunston et al., 1989; Daghyani et al., 1995). Consequently, the results reported in the literature indicated a strong dependence of the fracture energy on the adhesive layer thickness, somewhat unexpected as these parameters are believed to be material properties.

The purpose of this paper is to present an analytical solution of the elastic-plastic deformation at the tip of a crack embedded in a constrained layer, valid for both small scale yielding and large scale yielding. Plastic yielding is assumed to be confined within the layer only and the substrates are assumed to remain elastic. The length of the plastic zone is determined using the concept of cohesive crack by Dugdale (1960) and Barrenblatt (1962), and the calculated plastic zone length is found to be in good agreement with finite element results. An energy kinetics model is proposed to account for the plastic dissipation under large scale yielding, with the predictions being in good agreement with the observed dependence of fracture toughness on layer thickness.

2. Elastic solutions of a cracked constrained layer

The problem to be considered is shown in Figure 1. A layer of height t_A and elastic properties μ_1, ν_1, which is sandwiched between two elastic half-planes with elastic properties μ_2, ν_2, contains a crack of length $2a$. Here μ_i and $\nu_i (i = 1, 2)$ are, respectively, the shear modulus and Poisson's ratio. The crack is lying in the mid-plane of the layer and parallel to the interface. In the case of the layer being elastic, by using the superposition principle it is easy to demonstrate that the elastic problem depicted in Figure 1 is equivalent to solving a perturbation problem

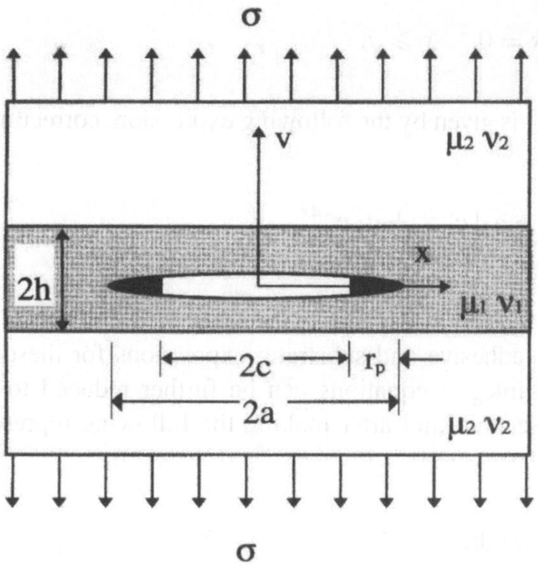

Figure 1. A crack embedded in an elastic-plastic layer which is sandwiched between elastic substrates.

where the cracks faces are subjected to pressure $\sigma_{yy}(x,0) = -p(x)$. In the following we begin by considering the elastic analysis and then concentrate on the plastic deformation behaviour near crack tip. The cohesive crack concept by Dugdale (1960) and Barrenblatt (1962) for cracks in homogeneous materials will be adapted to analyse cracks in constrained layers.

Elastic analysis of the perturbation problem involves finding two Airy stress functions (Φ_1 and Φ_2) satisfying the following mixed-boundary value conditions along $y = 0$ and continuity conditions along $y = \pm t_A/2$ (see Sih and Chen, 1981),

$$\sigma_{yy}(x,0) = -p(x), \quad |x| < a, \tag{1a}$$

$$u^{(1)}(x,0) = 0, \quad |x| \geqslant a, \tag{1b}$$

$$\tau_{yy}(x,0) = 0, \tag{1c}$$

$$\sigma_{yy}^{(1)}(x,\pm h) = \sigma_{yy}^{(2)}(x,\pm h), \tag{1d}$$

$$\tau_{xy}^{(1)}(x,\pm h) = \tau_{xy}^{(2)}(x,\pm h), \tag{1e}$$

$$u^{(1)}(x,\pm h) = u^{(2)}(x,\pm h), \tag{1f}$$

$$v^{(1)}(x,\pm h) = v^{(2)}(x,\pm h), \tag{1g}$$

where $h = t_A/2$. Assuming that the traction acting on the crack faces is an even function, $p(x) = p(-x)$, the Airy stress functions corresponding to the adhesive layer and the substrates can be expressed in terms of Fourier cosine transforms, from which the stress and displacement components can be determined (details are given in Appendix A). After some lengthy manipulations the problem can be reduced to a single unknown $A(s)$ that satisfies a pair of dual integral equations,

$$\int_0^\infty sF(sh)A(s)\cos(sx)\,\mathrm{d}s = \frac{\pi}{2}p(x), \quad x < a, \tag{2a}$$

$$\int_0^\infty A(s)\cos(sx)\,\mathrm{d}s = 0, \quad x \geqslant a, \tag{2b}$$

where the function $F(x)$ is given by the following expression, correcting a typographical error in Sih and Chen (1981),

$$F(x) = \frac{a_1 + [a_2x^2 + a_3]\,\mathrm{e}^{-2x} + a_4\,\mathrm{e}^{-4x}}{a_1 - a_2x\,\mathrm{e}^{-2x} - a_4\,\mathrm{e}^{-4x}}, \tag{3}$$

where constants a_1, a_2, a_3, and a_4 depend on the shear modulus ratio μ_1/μ_2 and the Poisson's ratios, ν_1 and ν_2 of the adhesive and substrate; expressions for these constants are given in Appendix B. The dual integral equations can be further reduced to a single Fredholm integral equation of the second kind after making the following representation (Sneddon and Lowengrub, 1969),

$$A(s) = \int_0^a \Phi(t)J_0(st)\,\mathrm{d}t, \tag{4}$$

where $J_0(x)$ is the zero-th order Bessel function of the first kind. The dual integral equation (2) can be reduced to a single Abel integral equation, which yields

$$\Phi(t) + t\int_0^a \Phi(x)L(t,x)\,\mathrm{d}x = t\int_0^t \frac{p(x)}{\sqrt{t^2 - x^2}}\,\mathrm{d}x, \tag{5}$$

where

$$L(t,x) = \int_0^\infty s[F(sh) - 1]J_0(st)J_0(sx)\,\mathrm{d}s. \tag{6}$$

To facilitate parametric studies, the integral equation (5) is normalised by introducing $x_1 = x/a$, $t_1 = t/a$, and $\phi = \Phi/(\pi a/2)$,

$$\phi(t_1) + t_1\int_0^1 \phi(x_1)l(t_1,x_1)\,\mathrm{d}x_1 = \frac{2t_1}{\pi}\int_0^{t_1} \frac{p(x_1)}{\sqrt{t_1^2 - x_1^2}}\,\mathrm{d}x_1, \tag{7}$$

where

$$l(t_1,x_1) = \left(\frac{a}{h}\right)^2 \int_0^\infty z[F(z) - 1]J_0(zt_1a/h)J_0(zx_1a/h)\,\mathrm{d}z \tag{8}$$

the stress intensity factor is now given by

$$K = \phi(1)\sqrt{\pi a}. \tag{9}$$

In the case of uniform traction, $p(x) = \sigma$, the right-hand side of the integral equation (5) becomes σt_1. In the special case of homogeneous materials ($\mu_1 = \mu_2, \nu_1 = \nu_2$), we have $F(s) = 1.0$, and hence it is easy to verify that the stress intensity factor for a central crack in a single material, $K = \sigma\sqrt{\pi a}$, is recovered.

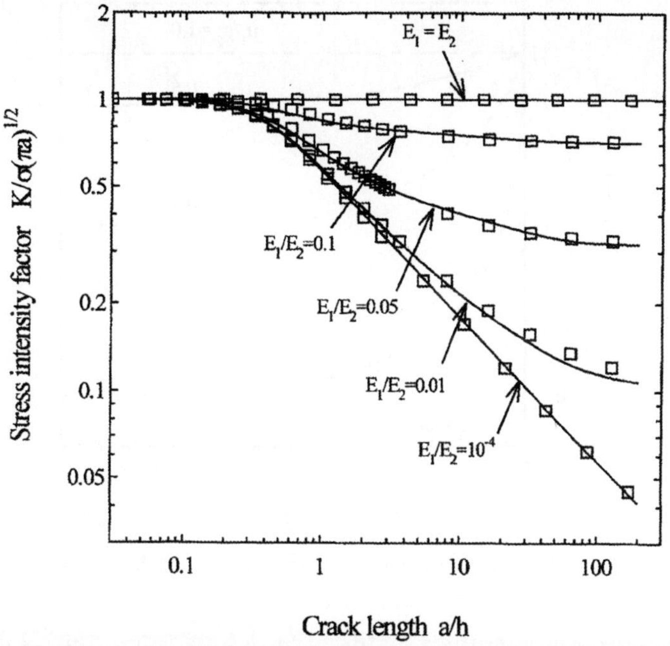

Figure 2. Stress intensity factor of a crack in an elastic constrained layer. Symbols denote the numerical solutions of integral equation and curves represent Equation 11.

It is clear from equation (7) that for given elastic modulus ratio (μ_1/μ_2) and Poisson's ratios, the stress intensity factor depends parametrically on the crack length to layer thickness ratio (a/h),

$$K = \sigma\sqrt{\pi a}Y\left(\frac{h}{a}, \frac{\mu_1}{\mu_2}, \nu_1, \nu_2\right). \tag{10}$$

The integral equation (5) can be solved using the Gauss–Legendre method (Press et al., 1992). A convergence study showed that for $a/h < 100$ an accuracy of 0.1 percent in the stress intensity factor can be achieved with $N = 60$, where N is the order of Gauss–Legendre polynomials. The numerical results of the integral equation are shown in Figure 2, which indicate two distinctly different regimes, denoted here as the 'short crack' and the 'long crack' regimes. For crack lengths far shorter than the thickness of the layer, the crack field (characterised by K) is not affected by the constraint imparted by the substrates. However, in the 'long crack' limit, viz crack length is much greater than layer thickness ($a/h \gg 1$), the stress intensity factor is strongly influenced by the ratio of the Young's moduli of the adhesive to the substrate. An approximate formula (Wang, 1997b) constructed based on the asymptotic solutions for short and long crack limits is also shown in the figure for comparison,

$$K = \sigma\sqrt{\pi a}\left[\beta\frac{E_1}{E_2} + \frac{h}{\pi a}\left(1 - \beta\frac{E_1}{E_2}\right)\tanh\frac{\pi a}{h}\right]^{1/2}, \tag{11}$$

where

$$\beta = 1 + [1 - (E_1/E_2)^{1/3}]\, e^{-Ba/h} \tag{12}$$

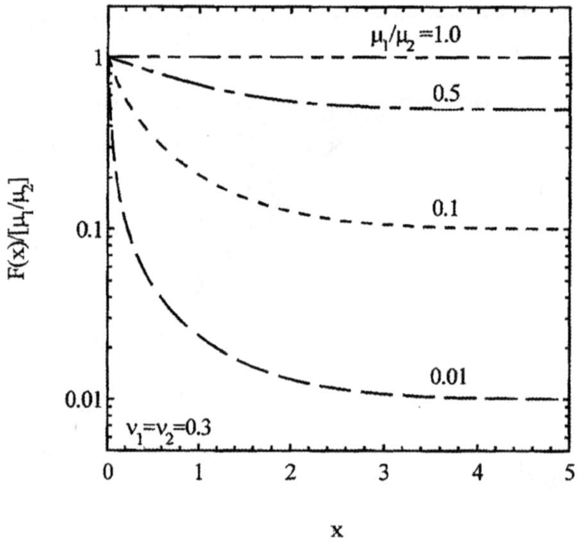

Figure 3. Shape of function $F(s)$ for various modulus ratios.

with constant B being determined by curve-fitting the numerical results of the integral equation, $B = 0.048$.

In the long crack limiting case, Sih and Chen (1981) reported that the stress intensity factor can be expressed as

$$K = \frac{\mu_1}{\mu_2}\frac{1-\nu_2}{1-\nu_1}\sigma\sqrt{\pi a} \tag{13}$$

upon equating $F(s)$ in equation (2) to the following limiting value,

$$F(0) = \lim_{x\to 0} F(x) = \frac{a_1 + a_3 + a_4}{a_1 - a_4} = \frac{\mu_2}{\mu_1}\frac{1-\nu_1}{1-\nu_2}. \tag{14}$$

It should be pointed that such simplification is incorrect, because $F(x)$ approaches rapidly from $F(0)$ to 1.0 as x increases from 0 to ∞, provided $h \neq 0$, as illustrated in Figure 3. Consequently, such an approximation would significantly overestimate (or underestimate) the integral on the left-hand side of equation (2a) if $F(0)$ is greater (or less) than unity, thus leading to erroneous solution of the stress intensity factor. The function $A(s)$ determined in this way will not capture the boundary layer behaviour near the crack tip. Furthermore, it is easy to see from equation (8) that such an approximation would cause the kernel $l(t_1, x_1)$ to diverge, except when $F(0) = 1.0$ which corresponds to a homogeneous material system.

The results of the stress intensity factor for this limiting case are plotted in Figure 4 for $h/a = 10^{-3}$ (no solution is possible for $h = 0$), together with the results obtained by Sih and Chen. Also shown in the figure is the following solution derived based on energy method (Trantina, 1972) or the J-integral theory (Wang, 1997) in plane strain,

$$K = \sqrt{\frac{\mu_1}{\mu_2}\frac{1-\nu_2}{1-\nu_1}}\sigma\sqrt{\pi a} = \sqrt{\frac{\mu_1}{\mu_2}\frac{1-\nu_2}{1-\nu_1}}K_N, \tag{15}$$

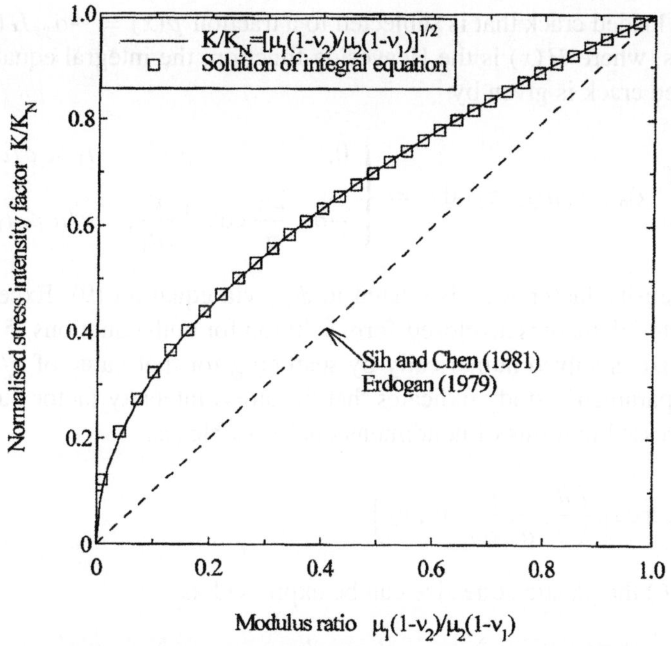

Figure 4. Normalised stress intensity factor for the long crack limit.

where $K_N = \sigma\sqrt{\pi a}$ denotes the nominal stress intensity factor. It is clearly seen that except for the homogeneous case, the formula given by Sih and Chen (1981) and Erdogan (1979) significantly underestimates the actual stress intensity factor of a long crack in a constrained layer, while equation (15) correlates very well with the numerical results. It should be pointed that, although the integral (8) converges for any a/h ratios, computation of the kernel $l(t_1, x_1)$ can be extremely expensive for large a/h values, as the oscillating frequency of the Bessel functions in equation (8) increases rapidly with a/h.

3. A strip yield model

The length of the plastic zone can be determined by superimposing the solutions of two elastic problems, as advocated by Dugdale (1960). This is done by cancelling the stress intensity factor at the tip of a fictitious crack of length $2a$, which is subjected to the remote applied load and a cohesive stress over a region ($c \leqslant |x| < a$) near the crack tip. The length over which the cohesive stress is required to cancel the stress intensity factor at the end of this fictitious crack is defined as the plastic zone length, $r_p = a - c$. This condition can be expressed as

$$K + K_{\sigma_{ys}} = 0, \tag{16}$$

where K and $K_{\sigma_{ys}}$ are, respectively, the stress intensity factors corresponding to a crack subjected to the remote applied stress and a cohesive stress acting on the crack faces. In the present case, plastic yielding is assumed to be contained in the adhesive layer while the substrates are assumed to remain elastic.

For a partially loaded crack that is subjected to a traction $p(x) = -\sigma_{ys}H(x_1 - c/a)$ acting on the crack faces, where $H(x)$ is the Heaveside function, the integral equation representing the partially loaded crack is given by,

$$\phi_{\sigma_{ys}}(t_1) + t_1 \int_0^1 \phi_{\sigma_{ys}}(x_1) l(t_1, x_1)\,\mathrm{d}x_1 = \begin{cases} 0, & t_1 < c/a \\ -\sigma_{ys}\dfrac{2t_1}{\pi}\cos^{-1}\dfrac{c}{at_1}, & c/a \leqslant t_1 < 1 \end{cases} \tag{17}$$

and the stress intensity factor $K_{\sigma_{ys}}$ is related to $\phi_{\sigma_{ys}}$ via equation (9). Except in the case of homogeneous material, there is no closed-form solution for both equations (5) and (17). Hence the plastic zone size is solved numerically by searching for that value of c/a which satisfies equation (16). A parametric study indicates that the stress intensity factor due to the cohesive stress can be expressed in terms of nondimensional variables as

$$K_{\sigma_{ys}} = -\sigma_{ys}\sqrt{\pi a}Y_0\left(\frac{h}{a}, \frac{c}{a}, \frac{\mu_1}{\mu_2}, \nu_1, \nu_2\right). \tag{18}$$

From equation (16) the plastic zone size can be expressed as

$$\frac{r_p}{a} = F_0\left(\frac{\sigma}{\sigma_{ys}}Y, \frac{h}{a}, \frac{\mu_1}{\mu_2}, \nu_1, \nu_2\right) = F_1\left(\frac{\sigma}{\sigma_{ys}}, \frac{h}{a}, \frac{\mu_1}{\mu_2}, \nu_1, \nu_2\right), \tag{19}$$

where function F_1 in general can only be determined numerically, except in the limiting case of small scale yielding where the following conventional relationship is recovered,

$$r_{p,K} = \frac{8}{\pi}\left(\frac{K}{\sigma_{ys}}\right)^2. \tag{20}$$

It has been found that, for a cracked layer with elastic modulus lower than that of the substrate, (i) the region in which the stress/strain field is dominated by the K-field may be comparable or even smaller than the layer thickness (Wang, 1997a), and (ii) outside the singular field the y-stress ahead the tip of a crack in a constrained layer is much larger than that for a homogeneous material subjected to the same stress intensity factor. In fact a plateau in the y-stress has been observed from finite element analysis (Wang, 1997a). The spread of plastic yielding can thus be rather ‘abrupt’ upon crossing over this plateau region, i.e., the plastic zone grows faster than K^2.

For convenience, let us now define the size of the singular region in which the stress/strain field is dominated by the K-field as r_K [e.g., the singularity approximation results in roughly a 2 percent underestimate of σ_{yy} and a 20 percent overestimate of σ_{xx}, see Anderson (1995)], and the region in which the stress/strain field is dominated by the nominal field K_N as R. In the special case of homogeneous materials, we have $r_K = R$. For the conventional small scale yielding to hold, the following condition must be met: $r_p \ll r_K$ and $r_p \ll h \ll a$, so that there exists a similitude between different geometries. However, with the increase of the stress intensity factor, the plastic zone will come into contact with the stiff elastic substrates and eventually spread laterally along the adhesive layer (Varias et al., 1991), beyond the crack-tip K-field. Consequently the problem ought to be considered as large scale yielding, within which two more cases can be distinguished: *case A*: $h \ll r_p \ll R$ and *case B*: $h \ll R \ll r_p$. In *case A*, the growth of the plastic zone has been shown to scale with $r_p = K_N^2/\pi(\alpha\sigma_0)^2$

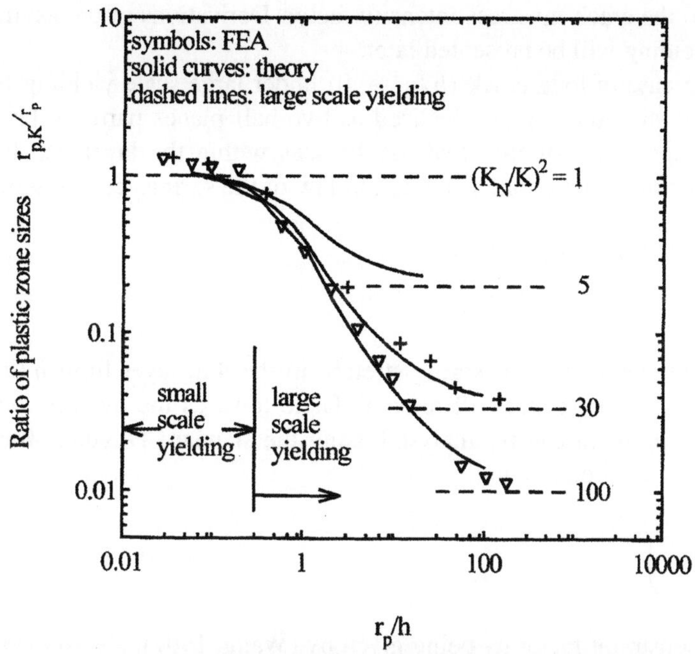

Figure 5. Transition from small scale to large scale yielding.

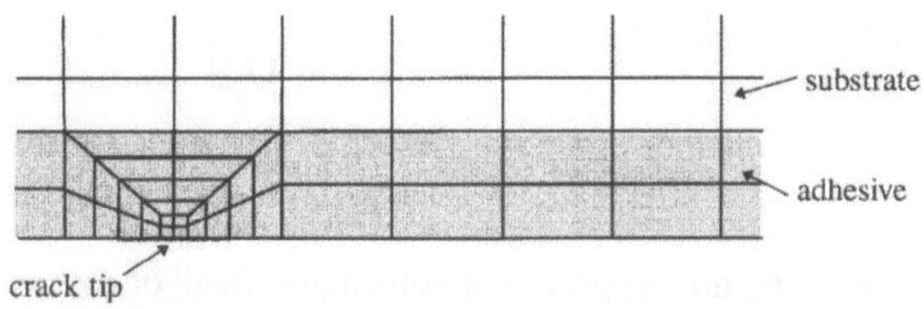

Figure 6. Finite element mesh near crack tip.

(Varias et al., 1991), where α denotes a plasticity constraint factor. This means that for layers with elastic modulus lower than the substrates, there exists a rapid transition from small scale yielding to large scale yielding, because the 'shielding' effect of stiff elastic substrates leads to $K \ll K_N$.

The numerical results of the plastic zone length computed using the strip yield model are shown in Figure 5 for three different modulus ratios. Here the demarcation between small scale yielding and large scale yielding is defined as when the small scale yielding assumption will underestimate the plastic zone size by 20 percent. To verify the strip yield model, finite element analysis has been carried out using a general purpose code (ABAQUS, 1995). The finite element mesh near the crack tip region is shown in Figure 6. The adhesive layer was assumed to be elastic-perfectly plastic and associated plasticity theories were employed with the von Mises yield criterion. As seen in the figure, the predictions of the strip yield model are in good agreement with the finite element results. The results shown in Figure 5 clearly reveal that the plastic zone size is no longer uniquely characterised by the stress intensity factor except when the plastic zone length is much smaller than the layer thickness. The transition from small scale yielding to large scale yielding depends on the ratio of the nominal stress intensity factor to the crack tip stress intensity factor. Derivations of the asymptotic solutions for large scale yielding will be presented later.

In the limiting case of long crack ($h/a \to 0$) under large scale yielding ($r_p/h \gg 1$), the entire sandwiched structure can be idealised as two half planes partially bridged by elastic-perfectly plastic springs, assuming that the stresses within the layer can be considered as constant in the thickness direction. The traction law of the springs can be written as

$$\sigma_b = \begin{cases} E_1 u_y/h & u_y \leqslant \sigma_{ys} h/E_1 \\ \sigma_{ys} & u_y > \sigma_{ys} h/E_1, \end{cases} \tag{21}$$

where $2u_y$ denotes the amount of spring stretch. In the thin layer limit $h/a \to 0$, the contribution from those springs that still remain elastic towards the overall compliance can be neglected and the problem can be analysed using the standard Dugdale–Barrenblatt model, for which solutions are well known,

$$r_p = \frac{\pi}{8}\left(\frac{K_N}{\alpha_1 \sigma_{ys}}\right)^2 \tag{22}$$

with the plastic constraint factor α_1 being given by (Wang, 1997),

$$\alpha = \begin{cases} \dfrac{1}{\sqrt{1 - \nu_1 + \nu_1^2}} \approx 1.134 & \text{plane stress} \\ & (\nu_1 = 1/3) \\ \dfrac{1 - \nu_1}{1 - 2\nu_1} = 2.0 & \text{plane strain} \end{cases} \tag{23}$$

assuming that no post-yielding nonproportional deformation will occur within the plastic zone so that the ratio of the effective cohesive stress to the uniaxial yield stress can be considered as the same as that prior to plastic yielding. For comparison, Equation (22) is also plotted in Figure 5 (dashed line). The large scale yielding solution given by Equation (22) is consistent with the approximate solution obtained by Varias et al. (1991), assuming that the plastic zone size is equal to the distance over which the elastic equivalent stress is equal to the material's yield stress. After modifying their original formula to account for the stress redistribution as a result of plastic yielding, the plastic length is given by

$$r_p = \frac{1}{\pi}\left(\frac{K_N}{\alpha_2 \sigma_{ys}}\right)^2 \tag{24}$$

with the plastic constraint factor α_2 being given by,

$$\alpha_2 = \left[\frac{2}{A^2 + B^2 + C^2}\right]^{1/2}, \tag{25}$$

where A, B, C are given in Appendix C. A graphical representation of the dependence of α_1 on the elastic property mismatch is shown in Figure 7. In the limiting case of $E_1/E_2 \to 0$, the constraint factor given by Equation (25) is identical to (23).

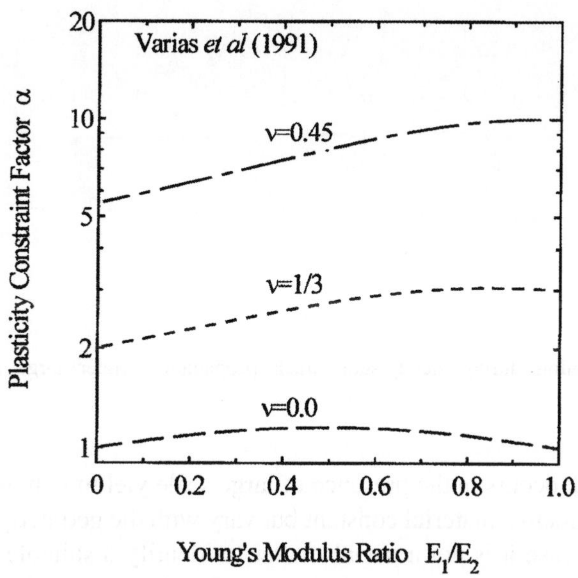

Figure 7. Variation of constraint factor versus Young's modulus ratio.

4. Energy dissipation under steady-state large scale yielding

From the strip yield model, the crack tip opening displacement (δ) can be determined,

$$\delta = \frac{4a}{E_1} \int_{c/a}^{1} \frac{\phi(t_1) + \phi_{\sigma_{ys}}(t_1)}{\sqrt{t_1^2 - (c/a)^2}} \, dt_1 . \tag{26}$$

An extensive numerical study confirms that, as expected, the relationship between J-integral and the crack tip opening displacement δ is independent of the plastic zone length; no details will be given here. This can be easily verified by taking advantage of the path-independence of J-integral. However, although the relationship between J and δ holds for both small scale yielding and large scale yielding ($r_p \gg h$), the energy dissipation during fracture under small scale yielding and large scale yielding conditions differs significantly due to the break down in similitude.

When fracture due to crack propagation is concerned, the critical energy release rate per unit of crack advance is (Wei and Hutchinson, 1997),

$$G_c = \Gamma_0 + \Gamma_p , \tag{27}$$

where Γ_0 is the work consumed by the near-tip fracture process in creating unit area of new crack surfaces, and Γ_P denotes the sum of the energy dissipated by plastic deformation and the residual elastic energy stored in the wake. In general it is not possible to compute Γ_P directly based on traditional mechanical properties of the material, as the calculation of the energy dissipation within the plastic zone is extremely complicated. Instead, the traditional framework of fracture mechanics relies on the assumption of small-scale yielding at the crack tip, so that Γ_P can be indirectly measured for a given material using one particular specimen configuration. The values of Γ_P and hence Gc are independent of the geometry or loading conditions under which these values are measured.

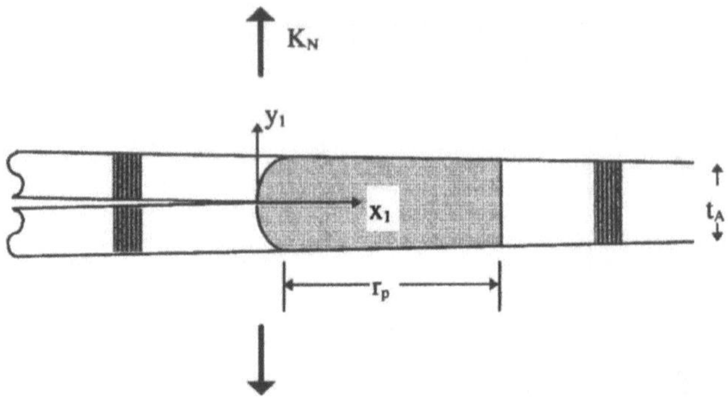

Figure 8. Energy dissipation during steady-state crack propagation under large scale yielding condition ($r_p \gg t_A$).

When crack growth occurs in the presence of large scale yielding, however, the fracture energy may no longer remain a material constant but vary with the geometry or loading condition of a structure. In this case it is not a trivial matter to identify a suitable near-tip deformation measure, whose critical value is a material constant. In this regard, criteria based on a critical crack-tip opening displacement or the near-tip opening angle have often been used. Since the similitude requirement is not exactly met under large scale yielding conditions, these near-tip fracture parameters are not strictly material constants, although they are often viewed as such. For a crack embedded in a constrained layer, the plastic constraint imparted by the stiff elastic substrates differs significantly from that existing in a homogeneous material, especially under large scale yielding. Numerous experimental studies (Kinloch,1987; Chai,1988; 1992) have found that the fracture energy of joints bonded with ductile adhesives exhibit a more or less linear dependence on the adhesive layer thickness, clearly demonstrating the failure of crack tip opening displacement as a fracture criterion.

4.1. MODE I FRACTURE

Let us now consider the critical condition for steady-state propagation of a long crack under large scale yielding. A semi-infinite crack advances under steady-state conditions along the mid-plane of layer as depicted in Figure 8. Plane strain deformations are considered, and the substrates are assumed to remain elastic. The extent of plastic deformation within the layer is much greater than the layer thickness. In steady-state, a material point at (x_1, y_1) experiences a history identical to that of the cumulative history of a spatial sequence of points on the straight line connecting (∞, y_1). The energy dissipated per unit of crack advance due to plasticity (Γ_P) is equal to the work required to convert (plastically deform) a unit width element at $x_1 \to \infty$ to a unit width element at $x_1 \to -\infty$

$$\Gamma_p = \int_{-t_A/2}^{t_A/2} W \, \mathrm{d}y_1, \tag{28}$$

where W denotes the strain energy density defined as

$$W = \int_0^{\varepsilon} \sigma_{ij}(x_1 \to -\infty) \, \mathrm{d}\varepsilon_{ij}. \tag{29}$$

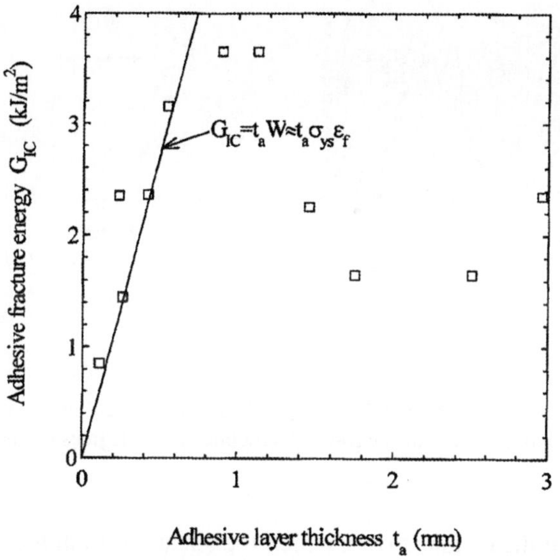

Figure 9. Energy dissipation during steady-state mode I cracking under large scale yielding condition ($\sigma_{ys} = 60$ MPa, $\varepsilon_f = 0.1$) (from Kinloch and Shaw, 1981).

In the present case, it is not unreasonable to assume that both the stresses and strains at $x_1 \to -\infty$ are uniform in y_1, with the only nonzero stress component in the layer being σ_{yy}. Therefore the energy dissipation due to plasticity can be further simplified, noting $\varepsilon_{xx} = \varepsilon_{zz} = \gamma_{xy} = \gamma_{yz} = \gamma_{zx} = 0$,

$$\Gamma_P = t_A \int_0^{\varepsilon_f} \sigma_{yy} \, d\varepsilon_{yy}, \tag{30}$$

where the integral on the right-hand side of the above equation represents the critical strain energy density (the total area below the tensile stress-strain curve) of the layer material determined under constrained conditions ($\varepsilon_{xx} = \varepsilon_{zz} = 0$). Therefore the critical energy release rate for a constrained layer containing a semi-infinite crack under large scale yielding conditions is given by

$$G_c = \Gamma_0 + t_A \int_0^{\varepsilon_f} \sigma \, d\varepsilon, \tag{31}$$

which suggests a linear dependence of the work of fracture on the layer thickness, consistent with many experimental findings of adhesively bonded joints under mode I (Kinloch,1987). For instance, the experimental results reported by Kinloch and Shaw (1981) are replotted in Figure 9 together with the predictions made by equation (31), indicating a very good correlation in the small thickness regime where large scale yielding is most likely to occur prior to final fracture.

4.2. Shear mode fracture

Shear cracking: at the tip of the main crack microcracks will form (see Anderson, 1995), at an angle of 45°. The spacing between the ligaments is approximately equal to the layer thickness

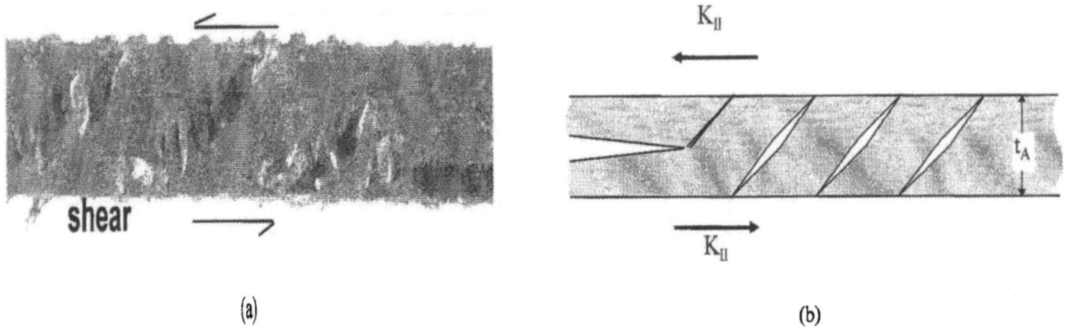

Figure 10. Shear cracking in a constrained layer (a) adhesively bonded joint under mode II loading, (b) steady-state cracking due to advancement of micro-cracks inclined at 45 degrees to the main crack.

as it has been found that the distance over which the shear stress in a constrained layer attains its peak value is approximately equal to the layer thickness (Wang and Heller, 1997). An example of shear cracking in the adhesive layer of a bonded joint is shown in Figure 10. Since the formation of the macro-cracks is mainly driven by the tensile stress, only very little amount of plastic deformation hence energy dissipation will occur. Therefore the major source of energy dissipation is due to the subsequent plastic deformation of the un-cracked ligament, which is effectively under simple compression along the ligament direction. This is probably why the shear failure strains of many ductile adhesives can be as high as 100 percent or even more (Kinloch, 1987). For many structural adhesives, the failure strain in compression is much greater than the tensile failure strain, giving rise to much higher mode II fracture toughness than mode I (Wang, 1996).

For simplicity, let us consider the shear deformation of an inclined ligament as depicted in Figure 11(a). Due to the extremely complex stress distribution within the ligament and accompanied large strain deformation during plastic straining of the ligament, it is almost impossible to precisely quantify analytically the plastic energy dissipation. To this end, an elastic-plastic finite element analysis has been conducted using a general purpose finite element software (ABAQUS, 1995). The finite element analysis is performed using large strain, geometrically nonlinear, formulation. The adhesive is assumed to be a linear kinematic hardening law with a plastic modulus $E_p = E/2000$, approaching an elastic-perfectly plastic limit. The force-displacement results are presented in Figure 11(b). It can be seen that during the plastic deformation, the average shear stress is approximately equal to the yield stress of the material. To further illustrate this point, the distribution of shear stress for two different deformation states ($u/t_A = 1$ percent and $u/t_A = 100$ percent) are shown in Figures 12 (a) and (b). The shear stress appears to be more or less uniform, equal to the yield stress, over a large area.

The area below the force-displacement curve shown in Figure 11(b) is equal to the total amount of energy dissipated during plastic deformation, provided the ligament does not fail until $u/t_A = 100$ percent. Tensile stress starts to develop when u/t_A exceeds unity. Therefore the shear mode fracture energy of bonded joints is given by

$$G_{\mathrm{IIc}} = G_0 + t_A \int_0^{\gamma_{\max}} \tau \, \mathrm{d}\gamma, \tag{32}$$

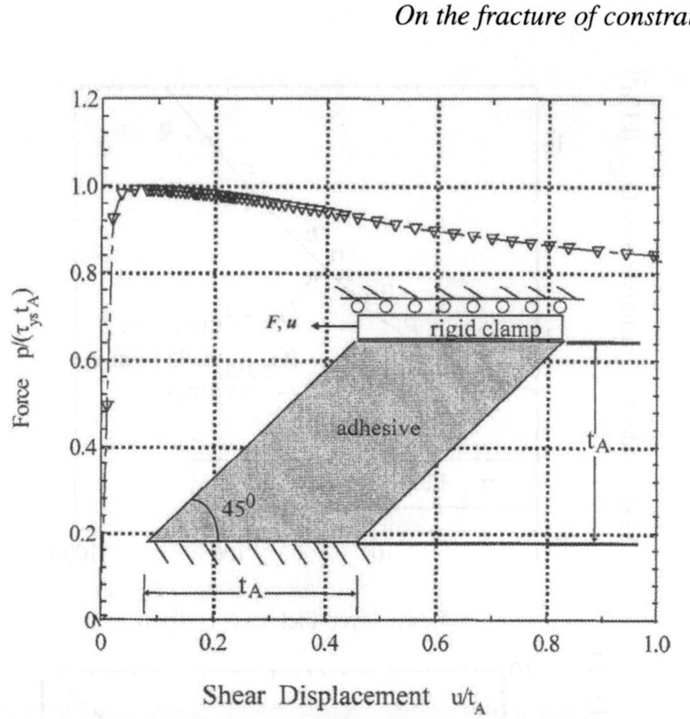

Figure 11. The force-displacement (u/t_A) relation, determined using finite element method, of an un-cracked ligament subjected to simple shear.

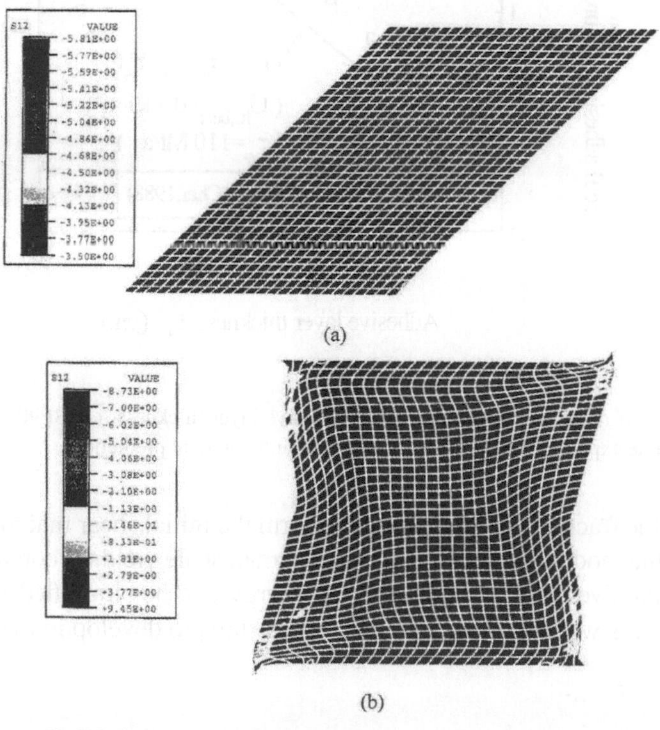

Figure 12. Distributions of shear stress at (a) $u/t_A = 0.01$ and (b) $u/t_A = 1$.

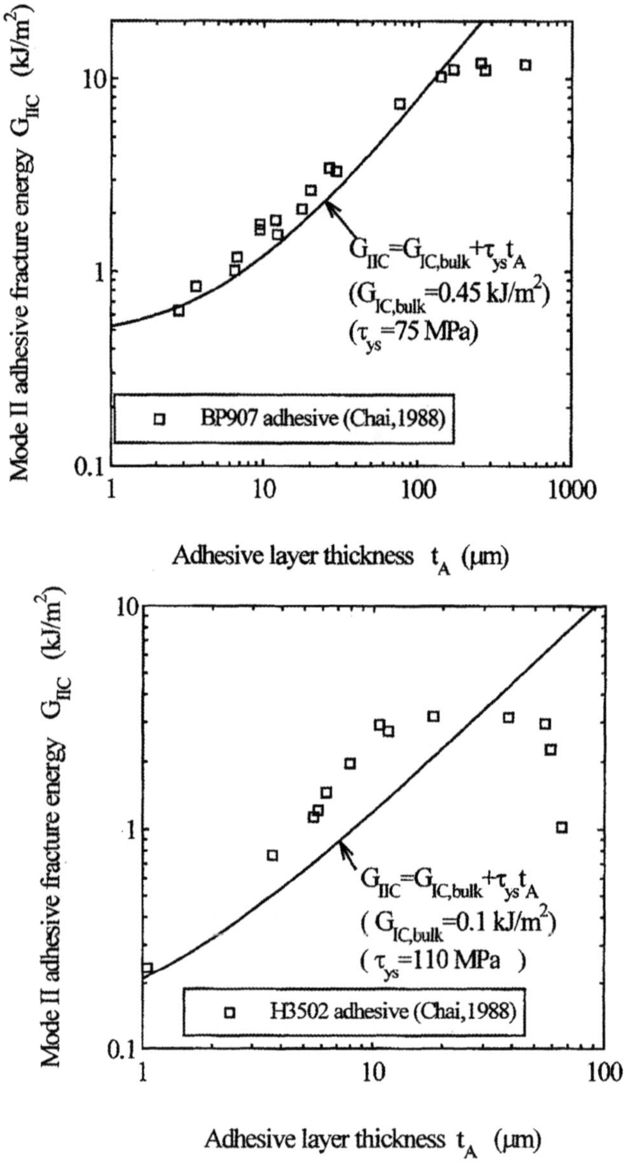

(b)

Figure 13. Dependence of mode II fracture energy on adhesive layer thickness. (a) BP907 adhesive, (b) h3502 adhesive. Symbols denote experimental data, and curves denote theoretical predictions.

where G_0 denotes the fracture energy required to form the initial shear macro-cracks, which is normally equal to the mode I fracture energy under small scale yielding condition. For elastic-perfectly plastic adhesives, the mode II fracture energy can be simplified to, assuming that ligament failure occurs when $u/t_A = 1$ (tensile stress starts to develop in the ligament),

$$G_{\mathrm{IIc}} = G_{\mathrm{IC,bulk}} + \tau_{ys} t_A. \tag{33}$$

As seen in Figure 13, the present theory agrees reasonably well with experimental results obtained by Chai (1988) for one brittle adhesive (H3502, $\tau_{ys} = 110$ MPa) and one ductile adhesive (BP907, $\tau_{ys} = 75$ MPa). The slight underestimations could be ascribed to possible friction between adherends when tested under shear mode, which has neglected in the theory.

5. Conclusions

The spread of plastic yielding of a crack embedded in a constrained layer has been analysed using integral transform technique. An approximate solution has also been obtained, which is shown to be in close agreement with the finite element results. The transition from small scale yielding and to large scale yielding normally occurs at a plastic zone length to layer thickness much less than unity. The smaller the ratio of the Young's moduli of the adhesive to that of the substrates, the sharper the transition from small scale yielding to large scale yielding. A theory has been proposed to model the thickness-dependence of the fracture energy of adhesively bonded joints, which appears to correlate well with experimental results.

Appendix A: Displacements and stresses in terms of stress function

Making use of the Fourier transform (Sneddon, 1951), the two-dimensional biharmonic equation can be solved to yield

$$\Phi(x, y) = \frac{2}{\pi} \int_0^{\infty} G(s, y) \cos(sx) \, ds, \tag{A1}$$

where the symmetry with respect to $x = 0$ has been used. Here the function $G(s, y)$ satisfies the following equation,

$$\left(\frac{d^2}{dy^2} - s^2 \right)^2 G(s, y) = 0, \tag{A2}$$

whose general solution can be written in the form

$$G(s, y) = [A(s) + B(s)y] \cosh(sy) + [C(s) + D(s)y] \sinh(sy), \tag{A3}$$

where the functions $A(s)$, $B(s)$, $C(s)$ and $D(s)$ are yet to be determined from the boundary conditions.

The corresponding stress components are given by

$$\sigma_{xx}(x, y) = \frac{2}{\pi} \int_0^{\infty} \frac{\partial^2 G(s, y)}{\partial y^2} \cos(sx) \, ds, \tag{A4}$$

$$\sigma_{yy}(x, y) = -\frac{2}{\pi} \int_0^{\infty} s^2 G(s, y) \cos(sx) \, ds, \tag{A5}$$

$$\tau_{xy}(x, y) = \frac{2}{\pi} \int_0^{\infty} s \frac{\partial G(s, y)}{\partial y} \sin(sx) \, ds. \tag{A6}$$

The components of the displacement vector in plane strain are

$$u(x, y) = \frac{1}{\pi \mu} \int_0^{\infty} \left[(1 - \nu) \frac{\partial^2 G(s, y)}{\partial y^2} + \nu s^2 G(s, y) \right] \frac{\sin(sx)}{s} \, ds, \tag{A7}$$

$$v(x, y) = \frac{1}{\pi\mu}\int_0^\infty \left[(1-\nu)\frac{\partial^3 G(s, y)}{\partial y^3} - (2+\nu)s^2\frac{\partial G(s, y)}{\partial y}\right]\frac{\cos(sx)}{s^2}\,\mathrm{d}s. \tag{A8}$$

Appendix B: Expressions of $a_i (i = 1, 2, 3, 4)$

The following expressions represent the parameters used in equation (3).

$$a_1 = (3-4\nu_1) + (10 - 12\nu_1 - 12\nu_2 + 16\nu_1\nu_2)\frac{\mu_1}{\mu_2} + (3-4\nu_2)\left(\frac{\mu_1}{\mu_2}\right)^2, \tag{B1}$$

$$a_2 = 4 + (8 - 16\nu_2)\frac{\mu_1}{\mu_2} - (12 - 16\nu_2)\left(\frac{\mu_1}{\mu_2}\right)^2, \tag{B2}$$

$$a_3 = (10 - 24\nu_1 + 16\nu_1^2) - (4 - 8\nu_1 - 8\nu_2 + 16\nu_1\nu_2)\frac{\mu_1}{\mu_2} - (6 - 8\nu_2)\left(\frac{\mu_1}{\mu_2}\right)^2, \tag{B3}$$

$$a_4 = (3-4\nu_1) - (6 - 4\nu_1 - 4\nu_2)\frac{\mu_1}{\mu_2} + (3-4\nu_2)\left(\frac{\mu_1}{\mu_2}\right)^2. \tag{B4}$$

Appendix C: Expressions for A, B, C

Following Varias et al. (1991), under plane strain conditions the stresses and strains in the adherends are

$$\sigma_{xx}^{(2} = \sigma_{yy}^{(2)} = \sigma, \qquad \sigma = \frac{K_N}{\sqrt{2\pi r}}, \tag{C1}$$

$$\varepsilon_{zz}^{(2)} = 0 \tag{C2}$$

and the continuity conditions across the interface are

$$\varepsilon_{xx}^{(1)} = \varepsilon_{xx}^{(2)}, \qquad \sigma_{yy}^{(1)} = \sigma_{yy}^{(2)}, \tag{C3}$$

$$\varepsilon_{zz}^{(1)} = 0, \tag{C4}$$

from which the stresses in the adhesive layer can be expressed in terms of the stress in the adherends

$$\sigma_{xx}^{(1)} = \left[\frac{1}{1-\nu_1^2}\frac{E_1}{E_2}(1 - \nu_2 - 2\nu_2^2) + \frac{\nu}{1-\nu}\right]\sigma, \tag{C5}$$

$$\sigma_{yy}^{(1)} = \sigma, \tag{C6}$$

$$\sigma_{zz}^{(1)} = \left[\frac{\nu}{1-\nu_1^2}\frac{E_1}{E_2}(1 - \nu_2 - 2\nu_2^2) + \frac{\nu}{1-\nu}\right]\sigma. \tag{C7}$$

After making use of the von Mises yield criterion, the following plasticity constraint factor is obtained

$$\alpha_2 = \left[\frac{2}{A^2 + B^2 + C^2}\right]^{1/2}, \tag{C8}$$

where

$$A = \frac{1}{1-\nu_1^2}\left[(1-\nu_1-2\nu_1^2) - \frac{E_1}{E_2}(1-\nu_2-2\nu_2^2)\right], \tag{C9}$$

$$B = \frac{1}{1-\nu_1^2}\left[(1-\nu_1-2\nu_1^2) - \frac{E_1}{E_2}\nu_1(1-\nu_2-2\nu_2^2)\right], \tag{C10}$$

$$C = \frac{1}{1+\nu_1}\frac{E_1}{E_2}(1-\nu_2-2\nu_2^2). \tag{C11}$$

Acknowledgment

The author would like to thank Dr. L.R.F. Rose for useful discussions.

References

Anderson, T.L. (1995). *Fracture Mechanics, Fundamentals and Applications*, 2nd Edition, CRC Press Inc, Boca Raton.

Barenblatt, G.I. (1962). The mathematical theory of equilibrium cracks in brittle fracture. *Advances in Applied Mechanics* **7**, 55–129.

Becker, T.L. Jr., McNaney, J.M., Cannon, R.M. and Ritchie, R.O. (1997). Limitations of the use of the mixed-mode delaminating beam test specimen: Effects of the size of the region of K-dominance. *Mechanics of Materials* **25**, 291–308.

Chai, H. (1986). Bond thickness effect in adhesive joints and its significance for mode I interlaminar fracture of composites. *Composite Materials: Testing and Design (Seventh Conference), ASTM STP 893* (Edited by J.M. Whiney), American Society for Testing and Materials, Philadelphia, 209–231.

Chai, H. (1988). Shear fracture. *International Journal of Fracture* **37**, 137–159.

Chai, H. (1993). Observation of deformation and damage at the tip of cracks in adhesive bonded loaded in shear and assessment of a criterion for fracture. *International Journal of Fracture* **60**, 311–326.

Daghyani, H.R., Lin, Y. and Mai, Y.-W. (1995). Mode-I fracture behaviour of adhesively joints. Part I. Relationship between fracture energy and bond thickness. *Journal of Adhesion* **53**, 149–162.

Dalgleish, B.J., Lu, M.C. and Evans, A.G. (1988). The strength of ceramics bonded with metals. *Acta Metallurgica* **36**, 2029–2035.

Dugdale, D.S. (1960). Yielding of steel sheets containing slits. *Journal of the Mechanics and Physics of Solids* **8**, 100.

Erdogan, F. (1972). Fracture problems in composite materials. *Engineering of Fracture Mechanics* **4**, 811–840.

Hunston, D.L., Kinloch, A.J. and Wang, S.S. (1989). Micromechanics of fracture in structural adhesive bonds. *Journal of Adhesion* **28**, 103–114.

Hutchinson, J.W. and Suo, Z. (1992). Mixed mode cracking in layered materials. *Advances in Applied Mechanics* **29**, 63–191.

Kinloch, A.J. (1987). *Adhesion and Adhesives, Science and Technology*, Chapman & Hall, London.

Kinloch, A.J. and Shaw, S.J. (1981). The fracture resistance of a toughened Epoxy adhesive. *Journal of Adhesion* **12**, 59–77.

Knauss, W.G. (1966). Stresses in an infinite strip containing a semi-infinite crack. *Journal of Applied Mechanics* **33**, 356–362.

Paris, P.C. and Sih, G.C. (1965). In *Fracture Toughness Testing and its Applications*, ASTM STP 381, 30–81.

Press, W.H., Teukolsky, S.A., Vetterling, W.T. and Flannery, B.P. (1992). *Numerical Recipes in Fortran 77*, Cambridge University Press.

Rice, J.R. (1967). Stresses in an infinite strip containing a semi-infinite crack. *Journal of Applied Mechanics* **34**, 248–249.

Sih, G.C. and Chen, E.P. (1981). Composite material with a cracked layer and fibre, in *Cracks in Composite Materials*, Martinus Nijhoff Publishers, Hague.

Sneddon, I.N. (1951). *Fourier Transforms*, McGraw-Hill Book Company, Inc., New York.

Sneddon, I.N. and Lowengrub, M. (1969). *Crack Problems in the Classical Theory of Elasticity*, John Wiley and Sons, Inc., New York.

Trantina, G.G. (1972). Fracture mechanics approach to adhesive joints. *Journal of Composite Materials* **6**, 192–207.

Tvergaard, V. and Hutchinson, J.W. (1996). On the toughness of ductile adhesive joints. *Journal of the Mechanics and Physics of Solids* **44**, 789–800.

Varias, A.G., Suo, Z. and Shih, C.F. (1991). Ductile failure of a constrained metal foil. *Journal of the Mechanics and Physics of Solids* **39**, 963–986.

Varias, A.G., Suo, Z. and Shih, C.F. (1992). Mode mixity effect on the damage of a constrained ductile layer. *Journal of the Mechanics and Physics of Solids* **40**, 485–509.

Wang, C.H. (1996). Fracture of interface cracks under combined loading. *Journal of Engineering Fracture Mechanics.* **56**(1), 77–86.

Wang, C.H. (1997a). Analysis of cracks in constrained layers. *International Journal of Fracture* **83**, 1–17.

Wang, C.H. (1997b). Strip yielding in a constrained layer. *Advances in Fracture Research* (proceedings of ICF9), **1**, Pergamon Press, 587–594.

Wang, C.H. and Heller, M. (1997). Stress concentrations in bonded lap joints. Submitted to *Journal of Strain Analysis* for publication.

Wang, S.S., Mandell, J.F. and McGarry, F.J. (1978). An analysis of the crack tip stress field in DCB adhesive fracture specimens. *International Journal of Fracture* **14**(1), 39–58.

Wei, Yueguang and Hutchinson, J.W. (1997). Nonlinear delamination mechanics for thin films. *Journal of the Mechanics and Physics of Solids* **45**(7), 1137–1159.

International Journal of Fracture **93:** 247–259, 1998.

Fracture size effects and polycrystalline inhomogeneity

Dedicated to Mel and Max Williams

KHALED ABDEL-TAWAB[1] and GREGORY J. RODIN[2]
[1]*Center for Advanced Computational Technology, University of Virginia, NASA Langley Research Center, Hampton, VA 23681, U.S.A.*
[2]*Center for Solids Structures and Materials, The University of Texas at Austin, Austin, TX 78712, U.S.A.*

Received 1 August 1997; accepted in revised form 12 March 1998

Abstract. Fracture toughness testing is prone to exhibit various size effects that have to be understood in order to properly interpret the results. It is shown that the size effect associated with polycrystalline inhomogeneity is weak. Instead polycrystalline inhomogeneity induces statistical effects controlled by a small number of grains adjacent to the crack tip.

Key words: Size effects, statistical effects, fracture toughness, brittle fracture, polycrystalline inhomogeneity.

1. Introduction

Linear elastic fracture mechanics is a meaningful theory for predominantly homogeneous isotropic linear elastic specimens with sharp cracks. The term predominantly means that the crack tip process zone, inside which the material cannot be regarded as homogeneous isotropic linear elastic, constitutes a small fraction of the specimen. It is common to refer to such specimens as K-dominant because they conform with the assumptions under which the K-fields of linear elastic fracture mechanics were derived by Professor M.L. Williams in his celebrated paper (Williams, 1957).

The best understood case of K-dominance involves metal specimens tested at room temperature (e.g. Broek, 1982). In those specimens, the process zone is dominated by plastic yielding and as a result small-scale yielding is usually regarded as a necessary and sufficient condition for K-dominance. This condition is realized in specimens whose minimum size

$$L \geqslant L_y = 2.5 \left(\frac{K_{\mathrm{Ic}}}{\sigma_y} \right)^2 . \tag{1}$$

Here K_{Ic} is the fracture toughness and σ_y is the yield stress. The numerical factor in (1) implies that

$$L_y = 25 r_y, \tag{2}$$

based on the estimate for the plastic zone radius

$$r_y = 0.1 \left(\frac{K_{\mathrm{Ic}}}{\sigma_y} \right)^2 . \tag{3}$$

Thus, if (1) holds, K_{Ic} is a material property insensitive to specimen details. However, if (1) does not hold, the apparent fracture toughness K_c depends on L and possibly other test variables. In practice, such a dependence is usually manifested in the form of the so-called thickness effect, which is particular size effect.

In contrast to smooth specimens used for deformation testing, in which size effects are possible but not that common, fracture specimens are prone to size effects because the fracture process is controlled by both area and volume specific material constants which cannot be scaled without introducing a length dimension. Further, due to significant stress variations ahead of the crack tip, the number of deformation mechanisms and consequently the number of material constants controlling the fracture process are expected to be large. Thus scaling fracture laws must reflect the complexity of deformation behavior and only under rather special conditions are expected to be simple. From this perspective, it is clear that semi-empirical scaling fracture laws based on simple dimensional (Bazant, 1993) or geometrical (Carpinteri, 1994) arguments are of limited use. Instead, fracture size effects should be approached in a manner similar to that proposed by Frost and Ashby (1982) for deformation mechanisms. That is, it is imperative to understand fracture size effects associated with the individual deformation mechanisms first and only then attempt construction of general scaling laws.

Our interest in size effects was motivated by experimental studies on ice conducted during the past ten years by Dempsey and co-workers (Dempsey et al., 1989a, b; Bently et al., 1989; Adamson et al., 1995; Adamson et al., 1997; LeClair et al., 1997) who obtained a unique set of fracture toughness data that involves specimens with L in the range between 10 cm and 80 m; results of those studies are summarized by Dempsey (1991; 1996). Due to the complex ice microstructure, rate effects, and significant temperature variations, the above data exhibit size effects that defy simple explanations. In this paper, we focus on the size effect associated with polycrystalline inhomogeneity. This size effect has to be considered because for ice a typical grain diameter, d, is several millimeters and as a result a typical laboratory fracture specimen is characterized by $L \approx 10d$ Dempsey (1991). Note that due to grain randomness the size effect associated with polycrystalline inhomogeneity has to be described in statistical rather than deterministic terms.

The basic problem considered in this paper is as follows. Suppose that fracture toughness tests are conducted on specimens that satisfy (1) and in addition rate effects are negligible. However, the coarse grain structure of those specimens indicates that the material behavior of the specimens cannot be regarded as homogeneous. Our objective is to evaluate the effect of inhomogeneity and to provide recommendations on how to interpret the fracture toughness tests data.

The stated problem is hardly ever mentioned in connection with room temperature fracture toughness testing of metals. One explanation for that is that for most metals $r_y \gg d$. This inequality implies that the plastic zone is surrounded by a large number of grains, presumably large enough so that one can assume that collectively those grains behave as a homogeneous material. However, one may expect that in brittle polycrystals (small r_y) formed by coarse grains (large d) this assumption may not hold. For such polycrystals the dimensionless number

$$\mathrm{GB} = \frac{d}{r_y}, \tag{4}$$

GB stands for grain brittleness, must be large. This number can vary significantly from one material to another and, indeed, it is large for materials commonly regarded as brittle and

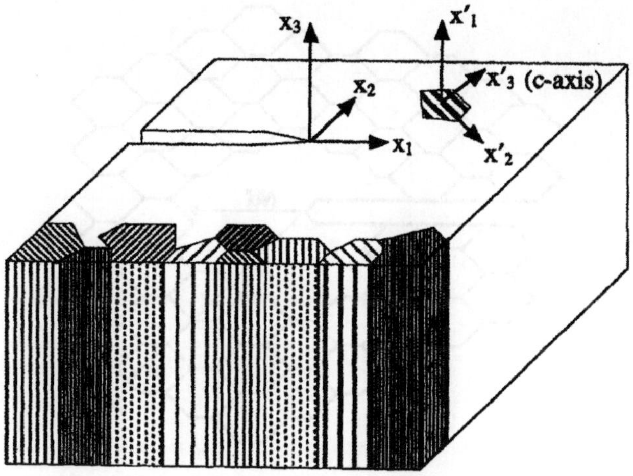

Figure 1. Fracture specimen with columnar microstructure.

small for materials commonly regarded as ductile. The following numerical examples are quite revealing. For alumina with $d = 10\,\mu\text{m}$ ($K_{\text{Ic}} = 4\,MPa\sqrt{\text{m}}$ and $\sigma_y = 5\,\text{GPa}$; Ashby and Jones, 1982) GB = 156. For fresh-water ice ($d = 3\,\text{mm}$, $K_{\text{Ic}} = 120\,\text{KPa}\sqrt{\text{m}}$ and $\sigma_y = 1\,\text{MPa}$; Bentley et al., 1989) GB = 2. In contrast, for high-strength aluminum alloys with grain diameter $d = 10\,\mu\text{m}$ ($K_{\text{Ic}} = 23\,\text{MPa}\sqrt{\text{m}}$ and $\sigma_y = 600\,\text{MPa}$; Ashby and Jones, 1982) GB = 0.07. Thus the stated problem is only relevant to polycrystals characterized by large GB values – otherwise the size effect due to polycrystalline inhomogeneity becomes insignificant in comparison to plastic yielding near the crack tip.

The rest of this paper is divided into two parts. In Section 2, we develop statistical asymptotic and numerical estimates that allow us to characterize polycrystalline inhomogeneity in quantitative terms. In Section 3, we discuss implications of those estimates for interpretation of fracture toughness data.

2. Analysis

2.1. Background

In this section, we consider polycrystalline fracture specimens formed by columnar grains. This two-dimensional microstructure is relevant to S2 ice and it is ideally suited for analysis; later in this paper we demonstrate that the principle conclusions of our work can be also applied to three-dimensional microstructures. An idealized fracture specimen of S2 ice is shown in Figure 1. We assign to the specimen global Cartesian coordinates such that the x_3 axis is along the thickness direction and the negative x_1 axis is along the crack line. On the microscopic scale, the specimen is formed by columnar grains aligned along the thickness direction. The grains are hexagonal (transversely isotropic) crystals whose c-axes are randomly oriented in the x_3 coordinate plane. Thus a sufficiently large sheet of S2 ice should be transversely isotropic with the symmetry axis along the x_3 axis. For a reference grain, local coordinates are such that the x_3' axis coincides with the c-axis and the x_1' axis coincides with the global x_3 axis (Figure 1). If the specimen shown in Figure 1 is subjected to in-plane loading, it should realize predominantly plane strain fields both on the microscopic and macroscopic

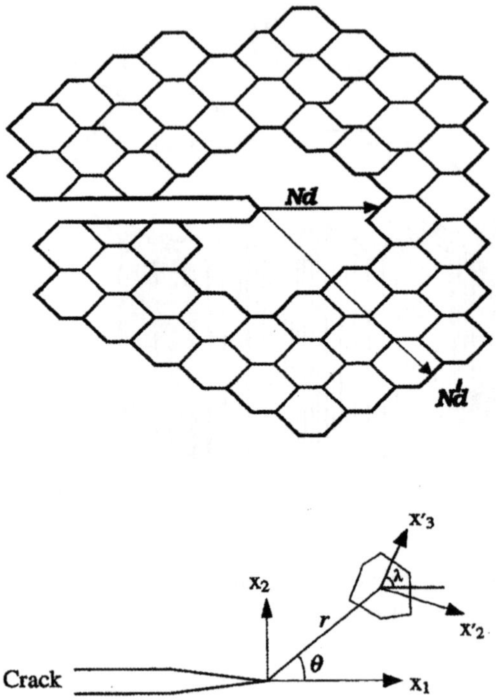

Figure 2. Analysis of ΔK induced by remote grains: (a) specimen; (b) reference grain.

scales. Furthermore, one can show that the macroscopic anisotropy is such that it does not affect calculations of the stress intensity factors.

The objective in this section is to establish how the columnar microstructure affects the relationship between the macroscopic and the microscopic stress intensity factors (SIFs) in specimens similar to the one shown in Figure 1. To this end we suppose that a typical specimen is subjected to boundary conditions consistent with the plain strain Mode I fields controlled by the stress intensity factor K. The induced microscopic Mode I SIF is denoted by K^0 and the difference $K^0 - K$ by ΔK.

2.2. Asymptotic Analysis

In this sub-section we derive an asymptotic statistical estimate for ΔK induced by grains that are at least N grain diameters away from the crack tip; N is sufficiently large but arbitrary otherwise. To suppress the microscopic SIF induced by the remaining grains we replace those grains with a homogeneous material whose properties represent the macroscopic response of the columnar microstructure (Figure 2a). Further, in order to simplify the calculations, we assume that the specimen is infinite.

We compute ΔK in three steps. First, we replace the grains with the equivalent transformation strains. Second, we calculate δK, the Mode I SIF induced by the equivalent transformation strain of a reference grain. Third, using methods of probability theory, we calculate ΔK as the sum over δK's.

The equivalent transformation strain $\boldsymbol{\gamma}$ of a reference grain with stiffness tensor $\hat{\mathbf{C}}$ is determined from Eshelby's equivalence condition (Eshelby, 1957; Mura, 1982)

$$\hat{\mathbf{C}}(\boldsymbol{\varepsilon} + \mathbf{S}\boldsymbol{\gamma}) = \mathbf{C}(\boldsymbol{\varepsilon} + \mathbf{S}\boldsymbol{\gamma} - \boldsymbol{\gamma}). \tag{5}$$

In (5) $\boldsymbol{\varepsilon}$ is the ambient strain field, **S** is Eshelby's tensor (in general an integral operator), and **C** is the macroscopic stiffness tensor. For our purposes, integral equation (5) does not have to be solved exactly since ultimately we are interested in the stress field induced by $\boldsymbol{\gamma}$ at a distance that significantly exceeds the grain size. The following assumptions allow us to approximate (5) with an algebraic equation:

- The reference grain is a circular cylinder with diameter d.
- The elastic interactions of the reference grain with other grains are neglected.
- The variation of $\boldsymbol{\varepsilon}$ within the grain is neglected.

With the assumptions, $\boldsymbol{\gamma}$ is approximated as

$$\boldsymbol{\gamma} = \frac{K}{\sqrt{2\pi r}}(\mathbf{C} - \mathbf{CS} + \hat{\mathbf{C}}\mathbf{S})^{-1}(\mathbf{C} - \hat{\mathbf{C}})\mathbf{C}^{-1}\boldsymbol{\tau}(\theta). \tag{6}$$

In (6) **S** is Eshelby's tensor of a long cylinder (Mura, 1982), $\boldsymbol{\tau}$ is the normalized stress field corresponding to the Mode I plane strain solution, and r and θ are the polar coordinates of the grain center (Figure 2b). Note that in (6) both **S** and $\boldsymbol{\tau}$ must be derived as solutions of anisotropic elasticity. Nevertheless, one can prove that the symmetry of **C** is such that δK can be calculated using solutions of isotropic elasticity. To do this we simply redefine **C** as an isotropic stiffness tensor whose constants E and ν coincide with the corresponding in-plane constants of the original transversely isotropic **C**. Also both **S** and $\boldsymbol{\tau}$ are redefined as the solutions corresponding to the new **C** (equations for both **S** and $\boldsymbol{\tau}$ are listed in Appendix A).

Following Hutchinson (1974) and Lambropoulos (1986), the reference grain induces the SIF

$$\delta K = \frac{\sqrt{\pi}}{16\sqrt{2}}\frac{E}{1-\nu^2}d^2 r^{-(3/2)}\boldsymbol{\mu}\cdot\boldsymbol{\gamma}. \tag{7}$$

The tensor $\boldsymbol{\mu}$ is a function of θ only whose explicit expression is given in Appendix A. Equations (6) and (7) can be conbined in the form

$$\delta K = \frac{K}{32n^2}\mu_{ij}M_{ijkl}\tau_{kl} = \frac{K}{32n^2}\boldsymbol{\mu}\cdot\mathbf{M}\cdot\boldsymbol{\tau} \tag{8}$$

with

$$n = \frac{r}{d} \quad \text{and} \quad \mathbf{M} = \frac{E}{1-\nu^2}(\mathbf{C} - \mathbf{CS} + \hat{\mathbf{C}}\mathbf{S})^{-1}(\mathbf{C} - \hat{\mathbf{C}})\mathbf{C}^{-1}.$$

Upon superposition of δK's we obtain

$$\Delta K = \sum_{n\geqslant N}\delta K = \sum_{n\geqslant N}\frac{K}{32n^2}\boldsymbol{\mu}(\theta_n)\cdot\mathbf{M}(\lambda_n)\cdot\boldsymbol{\tau}(\theta_n). \tag{9}$$

This notation emphasizes that the dependencies on the angles θ and λ; the latter is defined as the angle between the c-axis of the reference grain and the x_1 axis (Figure 2b). Since the c-axes are randomly oriented in the specimen plane, we regard λ as a random variable uniformly distributed over the interval $[0, \pi]$. With this provision, (9) can be evaluated using the central limit theorem of probability theory (Ross, 1994; also see Appendix B). To this end we rewrite (9) as

$$\Delta K = \sum_{n=N}^{\infty} \Delta_n K = \sum_{n=N}^{\infty} \frac{K}{32n^2} \sum_{m=1}^{n_c} \boldsymbol{\mu}(\theta_m) \cdot \mathbf{M}(\lambda_m) \cdot \boldsymbol{\tau}(\theta_m); \quad n_c \approx 2\pi n. \tag{10}$$

In the double sum in (10),the inner sum is over the grains that a form a quasi-circular ring with circumference $n_c d$ and center at the crack tip, and the outer sum is over the rings. To evaluate the inner sum we observe that the random variable $\Delta_n K$ has zero mean simply because the integral

$$\int_{-\pi}^{\pi} \mu_{ij}(\theta)\tau_{kl}(\theta)\, \mathrm{d}\theta = 0.$$

Then the central limit theorem applied to the inner sum yields

$$\text{Probability} \left\{ \frac{\sum_{m=1}^{n_c} \boldsymbol{\mu}(\theta_m) \cdot \mathbf{M}(\lambda_m) \cdot \boldsymbol{\tau}(\theta_m)}{\sqrt{\sum_{m=1}^{n_c} \sigma[\boldsymbol{\mu}(\theta_m) \cdot \mathbf{M}(\lambda_m) \cdot \boldsymbol{\tau}(\theta_m)]^2}} \leqslant \omega \right\} \to \boldsymbol{\Phi}(\omega) \quad \text{as } n_c \to \infty. \tag{11}$$

In (11) ω is an arbitrary number, $\Phi(\omega)$ is the cumulative distribution function of the normal random variable, and σ is the standard deviation. The arrow in (11) emphasizes that the central limit theorem is an asymptotic estimate valid for large n_c only. An alternative form of (11) is

$$\frac{\Delta_n K}{K} \to \mathcal{N} \left\{ 0, \frac{1}{32n^2} \sqrt{\sum_{m=1}^{n_c} \sigma[\boldsymbol{\mu}(\theta_m) \cdot \mathbf{M}(\lambda_m) \cdot \boldsymbol{\tau}(\theta_m)]^2} \right\}, \tag{12}$$

which states that asymptotically $\Delta_n K/K$ is a normal random variable identified via its mean and standard deviation (see Appendix B). The sum in (12) can be calculated asymptotically as

$$\sum_{m=1}^{n_c} \sigma[\boldsymbol{\mu}(\theta_m) \cdot \mathbf{M}(\lambda_m) \cdot \boldsymbol{\tau}(\theta_m)]^2 = \sum_{m=1}^{n_c} \frac{1}{\pi} \int_0^{\pi} [\boldsymbol{\mu}(\theta) \cdot \mathbf{M}(\lambda) \cdot \boldsymbol{\tau}(\theta)]^2\, \mathrm{d}\lambda \to n\eta^2,$$

with

$$\eta = \sqrt{\frac{1}{\pi} \int_0^{\pi} \int_{-\pi}^{\pi} [\boldsymbol{\mu}(\theta) \cdot \mathbf{M}(\lambda) \cdot \boldsymbol{\tau}(\theta)]^2\, \mathrm{d}\lambda\, \mathrm{d}\theta}, \tag{13}$$

so that (12) can be rewritten as

$$\frac{\Delta_n K}{K} \to \mathcal{N} \left(0, \frac{\eta}{32n^{3/2}} \right). \tag{14}$$

Finally, by applying the addition theorem for normal random variables (Ross, 1994) (Appendix B) to (14), we obtain

$$\frac{\Delta K}{K} = \sum_{n=N}^{\infty} \frac{\Delta_n K}{K} \to \mathcal{N} \left(0, \frac{\eta}{32\sqrt{2N}} \right). \tag{15}$$

Table 1. Evaluation of the size effect for various levels of anisotropy.

ψ	η	E GPa	ν	N
1	9.7	9.7	0.30	6
2	12.6	7.0	0.36	8
5	21.6	4.4	0.41	13
10	57.3	3.1	0.44	33
20	465.0	1.3	0.47	266
100	1030.0	0.9	0.48	588

Note that in (15) the sum $\sum_N^\infty n^{-3}$ is replaced with its asymptotic equivalent $2N^{-2}$.

To evaluate (15) for a given set of material constants, first, we must calculate the parameter η from (13). Altough integration in (13) involves triginometric functions only, an explicit expression for η in terms of the elastic constants is too long. Therefore we compute η using *Mathematica* which allows us to combine symbolic integration with numerical algebraic operations.

For S2 fresh-water ice at $T = -10°$ C, the elastic constants are (Hobbs, 1974)

$$\begin{aligned} &\hat{C}_{1'1'1'1'} = 13.6\,\text{GPa}, \quad \hat{C}_{1'1'2'2'} = 6.7\,\text{GPa}, \quad \hat{C}_{1'1'3'3'} = 4.2\,\text{GPa}, \\ &\hat{C}_{3'3'3'3'} = 14.6\,\text{GPa}, \quad \hat{C}_{1'3'1'3'} = 3.1\,\text{GPa}, \\ &E = 9.7\,\text{GPa}, \quad \nu = 0.3. \end{aligned} \tag{16}$$

For these constants we compute $\eta = 9.7$. Upon substitution of $\eta = 9.7$ in (15) we calculate that for $N = 6$ the probability that $|\Delta K/K| \leqslant 0.1$ is above 0.99; for $N = 9$ the probability that $|\Delta K/K| \leqslant 0.1$ is above 0.9999. These estimates clearly indicate that, in S2 fresh-water ice specimens, remote grains cannot induce significant ΔK and consequently the size effect is weak.

It is possible that in specimens made of materials that are more anisotropic than S2 fresh-water ice remote grains can induce significant ΔK. To examine this possibility, we evaluate (13) and (15) for a family of single crystal constants in which all single crystal constants are the same as in (16) but $\hat{C}_{1'3'1'3'}$ is divided by a factor $\psi > 1$. This stiffness reduction can be associated with the presence of salts in S2 saline ice. Computational results for $1 \leqslant \psi \leqslant 100$ are presented in Table 1. In this table, the polycrystalline elastic constants E and ν are calculated using a self- consistent analysis as formulated by Wang and Schapery (1995). Note that the self-consistent analysis is in excellent agreement with experimental data for S2 fresh-water ice. The last column in Table 1 is the minimum value of N chosen such that $|\Delta K/K| \leqslant 0.1$ with a probability that exceeds 0.99. Based on experimental data for the single crystal elastic constants (Simmons and Wang, 1971), we conclude that for real materials $\psi \leqslant 20$. This means that in fracture specimens made of strongly anisotropic materials remote grains ΔK and consequently the size effect become insignificant if N is in the hundreds.

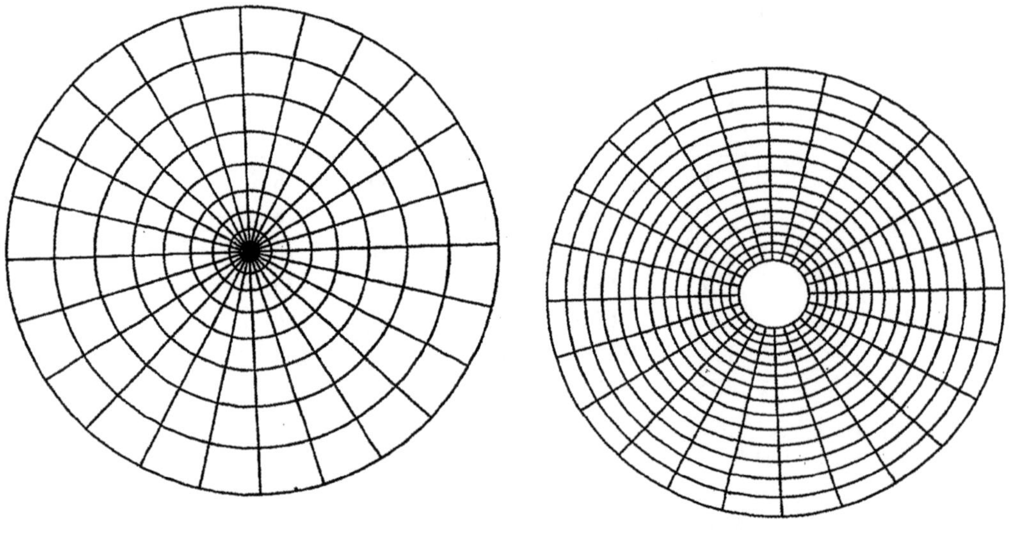

Figure 3. Finite element mesh: (a) grain that contains the crack tip; (b) remaining grains

2.3. NUMERICAL ESTIMATES

In this sub-section we present a set of numerical experiments that not only reinforce the notation that $\Delta K \ll K$ for remote grains but also suggest that ΔK is primarily controlled by a very small number of grains adjacent to the crack tip. The microstructure of a typical specimen analyzed in this sub-section includes a circular grain whose center is at the crack tip and quadrilateral grains. In a typical numerical experiment, the c-axis of the grain that contains the crack tip is fixed while the c-axes of the remaining grains are either randomly oriented in the specimen plane or aligned with the c-axis of the grain that contains the crack tip. The specimens are loaded by prescribing displacements consistent with the Mode I field. The specimens are analyzed with the ABAQUS finite element program supplemented with a post-processing sub- routine that implements a contour integral method (Soni and Stern, 1976) to compute $K_{\rm I}^0$ and $K_{\rm II}^0$. The finite element mesh (Figure 3) is constructed such that the grain that contains the crack tip is modeled with ten layers of elements and its radius is 0.14 of the mesh radius. The finite element mesh dictates the shape of the remaining grains, which are represented by contiguous element sets such that the area of each set is approximately equal to the area of the circular grain. The mesh shown in Figure 3 provides convergent solutions in the sense that neither smaller element size nor larger specimen size, as compared to the grain diameter, significantly affect computed $K_{\rm I}^0$ and $K_{\rm II}^0$.

In Figure 4, $K_{\rm I}^0$ normalized by K is plotted as a function of the c-axis orientation angle λ of the grain that contains the crack tip. The solid curve corresponds to S2 fresh-water ice. The dashed and dotted lines correspond to the artificial elastic constants obtained by choosing $\psi = 2$ and $\psi = 5$, respectively; numerical results appear to be unreliable for $\psi > 5$. The remarkable feature of the curves in Figure 4 is that each point on these curves represents four numerical experiments. In the first experiment, the c-axes of all grains are aligned, i.e. the specimen is anisotropic homogeneous. In the remaining experiments, the c-axes of the quadrilateral grains are chosen at random. This degree of coincidence is a strong indication that the grains other than the one that contains the crack tip cannot significantly affect $K_{\rm I}^0$ and

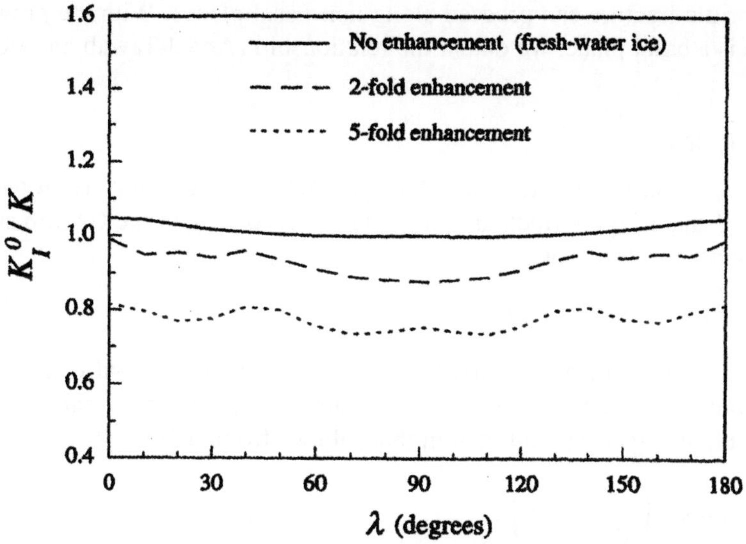

Figure 4. The normalized microscopic stress intensity factor as a function of the angle λ.

therefore the size effect is weak. The same statement applies to $K_{\rm II}^0$ which in all cases was a small fraction of K. Note that for S2 fresh-water ice $K_{\rm I}^0$ is so close to K that for all practical purposes elastic anisotropy can be neglected. As the strength of elastic anisotropy increases, $\psi > 1$, one observes modest shielding, $K_{\rm I}^0 < K$, which may be insignificant in comparison to the fracture toughness anisotropy effect discussed in Section 3.

The microscopic geometry considered in this subsection is biased against grain boundary effects. It is clear that if the crack tip were located close to a grain boundary or a triple-point junction its behavior would be influenced by more than one grain. Nevertheless the numerical experiments considered in this subsection strongly suggest that the number of grains that can control the crack tip fields is very small.

3. Discussion

In the preceeding section, we established that in S2 fresh-water ice fracture specimens remote grains cannot significantly influence the SIFs transmitted to the crack tip and therefore the size effect associated with polycrystalline inhomogeneity is weak. For materials with strongly anisotropic single crystals, a modest size effect is possible but it may be insignificant for all practical purposes. Furthermore, it appears that the crack tip fields in materials whose GB is large are controlled by a very small number of grains adjacent to the crack tip – the remaining grains serve merely as a loading device. This conclusion agrees with the notion that fracture toughness data for brittle polycrystals are characterized by significant statistical effects. To examine this issue further, let us suppose that the fracture toughness data are obtained using S2 fresh-water ice specimens like those analyzed in Section 2.3. For those specimens, the outcome of each test should be a strong function of λ because the crack can grow only along a specific crystallographic plane(s). For S2 fresh-water ice, there is no reliable experimental evidence that brittle fracture can proceed along non-basal planes. With the provision that the cleavage plane is a basal plane, we obtain the relationship (Abdel-Tawab and Rodin, 1993b)

$$K_c(\lambda) = \frac{1}{1+\sin\lambda} K_c(\tfrac{1}{2}\pi). \tag{17}$$

Equation (17) is derived by assuming that the material is elasticly isotropic and that K_c corresponds to K that triggers crack growth initiation according to the Griffith criterion stated for the cleavage plane. Note that

$$K_c(\tfrac{1}{2}\pi) \leqslant K_c(\lambda) \leqslant K_c(0) = 2K_c(\tfrac{1}{2}\pi),$$

with the lower and upper bounds realized, respectively, when the basal plane is aligned with and perpendicular to the crack. If λ is a uniformly distributed random variable over the interval $[0, \pi]$, the cumulative distribution function that follows from (17) is

$$F(K_c) = \frac{2}{\pi} \arcsin\left(\frac{K_c}{K_{\min}} - 1\right).$$

Here we use $K_{\min}$ instead of $K_c(\frac{1}{2}\pi)$ to emphasize that when λ is not resolved $K_c(\frac{1}{2}\pi)$ does appear as $K_{\min}$. Of course, if λ is resolved the statistical effect disappears.

Although the calculations in Section 2 involved columnar grains, the conclusion that the size effect is insignificant should also hold for three-dimensional grains. In this regard, we observe that the columnar microstructure can be viewed as a hypothetical three-dimensional microstructure in which the grains have the same crystallographic orientations within each column. This order implies that in the hypothetical microstructure the grains within each column interact with the crack in a cooperative manner while in the real microstructure the crack-grain interactions involve cancellations. The weakness of the three-dimensional versus two-dimensional elastic interactions is a fairly general property that can be explained in terms of decay rates of the fundamental solutions (Kachanov, 1993).

In our previous study (Abdel-Tawab and Rodin, 1993a) we attempted to evaluate the effect of polycrystalline inhomogeneity by calculating the jump in the J-integral value across the process zone boundary. The scaling for the standard deviation predicted in that paper was $N^{-(1/2)}$ as opposed to N^{-1} in (15). First, the grains other than those adjacent to the process zone were excluded. Second, in terms of the analysis presented in Section 2.2, in the previous work the strength of polycrystalline inhomogeneity was tied to the randomness of the equivalent transformation strain field; presently, polycrystalline inhomogeneity is tied directly to ΔK.

In conclusion, let us mention that there may be a significant microstructural size effect in the presence of damage distributed throughout the specimen. Such damage, for example, can be in the form of grain boundary sliding which creates a network of Mode II microcracks through which the main crack and specimen boundary can interact (Hsia et al., 1991; Onck and van der Giessen, 1997, 1998). Grain boundary sliding is expected to be quite significant at temperatures very close to melting.

Appendix A

In this appendix, we list expressions for the tensors $\boldsymbol{\tau}$, $\boldsymbol{\mu}$, and $\mathbf{S}$ introduced in Section 2.2. For compactness, we use Voight's notation in which symmetric second-rank tensors are represented as 6×1 vectors and fourth-rank tensors as 6×6 matrices. The components of $\boldsymbol{\tau}$ and $\boldsymbol{\mu}$ are defined with respect to the global coordinates and the components of $\mathbf{S}$ are defined with respect to the local coordinates as defined in Figure 1 and 2

$$\{\tau\} = \left\{ \begin{array}{c} \tau_{11} \\ \tau_{22} \\ \tau_{33} \\ \tau_{12} \\ \tau_{13} \\ \tau_{23} \end{array} \right\} = \frac{1}{4} \left\{ \begin{array}{c} 3\cos\frac{\theta}{2} + \cos\frac{5\theta}{2} \\ 5\cos\frac{\theta}{2} - \cos\frac{5\theta}{2} \\ 8\nu\cos\frac{\theta}{2} \\ -\sin\frac{\theta}{2} + \sin\frac{5\theta}{2} \\ 0 \\ 0 \end{array} \right\},$$

$$\{\mu\} = \frac{1}{4} \left\{ \begin{array}{c} \cos\frac{3\theta}{2} + 3\cos\frac{7\theta}{2} \\ 7\cos 3\frac{\theta}{2} + 3\cos\frac{7\theta}{2} \\ 0 \\ -3\sin\frac{3\theta}{2} + 3\sin\frac{7\theta}{2} \\ 0 \\ 0 \end{array} \right\},$$

$$[\mathbf{S}] = \frac{1}{8(1-\nu)} \begin{bmatrix} 0 & 0 & 0 & 0 & 0 & 0 \\ 4\nu & 5-4\nu & -1+4\nu & 0 & 0 & 0 \\ 4\nu & -1+4\nu & 5-4\nu & 0 & 0 & 0 \\ 0 & 0 & 0 & 2-2\nu & 0 & 0 \\ 0 & 0 & 0 & 0 & 2-2\nu & 0 \\ 0 & 0 & 0 & 0 & 0 & 3-4\nu \end{bmatrix}.$$

Appendix B

The purpose of this appendix is to summarize basic results from probability theory (Ross, 1994).

For a continuous random variable X, prescribed in terms of its probability density function $f(x)$, where x are the values that X can take, the cumulative distribution function is defined as

$$F(\omega) = \text{Probability}\,\{X \leqslant \omega\} = \int_{-\infty}^{\omega} f(x)\,\mathrm{d}x.$$

The mean value and the standard deviation of X are defined as

$$\overline{X} = \int_{-\infty}^{\infty} x f(x)\,\mathrm{d}x \quad \text{and} \quad \sigma = \sqrt{\int_{-\infty}^{\infty} (x - \overline{X})^2 f(x)\,\mathrm{d}x},$$

respectively. If

$$f(x) = \frac{1}{\sqrt{2\pi}\,\beta} \exp\left[-\frac{(x-\alpha)^2}{2\beta^2}\right],$$

X is a normal random variable with $\overline{X} = \alpha$ and $\sigma = \beta$. In this case, we also write

$$X = \mathcal{N}(\alpha, \beta).$$

If $\alpha = 0$ and $\beta = 1$, X is a standard normal variable, and its cumulative distribution function is

$$\Phi(\omega) = \frac{1}{\sqrt{2\pi}} \int_{-\infty}^{\omega} \exp\left(-\frac{x^2}{2}\right) \mathrm{d}x = \tfrac{1}{2}\,\mathrm{erf}\left(\frac{\omega}{\sqrt{2}}\right) + \tfrac{1}{2},$$

where erf denotes the error function. The function $\Phi(x)$ can be used to calculate the probability for an arbitrary normal random variable

$$\text{Probability}\,[\mathcal{N}(\alpha, \beta) - \alpha < \beta\omega] = \Phi(\omega) \quad \forall\omega.$$

Also

$$\text{Probability}\,[|\mathcal{N}(\alpha, \beta) - \alpha| < \beta\omega] = 2\Phi(\omega) - 1 \quad \forall\omega > 0.$$

ADDITION THEOREM. *Let X_i be a set of n independent normal random variables, $X_i = \mathcal{N}(\alpha_i, \beta_i)$, and $Y = \sum_{i=1}^{n} X_i$. Then $Y = \mathcal{N}(\alpha, \beta)$ with $\alpha = \sum_{i=1}^{n} \alpha_i$ and $\beta = \sqrt{\sum_{i=1}^{n} \beta_i^2}$.*

CENTRAL LIMIT THEOREM. *Let X_i be a sequence of independent random variables having respective means $\overline{X}_i$ and standard deviations σ_i. If (a) the X_i are uniformly bounded and (b) $\sum_{i=1}^{\infty} \sigma_i \to \infty$, then*

$$\text{Probability}\left[\frac{\sum_{i=1}^{n}(X_i - \overline{X}_i)}{\sqrt{\sum_{i=1}^{n} \sigma_i^2}} < \omega\right] \to \Phi(\omega) \quad \text{as } n \to \infty.$$

Acknowledgements

We with to thank Professor John P. Dempsey of Clarkson University for his encouragement and technical expertise. We are grateful to Hibbitt, Karlson and Sorensen, Inc. for providing us with an academic license to use the program ABAQUS. This work was funded by ONR through the grant N00014-92J1332, Program Director Dr. Y.D.S. Rajapakse.

References

Abdel-Tawab, K. and Rodin, G.J. (1993a). On the relevance of linear elastic fracture mechanics to ice. *International Journal of Fracture* **62**, 171–181.

Abdel-Tawab, K. and Rodin, G.J. (1993b). An interpretation of results of the fracture toughness tests on ice. *Ice Mechanics 1993* (Edited by J.P. Dempsey, Z.P. Bazant, Y.D.S. Rajapakse and S.S. Sunder) AMD-Vol. 163, The American Society of Mechanical Engineers, New York, 49–60.

Adamson, R.M., Dempsey, J.P., DeFranco, S.J. and Xie, Y. (1995). Large-scale in-situ fracture experiments–Part I: Experimental aspects. *Ice Mechanics 1995* (Edited by J.P. Dempsey and Y.D.S. Rajapakse), AMD-Vol. 207, The American Society of Mechanical Engineers, New York, 107–128.

Adamson, R.M., Schapiro, L.H. and Dempsey, J.P. (1997). Core-based SCB fracture of aligned first-year sea ice (Phases III and IV). *Journal of Cold Regions Engineering* **11**, 30–44.

Ashby, M.F. and Jones, D.R.H. (1980). *Engineering Materials I*. Pergamon Press, New York.

Bazant, Z.P. (1993). Scaling laws in mechanics of failure. *ASCE Journal of Engineering Mechanics* **119**, 1828–1844.

Bentley, D.L., Dempsey, J.P., Sodhi, D.S. and Wei, Y. (1989). Fracture toughness of S2 columnar freshwater ice: Large scale DCB tests. *Cold Regions Science and Technology* **17**, 7–20.

Broek, D. (1982). *Elementary Engineering Fracture Mechanics*. Martinus Nijhoff Publishers, The Hague.

Carpinteri, A. (1994). Fractal nature of material microstructure and size effects on apparent mechanical properties. *Mechanics of Materials* **18**, 89–101.

Dempsey, J.P. (1991). The fracture toughness of ice. *Ice-Structure Interaction* (Edited by S. Jones, R.F. McKenna, J. Tilloston, and I. Jordaan), IUTAM-IAHR Symposium, St. John's, Newfoundland, Canada, Springer-Verlag, Heidelberg, 109–145.

Dempsey, J.P. (1996). Scale effects on the fracture of ice. *The Johannes Weertman Symposium* (Edited by R.J. Arsenault, D. Cole, T. Gross, G. Kostorz, P.K. Liaw, S. Parameswaran and H. Sizek), The Minerals, Metals and Materials Society, Warrendale Pennsylvania, 351–361.

Dempsey, J.P., Wei, Y., DeFranco, S., Ruben, R. and Frachetti, R. (1989a). Fracture toughness of S2 columnar freshwater ice: Crack length and specimen size effects – Part I. *Proceedings of the 8th International Conference on Offshore Mechanics and Arctic Engineering* IV, The Hague, The Netherlands, 83–89.

Dempsey, J.P., Wei, Y., DeFrance, S., Ruben, R. and Frachetti, R. (1989b). Fracture toughness of S2 columnar freshwater ice: crack length and specimen size effects – Part II. *Proceedings of the 10th International Conference on Port and Ocean Engineering Under Arctic Conditions* IV, Lulea, Sweden, 199–207.

Eshelby, J.D. (1957). The determination of the elastic field of an ellipsoidal inclusion and related problems. *Proceedings of the Royal Society* **A241**, 376–396.

Frost, H.J. and Ashby, M.F. (1982). *Deformation-Mechanism Maps*. Pergamon Press, Oxford.

Hobbs, P.V. (1974). *Ice Physics*. Clarendon Press, New York.

Hsia, K.J., Argon, A.S. and Parks, D.M. (1991). Effects of grain boundary sliding on creep-constrained boundary deformation and creep deformation. *Mechanics of Materials* **11**, 43–62.

Hutchinson, J.W. (1974). On steady quasi-static crack growth. Report AFOSR-TR74-1042.

Kachanov, M. (1993). Elastic solids with many cracks and related problems. *Advances in Applied Mechanics* (Edited by J.W. Hutchinson and T. Wu), Academic Press, New York, 259–445.

Lambropoulos, J.C. (1986). Shear, shape and orientation effects in transformation toughening. *International Journal of Solids Structures* **22**, 1083–1106.

LeClair, E.S., Adamson, R.M. and Dempsey, J.P. (1997). Core-based SCB fracture of aligned first-year sea ice (Phase VI). *Journal of Cold Regions Engineering* **11**, 45–58.

Mura, T. (1982). *Micromechanics of Defects in Solids*. Martinus Nijhoff Publishers, The Hague.

Onck, P. and van der Giessen, E. (1997). Influence of microstructural variations on steady state creep and facet stresses in 2-D freely sliding polycrystals. *International Journal of Solids Structures* **34**, 703–726.

Onck, P. and van der Giessen, E. (1998). Private communication.

Ross, S. (1994). *A First Course In Probability*. Macmillan College Publishing Company, New York.

Simmons, G. and Wang, H. (1971). *Single Crystal Elastic Constants and Calculated Aggregate Properties: A Handbook*. MIT Press, Cambridge Massachusetts.

Soni, M.L. and Stern, M. (1976). On the computation of stress intensity factors in fiber composite media using a contour integral method. *International Journal of Fracture* **12**, 331–344.

Wang, L. and Schapery, R.A. (1995). Prediction of elastic and viscoelastic properties of columnar ice. *Ice Mechanics 1995* (Edited by J.P. Dempsey and Y.D.S. Rajapakse), AMD-Vol. 207, The American Society of Mechanical Engineers, New York, 33–48.

Williams, M.L. (1957). On the stress distribution at the base of a stationary crack. *Journal of Applied Mechanics* **25**, 109–114.

International Journal of Fracture **93:** 261–280, 1998.

A dislocation barrier model for fatigue crack growth threshold

H.W. LIU
Emeritus Professor of Syracuse University, 1040 Continentals Way, Belmont, CA 94002, U.S.A.

Received 2 July 1997; accepted in revised form 18 June 1998

Abstract. The primary mechanism of fatigue crack growth is crack-tip dislocation emission followed by the glide of the emitted dislocations. Both of these two processes are controlled by the crack-tip resolved shear stress field, which is characterized by the resolved shear stress intensity factor, $K_{R\tau}$.

A dislocation barrier model for fatigue crack growth threshold is constructed. The model assumes that a fatigue crack stops growing when crack-tip slip bands are incapable of penetrating the primary dislocation barrier. The derived and deduced threshold behaviors agree with the observed constant threshold $K_{\max,\text{th}}$ in the low R region and constant threshold ΔK_{th} in the high R region. $K_{\max,\text{th}}$ is the $K_{\max}$ at the threshold. The constant $K_{\max,\text{th}}$ is related to the resistance of the primary dislocation barrier, which in most of cases is grain boundary; and the constant ΔK_{th} is related to the resistance of secondary barriers. Furthermore, the analysis shows that $K_{\max,\text{th}}$ is proportional to $\sqrt{d}$, where d is the grain size. The relation has been observed in steels. The model also helps to explain the characteristics of, and the transition from, microstructure-sensitive to microstructure-insensitive growth.

Key words: Fatigue crack growth, fatigue threshold, dislocation model, dislocation barriers, R-ratio, grain size, microstructure sensitive, microstructure insensitive, facet.

1. Introduction

Cooke and Beevers (1973) found that the $K_{\max,\text{th}}$'s of five medium carbon steels are constant in the low R region. $K_{\max,\text{th}}$ is the $K_{\max}$ at the threshold of fatigue crack growth. Constant $K_{\max,\text{th}}$ has also been oberved in 316 stainless steel (Priddle, 1978) and ferritic steel (Masounave and Baïlon, 1976). On the other hand, the ΔK_{th} in the high R region is constant. The constant $K_{\max,\text{th}}$ in the low R region together with the constant ΔK_{th} in the high R region have been observed in a number of materials: 2024-T3 aluminum, A533 and T1 steels (Schmidt and Paris, 1973), Al-Li 8090 alloy (Greene, Holtz, Sadannada and Vasudevan, 1995), and titanium aluminides (Sandananda and Vasudevan, 1995). Furthermore, ΔK_{th} is proportional to $\sqrt{d}$ in ferritic steel (Masounave and Baïlon, 1976), ferritic-pearlitic steel (Ritchie, 1979), 0.07%C – 0.46Mn steel (Beevers, 1977), and low carbon steel (Taira, Tanaka, and Hoshina, 1979), where d is the grain size.

The primary mechanism of fatigue crack growth in metals and alloys is crack-tip dislocation emission followed by the glide of the emitted dislocations. Both crack-tip dislocation emission and the subsequent dislocation glide are driven by the dislocation glide force field. Dislocation glide force is linearly proportional to resolved shear stress. The crack-tip resolved shear stress field or dislocation glide force field is characterized by the resolved shear stress intensity factor, $K_{R\tau}$.

This paper introduces a dislocation barrier model to analyze fatigue crack growth threshold. The analysis of the model indicates that $K_{\max,\text{th}}$ in the low R ratio region is constant and is proportional to $\sqrt{d}$. Based on the model, it can also be deduced that ΔK_{th} is constant in the high R region. The effects of crack closure are reviewed and discussed. The model also helps to understand the characteristics of, and the transition from, microstructure-sensitive to microstructure-insensitive fatigue crack growth.

2. Resolved shear stress intensity factor

An applied stress exerts a force, $\mathbf{F}$, on a dislocation, which consists of a component in the slip plane and a component normal to the slip plane. The component in the slip plane is a glide force, $\mathbf{F^g}$. The force normal to the plane is $\mathbf{F^n} = \mathbf{n}(\mathbf{F}\cdot\mathbf{n})$, where $\mathbf{n}$ is the unit normal vector of the plane. The glide force is

$$\mathbf{F^g} = \mathbf{F} - \mathbf{F^n} = \mathbf{n} \times (\mathbf{F} \times \mathbf{n}). \tag{1}$$

With the force per unit length of a dislocation, $\mathbf{F}$ (Hirth and Lothe, 1968), the expression of $\mathbf{F^g}$ can be derived.

$$\mathbf{F^g} = (\mathbf{n} \times \boldsymbol{\xi})(\boldsymbol{\sigma}\cdot\mathbf{n}\cdot\mathbf{b}) = (\mathbf{n} \times \boldsymbol{\xi})\tau_R b, \tag{2}$$

$$F^g = \boldsymbol{\sigma}\cdot\mathbf{n}\cdot\mathbf{b} = \sigma_{ij} n_i b_j = \tau_R b, \tag{3}$$

and

$$\tau_R = \boldsymbol{\sigma}\cdot\mathbf{n}\cdot\frac{\mathbf{b}}{b} = \sigma_{ij} n_i \frac{b_j}{b} \tag{4}$$

where $\boldsymbol{\sigma}$ is the applied stress, $\mathbf{b}$ is the Burgers vector, and $\boldsymbol{\xi}$ is a unit vector tangent to the dislocation. $(\boldsymbol{\sigma}\cdot\mathbf{n}\cdot\mathbf{b})$ is the resolved shear stress, τ_R, of the slip system multiplied by the magnitude, b, of the Burgers vector. $(\mathbf{n} \times \boldsymbol{\xi})$ is the direction of the force. An applied stress may have all nine components, but only the resolved shear stress exerts a glide force. The resolved shear stress is linearly proportional to the dislocation glide force. Thus, the yield criterion of single crystals can be stated as 'the dislocation glide force at the yield initiation of a single crystal is constant.' Nabarro (1967) and Friedel (1967).

A stage-I fatigue crack is a shear crack and it grows by a shear mechanism. Even though a stage-II crack is on a tensile plane, the crack grows by alternate shear on two intersecting slip planes. Undoubtedly, crack-tip shear slip or shear decohesion plays a major role in fatigue crack growth as shown by McEvily and Boettner (1963), Laird (1967), Tomkins (1968), Pelloux (1969), Neumann (1974), and Vehoff and Neumann (1979). Shear decohesion is caused by dislocation emission from a crack tip and the subsequent glide motion of the emitted dislocations. Both dislocation emission and glide are the result of a relative shear displacement in the direction of slip between the atoms of two neighboring slip planes. Therefore, both dislocation emission and glide are controlled by the shear stress field in the slip plane in the slip direction, i.e., by the resolved shear stress field of the slip system.

A crack in a crystal is often inclined and tilted relative to an applied stress. Thus, a crack-tip stress field, $\sigma_{ij}(r, \theta)$, may consist of all three modes of crack-tip stress fields,

$$\sigma_{ij}(r, \theta) = \sigma_{\mathrm{I},ij}(r, \theta) + \sigma_{\mathrm{II},ij}(r, \theta) + \sigma_{\mathrm{III},ij}(r, \theta) \tag{5}$$

where $\sigma_{\mathrm{I},ij}$, $\sigma_{\mathrm{II},ij}$, and $\sigma_{\mathrm{III},ij}$ are, respectively, the mode-I, mode-II, and mode-III crack-tip stress fields.

A crystal has a number of slip systems. Equation (4) can be used to calculate the resolved shear stress fields of the slip systems of the crystal at a crack tip. For a slip system defined by **n** and **b**,

$$\tau_R(r) = \sigma_{ij}(r, \theta) n_i \frac{b_j}{b} = \frac{K_{R\tau}}{\sqrt{2\pi r}}. \tag{6}$$

$\tau_R(r)$ is the resolved shear stress field of the slip system, and $\sigma_{ij}(r, \theta)$ is the crack-tip stress field given by (5). $K_{R\tau}$ is the resolved shear stress intensity factor. $K_{R\tau}$ is a function of $K_{\rm I}$, $K_{\rm II}$, $K_{\rm III}$, and θ, which is the conventional crack-tip polar coordinate. Equation (6) indicates that a crack-tip resolved shear stress field is characterized by $K_{R\tau}$. In other words, the same value of $K_{R\tau}$ gives the same resolved shear stress field. While a single stress intensity factor $K_{\rm I}$, $K_{\rm II}$, or $K_{\rm III}$ characterizes an *entire* mode-I, mode-II, or mode-III crack-tip stress field, $K_{R\tau}$ characterizes the resolved shear stress field of only a *single* slip system. The number of $K_{R\tau}$'s at a crack tip is equal to the number of slip systems of the crystal. $K_{R\tau}$ should be useful in analyzing fatigue crack growth, especially when a crack grows by a shear process. A detailed discussion of the characteristics of $K_{R\tau}$ is given by Chen (1992).

3. A dislocation barrier model for fatigue crack growth threshold

The physical mechanism of the threshold and the relations between $\Delta K_{\rm th}$, microstructure, load ratio, R, and environment are rather complicated. The near threshold fatigue crack growth in steels was reviewed by Ritchie (1979). While crack closure and crack-tip residual stress may contribute to the cause of threshold, a strong dislocation barrier can also stop fatigue crack growth. This paper presents a model of threshold based on the stopping of crack growth by a dislocation barrier. The derived relation is examined in view of the empirical observations of the effects of grain size and R-ratio.

Cooke and Beevers (1973) have shown that fatigue crack growth threshold behavior is strongly influenced by R-ratio. Table 1 shows their data of five medium carbon steels of three different microstructures: mixed ferrite and pearlite, pearlite, and mixed pearlite and martensite. In the low R region, the $K_{\rm max,th}$ of each steel is constant and is independent of R. The numbers within parentheses are averages. The deviations from the averages are rather small. If $K_{\rm max,th}$ is constant, $\Delta K_{\rm th}$ is proportional to $(1 - R)$. $K_{\rm max,th}$ is the $K_{\rm max}$ at the threshold. In the low R region, the relation, $\Delta K_{\rm th} \sim (1 - R)$, and, thus, the constant $K_{\rm max,th}$ have also been observed in 316 stainless steel (Priddle, 1978) and 7075 and 2L93 aluminum alloys (Kirby and Beevers, 1979).

Masounave and Baïlon (1976) found the following empirical relation for a ferritic steel.

$$\frac{\Delta K_{\rm th}}{(1 - R)} = 3.8 + 1.14 \times 10^3 \sqrt{d} = K_{\rm max,th}. \tag{7}$$

K is in MN m$^{-3/2}$, and d is grain size in meters. Note that $\Delta K_{\rm th}/(1 - R) = K_{\rm max,th}$. At a given d, $K_{\rm max,th}$ is, thus, constant. Equation (7) is consistent with the data in Table 1.

Beevers (1977) has shown that $\Delta K_{\rm th}$ for 0.07%C – 0.46 Mn steel increases with $\sqrt{d}$.

$$\Delta K_{\rm th} = 3.8 + 1.4 \times 10^3 \sqrt{d}. \tag{8}$$

K is in MN m$^{-3/2}$, and d is in meters. Taira, Tanaka, and Hoshina (1979) found that the $\Delta K_{\rm th}$ of a low carbon steel is proportional to $\sqrt{d}$ at three different R ratios, 0.5, 0, -1.

Table 1. ΔK_{th} and $K_{\text{max,th}}$ of Medium Carbon Steels. Cooke and Beevers (1973)

Microstructure	R-ratio	ΔK_{th} (MN/m$^{3/2}$)	$K_{\text{max,th}}$ (MN/m$^{3/2}$)
Ferrite–Pearlite	0.26	6.61	8.95
	0.27	6.60	9.04
	0.34	5.45	8.26
	0.51	4.42	9.02
			(8.82)
Pearlite	0.34	5.63	8.53
Pearlite	0.26	7.25	9.81
	0.34	6.10	9.25
			(9.53)
Martensite–Pearlite	0.34	8.16	12.36
	0.50	6.13	12.61
			(12.48)
Martensite–Pearlite	0.27	8.50	11.62
	0.36	7.67	12.00
	0.37	7.56	12.02
	0.37	7.60	12.08
	0.50	6.00	12.00
			(11.94)

Ritchie (1979) has shown a linear relation between ΔK_{th} and ferrite grain size for two low-strength ferritic–pearlitic steels. The data is re-plotted in Figure 1 in the form of ΔK_{th} versus $\sqrt{d}$. The linear correlation between ΔK_{th} and $\sqrt{d}$ is quite reasonable.

$K_{\text{max,th}}$ is constant in the low R region, ΔK_{th} is constant in the high R region, and R_T is the transition point as shown schematically in Figure 2. Such threshold behavior has been observed in a number of materials: 2024-T3 aluminum, A533 and T1 steels (Schmidt and Paris, 1973), titanium aluminides (Sadananda and Vasudevan, 1995), and Al-Li 8090 alloy (Greene, Holtz, Sadannada and Vasudevan, 1995). The analysis of this paper shows that the constant $K_{\text{max,th}}$ and the constant ΔK_{th} are related to the intrinsic slip resistance of the primary and secondary dislocation barriers, and the constant $K_{\text{max,th}}$ and constant ΔK_{th} are denoted by $K^{\text{I}}_{\text{max,th}}$ and $\Delta K^{\text{I}}_{\text{th}}$ respectively as shown in the figure.

Before the construction of the model for the threshold in the low R region, the physical processes of slip band formation and microcrack nucleation and their relation to fatigue crack growth are reviewed. The formation of slip bands and the nucleation of microcracks at a stress below fatigue limit are well known. Taira, Tanaka, and Hoshina (1979) measured the numbers of cycles at slip band formation, N_s, microcrack nucleation, N_c, and fatigue fracture, N_f, of smooth low-carbon steel specimens. Figure 3 shows the measured N_s, N_c, and N_f. Below the fatigue limit, micro cracks were formed, but the cracks became nonpropagating. It was also observed that below the fatigue limit, micro cracks grew parallel to a slip plane, and most

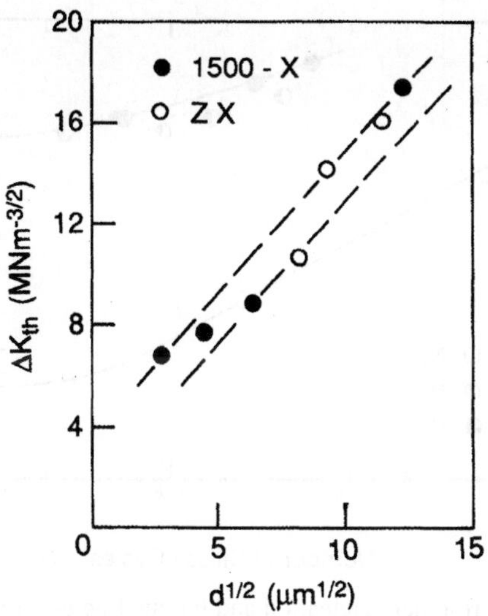

Figure 1. ΔK_{th} vs. $d^{1/2}$. $R = 0.05$. Low strength steels. $\sigma_Y < 500\,\mathrm{MN\,m^{-2}}$. Data from Ritchie (1979).

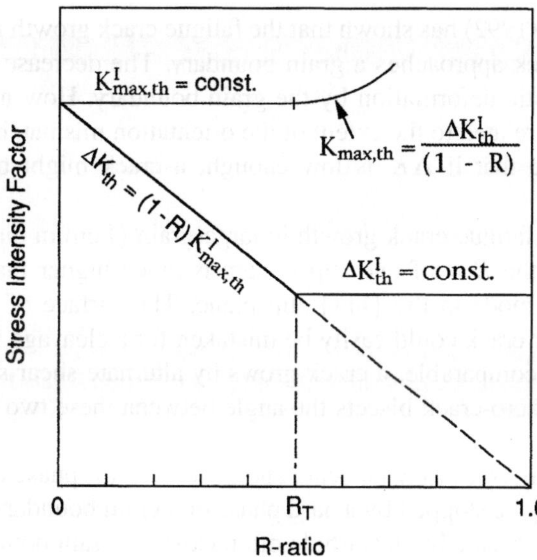

Figure 2. A schematic figure showing threshold behavior as influenced by R-ratio.

slip bands were blocked by grain boundaries. In other words, a grain boundary is capable of stopping slip bands and microcracks.

Kunio, Shimizu, and Yamada (1969) examined slip band formation, microcrack nucleation, and fatigue fractures of smooth ferritic-martensitic steel specimens. A fatigue crack is often nucleated in a soft ferrite grain. If the applied stress is too low, the nucleated crack is stopped by the surrounding hard martensite and becomes nonpropagating. Yang (1979) measured fatigue crack growth rate in a ferrite as the crack grew toward a martensite and found that the crack growth rate decreased as the crack-tip plastic zone impinged upon the martensite. The martensite constrains the cyclic plastic deformation ahead of the crack tip. The decrease of the crack growth rate is related to the constraint of the cyclic deformation.

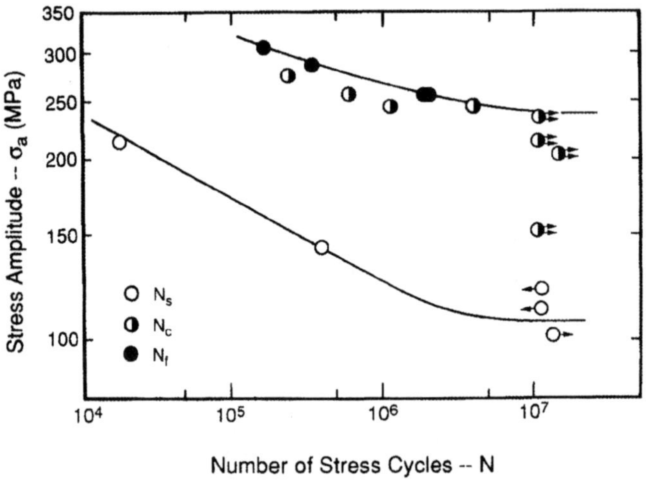

Figure 3. Slip-band formation, micro-crack nucleation and fatigue fractures of a low-carbon steel. Taira, Tanaka, Hoshina (1979).

More recently, Chen (1992) has shown that the fatigue crack growth rate in 7079 aluminum alloy decreases as a crack approaches a grain boundary. The decrease is caused by the constraint of the cyclic plastic deformation by the grain boundary. How much the fatigue crack growth rate decreases is related to the extent of the orientation mismatch between neighboring grains. It is conceivable that if ΔK is low enough, a crack might be stopped by a grain boundary.

Chen (1992) studied fatigue crack growth in large-grain (1 cm in diameter or larger) 7079 aluminum alloy. When the $K_{R\tau}$ of one slip system is much higher than that of all others, a crack grows in a shear mode in the {111} slip plane. The surface of a shear crack is 'flat' and shiny. A shiny shear crack could easily be mistaken for a cleavage facet. When the $K_{R\tau}$'s of two slip systems are comparable, a crack grows by alternate shear slip in these two {111} slip planes, while the macro-crack bisects the angle between these two planes, giving a {100} 'tensile' crack.

In summary: A fatigue crack may nucleate and grow in a soft phase of a steel, but the crack as well as slip bands may be stopped by a hard phase or a grain boundary. Ferrite and austenite are soft, carbide and martensite hard. Carbide often exists in grain boundaries of steels. If the effects of carbide and orientation mismatch are combined, a grain boundary in a steel would be a very strong dislocation barrier. Close to threshold, the $K_{R\tau}$'s of a crack-tip grains are very low. Therefore, only one slip system of the grain is activated, and the crack at the grain grows by a shear mode. When K is too low, crack-tip slip is unable to penetrate martensite, carbide, or a grain boundary, and the crack stops propagating. This could be the underlying cause of the fatigue crack growth threshold.

Observing that the crack-tip slip band of a nonpropagating crack is stopped by grain boundary, Taira, Tanaka and Nakai (1978) constructed a model to study the grain size effect on ΔK_{th}. They used the continuously distributed dislocation theory of Bilby, Cottrell, and Swindon to analyze an array of climbing edge dislocations under a tensile load. The array is stopped by a grain boundary. The analysis gives a linear relation between ΔK_{th} and $\sqrt{d}$, and the relation agrees qualitatively with those observed by Masounave and Baïlon (1976), Beevers (1977), and Taira, Tanaka, and Hoshina (1979). However, their model gives a mode-I stress field at

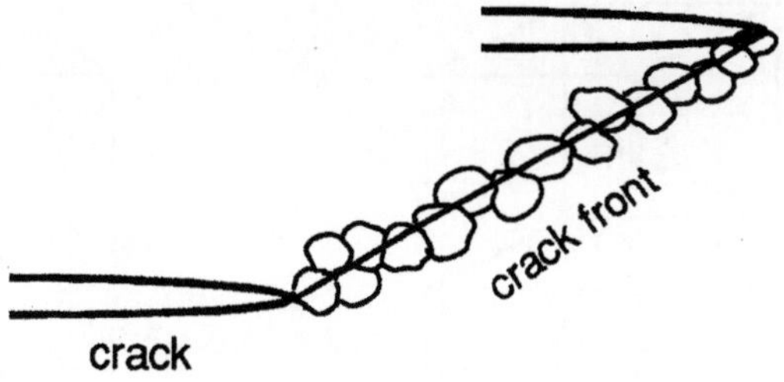

Figure 4. A number of grains lie along a crack front. Close to threshold, the grain that has the maximum $K_{R\tau}$ yields first, and the crack front at the grain starts to grow first.

the tip of the array of climbing edge dislocations, while the slip band is a shear band. In this paper, the Taira–Tanaka–Nakai model is modified and expanded to reflect the shear stress field of the slip band.

A number of grains lie along a crack front in a polycrystal, as shown schematically in Figure 4. The slip system that coincides with the plane and the direction of the maximum crack-tip shear stress, $\tau_{\theta r}$, has the maximum $K_{R\tau}$. In the state of plane strain, the slip system with the slip plane at $\theta = 70°$ and **b** in the plane of the plate has the maximum τ_R and maximum $K_{R\tau}$.

$$\tau_{R,\max} = \tau_{\theta r(\theta=70°)} = 0.39\frac{K_{\mathrm{I}}}{\sqrt{2\pi r}} = \frac{K_{R\tau}}{\sqrt{2\pi r}}$$

or

$$K_{R\tau} = 0.39 K_{\mathrm{I}}, \qquad (9)$$

This is the maximum $K_{R\tau}$ of the optimally oriented single slip system. At any other orientation, $K_{R\tau}$ is lower.

Figure 5a shows two slip systems on two slip planes oriented symmetrically to the {100} crack plane in a bcc crystal. The slip planes are outlined by dashed lines. Only the Burgers vector, **b**, of the upper cubic cell is shown. These two slip systems have the same $K_{R\tau}$. The shear stress on the slip plane is

$$\tau_{\theta r(\theta=45°)} = 0.33\frac{K_{\mathrm{I}}}{\sqrt{2\pi r}}.$$

The resultant of the shear stress in the direction of **b** is τ_R.

$$\tau_R = 0.27\frac{K_{\mathrm{I}}}{\sqrt{2\pi r}}$$

or

$$K_{R\tau} = 0.27 K_{\mathrm{I}}. \qquad (10)$$

This is the $K_{R\tau}$ of the optimally oriented two-slip system shown in Figure 5a, and its $K_{R\tau}$ is lower than that of the optimally oriented single-slip system given by (9). The K_{I} necessary to emit and move dislocations on the two-slip system is nearly 50 percent higher than that on the single slip system.

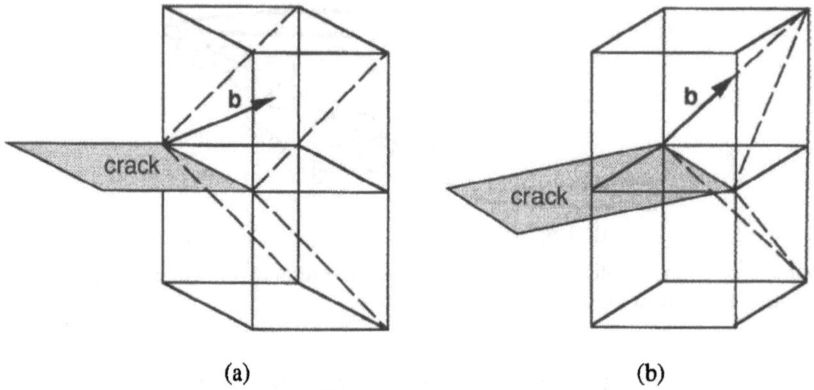

Figure 5. Two symmetrically oriented two-slip systems. (a) bcc crystal, $\{110\}\langle 111\rangle$ slip system. (b) fcc crystal, $\{111\}\langle 110\rangle$ slip system.

A similar calculation can be made for the fcc crystal in Figure 5b. The $K_{R\tau}$ is

$$K_{R\tau} = 0.31K_{\mathrm{I}}. \tag{11}$$

For the $\{111\}\langle 110\rangle$ slip system of a fcc crystal, the difference between the $K_{R\tau}$'s of a single slip system and a two-slip-system is nearly 25 percent. Therefore, close to threshold, crack-tip shear decohesion occurs on a single slip system, and a crack grows by shear mode. Only when ΔK is increased by 50 percent or 25 percent, two slip systems are activated, and a crack may then grow by a 'tensile' mode.

When a cracked polycrystal is stressed, the grain having the maximum $K_{R\tau}$ slips first, forming a slip band (see Figure 6a). As $K_{R\tau}$ increases, the slip band extends to the grain boundary, GB, and spans an entire grain as illustrated in Figure 6b. d is the grain diameter. The crack stops at the grain boundary because of the abrupt change of crack path, which increases the local resistance to crack growth.

To simplify the calculation, the crack and the slip band are assumed to be collinear, and the crack becomes a shear crack. The applied shear stress τ_R gives rise to $K_{R\tau}$. The magnitude of $K_{R\tau}$ is given by (9).

The slip band is equivalent to a shear strip yield zone. A shear stress equal to the shear yield stress, τ_{RY}, is applied to the upper and lower 'surfaces' of the strip yield zone as indicated by the arrows in Figure 6c. The applied τ_{RY} gives rise to $K'_{R\tau}$. For the strip yield zone fully extended to the grain boundary as shown in Figures 6b and 6c,

$$K'_{R\tau} = \left(\frac{8}{\pi}\right)^{1/2} \tau_{RY}\sqrt{d}. \tag{12}$$

At a $K_{R\tau}$ higher than $K'_{R\tau}$, a fully realized slip band would extend beyond GB. However, the slip band is stopped by GB. When the slip band is stopped and constrained, a stress singularity, $K_{R\tau,\mathrm{net}}$, exists at the tip of the slip band.

$$K_{R\tau,\mathrm{net}} = (K_{R\tau} - K'_{R\tau}) = K_{R\tau} - \left(\frac{8}{\pi}\right)^{1/2} \tau_{RY}\sqrt{d}. \tag{13}$$

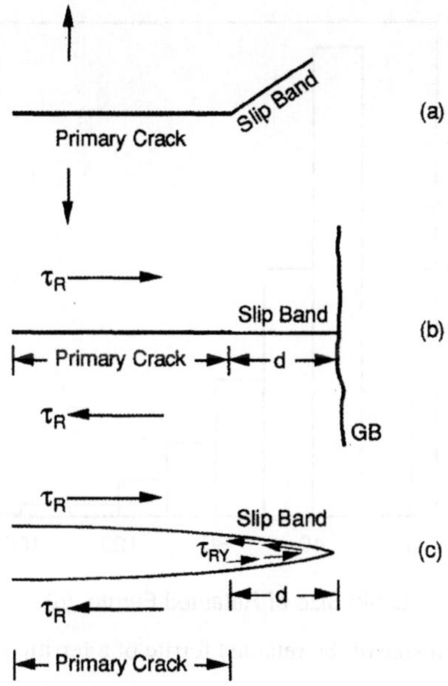

The arrows on the internal surfaces of d indicate shear yield stress, τ_{RY}.

Figure 6. The crack-tip shear band is stopped by a grain boundary, GB; *d* is the grain diameter.

During a cyclic load, the maximum $K_{R\tau}$ of the grain having the optimum orientation is $0.39K_{\mathrm{I,max}}$. See (9). The simplification of Figure 6b introduces an error of a constant factor, γ. Thus, (13) becomes

$$K_{R\tau,\mathrm{net}} = 0.39K_{\mathrm{I,max}} - \gamma\left(\frac{8}{\pi}\right)^{1/2}\tau_{RY}\sqrt{d}. \tag{14}$$

Below the threshold, the stress field at the tip of the slip band is not high enough to penetrate the grain boundary, and the crack becomes dormant. Above the threshold, grain boundary resistance is overcome, cyclic slip extends beyond the boundary, and the crack starts to grow in a shear mode. At the threshold, $K_{\mathrm{I,max}}$ becomes $K_{\mathrm{max,th}}$, and $K_{R\tau,\mathrm{net}} = K_{c,\mathrm{GB}}$. $K_{\mathrm{max,th}}$ is the $K_{\mathrm{I,max}}$ at the threshold, and $K_{c,\mathrm{GB}}$ is the resistance of the grain boundary. If $K_{c,\mathrm{GB}}$ is a constant, (14) becomes

$$K_{\mathrm{max,th}} = \frac{\Delta K_{\mathrm{th}}}{(1-R)} = \frac{K_{c,\mathrm{GB}}}{0.39} + \frac{\gamma}{0.39}\left(\frac{8}{\pi}\right)^{1/2}\tau_{RY}\sqrt{d} = K^{\mathrm{I}}_{\mathrm{max,th}}. \tag{15}$$

$K^{\mathrm{I}}_{\mathrm{max,th}}$ and ΔK_{th} are proportional to $\sqrt{d}$. For a given material, *d* is constant, thus, $K^{\mathrm{I}}_{\mathrm{max,th}} =$ constant, and $\Delta K_{\mathrm{th}} = (1-R)K^{\mathrm{I}}_{\mathrm{max,th}}$. Both of these two relations are shown schematically in Figure 2. As *d* becomes infinitesimal, $K_{\mathrm{max,th}} = K_{c,\mathrm{GB}}/0.39$. $K_{c,\mathrm{GB}}$ can be obtained by the extrapolation of the data of $K_{\mathrm{max,th}}$ vs. $\sqrt{d}$. $K_{\mathrm{max,th}}$ is a measure of the intrinsic slip resistance of the grain boundary, and, thus, the quantity is denoted by $K^{\mathrm{I}}_{\mathrm{max,th}}$.

The constants of (15) can be evaluated. Equation (15) has the same form as the empirical equation, (7), of Masounave and Baïlon. The value of $K_{c,\mathrm{GB}}$ of the steel can be obtained from the empirical constant of (7). $K_{c,\mathrm{GB}} = 1.5\,\mathrm{MN\,m^{-3/2}}$. The values of γ and τ_{RY} are needed to

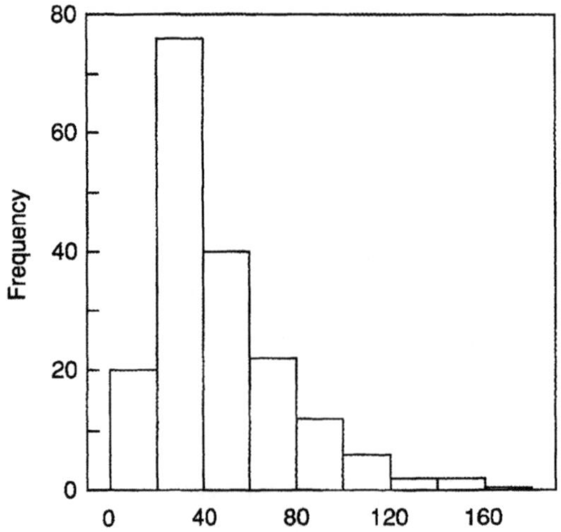

Figure 7. The variation of the grain size of the retained ferrite of a ferritic-martensitic steel. Yamada (1970).

evaluate the constant, $\left(\frac{\gamma}{0.39}\left(\frac{8}{\pi}\right)^{1/2}\tau_{RY}\right)$, of the second term of (15). Figure 3 shows the stress levels of slip band formation and microcrack nucleation of a low carbon steel. Perhaps, the steel used by Masounave and Baïlon is close to that of Figure 3, and these values can be used as τ_{RY}. The lower limits of the stress amplitudes for slip band formation, σ_s, and microcrack nucleation, σ_c, are 107 and 160 MPa, respectively. If we take $\sigma_s/2$ as τ_{RY} and assume $\gamma = 1$, then the constant is 220 MPa, a factor of five lower than the empirical value of (7), 1140 MPa. However, if an applied stress is between the static and cyclic yield stresses, and if the material is cyclically hardened and becomes elastic, slip bands may appear, but no micro crack is formed. Only persistent cyclic plastic deformation causes crack nucleation. If this is, indeed, the case, τ_{RY} (cyclic) should be close to $\sigma_c/2$. If we take $\sigma_c/2$ as τ_{RY} (cyclic), the constant becomes 332 MPa, and the factor of difference is reduced from 5 to 3.4.

The d of (7) is the average or nominal grain size. Yamada (1970) examined the ferrite grain size of a ferritic-martensitic steel. The data of one of the heat treatments is shown in Figure 7. The grain size of the steel varies widely. Along a crack front, there are a number of grains having the optimum orientation, some small and some large. If a grain is small, $K_{R\tau,\mathrm{net}} > K_{c,\mathrm{GB}}$, the grain boundary resistance ahead of the slip band is overcome, and the crack in the grain grows until the crack front is stopped by less favorably oriented grains or by a large grain. If the grain is large enough, $K_{R\tau,\mathrm{net}} < K_{c,\mathrm{GB}}$, the grain boundary is not penetrated, and the crack segment is stopped. Therefore, at threshold, a crack front is stopped by large grains. Large ferrite grains could be four times the average size, as indicated by the data of Figure 7. The difference between the maximum and the average grain sizes could bring the difference between the theoretical and the empirical constants to a factor of 1.7, which is quite reasonable at this stage of quantitative analysis of microstructural effects.

For a given material, (15) indicates that K_{max} at the threshold is constant as observed by Cooke and Beevers (1973) in five medium carbon steels of three different microstructures. At a given R-ratio, ΔK_{th} is proportional to $\sqrt{d}$. Equation (15) can, thus, be written in the form

$$\Delta K_{\mathrm{th}} = \alpha + \beta\sqrt{d} \qquad (16)$$

which agrees with the measurements of Beevers (1977) on 0.07%C – 0.46%Mn steel, that of Taira, Tanaka, and Hoshina (1979) on low carbon steels, and that of Ritchie (1979) shown in Figure 1 on low-strength ferritic-pearlitic steels. The dislocation barrier model agrees with the empirical relation between ΔK_{th} and grain size in the low R region.

Fatigue crack growth is caused by shear slip, and shear slip is resisted by dislocation barriers. In order to grow a crack in a crystal, the glide force has to overcome all dislocation barriers. The $K^{I}_{max,th}$ of (15) is derived from the physical process of dislocation penetration of grain boundary, the primary barrier. Once the grain boundary is penetrated, following the crack growth process step by step, the glide force has to overcome the intragrain secondary barriers, so that the crack-tip cyclic slip or shear decohesion can continue to occur. There must be a maximum ΔK below which the cyclic slip is stopped by secondary barriers, and, hence, the crack ceases to grow. This maximum is ΔK^{I}_{th}. ΔK^{I}_{th} is the threshold for growing a crack without grain boundaries as dislocation barriers in front of the slip bands. It is also a measure of the intrinsic slip resistance of the secondary barriers. As such, ΔK^{I}_{th} should be a constant.

Grain boundary penetration is a one-step process, thus, the threshold is controlled by the maximum K, K_{max}. After grain boundary penetration, the crack grows by repeated slip cycle by cycle. Therefore, the threshold is controlled by the cyclic range of K, ΔK. In order to grow a crack, K_{max} has to be higher than $K^{I}_{max,th}$, and ΔK has to be higher than ΔK^{I}_{th}. At the threshold in the low R region, $K_{max} = K^{I}_{max,th} =$ constant, and $\Delta K > \Delta K^{I}_{th}$. Therefore, the threshold is controlled by K_{max} and defined by $K^{I}_{max,th}$. As R increases, ΔK_{th} decreases with $(1 - R)$ in order to keep $K^{I}_{max,th}$ constant. Beyond R_T, as ΔK^{I}_{th} stays constant, K_{max} increases rapidly with $1/(1 - R)$, and $K_{max} > K^{I}_{max,th}$. Thus, in the high R region, $\Delta K = \Delta K^{I}_{th}$, and $K_{max} > K^{I}_{max,th}$. The threshold is, therefore, controlled by ΔK and defined by ΔK^{I}_{th}. The dislocation barrier model conforms to all of these observed threshold behaviors.

The model is also consistent with the analysis of the fatigue limit of smooth specimens based on the resistance of the grain boundaries to shear slip. Figure 3 shows the data of slip-band formation, micro-crack nucleation, and fatigue fracture of a low carbon steel. There are slip bands and micro cracks in the smooth specimens tested at stress levels below the fatigue limit. Nonpropagating microcracks and slip bands are stopped by grain boundaries. The fatigue limit is the stress at the threshold of the micro cracks. The same model of dislocation barrier to shear slip is applicable to both fatigue threshold in cracked specimens and fatigue limit of smooth specimens. The analysis of fatigue limit is summarized in a companion paper (Liu, 1998).

If the grain diameter is very large, the size of the strip yield zone at threshold is smaller than d, and

$$K^{I}_{max,th} = \frac{\Delta K_{th}}{(1 - R)} = \frac{K_{c,GB}}{0.39} + \frac{\gamma}{0.39}\left(\frac{8}{\pi}\right)^{1/2} \tau_{RY}\sqrt{r_{SY}} \tag{17}$$

where r_{SY} is the size of the strip yield zone at threshold. In this case, $K^{I}_{max,th}$ as well as ΔK_{th} are independent of grain size.

According to (15), $K^{I}_{max,th}$ and ΔK_{th} are linearly proportional to τ_{RY}. This seems to contradict the inverse relation between ΔK_{th} and the tensile yield strength, σ_Y, of polycrystalline low carbon steels as observed by Taira, Tanaka, and Hoshina (1979). As d decreases, σ_Y increases and ΔK_{th} decreases. The changes in both σ_Y and ΔK_{th} are caused by the change in grain size. τ_{RY} is the resolved shear yield strength of a 'single' crystal without the strengthening effect of grain boundaries. τ_{RY} changes with dislocation density and with the densities and sizes of solutes and precipitates but stays constant as d changes.

4. Crack closure

Elber (1970) has observed crack closure during a fatigue cycle. Crack closure has been studied experimentally by Lindley and Richards (1974), Sharpe and Grandt (1976), Schmidt and Paris (1973), and numerically with finite element method by Newman and Armen (1974), Ohji et al. (1975), and Newman (1976). Overwhelming evidence indicates that a crack tip does close during a portion of a fatigue cycle.

Schmidt and Paris (1973) found that the threshold $K_{\text{max,th}}$ is constant in the low R region, and the threshold ΔK_{th} is constant in the high R region for 2023-T3 aluminum, and A533 and T1 steels. They explained the observed threshold behavior on the basis of crack closure. Using side-face-mounted strain gages, they determined the stress intensity necessary to open the crack, K_{op}. The range of stress intensity factor above K_{op} necessary to produce fatigue crack growth is the threshold range and is called ΔK_O. Schmidt and Paris assumed that both K_{op} and ΔK_O are constant, i.e., independent of R and K_{min}. Consequently,

$$K_{\text{max,th}} = K_{\text{op}} + \Delta K_O = \frac{\Delta K_{\text{th}}}{(1-R)} = K^{\text{I}}_{\text{max,th}} = \text{constant}. \tag{18a}$$

This relation is applicable in the low R region, where $K_{\text{op}} > K_{\text{min}}$, and the crack does close. Above the transition point, R_T, in the high R region, $K_{\text{op}} < K_{\text{min}}$, the crack does not close. Therefore,

$$\Delta K_{\text{th}} = \Delta K_O = \Delta K^{\text{I}}_{\text{th}}. \tag{18b}$$

A crack-tip stress-strain field is three dimensional (Liu, 1975). Close to the broad side face of a plate specimen, the crack-tip field is plane stress; in the interior of a thick plate, it is plane strain. Yang (1979) studied crack closure during an unloading cycle with two-dimensional finite element models for plane strain and plane stress states for the same specimen geometry and with the same stress-strain relation. First, a cracked specimen is loaded to K_{max}, and then the nodal force at the crack tip is released. The 'crack tip' opens up, and the tip advances to the next nodal point. After the nodal force is released, the applied load is reduced incrementally. The crack tip starts to close during the unloading cycle when the applied K becomes low enough. During the subsequent reloading cycle, the crack opens up gradually. Upon reloading to K_{max}, the nodal force at the 'new' crack tip is released. This process of nodal force release and crack tip advancement is repeated for four cycles in order to 'stabilize' the crack-tip cyclic stresses and strains. The K at the instant of crack closure during the fourth unloading cycle is recorded as K_{cl}. After unloading from the same $K_{\text{max}} = 16.6\,\text{MPa}\sqrt{\text{m}}$, the crack tip in the plane stress state closes at $K_{\text{cl}} = 10.8\,\text{MPa}\sqrt{\text{m}}$, and the crack tip in the plane strain state closes at $K_{\text{cl}} = 7.2\,\text{MPa}\sqrt{\text{m}}$. The K_{cl} at which the crack tip in the plane stress state closes is higher than that in the plane strain state. The results of the two-dimensional calculations suggest the following three-dimensional effect: During the unloading cycle of a cracked thick plate, the crack tip in the plane stress state in the surface layer closes first, and the crack tip in the plane strain state in the interior remains open until a much lower K_{cl}.

Crack closure has been detected with crack-mouth displacement gages, side-face-mounted or back-face-mounted strain gages, potential drop, and laser interferrometry. None of the these techniques can distinguish the plane stress and plane strain crack-tip opening or closing. Therefore, these techniques cannot pinpoint the crack-tip opening or closing in the interior of a thick plate. There is a distinct possibility that while the crack tip in the surface layer is closed, the crack tip in the interior is 'wedged open' by the closure of the surface layer, or, at least, the K_{op} in the interior is much lower than that in the surface layer. Near the threshold, only one slip system is activated, and a crack grows in a shear mode. A slip band is shown in Figure 6a. As the applied load increases, the shear slip in the slip band causes shear decohesion and shear mode crack growth. During the following unloading cycle, the newly formed shear crack tip may retract somewhat by reverse slip but not enough to pass its initial position and cause crack closure. However, closure may occur at the boundaries between the grains along a crack front.

Both the crack closure model and the dislocation barrier model give the same conclusion regarding the threshold behavior: $K_{max,th}$ is constant in the low R region, and ΔK_{th} is constant in the high R region. Schmidt and Paris derived the constant $K_{max,th}$ based on the assumption that both ΔK_O and K_{op} are constant in the entire range of R and the fact that $K_{max,th} = (K_{op} + \Delta K_O)$ as defined. The same ΔK_O is related to the same physical cause of threshold in the entire range of R. According to the dislocation barrier model, the constant $K^{I}_{max,th}$ and the grain size dependency in the low R region are derived directly from the strength of the resistance of the primary dislocation barrier, and the constant ΔK^{I}_{th} in the high R region is related to the resistance of the secondary barriers. Two different physical causes of threshold in the low and high R regions. The strength of the grain boundary of a ferritic steel is likely related to the amount of carbon in the steel. If this is true, then, according to the dislocation barrier model, the grain boundary strength and, thus, $K^{I}_{max,th}$ should increase with carbon. According to the crack closure model, conversely, an increase in carbon increases the yield strength, and, thereby, may very well reduce crack closure and $K^{I}_{max,th}$. Empirical evidence is needed.

The dislocation barrier model is valid if K_{max} is capable of characterizing the crack-tip stresses at K_{max}. This will be the case if K_{op} is independent of R as assumed by Schmidt and Paris or if the interior crack tip is wedged open so that the state of the crack-tip stress is elastic during unloading. It would be of great interest to study the effect of crack closure on crack-tip stresses. The crack configuration is three dimensional. Near the broad side surface, a shear lip is formed, and the crack front on the shear lip lags behind. The crack in the interior grows in a shear mode, and the crack tip tunnels ahead. The problem of the three-dimensional crack configuration is rather complicated. However, a three-dimensional finite element model for a flat crack with a straight crack front may help to clarify some aspects of this complicated problem.

Empirically, a carburized steel specimen or a sharp side groove in a very thick plate may reduce or even eliminate the complication of the wedging effect of plane-stress crack-closure. If the plane stress closure is reduced or eliminated, K_{op} and $K_{max,th}$ will decrease by the same amount according to the crack closure model. On the other hand, according to the dislocation barrier model, the decrease in $K_{max,th}$ is not necessarily equal to the decrease in K_{op}; it is dependent on the resolved shear stress intensity factor. The mechanics of the problem need to be worked out in order to make quantitative assessment. The correct model describes what actually happens and is consistent with other observed crack growth phenomena, and the model, thus, offers an opportunity for further development.

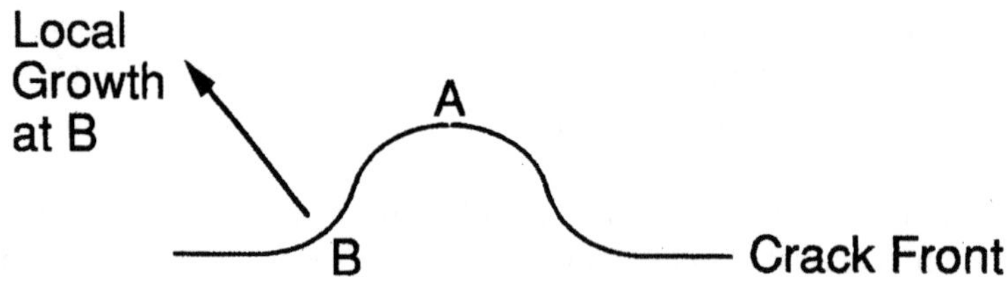

Figure 8. A crack front bulge. The local K at the concave crack front at A is ower than the applied K, while that at the convex crack front at B is higher. The crack is dormant at A and grows laterally at B.

5. Microstructure-sensitive and -insensitive fatigue crack growths

A number of grains lie along a crack front as illustrated in Figure 4. At a ΔK slightly above ΔK_{th}, the crack front at a grain having a higher $K_{R\tau,\text{net}}$ grows faster. Consequently, the local crack front at the grain bulges out as illustrated in Figure 8. Once a bulge is formed, the local K at the concave crack front, A, is lower than the nominal K, while the local K at the convex crack front, B, is higher. Therefore, the crack front at A slows down or stays dormant, while the crack front at B grows to the left and upward. Immediately above ΔK_{th}, $\mathrm{d}a/\mathrm{d}N$ is related to the rate of the total lateral growth, which, in turn, is related to bulge density. Bulge density is equal to the fraction of the crack-tip grains whose $K_{R\tau,\text{net}} > K_{c,\text{GB}}$. The fraction is a function of the orientations of the slip planes and Burgers vectors of the grains relative to the plane and direction of the maximum crack-tip shear stress. It is also a function of the statistical distribution of grain size. Immediately above ΔK_{th}, the fraction and, thereby, $\mathrm{d}a/\mathrm{d}N$ increase rapidly with ΔK.

The local crack increment per cycle, Δa, varies from grain to grain depending on the orientation of the grain and the geometry of the local crack front. The measured $\mathrm{d}a/\mathrm{d}N$ is the average of the Δa's. The same set of ΔK and R gives the same cyclic elastic-plastic crack-tip stresses and strains, which, in turn, give the same bulge density and the same statistical distribution of Δa's. Thus, at the same set of ΔK and R, $\mathrm{d}a/\mathrm{d}N$ is the same. In other words, $\mathrm{d}a/\mathrm{d}N$ is a function of ΔK and R.

$$\frac{\mathrm{d}a}{\mathrm{d}N} = f(\Delta K, R). \tag{19}$$

The relation that the same set of ΔK and R gives the same $\mathrm{d}a/\mathrm{d}N$, in a sense, indicates that the fatigue crack growth behavior is homogeneous. Bulge density increases with ΔK. As ΔK increases, more slip systems in a grain are activated. Multiple slip causes rapid strain hardening. As r_p encloses more grains, grain boundary strengthening increases. The bulge density, the deformation behavior of the crack-tip shear decohesion process, and the statistical distribution of the Δa's change with ΔK. Just above ΔK_{th}, $\mathrm{d}a/\mathrm{d}N$ is also related to grain size and to the strengths of dislocation barriers. Therefore, fatigue crack growth is strongly microstructure sensitive. The same crack growth behavior and the same statistical distribution of the Δa's are limited to the same set of ΔK and R. Therefore, the crack growth behavior in the microstructure sensitive region is said to be *limitedly homogeneous*. Equation (19) is deduced from this limited homogeneity. In essence, this is also the concept of similitude. The physical meaning of similitude is clearly spelled out by the above analysis.

A typical fatigue crack growth curve has three regions. In the low ΔK region immediately above ΔK_{th}, region 1, fatigue crack growth is microstructure sensitive; and in the intermediate

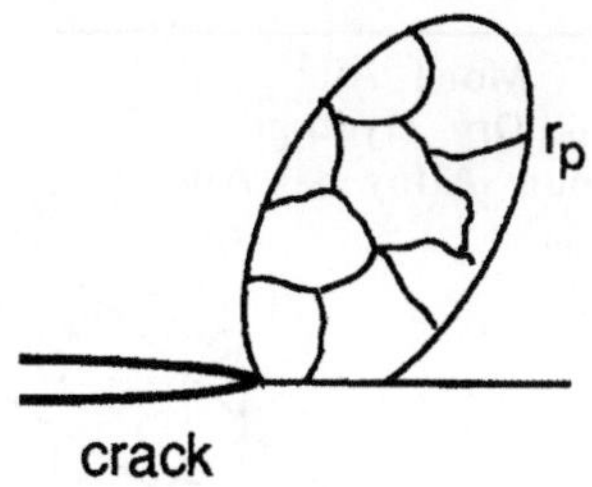

Figure 9. A large crack-tip plastic zone encloses a number of grains.

ΔK region, region 2, it is microstructure insensitive. The transition takes place at $\Delta K_{1,2}$. Above $\Delta K_{1,2}$, the crack-tip plastic zone, r_p, encloses a large number of grains as illustrated in Figure 9, and more than one slip system of a crack-tip grain is activated. Shear decohesion takes place alternately on two slip planes, and striations are formed. The local crack surface in a grain is close to the plane bisecting the angle between these two slip planes, and the macro-crack appears as a 'tensile' crack. The effects of grain boundary strengthening and multiple-slip strain hardening are fully realized. The deformation behavior becomes *uniformly homogeneous*. In other words, the relation between the imposed stress and the induced deformation at geometrically similar points, x_i/r_p, averaged over an entire crack front, does not change with a further increase in ΔK and r_p. While shear decohesion may still occur on the active slip planes of an individual crack-tip grain, the average amount of shear decohesion along an entire crack front is controlled by the uniformly homogeneous deformation within r_p.

A fatigue crack may also grow by a combination of shear decohesion and fracture. When brittle particles are much smaller than the crack increment per cycle, the fracture behavior of these particles is uniformly homogeneous. *When and only when a crack growth process is uniformly homogeneous, fatigue crack growth is completely microstructure insensitive.* If the fatigue crack growth process is indeed uniformly homogeneous, a more specific relation between $\mathrm{d}a/\mathrm{d}N$ and ΔK can be deduced rather rigorously. Liu (1991, 1995) has shown that, at any ΔK, the cyclic elastic-plastic stresses and strains within the dimensionless increment, $\Delta a/r_p$, are the same if the condition of small scale yielding prevails. If the crack growth behavior is uniformly homogeneous, the cycle increment, ΔN, needed to propagate through the crack increment, Δa, is the same regardless of the size of r_p or the level of ΔK. Thus, $\mathrm{d}a/\mathrm{d}N$ is proportional to r_p, which in turn is proportional to ΔK^2.

$$\frac{\mathrm{d}a}{\mathrm{d}N} = f(R)\Delta K^2. \tag{20}$$

If crack closure does occur, and if the deformation and fracture behaviors are completely microstructure insensitive, according to a plane strain analysis, K_{op} as well as K_{cl} should be linearly proportional to ΔK. Thus, even if crack closure occurs, (20) is still valid. If the fatigue crack growth in region 2 is weakly dependent on microstructure, the slope of the crack growth curve should be much less than that of region 1 and still close to 2. Liu, Xu, and Wen (1995) have shown that this is, indeed, the case for fatigue crack growth in 7075-T6 aluminum alloy. The weakly microstructure sensitive growth is caused by inclusion fractures.

As discussed earlier, both ΔK_{th} and $\Delta K_{1,2}$ are related to the crack-tip plastic zone sizes at the threshold and the transition point, and both of these two sizes are related to grain size.

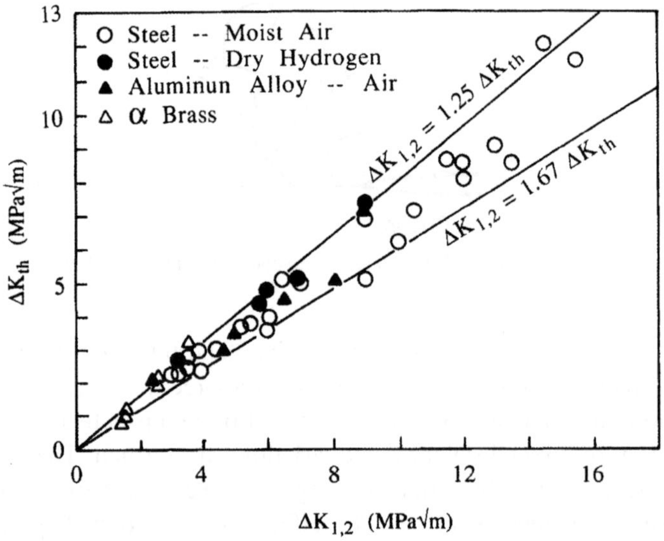

Figure 10. ΔK_{th} versus $\Delta K_{1,2}$. Liu and Liu (1982).

Therefore, these two quantities must be related to each other. Figure 10 shows the plot of ΔK_{th} versus $\Delta K_{1,2}$ of nearly forty materials. Almost all of the data points are between the lines

$$\Delta K_{1,2} = 1.25\Delta K_{\text{th}} \quad \text{and} \quad \Delta K_{1,2} = 1.67\Delta K_{\text{th}} \tag{21}$$

The average of the constants of proportionality is 1.46, which indicates that the ratio of the plastic zone sizes at $\Delta K_{1,2}$ and ΔK_{th} is close to 2. At ΔK_{th}, the slip band is about the size of large grains, while at $\Delta K_{1,2}$, r_p is twice that. It seems reasonable that when r_p is twice the size of large grains, the crack growth becomes uniformly homogeneous, and the transition from microstructure-sensitive crack growth to microstructure-insensitive growth occurs.

Close to ΔK_{th} in the microstructure sensitive region, the slope of a fatigue crack growth curve is very steep because the proportion of crack-tip grains participating fully in the crack growth process increases rapidly with ΔK. At a higher ΔK, if fatigue crack growth is completely microstructure insensitive, the crack growth process is uniformly homogeneous, and the slope of the fatigue crack growth curve should decrease to 2. Even if the fatigue crack growth process is weakly dependent on microstructure, the slope should be close to 2 and much less than that in the low ΔK region immediately above the threshold. This explains the general shape of the fatigue growth curve in the low and intermediate ΔK regions.

The classification of microstructure-sensitive and -insensitive growths is based on empirical observations. The classification of limited and uniformly homogeneous crack growths is deduced from macro-mechanics. As shown, these two classifications can be reconciled and are consistent with each other. Liu (1991) gave a comprehensive macro-mechanical analysis of fatigue crack growth, and Liu (1995) explained the physical concept of limited and uniform homogeneities in detail.

Close to ΔK_{th} in the microstructure-sensitive region, only one slip system is activated, and a fatigue crack grows by the shear mode. If the stacking fault energy is low or if there is only a limited number of slip systems, a shear fatigue crack surface may be smooth, shiny and crystallographic, and the crack growth facets are planar transgranular very much like a

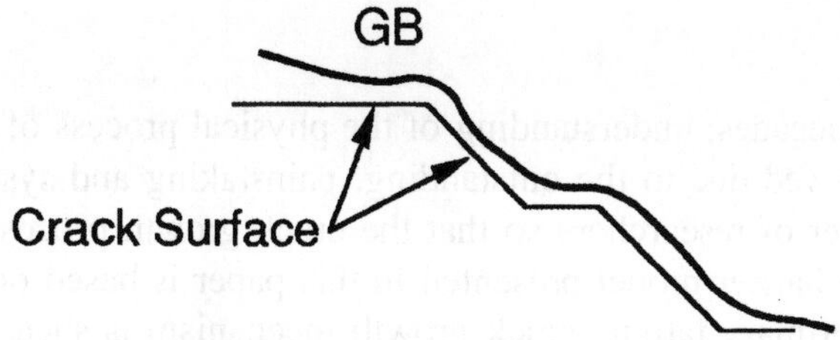

Figure 11. A crack surface is separated from a grain boundary by a thin layer of crystal.

cleavage facet. Such flat, smooth, shiny, {111}-plane shear crack surfaces have been observed by Chen (1992) in large-grain aluminum specimens.

In plain low carbon steels, facets are intergranular. The intergranular facets may be the manifestation of strong grain boundary resistance to slip penetration. Hydrogen-assisted cracking is caused by hydrogen-impaired resistance to shear slip, and the cracking process is a shear slip process. Hydrogen assisted cracking is often 'intergranular'. However, Birnbaum (1990) found that there is a thin layer of crystal between a crack surface and a grain boundary as illustrated in Figure 11. If a grain boundary is a strong dislocation barrier, the shear decohesion process is confined to one side of the boundary. The crack grows by shear slip alternately between two active slip planes, and the zigzag crack path traces these two sets of parallel planes, but the boundary and the crack are separated by a thin layer of crystal (see Figure 11). The observed intergranular fatigue facets in low carbon steels may have been formed by the same process. However, if cross slip occurs easily, a crack path may not be as well defined as the one illustrated in the figure.

Just above threshold, fatigue crack growth facets are often present. As summarized by Ritchie (1979):

> Very close to threshold, the proportion of facets is generally small (~ 1 percent), increasing to a maximum of anywhere from 10 to 80 percent as ΔK is increased, and then gradually diminishing at higher ΔK values in the mid growth rate regime. The maximum proportion of facets appears to occur when the cyclic plastic-zone size approaches the grain size, whereas the disappearance of facets seems to occur when the maximum plastic-zone size exceeds the grain size. Such facets are intergranular in ferritic steels, and transgranular in austenitic stainless steels and alloys of titanium, aluminum, copper, and nickel.

Close to ΔK_{th}, only one slip system in a crack-tip grain is active, and a crack grows primarily by shear mode. Therefore, fatigue facets are likely shear surfaces. When the cyclic plastic zone exceeds the grain size, more than one slip system is activated, and a crack grows by shear slip alternately on two slip planes. The local crack plane bisects the angle between these two planes, and facets disappear. Facets are intergranular in ferritic steels because of the strong grain boundaries as discussed above. However, many questions remain unanswered, among them, whether the maximum proportion of facets is related to the stacking fault energy and the number of available slip systems of a crystal, as should be the case for shear facets, and why the proportion of facets changes with cyclic plastic zone size.

6. Discussion

During the past three decades, understanding of the physical process of fatigue crack growth has been greatly improved due to the outstanding, painstaking and systematic experimental work of a great number of researchers so that the crack growth process is gradually understood. The dislocation barrier model presented in this paper is based on the following crack growth process: The primary fatigue crack growth mechanism is shear slip. Close to ΔK_{th}, only one slip system of a crack-tip grain is activated, and a slip band is formed. When a slip band is stopped and constrained by a dislocation barrier, a stress singularity exists at the tip of the slip band. Without the penetration of dislocation barriers, a fatigue crack becomes dormant and nonpropagating. Slip is driven by resolved shear stress. At threshold, the net applied resolved shear stress intensity factor is equal to the resistance of the barrier. If this physical process is accepted, the deductive logic and calculation are rather simple and straightforward.

In spite of the simplified calculation, the results of the analysis agree surprisingly well with a number of observations on the effects of grain size and R-ratio. The model is also consistent with the observed transgranular and intergranular facets, the characteristics of, and the transition from, microstructure-sensitive to microstructure-insensitive crack growth.

The analysis raises a number of questions and points out the need of additional work. To substantiate and improve the model, critical empirical work is needed such as the study of the detailed process of microstructure-sensitive crack growth near the threshold, the verification of crack surface facets as hypothesized, and a study of the statistical distribution of grain size. Perhaps, very thick specimens made of large-grained materials may help to trace the details of the crack growth process. The model is two dimensional, but the physical problem is three dimensional. Needed also are more accurate calculation of $K'_{R\tau}$ and a better understanding of the effects of crack closure on the state of the crack-tip stress at $K_{\max}$. As we learn more about the crack growth process, the details of the model may have to be modified.

7. Summary and Conclusions

(1) Fatigue crack growth is caused by crack-tip shear slip or shear decohesion. Shear decohesion is the result of crack-tip dislocation emission followed by the glide of the emitted dislocations. Both dislocation emission and glide are controlled by the resolved shear stress field at the crack tip, which is characterized by the resolved shear stress intensity factor, $K_{R\tau}$.

(2) A dislocation barrier model of fatigue crack growth threshold is constructed. In the low R region, the model relates $K^{\text{I}}_{\max,\text{th}}$ and ΔK_{th} to grain size, d, and R.

$$K^{\text{I}}_{\max,\text{th}} = \frac{\Delta K_{\text{th}}}{(1-R)} = \frac{K_{c,\text{GB}}}{0.39} + \frac{\gamma}{0.39}\left(\frac{8}{\pi}\right)^{1/2} \tau_{RY}\sqrt{d}. \tag{15}$$

This theoretical relation agrees surprisingly well with the empirical one of Masounave and Baïlon for a ferritic steel,

$$\frac{\Delta K_{\text{th}}}{(1-R)} = 3.8 + 1140\sqrt{d}. \tag{7}$$

(3) Equation (15) indicates that ΔK_{th} is proportional to $\sqrt{d}$. This agrees with the data of 0.07%C – 0.46%Mn steel obtained by Beevers, that of low carbon steels obtained by Taira et al., and that of low strength steels obtained by Ritchie.

(4) Equation (15) also indicates that $K^{\mathrm{I}}_{\mathrm{max,th}}$ of a given material is constant as observed by Cooke and Beevers in five medium carbon steels of three different microstructures and by many others in a variety of materials.
(5) Based on the model, it can also be deduced that the $\Delta K^{\mathrm{I}}_{\mathrm{th}}$ in the high-R region is constant.
(6) $K^{\mathrm{I}}_{\mathrm{max,th}}$ is related to the intrinsic resistance of the primary dislocation barrier, and $\Delta K^{\mathrm{I}}_{\mathrm{th}}$ is related to the intrinsic resistance of secondary barriers.
(7) The model is consistent with the characteristics of microstructure-sensitive and microstructure-insensitive fatigue crack growths and helps to explain the transition from microstructure-sensitive to microstructure-insensitive growth.
(8) The dislocation barrier model assumes that the crack growth near the threshold is a shear growth and suggests that a fatigue facet is a shear surface. However, experimental verification of the shear facet is needed.

Acknowledgment

The author gratefully acknowledges the full and sustained support of his wife, which is indispensable to the success of this study.

References

Beevers, C.J. (1977). Fatigue crack growth characteristics at low stress intensities of metals and alloys. *Metal Sciençe* **11**, 362–367.

Birnbaum, H.K. (1990). Mechanism of hydrogen related fracture of metals. *Hydrogen Effects on Material Behavior* (Edited by N.R. Moody and A.W. Thompson), The Minerals, Metals & Materials Society, 639–658.

Chen, Qi, (1992). *Shear Fatigue Crack Growth Analyses in Large Grain Polycrystals*, Ph.D. Dissertation, Solid State Science and Technology, Syracuse University, Syracuse, NY.

Cooke, R.J. and Beervers, C.J. (1973). The effect of load ratio on the threshold stresses for fatigue crack growth in medium carbon steels. *Engineering Fracture Mechanics* **5**, 1061–1071.

Elber, W. (1970). Fatigue crack closure under cyclic tension loading. *Engineering Fracture Mechanics* **2**(1), 37–45.

Elber, W. (1970). The significance of fatigue crack closure. *Damage Tolerance in Aircrat Structures*, ASTM STP 486, American Society for Testing and Materials, 230–242.

Friedel, J. (1967). *Dislocations*, Addison Wesley Publishing Co., Inc., 53.

Greene, C.A., Holtz, R.L., Sadananda, K. and Vasudevan, A.K., The controlling parameters of fatigue crack growth behavior in Al-Li 8090 alloy, In *Micromechanics of Advanced Materials*, The Minerals, Metals, and Materials Society, 127–133.

Hirth, J.P. and Lothe, J. (1968). *Theory of Dislocations*, McGraw-Hill Book Company, New York, 87.

Kirby, B.R. and Beevers, C.J. (1979). Slow fatigue crack growth and threshold behaviour in air and vacuum of commercial aluminum alloys. *Fatigue of Engineering Materials and Structures* **1**, 203–215.

Kunio, T., Shimizu, M. and Yamada, K. (1969). Microstructural aspects of the fatigue behaviour of rapid-heat-treated steel. *Proceedings of 2nd International Conference on Fracture*, Chapman & Hall, England, 630–642.

Laird, C. (1967). The influence of metallurgical structure on the mechanisms of fatigue crack propagation. *Fatigue Crack Propagation*, American Society for Testing and Materials, ASTM STP 415, 131–168.

Lindley, T.C. and Richards, C.E. (1974). The relevance of crack closure to fatigue crack propagation. *Materials Science and Engineering* **14**, 281–293.

Liu, H.W. (1991). A review of fatigue crack growth analyses. *Theoretical and Applied Fracture Mechanics* **16**, 91–108.

Liu, H.W. (1995). A logic framework for fatigue crack growth analyses. *A Symposium on Micromechanics of Advanced Materials*, (Edited by S.N.G. Chu, P.K. Liaw, R.J. Arsenault, K. Sadananda, K.S. Chan, W.W. Gerberich, C.C. Chau, and T.M. Kung), The Minerals, Metals & Materials Society, 171–180.

Liu, H.W. (1998). A dislocation barrier model for fatigue limit – as determined by crack non-initiation and crack non-propagation. To be published.

Liu, H.W. and Ke, J.S. (1975). Moire method. *Experimental Techniques in Fracture Mechanics* **2** (Edited by Albert Kobyashi), Society for Experimental Stress Analysis, 111–165.

Liu, H.W. and Liu, Dai (1982). Near threshold fatigue crack growth behavior. *Scripta Metallurgica* **16**, 595–600.

Liu, H.W., Xu, Jian Guo and Wen, Wang You (1995). An application of the logic framework for FCG analyses – A study of microstructural effects on FCG in Al 7075-T651. *A Symposium on Micromechanics of Advanced Materials* (Edited by S.N.G. Chu, P.K. Liaw, R.J. Arsenault, K. Sadananda, K.S. Chan, W.W. Gerberich, C.C. Chau, and T.M. Kung), The Minerals, Metals & Materials Society, 181–190.

Masounave, J. and Baïlon, J.-P. (1976). Effect of grain size on the threshold stress intensity factor in fatigue of a ferritic steel. *Scripta Metallurgica* **10**, 165–170.

McEvily, A.J. Jr. and Boettner, R.C. (1963). On fatigue crack propagation in F.C.C. metals. *Acta Metallurgica* **11**, 725–743.

Nabarro, F.R.N. (1967). *Theory of Crystal Dislocations*, Oxford University Press, 84.

Newman, J.C., Jr. (1976). A finite element analysis of fatigue crack closure. *Mechanics of Crack Growth*, ASTM STP 590, American Society for Testing and Materials, 281–301.

Newman, J.C., Jr. and Armen, H. Jr. (1974). Elastic-plastic analyses of a propagating crack under cyclic loading, AIAA paper no. 74-366, Presented at the AIAA/ASME/SAE 15th Structure, Structural Dynamics, and Materials Conference, Las Vegas, Nevada, April 17–19.

Neumann, V.P. (1974). Fatigue crack propagation I – New experiments concerning the slip processes at propagating fatigue cracks. *Acta Metallurgica* **22**(9), 1155–1165.

Neumann, V.P. (1974). Fatigue crack propagation II – The geometry of slip processes at a propagating fatigue crack, *Acta Metallurgica* **22**(9), 1167–1178.

Ohji, K., Ogura, K. and Ohkubo, Y. (1975). Cyclic analysis of a propagating crack and its correlation with fatigue crack growth. *Engineering Fracture Mechanics* **7**, 457–464.

Pelloux, R.M.N. (1969). Mechanisms of formation of fatigue striations. *Transactions of American Society of Metals* **62**, 281–285.

Priddle, K.E. (1978). The influence of grain size on threshold stress intensity for fatigue cracks in AISI 316 stainless steel. *Scripta Metallurgica* **12**, 49–56.

Ritchie, R.O. (1979). Near-threshold fatigue-crack propagation in steels. *International Metals Reviews*, Review 245(5 and 6).

Sadananda, K. and Vasudevan, A.K. (1995). Fatigue crack growth behavior in Titanium Aluminides. *Materials Science and Engineering* **A192/193**, 490–501.

Schmidt, R.A. and Paris, P.C. (1973). Threshold for fatigue crack propagation and the effects of load ratio and frequency. *Progress in Flaw Growth and Fracture Toughness Testing,* ASTM STP 536, American Society for Testing and Materials, 79–94.

Sharpe, W.N., Jr. and Grandt, A.F. Jr. (1976). A preliminary study of fatigue crack retardation using laser interferrometry to measure crack surface displacements. *Mechanics of Crack Growth*, ASTM STP 590, American Society for Testing and Materials, 302–320.

Taira, S., Tanaka, K. and Nakai, Y. (1978). A model of crack-tip slip band blocked by grain boundary. *Mechanics Research Communication,* **5**(6), 375–381.

Taira, S., Tanaka, K. and Hoshina, M. (1979). Grain size effect on crack nucleation and growth in long-life fatigue of low-carbon steel. *Fatigue Mechanisms*, ASTM STP 675, American Society for Testing and Materials, 135–173.

Tomkins, B. (1968). Fatigue crack propagation – An analysis. *Philosophical Magazine* **18**, 1041–1066.

Vehoff, H. and Neumann, P. (1979). In situ SEM experiments concerning the mechanism of ductile crack growth. *Acta Metallurgica* **27**, 915–925.

Yamada, K. (1970). *Microstructural Aspects of the Fatigue Behavior of Rapid-Heat-Treated Steel,* Ph.D. dissertation, Faculty of Engineering, Keio University, Yokohama, Japan.

Yang, Chuang-Yeh (1979). *Modeling of Crack Tip Deformation with Finite Element Method and Its Applications*, Ph.D. Dissertation, Solid State Science and Technology, Syracuse University, Syracuse, NY.

International Journal of Fracture **93:** 281–314, 1998.

A calibrated fracture process zone model for thin film blistering

A. SHIRANI and K.M. LIECHTI
Mechanics of Solids, Structures and Materials, Department of Aerospace Engineering and Engineering Mechanics, The University of Texas at Austin, Austin, Texas 78712-1085 U.S.A.

Received 1 September 1998; accepted in revised form 15 September 1998

Abstract. The purpose of this work was to examine the feasibility of using fracture process zone models for extracting the adhesive fracture energy of thin films on a thick substrate from circular blister experiments that involve a substantial amount of inelastic deformation in the thin film. The interface produced by vapor depositing polyimide on aluminum formed the basis of the experiments that were conducted. The experiments were conducted in volume control while the pressure history and the corresponding three dimensional blister shape were measured. The analysis accounted for the nonlinear kinematics and material behavior of the polyimide film and included a traction-separation law for the interface. The traction-separation law for the interface was calibrated in an iterative manner by comparing measured pressure-volume responses and crack opening displacements. The adhesive fracture energy obtained from the selected traction-separation law was reasonable considering that fracture occurred in an interphase region. It was bounded by values that were obtained from elastic analyses (with updated kinematics) of the type performed by Gent and Lewandowski (1987) and Chu et al. (1992 a, b).

1. Introduction

In view of his pioneering contributions to adhesive fracture mechanics through the introduction of the pressurized blister experiment (Williams, 1969, 1970 and Burton, Jones and Williams, 1970), it is fitting to present this paper which deals with the blistering of thin films. In this situation, unless the delaminating layer has an exceptionally high yield strength (Nix, 1997), plasticity effects can be expected to enter into the determination of adhesive fracture toughness. This has even been true with relatively thick blisters (Liechti and Hanson, 1989) and sandwiched blisters (Liechti and Liang, 1992). As many will already know, our honoree also provided a pioneering analysis of an adhesively bonded cantilever in which plasticity effects were extracted (Chang, DeVries and Williams, 1972).

Hinkley (1983) was the first to explore the use of the pressurized blister for determining the degree of adhesion of thin films. An elastic analysis of a spherical cap was used to determine the adhesive fracture energy of polyimide on silicon. A more general membrane analysis by Hencky (1915) was the basis of Gent and Lewandowski's (1987) approach for determining the onset and subsequent quasi static propagation of circular blisters, given an elastic film and a constant adhesive fracture energy. Chu et al. (1992 a, b) extended the Gent and Lewandowski analysis so that, in constant flow rate experiments, only pressure-time histories following initiation needed to be measured in order to extract the adhesive fracture energy.

Further extension of the nonlinear kinematics of blisters was provided by Jensen (1991) who employed von Karman plate theory in an elastic analysis of stationary cracks. The numerical solutions for energy release rate and mode-mix spanned the range of behaviors from thick plates to membranes in a consistent manner for the first time. The effect of residual stresses was considered in a subsequent analysis by Jensen and Thouless (1993). Similar kinematics were used by Liechti and Shirani (1994) to analyze circular, island and peninsula blisters in a consistent manner. The latter two configurations had been introduced by Allen and Senturia (1989) and Dillard and Bao (1991), respectively, in order to circumvent bursting problems that had been encountered in circular blisters. In comparing the stress levels in the three configurations for an adhesive fracture energy of 100 J/m^2, Liechti and Shirani found that the extent of yielding was the largest in circular blisters and smallest in the peninsula blister. The amount of yielding was generally substantial, and the only configuration which did not suffer any yielding was a relatively thick peninsula blister. The addition of a thicker, higher strength layer extended the range of aspect ratios (crack length to film thickness) over which linear elastic fracture mechanics could potentially be applied.

Thus, in general, it was clear that significant amounts of yielding could be expected in blister experiments. Unless accounted for, the accompanying plastic dissipation would mask the true values of the adhesive fracture energies of different combinations of materials and surface treatments as Kim and Aravas (1988) showed in their analysis of steady-state, elastoplastic peeling. More recently, Wei and Hutchinson (1996) have made similar distinctions between the intrinsic toughness of an interface and plastic dissipation in transient thin film peeling by making use of a traction separation law to represent the interface. This so-called fracture process zone model approach to analyzing interfacial crack growth problems was pioneered by Needleman (1987, 1990) and has been used (Tvergaard and Hutchinson, 1993, 1994; Swadener and Liechti, 1996) to examine the role of plasticity in explaining the asymmetrical shielding effect (Liechti and Chai, 1992) that occurs under shear-dominant loadings.

A key element in the fracture process zone modeling approach is the identification of the traction separation law of the interface. The simplest form would be a Dugdale or Barenblatt zone with a constant traction. These models have been used in theories and analyses of viscoelastic fracture (Schapery, 1975 a, b, c; Knauss 1983) and gave rise to atomistically small cohesive zone sizes for the relatively brittle materials that were being considered. More general traction separation laws for damage-softened composites and adhesives and crazes have been extracted by Ungsuwarungsri and Knauss (1987, 1988a, b) by making use of slope measurements on double cantilever beam specimens. More recently, Swadener and Liechti (1996) used an iterative, hybrid experimental/numerical approach in which measurements and finite element predictions of near tip submicron crack opening displacements were matched in steady state interfacial crack growth along a glass/epoxy interface. Because of the steady state condition, it was possible to specify a particular cohesive zone size and the decay rate of the tractions with the result that the traction separation law itself became part of the solution.

The problem being considered here was transient, in the sense that we wished to capture the initiation and propagation of debonds in circular blisters. As will be seen in what follows, this has required a different approach to be taken for extracting the traction separation laws. The materials being considered here are polyimide deposited on aluminum. In this paper, we examine the sensitivity of the pressure volume response and blister shape to various choices of traction separation laws. A comparison is then made between the adhesive fracture energy value extracted in this way and values that were obtained using the Gent and Lewandowski (1987) and Chu et al. (1992 a, b) approaches, but making use of more generalized kinematics than were employed in the original baseline analysis of Hencky (1915). The traction separation laws associated with various polyimide/aluminum interfaces will be developed in a companion paper and linked to spectroscopic analyses of the fracture surfaces and the interfaces prior to fracture.

2. Experiments

The adhesion experiments consisted of several key components. First came the preparation of the aluminum surfaces and the deposition of the polyimide. Secondly, a pressurization system was developed along with a computerized data acquisition system. The third aspect of the experiment was the measurement of the three dimensional shape of the blister. Once pressure and shape could be measured, bulge tests were conducted in order to determine the stress strain behavior of the polyimide. These were then followed by crack initiation and propagation experiments.

3. Specimen preparation

This portion of the work was done by Professor Boerio's group in the Materials Science and Engineeering Department of the University of Cincinnati. Aluminum substrates were prepared by degreasing in acetone, detergent washing, wet grinding and finally mechanical polishing using 6 μm and 1 μm diamond compound. Substrates were then ultrasonically cleaned in ethanol followed by acetone. The final cleaning step consisted of a ten minute exposure to an O_2 plasma. This was crucial to the removal of all traces of organic compounds from the aluminum surface and resulted in much more reproducible data.

Samples for blister test evaluations were prepared in a similar fashion from circular aluminum disks having a diameter and thickness of 100 mm and 25 mm, respectively. A hole for pressurization was machined at the center of the substrate, occluded with a thin disk of Potassium Bromide (KBr) pressed in-situ prior to final polishing, plasma cleaning, and spin coating. The polyimide deposition technique for the experiments in this paper consisted of vapor deposition of monomer and polymer films, carried out by sublimation from independently heated crucibles in a modified NRC thermal evaporation unit. The system pressure during deposition was approximately 5×10^{-6} Torr as measured with an ionization gauge. A 200 A° film of PMDA/ODA was vapor deposited onto the clean aluminum substrates and then imidized. The film was then treated in an oxygen plasma to ensure adhesion of subsequent layers, and then a nominally 50 μm film of PMDA/ODA polyamic acid was spin coated over the vapor-deposited film and imidized. This produced a film with identical bulk mechanical properties to spin coated films.

The blister specimen is shown in Figure 1. The aluminum disk includes a central hole (a_0) of about 6.5 mm in diameter, which conveys fluid (deionized water) to the film. Each specimen was used for several tests with different crack lengths. A very thin layer of white paint was sprayed on top of the polyimide film so that a clear pattern of projected grid lines would be observed in order to measure the height of blister. The specimen was fastened to two columns and these two columns were mounted to a table for support.

4. Apparatus

Figure 1 is also a schematic view of the apparatus that was used in this study. The pressurization system consisted of a syringe pump and a pressure transducer. In order to minimize the effect of compliance, the components were designed to be as close as possible to one another. A syringe pump with flow rates ranging from 0.02 μl/hr to 500 ml/hr and with an accuracy of 0.5 percent with a 10 cc plastic syringe was used to control the volume flow rate of deionized

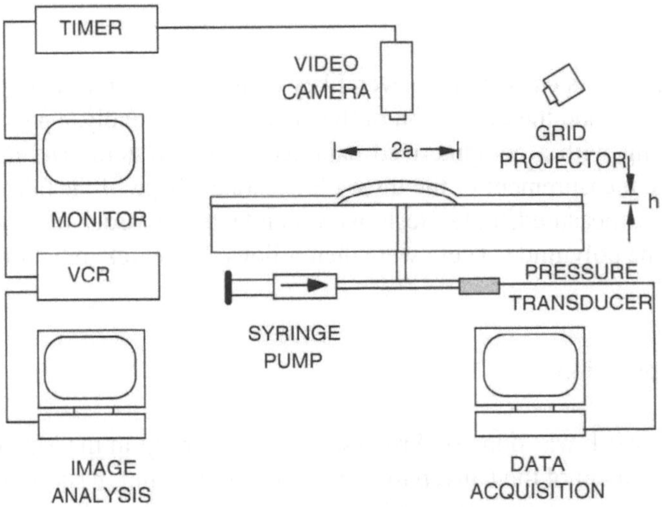

Figure 1. Blister test apparatus and circular blister geometry.

water at 20 ml/hr. In most of the experiments, especially for blister initiation from the pressure porthole, an applied pressure up to 1.05 MPa was required, thereby ruling out the use of glass syringes. The resulting pressure was measured with a Sensotec pressure transducer with a resolution of 0.25 percent of full range and a maximum capacity of 1.38 MPa. The applied pressure was recorded by a computerized data acquisition system with 16 bit resolution. A video camera and recording system along with a digital image analyzer were used to measure the dimensions of the crack front and the change of volume of the blisters.

One of the most challenging parts of this experiment was to remove or, at least, reduce the amount of the air trapped in the system. First the hole in the aluminum disk and then the pressure transducer were filled with deionized water and then by taking advantage of gravity, the circuit was closed. Nevertheless, as will be seen later, the volume measurement by the syringe pump could not be used due first to the compliance of the devices, especially the syringe which was plastic and second the compressibility of small amounts of air trapped in the system. As a result, it was necessary to measure the blister shape.

In most experiments, crack growth was slow enough so that cracks, once initiated and grown for about 2 mm, could then be arrested by unloading, leaving a new starter crack for the next experiment. In this way, a single specimen could be used to conduct several experiments, so that enough data was obtained for each processing mode.

5. Blister shape measurement

The projected grid method was adopted for measuring the blister shape. A Moiré method would have been easier for data reduction but the competing requirements of sufficient stand-off distance and fringe visibility could not be reconciled. Parallel lines were projected on the surface of interest. The lateral deflection, Δ, of the lines (Figure 2) is related to the out-of-plane displacement of the surface. The viewing from above provides a precise way of determining the shape of the crack front. A light source with grids of 20, 10 or 5 lines/mm was used to project a shadow of the lines on the specimen. The video system recorded the

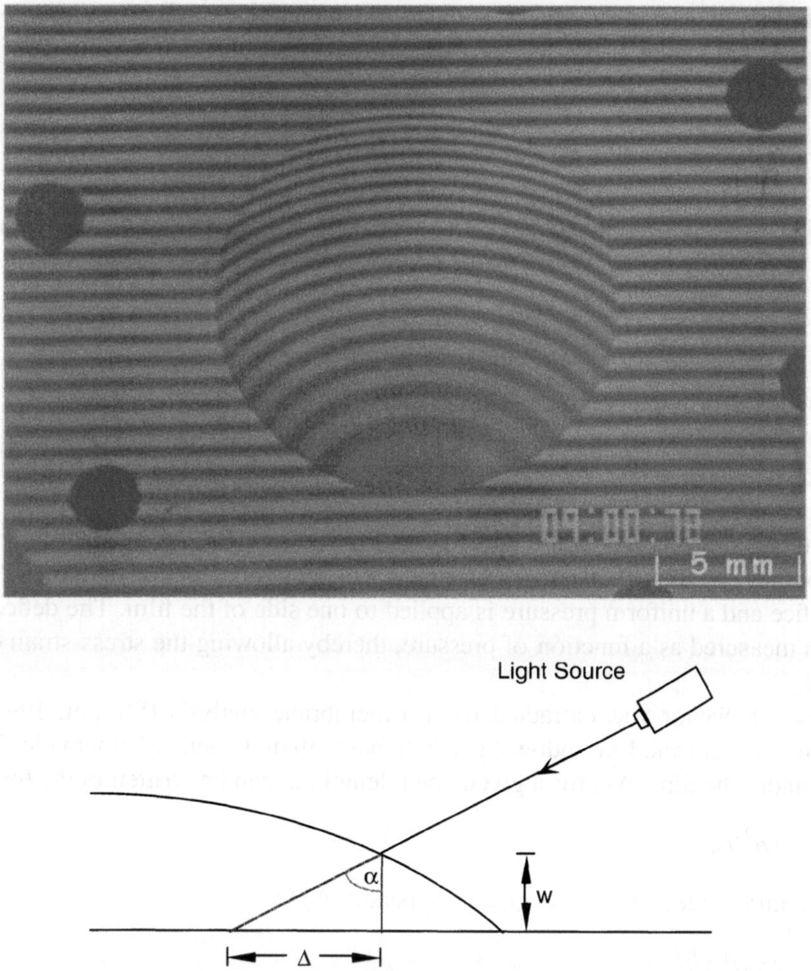

Figure 2. Projected grid measurement schematic and line pattern.

change in line spacing while the test was in progress. Prior to each test, a two-dimensional calibration of length/pixel was necessary for each magnification level.

The following procedure was adopted for determining the deflection, w, of the blister at any particular location. The image analysis system was used to digitize recorded frames and then scan the video image to obtain profiles of intensity of light versus pixel position. The location and order of the dark or bright lines were determined by identifying maxima and minima in the intensity profiles. The relative positions of the same bright or dark line in the undeformed and deformed configurations allowed the shift, Δ, to be determined. The deflection is related to the shift and the angle of illumination, α, through

$$w = \frac{\Delta}{\tan \alpha}. \tag{1}$$

The lines were projected at an angle of 30° giving rise to a resolution of 50 μm with a grating of 20 lines/mm by counting bright and dark fringes which could be located to within 5 mm. The overall accuracy of this technique was found to be 3 percent by using a calibrated 45° wedge (Shirani, 1997).

Once the film displacement was determined, the change in volume, Δv, was obtained by integrating the area under the deflected shape in the following manner. First, the change in volume is given by

$$\Delta v = 2\pi \int_0^a w(r) r \, dr, \tag{2}$$

where r is the distance from the center of the blister. In discrete form, (2) can be expressed as

$$\Delta v = 2\pi \sum_{i=1}^{n} \left(\frac{w_i - w_{i+1}}{2} \right) r_i \Delta r_i, \tag{3}$$

where i denotes the ith data point.

6. Bulge test

The bulge test is one the first techniques that was introduced for the study of thin film mechanical properties (Beams, 1959). In its original form, a circular film or membrane is clamped over an orifice and a uniform pressure is applied to one side of the film. The deflection of the film is then measured as a function of pressure, thereby allowing the stress-strain curve to be determined.

The elastic behavior was extracted from a membrane analysis (Shirani, 1997) that was similar to that of Gent and Lewandowski (1987), but with more general kinematics.The change of volume under the film, Δv, for a given crack length, a, can be written in the form

$$\Delta v = C_1 \pi a^2 w_0, \tag{4}$$

where the central deflection, w_0, is given (Appendix A) by

$$w_0 = C_2 \left(\frac{pa^4}{Eh} \right)^{1/3}, \tag{5}$$

where E is the Young's modulus and C_1 and C_2 are coefficients whose values depend upon the value of Poisson's ratio ν and the boundary conditions. For a Poisson's ratio of 0.3 and clamped boundaries, $C_1 = 0.519$ and $C_2 = 0.706$. Equations (4) and (5) result in

$$p = \kappa (\Delta v)^3, \tag{6}$$

where

$$\kappa = \left[(C_1 C_2 \pi a)^3 \frac{a}{Eh} \right]^{-1}.$$

Thus, for elastic deformations, the value of κ can obtained from pressure and volume measurements, thereby allowing the Young's modulus to be extracted.

Beyond the elastic limit, the stress in the film was derived using the spherical cap assumption and the condition of force equilibrium, resulting in the standard formula for stress in a thin-walled spherical pressure vessel

$$\sigma = \frac{pR}{2h}. \tag{7}$$

where p is the pressure, R is the blister radius and h is the film thickness.

With the reasonable assumption that, $w_0 \ll a$, for thin films, (7) can be written as

$$\sigma = \frac{pa^2}{4w_0 h}. \tag{8}$$

The strain in the film is similarly derived from geometrical considerations with the same assumptions. The strain is defined as the change in arc length divided by the original arc length with the following result

$$\varepsilon = \frac{2w_0^2}{3a^2}. \tag{9}$$

Thus, the stress-strain behavior can be extracted from (8) and (9) and measurements of pressure, central deflection and blister radius. A Ramberg Osgood fit was made to the stresses and strains derived from (8) and (9) using the form

$$\varepsilon = \frac{\sigma}{E}\left(1 + \frac{3}{7}\left(\frac{\sigma}{\sigma_0}\right)^{n-1}\right), \tag{10}$$

where n is the power law exponent and σ_0 is a fitting parameter, not to be confused with the yield strength σ_y, which was determined from the 0.2 percent offset strain level.

7. Analysis

In this section, we describe the steps that were taken to develop the analyses that were used to extract the adhesive fracture energies. The first was an entirely elastic analysis of quasi static debonding of a thin film from a rigid substrate following the approaches of Gent and Lewandowski (1987) and Chu et al. (1992 a, b) but incorporating more general kinematics than those that were employed in the analysis of Hencky (1915), which was the basis of both analyses. The second was a finite element analysis that included the interface as a separate constitutive entity, while the polyimide was taken to be elastoplastic and the aluminum rigid.

8. Elastic analysis

The details of the stress analysis of a clamped thin film with a residual stress σ_R in the film are given in Appendix A. It was based on the assumption that the deflection w and radial displacement u of the film were, respectively

$$w = w_0\left(1 - \frac{r^2}{a^2}\right)^2 \tag{11}$$

and

$$u = r(a - r)(c_1 + c_2 r). \tag{12}$$

The coefficients w_0, c_1 and c_2 were determined by minimizing the strain energy of the film with respect to c_1 and c_2 and then applying the principle of virtual displacements to determine w_0. As a result, the relationship between the pressure and the central displacement was

$$\frac{pa^4}{Eh} = e(v)h^2 w_0 + f\left(v, \frac{\sigma_R}{E}\right) a^2 w_0 + g(v) w_0^3, \tag{13}$$

where the functions e, f, and g are given in Appendix A. Acounting for only membrane effects yields (5).

Following the fracture analysis developed in Appendix B for bending and membrane effects and residual stresses, it was found that the adhesive fracture energy Γ_a was

$$\Gamma_a = \frac{1}{2\pi a}[\tfrac{3}{4}p - (\Delta v)\zeta]\left(\frac{\partial(\Delta v)}{\partial a}\right)_p - 2(\Delta v)^2 \frac{\zeta}{a}, \tag{14}$$

where

$$\left(\frac{\partial(\Delta v)}{\partial a}\right)_p = \left(\frac{6e(v)h^2 a^4 + 4f\left(v, \frac{\sigma_R}{E}\right) a^6 + \frac{10}{C_1^2 \pi^2} g(v)(\Delta v)^2}{e(v)h^2 a^5 + f\left(v, \frac{\sigma_R}{E}\right) a^7 + \frac{3}{C_1^2 \pi^2} g(v) a (\Delta v)^2}\right)(\Delta v) \tag{15}$$

and

$$\zeta = \frac{Eh}{2\pi C_1 a^4}\left[e(v)\frac{h^2}{a^2} + f\left(v, \frac{\sigma_0}{E}\right)\right]. \tag{16}$$

Equation (14) can be simplified by considering only membrane effects, thereby allowing these results to be compared with those predicted by Gent and Lewandowski (1987). Under these conditions,

$$\Gamma_a = C_3 p w_0, \tag{17}$$

where $C_3 = 0.619$ and is 4.6 percent lower than the value (0.649) from Gent and Lewandowski (1987). Thus the adhesive fracture energy during elastic blistering can be determined from measurements of the pressure and central deflection.

When a blister specimen is initially pressurized, its crack radius is that of the hole through which fluid is injected. The 'crack' is generally more like a bimaterial corner or a notch. As a result, the pressure level required to cause initial debonding is unusually high and subsequent growth is dynamic. Due to volume control, the crack arrests and subsequent repressurization results in slower growth from the sharp, dynamically arrested crack. It might be thought that initiation values of the adhesive fracture energy obtained from these subsequent experiments would be reasonable. However Chu et al. (1992 a, b) found that more consistent data could be obtained from measurements made during quasi-static growth rather than at initiation. The adhesive fracture energy was extracted on the basis of pressure vs. time data through an extension of the debonding analysis of Gent and Lewandowski (1987).

It was shown by Chu et al. (1992 a, b) that, in the absence of residual stresses and neglecting the effect of bending for very thin blisters, one can find a linear relationship between p^{-3} and time, t, in the form

$$p^{-3} = \frac{R_f}{C_1 C_2 \pi}\left(\frac{Eh}{K_r^{10}}\right)^{1/3}(t - t_c) + p_c^{-3}, \tag{18}$$

where t_c is the time when debonding begins, p_c is the critical pressure, R_f is the constant volume rate of fluid injection and C_1 and C_2 are the coefficients from the membrane analysis in (4) and (5). From (18), we can introduce the slope N_s of the plot p^{-3} vs. $(t - t_c)$ as

$$N_s = \frac{R_f}{C_1 C_2 \pi} \left(\frac{Eh}{K_r^{10}} \right)^{1/3}. \tag{19}$$

The constant K_r was obtained from (16) of Gent and Lewandowski (1987) which applies to the quasi-static debonding stage of the experiments. For this analysis

$$K_r = (C_4 Eh \Gamma_a^3)^{1/4}, \tag{20}$$

where

$$C_4 = \left(\frac{4}{5C_1C_2} \right)^3. \tag{21}$$

The value of N_s can be extracted from each experiment. Combining (19) and (20), one finds that

$$\Gamma_a = C_5 \left[\frac{R_f^2}{N_s^2 Eh} \right]^{1/5}, \tag{22}$$

where

$$C_5 = \frac{1}{\sqrt[3]{C_4}} \left[\frac{1}{(C_1 C_2 \pi)^2} \right]^{1/5}. \tag{23}$$

The Gent and Lewandowski (1987) analysis yielded $C_5 = 0.395$, whereas $C_5 = 0.431$ from the current analysis. Thus, based on Hencky's (1915) analysis, there is a 9 percent difference in Γ_a values, due to the inclusion of radial displacements here.

Nonetheless, it can be seen from (22) that the adhesive fracture energy Γ_a can be determined from measurements of the pressure history (N_s) and the rate of fluid injection (R_f). This greatly simplifies the data reduction process, because a height or volume measurement is not required, if the loading device compliance is low.

9. Elastoplastic debonding of the thin layers

In Liechti and Shirani (1994), it was shown that large scale yielding could occur in circular blister specimens with an adhesive fracture energy of 100 J/m^2. In some cases there was complete yielding of the delaminating film, so that J-dominance would no longer apply. In order to account for the large amount of yielding and still extract the adhesive fracture energy, the fracture process zone approach of Needleman (1987, 1990) and Tvergaard and Hutchinson (1993, 1994) was adopted for our study. This approach essentially involves attributing a traction-separation law to the interface and, because it allows crack growth to occur, the associated plastic dissipation from loading and unloading of points that are passed by the crack front is rigorously accounted for. As a result, the adhesive fracture energy is determined by the chosen traction-separation law. The dominant scale of the fracture process zone in many interface systems can indeed be measured in microns rather than nanometers. Thus, the traction-separation law for an interface should be regarded as a phenomenological characterization of the zone where the separation takes place along the interface and not necessarily a description of atomic separation.

Under mixed-mode conditions, traction-displacement relationships are introduced between the normal and tangential components δ_n and δ_t, respectively, of the relative displacement of the crack faces across the interface in the zone where the fracture processes are occurring and the corresponding tractions are T_n and T_t. When the components of the relative displacements reach the critical values, δ_n^c and δ_t^c, separation occurs. By introducing a single non-dimensional separation measure as

$$\lambda = \left[\left(\frac{\delta_n}{\delta_n^c} \right)^2 + \left(\frac{\delta_t}{\delta_t^c} \right)^2 \right]^{1/2}, \tag{24}$$

the tractions drop to zero when $\lambda = 1$. To specify the tractions T_n and T_t, a potential function is defined such that

$$\Phi(\delta_n, \delta_t) = \delta_n^c \int_0^\lambda \sigma(\lambda')\, d\lambda'. \tag{25}$$

The normal and tangential components of the traction acting on the interface in the fracture process zone are given by

$$\begin{aligned} T_n &= \frac{\partial \Phi}{\partial \delta_n} = \frac{\sigma(\lambda)}{\lambda} \frac{\delta_n}{\delta_n^c}, \\ T_t &= \frac{\partial \Phi}{\partial \delta_t} = \frac{\sigma(\lambda)}{\lambda} \frac{\delta_t}{\delta_t^c} \frac{\delta_n^c}{\delta_t^c}. \end{aligned} \tag{26}$$

Under purely normal separation, $\delta_t = 0$, (Mode I), the traction law becomes $T_n = \sigma(\lambda)$ where $\lambda = \delta_n/\delta_n^c$. Similarly, under purely tangential displacement, $\delta_n = 0$, (Mode II), the traction is $T_t = \sigma(\lambda)$ where $\lambda = \delta_t/\delta_t^c$. The maximum normal traction under pure normal separation is $\tilde{\sigma}$, and the peak shear traction is $\left(\frac{\delta_n^c}{\delta_t^c}\right)\tilde{\sigma}$. The work of separation of fracture energy per unit area of the interface is given by (25) with $\lambda = 1$.

The separation law of the interface contains a number of parameters: the adhesive fracture energy, Γ_a, the value of peak stress, $\tilde{\sigma}$, the critical displacement ratio δ_n^c/δ_t^c, along with others that define its shape. Earlier studies Tvergaard and Hutchinson (1993) and several cases that were considered in this study suggest that the details of the shape of the separation law are relatively unimportant. As a result, we considered an essentially triangular traction separation law so that

$$\Gamma_a \cong \tfrac{1}{2}\tilde{\sigma}\delta_n^c, \tag{27}$$

which shows that Γ_a and $\tilde{\sigma}$ are the two most important parameters for characterizing the fracture process in this model. The study on mixed mode interface toughness using this model (Tvergaard and Hutchinson, 1994) showed that the predictions do not seem to be very sensitive to other features of the traction-separation law such as the relative peak in the shear traction to normal traction as specified by the parameter δ_n^c/δ_t^c, which we therefore took to be unity. Use of the potential function ensures that the work of separation Γ_a is independent of the combination of normal and tangential displacements taking place in the process zone.

In this study, spring elements in the ABAQUS finite element code[1] were used to simulate the traction separation laws in the directions normal and tangential to the interface. The particular values of the parameters of the traction-separation law were chosen by making fits to pressure-volume and crack opening displacement data.

One point to note for circular blisters is that, for a given increment da of the crack radius, the area of debonding of the annulus increases as $2\pi a$ da. Since forces rather than tractions need to be prescribed for the springs, the peak force, $\tilde{F}$ had to be increased in order to maintain a constant value of $\tilde{\sigma}$. This was accomplished in a stepwise manner, where a new value of peak force was prescribed for every millimeter of crack extension.

The projected grid measurements indicated clearly that the crack front was circular. An axisymmetric analysis was therefore conducted considering the aluminum to be rigid and using four-node shell elements to represent the polyimide layer. The polyimide was modeled as a J_2 flow theory material with isotropic hardening. The Ramberg Osgood fit with parameters n and σ_0 derived from the bulge tests was used to represent the stress-strain curve.

10. Results

The results from the experimental and analytical procedures that were developed in the earlier sections are presented and compared here. First, data from bulge tests was used to determine the mechanical properties of the thin polyimide layers. Elastic and elastoplastic analyses of quasi static debonding are then compared, with special consideration being given to adjusting parameters in the fracture process zone model. This allowed us to determine the adhesive fracture energy of a particular polyimide/aluminum interface regardless of the amount of global plastic deformation in the delaminating layer.

11. Mechanical properties

Two approaches were taken to determine the material properties of thin polyimide layers using the pressure vs. change in volume response prior to debonding. For low pressures, the response (Figure 3(a)) was linear with the cube of the change in volume as suggested by (6). The value of E was determined from the slope κ and measured values of debond radius and film thickness with $C_1 = 0.519$ and $C_2 = 0.706$. As a result, E was found to be 2.18 GPa, which was in good agreement with the values from uniaxial tests (Shirani, 1997).

Higher pressures were used to find the Young's modulus, yield strength, σ_y, and the Ramberg Osgood parameters, σ_0 and n, from measurements of pressure and central deflection and (8) and (9). Loading and unloading tests were performed several times (Figure 4(a)) up to an applied pressure of 379.2 kPa, well before the initiation of delamination. The corresponding stress-strain behavior and the values of the parameters that were used in subsequent analysis are shown in Figure 3(b) and Table 1, respectively.

[1] We are grateful to Hiibbitt, Karlsson and Sorensen for providing ABAQUS under academic license.

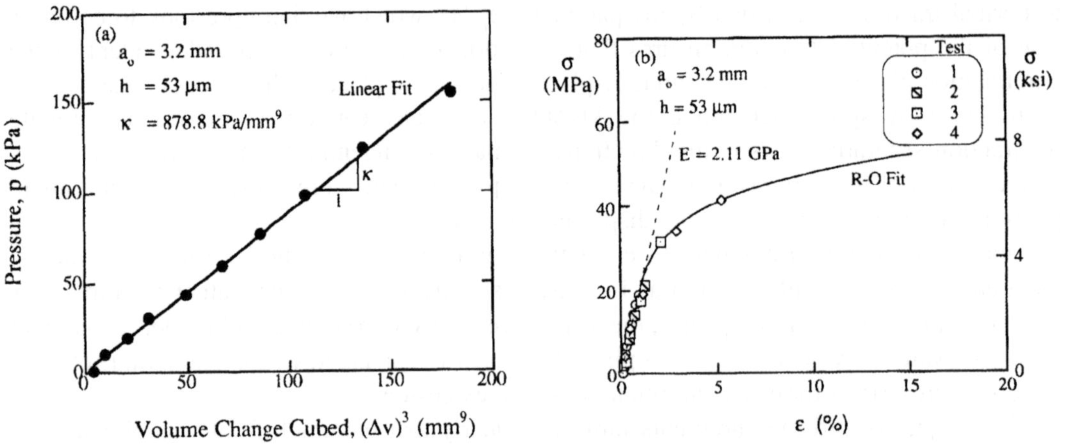

Figure 3. Data for extraction of (a) elastic and (b) Ramberg Osgood parameters.

Table 1. The mechanical properties of polyimide PMDA/ODA.

Young's Modulus E (GPa)	Yield Stress σ_y (MPa)	Ramberg Osgood Exponent n	Ramberg Osgood Stress σ_0 (MPa)
2.15	31	5.85	26

12. Pressure-time response

The pressure histories resulting from a nominal volume flow rate $R_f = 20$ ml/hr are shown in Figure 4. The results prior to the first delamination (bulge tests) are presented in Figure 4(a) while the response during the initiation and propagation of several cracks appears in Figure 4(b). For the initial debond, $a_0 = 3.2$ mm, the pressure level required to cause initial debonding was high and the crack growth was dynamic followed by crack arrest due to the volume control. In this particular specimen, the crack radius jumped from its initial value of 3.2 mm to 14.6 mm. The critical pressure, p_c, was 815 kPa, followed by a drop to about 170 kPa. The specimen was then unloaded, reloaded and unloaded several times, producing quasi-static growth for new starter crack radii of 14.6 mm, 18.0 mm and 20.2 mm. This data was the simplest to obtain and was used to extract the adhesive fracture energy using the approach of Chu et al. (1992 a,b) and (22).

13. Deformed shape

The deformed shapes of the blisters were obtained from the projected grid method which, for any pressure level, allowed the deflection at any radial position and, therefore, volume change to be extracted by making use of (1) and (3), respectively. Figure 5(a) shows the deformed shapes during the first delamination experiment prior to the onset of the dynamic growth. At zero pressure, there was some residual deformation of the polyimide layer due to yielding that had occurred from the previous bulge testing. For pressures from 174 kPa to 799 kPa there was a period of stable debonding that was followed by dynamic, unstable growth. The debonding was quite regular and resulted in circular crack fronts. The crack arrested at a radius

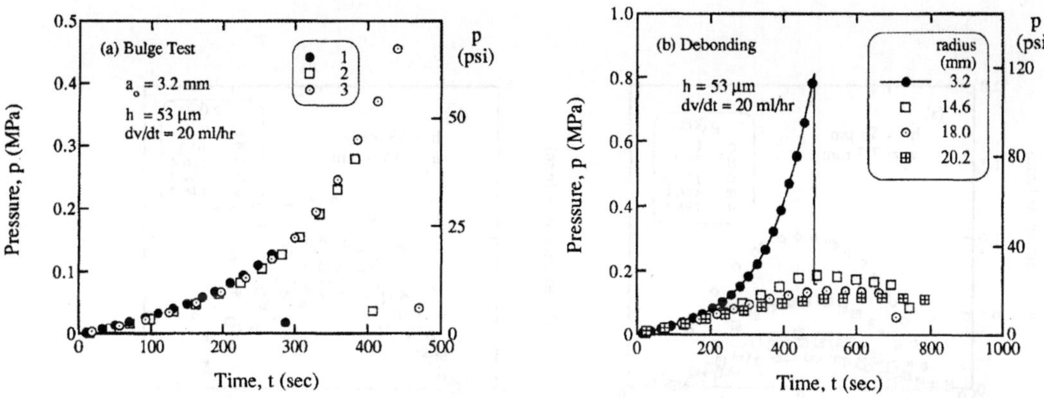

(a) (b)

Figure 4. Pressure histories (a) during bulge tests and (b) during initiation and propagation experiments.

of 14.6 mm, at which point the blister was depressurized and preparations were made for the next experiment (Figure 5(b)). The amount of permanent deformation after crack growth and unloading had increased significantly compared to the amount that had resulted from the bulge test. The sharp crack produced by the rapid arrest produced entirely stable growth upon reloading. Stable growth initiated at 176.4 kPa and was arrested after 3.4 mm of growth by reversing the fluid flow. The pressure dropped during the stable growth phase. The loading, growth and unloading sequence was repeated for the 18 and 20.2 mm cracks (Figures 5(c) and 5(d), respectively). Stable growth occurred in all experiments and the amount of residual deformation increased with increasing crack length while the maximum pressure dropped.

The projected grid method also yielded information on the crack front geometry. Crack fronts were generally quite circular. Any irregularities were usually traced to variations in adhesive layer thickness or surface treatment. As a result, crack radii were a very reasonable measure of the in-plane shape of the blisters. The crack radii histories are shown in Figure 6 for starter crack radii of 14.6, 18.0 and 20.2 mm. Each debond grew in a smooth, quasi-static manner and reached steady state speeds of 9.9, 6.5 and 4.7 μm/s, respectively from the 14.6, 18.0 and 20.2 mm radii.

The blister deflection and the crack radii in Figure 5 were used to determine the change in volume using (3). The latter is plotted along with the crack radii in Figure 6. It can be seen that the actual volume rate was not constant but did reach a steady state once steady state debonding was established. The rates were 14.3, 14.0 and 15.5 ml/hr for the 14.6, 18.0 and 20.2 mm radii cracks, respectively. These were considerably lower than the 20 ml/hr nominal rate, due to compliance in the plastic syringes that were used.

For all four cracks, the initial central deflection was not zero when the fluid pressure was zero, an indication that residual plastic strains were built up during loading and subsequent crack propagation. As an example of the global strain levels that were reached, we present results from the 20.2 mm crack in Table 2.

It can be seen that global strain levels of 10 percent were realized before the pressure reached its maximum value.

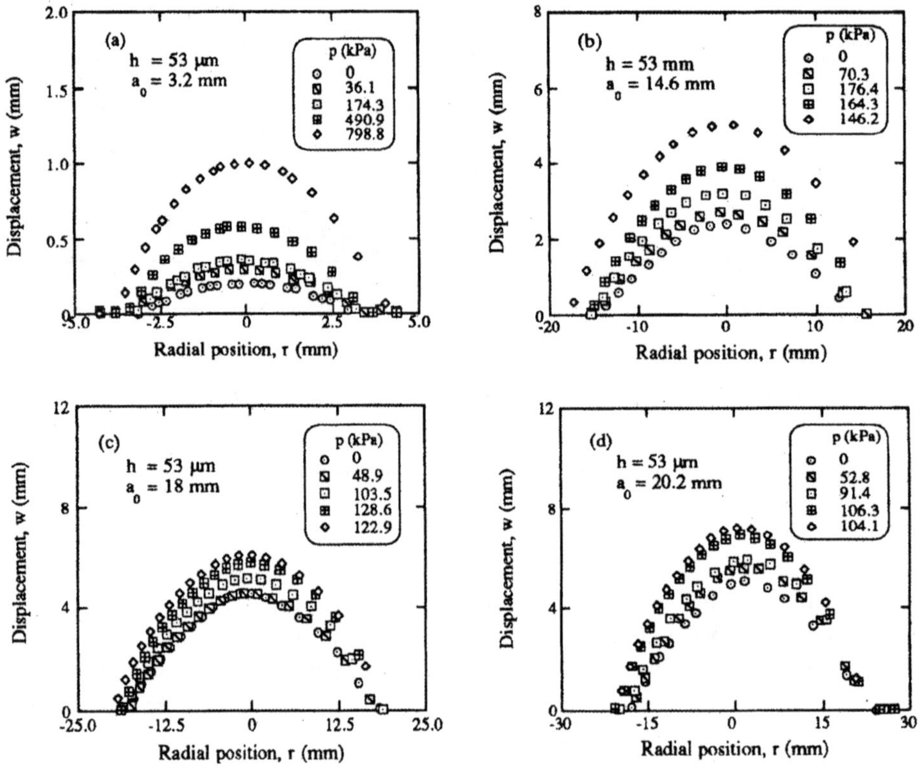

Figure 5. Deformed shapes of blisters for crack radii of (a) 3.25 mm, (b) 14.6 mm, (c) 18.0 mm and (d) 20.2 mm.

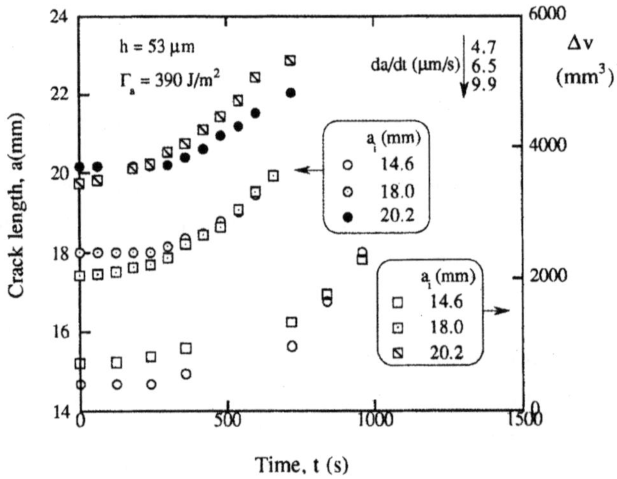

Figure 6. Crack radius and blister volume histories.

Table 2. Strain levels during the propagation of a 20 mm crack.

p(kPa)	a(mm)	w_0(mm)	ϵ(%)
62.6	20.21	6.93	7.84
85.4	20.62	7.43	8.66
97.1	21.19	8.23	10.06
97.3	22.05	8.48	9.86

14. Analysis

In this section, we compare the responses that were predicted by the various analyses that were conducted. Comparisons are made between elastic and elastoplastic material behavior and the effect of including the interface as a separate constitutive entity.

The inflation response while the blister was stationary is considered first. The predictions from elastic analyses ((8) of Gent and Lewandowski (1987) and (5) developed here) and elastic and elastoplastic finite element results with and without the interfacial elements are compared in Figure 7. The finite element analyses were designated elastic (EC) and elastoplastic (EPC) for clamped blisters so as to correspond to the conditions that were considered in the elastic analyses. The designations EI and EPI refer respectively to elastic and elastoplastic finite element analyses with interface elements. The stiffest response was that predicted by (8) of Gent and Lewandowski (1987), which differed from the prediction from (5) by as much as 19 percent because the latter included the radial displacement. The results from (5) and the clamped blister finite element analysis were within 2 percent, indicating that a suitable level of mesh refinement had been selected for the latter. The slight difference could be due to the fact that the finite element solution captured bending effects, whereas equation contained only membrane effects. Comparing the elastic and elastoplastic clamped blister analyses, it can be seen that plastic deformation affected the response at about 0.4 MPa for a crack length of 3.25 mm, as was observed in the bulge tests (Figure 4(a)). Even so, in Figure 7, the measured response still fell below the predicted response for the clamped elastoplastic blister by as much as 12 percent. The closest agreement between measurement and prediction is apparent for the elastoplastic analysis with interface elements. This is not due to the fact that the crack was actually growing so far in advance of the maximum load but rather because the springs that were modeling the interface stretched and provided some rotation at the crack front in the same way that an elastic foundation would. In an analysis of elastoplastic peeling, Kinloch and Williams (1994) have also shown that including the rotation is very important. The elastic finite element analysis with the same interface elements (EI) predicted a 50 percent lower peak stress than the corresponding elasoplastic analysis (EPI).

We now consider the initiation and propagation phase of the blistering (Figure 8). An adhesive fracture energy of 390 J/m^2 was chosen as the basis for the comparison. As a result, the pressure vs. central displacement response becomes hyperbolic (Gent and Lewandowski (1987) and (17) for elastic debonding. The highest peak pressure was predicted by (10) of Gent and Lewandowski (1987) due to the stiffer inflation response. However, once the crack was propagating, this analysis resulted in the lowest pressure levels at any given displacement. Equation (17) produced an intermediate response while the interface elements in the elastoplastic finite element analysis had been chosen in such a way that the initiation pressure

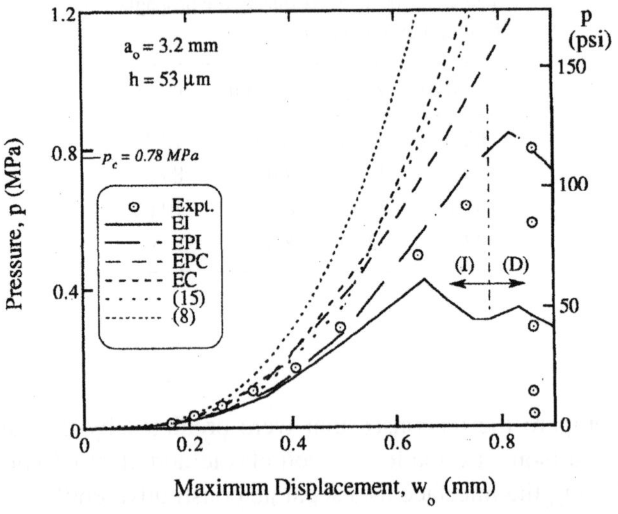

Figure 7. Comparison of predicted responses for a stationary blister.

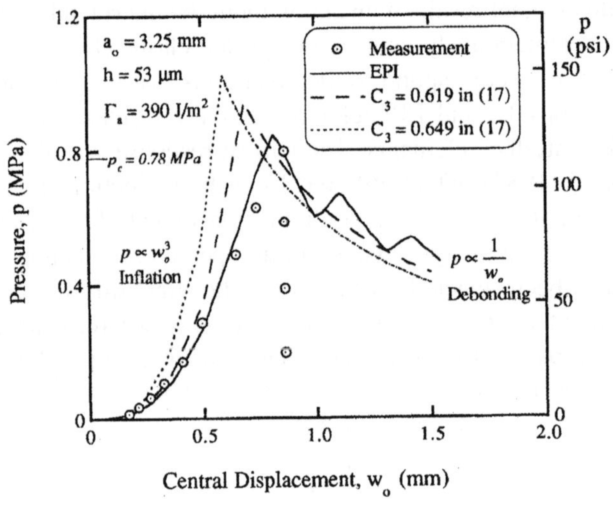

Figure 8. Comparison of predicted responses for a growing blister.

was matched. Since the blister grew dynamically from the initial radius, none of the analyses were able to capture that portion of the response. As will be seen later, once a sharp crack was established, propagation was quasi static and could be handled by the analysis.

15. Adhesive fracture energy

In this section, we compare the values of the adhesive fracture energy that were extracted using the Gent and Lewandowski (1987), Chu et al. (1992 a, b) and the fracture process zone approaches, embodied in (17), (22) and (27), respectively. Each approach has facets that need to be considered further than a straight forward application of the equations.

When (17) was used to obtain the adhesive fracture energy from measured values of pressure and central deflection, a range of values was obtained, due to resistance-curve type of

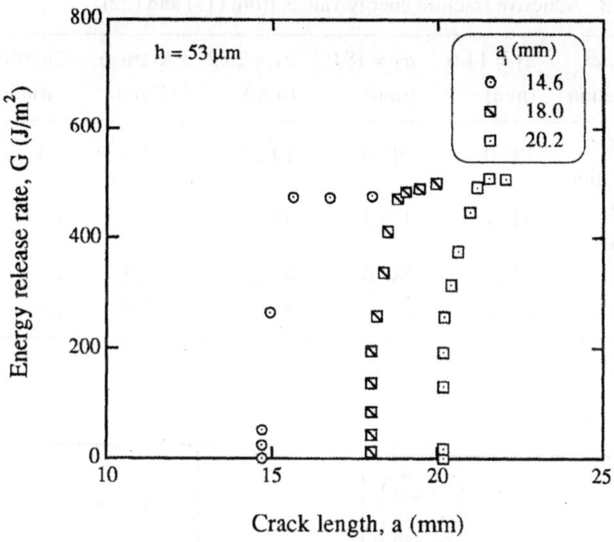

Figure 9. Resistance curves based on Equations (5) and (17) for inflation and debonding, respectively.

behavior (Figure 9). While the crack was stationary, (5) was used to determine J-integral values. Within the resolution limits of the crack radius and crack opening displacement measurements, the initiation values were quite consistent at about 193 J/m^2 (Table 3). There was more variation in the peak values that were obtained from the three experiments but the average was 454.4 J/m^2 with a coefficient of variation of 4.7 percent.

The adhesive fracture energy was next extracted from the time dependence of the blister pressure under a constant volume flow rate using (22). The method circumvents the problems associated with determining the adhesive fracture energy from the critical pressure. In addition, since it only requires the pressure history to be measured during debonding, experimental difficulties are minimized as long as the pressurization system is sufficiently stiff that the nominal volume flow rate is the actual flow rate seen by the blister. For a test carried out at a constant volumetric flow rate, (22) indicates that, after debonding begins, the cube of the reciprocal of the pressure should be a linear function of the debonding time if debonding is elastic. As it turned out, there was compliance in the syringe pump that was used in the experiments that were conducted here. Figure 6 showed that the rate of increase in the blister volume only became constant well after debonding had initiated. At that stage, the actual rate was 14.60 ml/hr, whereas the nominal rate was 20 ml/hr, an error of 27 percent, which is quite substantial. The major source of compliance was the polymer syringe that was used in the syringe pump. Using glass syringes was problematic, as their supporting lugs cracked easily. However, for future work, a metallic syringe would be better and possibly avoid having to make a local volume measurement.

The results for the three different experiments that were conducted after the initial debonding are shown in Figures 10(a) and 10(b). The former shows the full range of data following initiation, thereby including the growth that occurred while pressure was still increasing. This resulted in a consistent but nonlinear series of plots. When the early debonding data was ignored (Figure 10(b)), it can be seen that the results were quite linear once debonding was well established, the linearity suggesting that quasi static, elastic debonding was occurring. The slope of the linear portion was used to extract the fracture energy using (22) using $C_5 =$

Table 3. Adhesive fracture energy values from (17) and (22).

Figure/ Equation	$a_1 = 14.6$ (mm)	$a_2 = 18.0$ (mm)	$a_3 = 20.2$ (mm)	Average (J/m^2)	Coefficient of variation (%)
9/17 Initiation	193.0	193.0	193.0	193.0	0.0
9/17 Peak	432.7	475.5	455.1	454.4	4.7
10a/22	373.4	387.0	410.3	390.2	4.8
10b/22	314.1	305.9	317.5	312.5	1.9

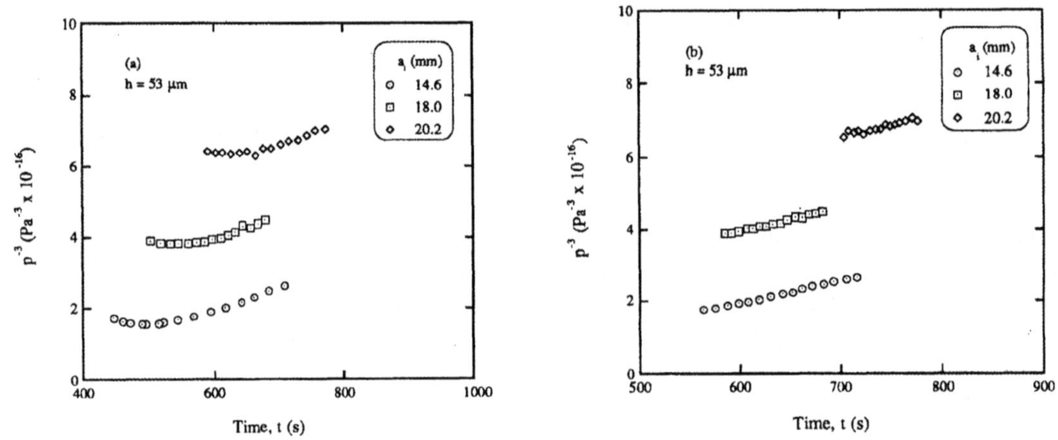

(a) (b)

Figure 10. Pressure-time data that was used in Equation (22). (a) all data, (b) data obtained during steady state debonding.

0.431 as discussed earlier and the measured rather than the nominal volume flow rates. The adhesive fracture energy values Γ_a, extracted from linear fits to the data in Figures 10(a) and 10(b) are shown in Table 3.

Incorporating all the data (Figure 10(a)) still gave rise to reasonable correlation to a linear fit ($R = 0.91$), lower slopes and therefore higher adhesive fracture energy values than were obtained using the data shown in Figure 10(b). The average value of the adhesive fracture energy taken from Figure 10(b) was 25 percent lower than the average taken from Figure 10(a) and the consistency of the adhesive fracture energy values extracted from the linear portion of the data was better. The difference suggests that there is a greater degree of dissipation at and immediately following initiation. This is consistent with the fact that the debonding rate in each experiment increased to a steady value (Figure 6). From a viscoplastic point of view, higher debonding rates should result in less yielding in the crack front region.

The values of adhesive fracture energy from the two fits to the pressure-time data were much lower than the peak values obtained from the resistance curves in Figure 9. At first

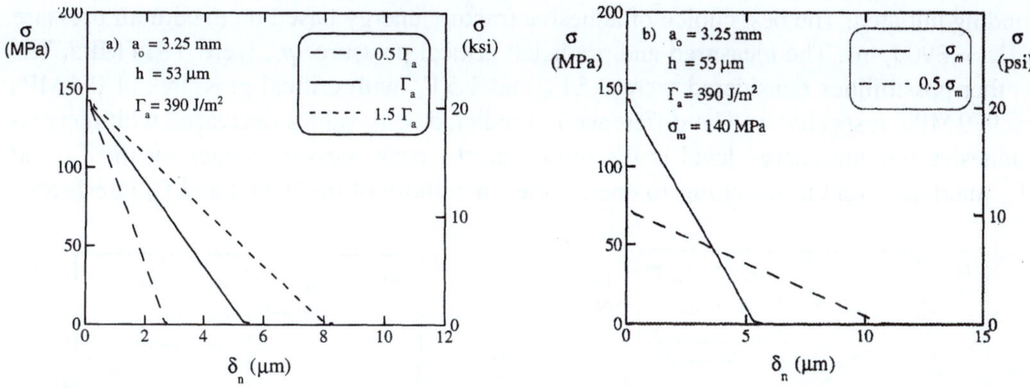

Figure 11. Various traction-separation laws that were considered.

sight, the differences are surprising because (17) was used in the derivation of (22). However, (17) is based on elastic analysis while the central displacement values that were used in it contained a plastic component, thereby yielding larger adhesive fracture energy values than purely elastic ones would have. The initiation values were consistently the lowest of all. Thus, based on elastic analyses, there was a lack of consistency in the values of adhesive fracture energy that were extracted.

We now give the details of the incorporation of the traction separation law in the elastoplastic finite element analyses. The responses of several nonlinear spring elements that were introduced along the interface are shown in Figures 11(a) and 11(b). In the former, the adhesive fracture energy Γ_a is varied while maintaining a fixed maximum traction level, $\tilde{\sigma}$. The situation is reversed in Figure 11(b). The yield strength of the polyimide was 31 MPa, about one fifth of the highest $\tilde{\sigma}$ level. The most appropriate choice of spring response was based on the measured maximum pressure, the overall pressure-volume change response and the measured crack opening displacements.

As will be seen, the baseline value of the adhesive fracture energy that was needed to match the measured pressure, volume and crack opening displacement (COD) responses was $\Gamma_a = 390\,\mathrm{J/m^2}$. This is much higher than the 5–10 $\mathrm{J/m^2}$ that is associated with breaking covalent bonds. The material-based length scale that is introduced is

$$R_0 = \frac{1}{3\pi(1-\nu^2)}\frac{E\Gamma_a}{\sigma_y^2}. \tag{28}$$

This reference length is an estimate of the height of the plastic zone size under mode I loading (Tvergaard and Hutchinson, 1993, 1994). For $\Gamma_a = 390\,\mathrm{J/m^2}$, and the mechanical properties of the polyimide given in Table 1, we find $R_0 = 115.3\,\mu\mathrm{m}$, about twice the film thickness. As a result, the process zones that are being modeled here are on a larger scale than those associated with atomic separation. In Figure 11(a), it can be seen that for $\Gamma_a = 390\,\mathrm{J/m^2}$, the maximum traction was 140 MPa, about five times the yield strength of the polyimide, while the crack tip opening displacement was 5.5 μm. This is a rather large displacement level, again emphasizing the point that the process zones considered here are large.

The measured pressure vs. volume response and the effect of changing Γ_a is shown in Figure 12. The data point with the highest pressure (Figure 12(a)) was taken just before debonding initiated. The best choice of adhesive fracture energy based on maximum pressure was $\Gamma_a = 390\,\mathrm{J/m^2}$. The measured and predicted critical pressure, p_c, were 0.814 MPa. The two other possibilities considered were $0.5\,\Gamma_a$ and $1.5\,\Gamma_a$ with critical pressures of 0.4 MPa and 0.979 MPa, respectively. The difference in predicted peak values decreased with increasing adhesive fracture energy level in the sense that the peak pressure values for the Γ_a and $1.5\,\Gamma_a$ interfaces were much closer to one another than those of the $0.5\,\Gamma_a$ and Γ_a interfaces.

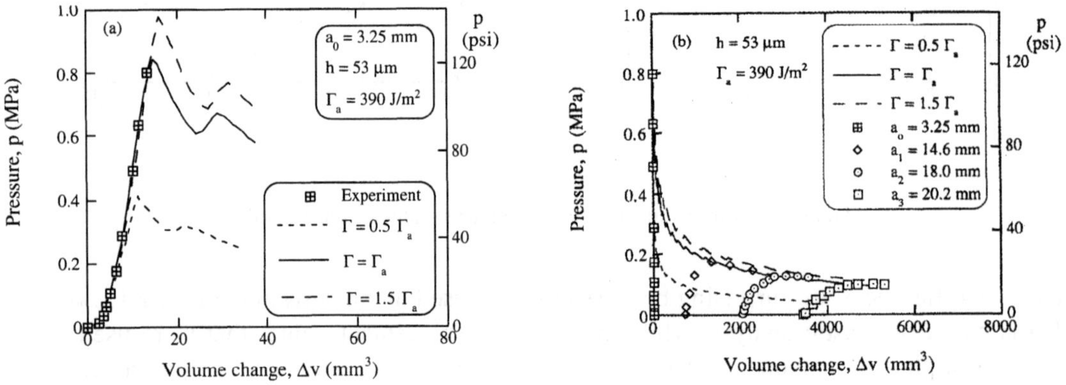

Figure 12. The effect of adhesive fracture energy on (a) peak pressures and (b) the quasi static debonding response.

The predicted pressure-volume responses for quasi-static debonding over the full range of volume change considered in the experiments is presented in Figure 12(b). The lack of smoothness of the predicted responses during debonding was due the periodic adjustment (every mm) of the spring forces in order to maintain the same traction separation law for all debond radii, as was discussed in the analyis section. Again, the interfacial springs representing $\Gamma_a = 390\,\mathrm{J/m^2}$ had been selected to match the steady state debonding response of all the subsequent experiments. Higher Γ_a values appear to have a diminishing effect as the responses of the Γ_a and $1.5\,\Gamma_a$ interfaces were much closer to one another than those of the $0.5\,\Gamma_a$ and Γ_a interfaces.

The next step that was taken was to compare the crack opening displacements (COD) at a fixed crack length. As expected from the traction separation laws (Figure 11(a)), the interface with the highest toughness produced the largest crack opening displacements outside the process zone (Figure 13). As can be seen from the inserted table, the cohesive zone size, d, became smaller with increasing adhesive fracture energy but was on the order of 200 μm. This is about four times the thickness of the polyimide layer and about twice R_0, the plastic zone height associated with $\Gamma_a = 390\,\mathrm{J/m^2}$.

The spatial and displacement resolution of the projected grid technique was not sufficient to resolve the actual response at the size scale shown in Figure 13. However, the effect can still be seen further away from the crack front (Figure 14). Comparisons between measured and predicted COD were made at three different volume levels. This means that the simulated cracks had grown to different radii at each volume level depending on the level of adhesion. Higher adhesion levels resulted in lower amounts of debonding. For $1.5\,\Gamma_a$ and $0.5\,\Gamma_a$, the crack radii were about 2 mm less and 3 mm greater than crack radius of an interface with an adhesive fracture energy of Γ_a. As a result, when viewed in the fixed reference frame of

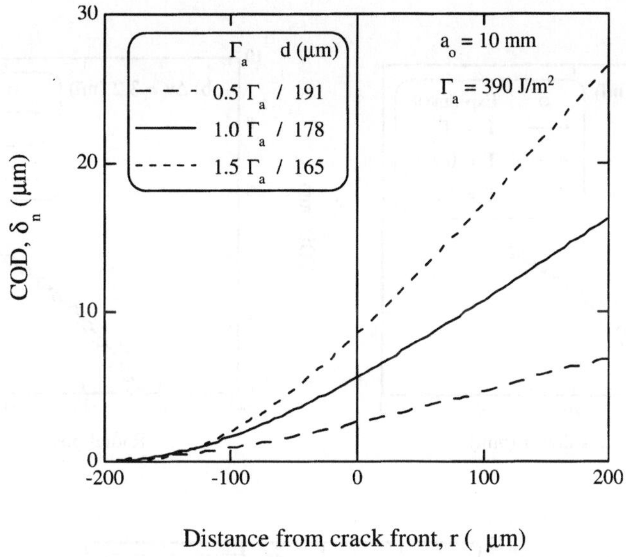

Figure 13. Crack opening displacements near the crack front for various degrees of adhesion.

Figure 14, the COD for the tougher interface were initially the lowest, in spite of the result in Figure 13, which was based on a fixed crack length. However, further away from the crack front, the COD associated with the toughest interface do eventually became the largest. Thus the interface with $\Gamma_a = 390\,\mathrm{J/m^2}$ best matched the crack radius and COD data.

Although the predictions were made for a single crack starting at a crack radius of 3.25 mm and growing to 22 mm without unloading and reloading at starter crack radii of 14.6, 18.0 and 20.2 mm (Figure 6) as was done in the experiments, it is instructive to compare the crack velocities at the same point in the crack growth history. The velocities at the three crack radii mentioned above are compared (Table 4) with those taken from the steady state portion of the measured crack growth histories shown in Figure 6. Again, it can be seen that the simulation with $\Gamma_a = 390\,\mathrm{J/m^2}$ provides the best match with the measured velocities.

The effect of changing $\tilde{\sigma}$ while keeping Γ_a constant (Figure 11(b)) is shown in Figure 15 where the pressure vs. volume changes are compared. The $\tilde{\sigma}$ values that were considered were 140 MPa and 70 MPa with an adhesive fracture energy of $390\,\mathrm{J/m^2}$. In Figure 15(a), where the responses are compared for the initial crack radius ($a = 3.25$ mm) that was used in the experiment, it can be seen that the lower $\tilde{\sigma}$ value gave rise to consistently lower pressure values during both the inflation and debonding phases. The more compliant response is due to the fact that the initial stiffness of the interfacial springs was lower (see insert to Figure 11(b)). The trend continues in Figure 15(b) where the measured steady state response was matched by the response of the interface with $\tilde{\sigma} = 140$ MPa whereas the interface with $\tilde{\sigma} = 70$ MPa produced consistently lower pressures. This can be understood in a straightforward manner from the point of view that the traction required to produce debonding is lower.

To summarize this portion of the work, it has been shown that it is possible to select a traction-separation law that fits measured pressure-central displacement and crack opening displacement responses in a consistent manner. It is an iterative process but it converges quickly. The cases where the chosen parameters did not provide fits to the measured responses were rather extreme for illustrative purposes and the iterative process was found to be quite sensitive to choices of traction-separation parameters.

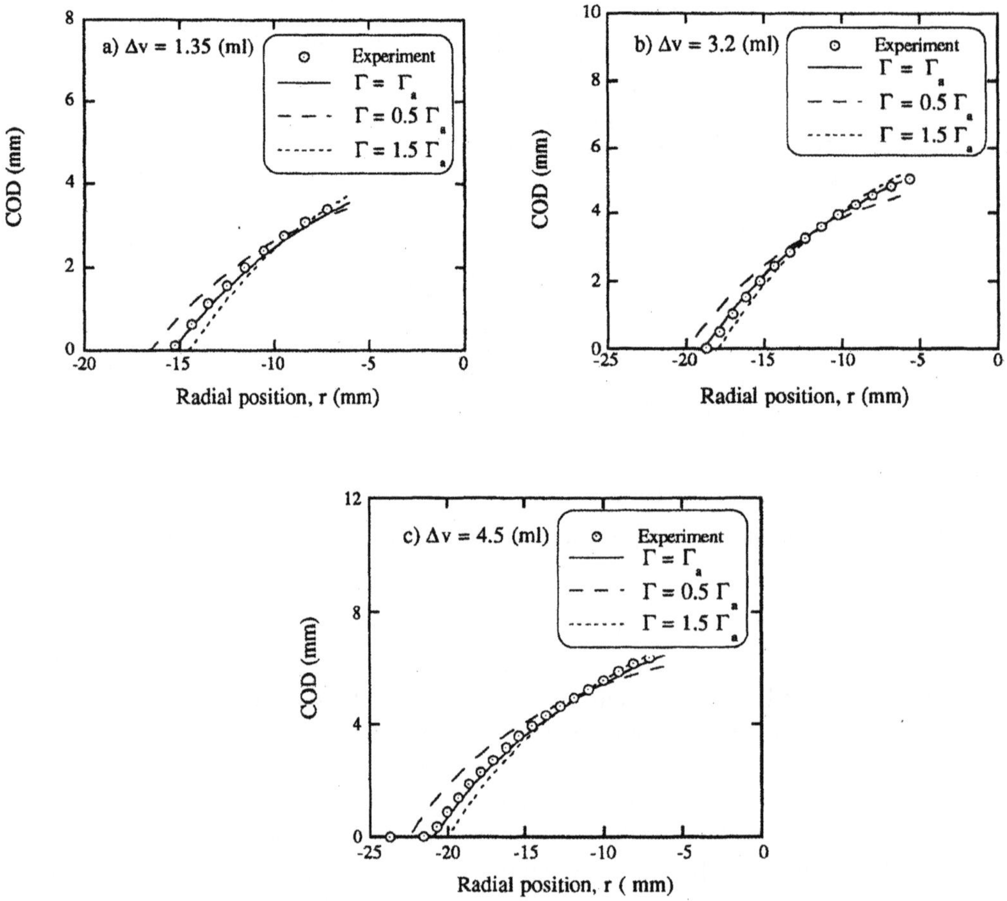

Figure 14. The effect of different degrees of adhesion on the COD.

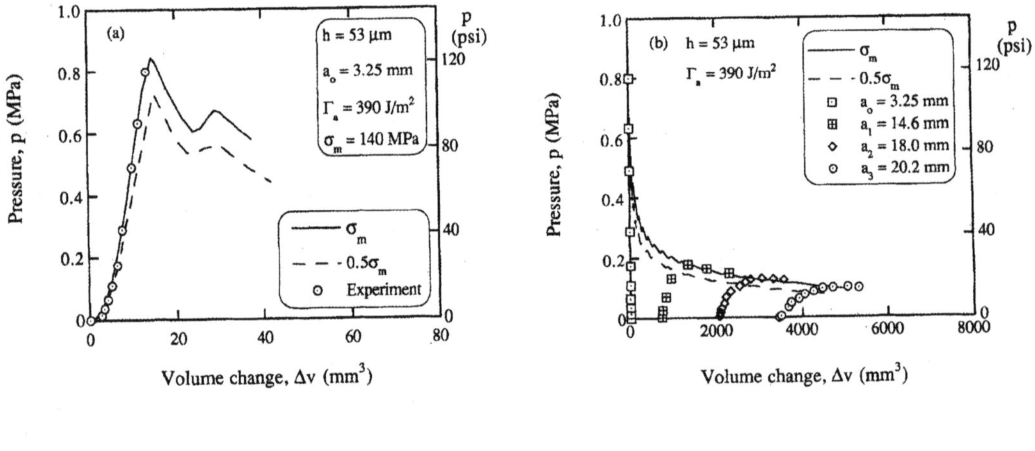

Figure 15. The effect of the value of the peak stress $\tilde{\sigma}$ on the pressure volume response.

Table 4. Measured and predicted crack velocities (μm/s).

Interface/Crack Radius	10 mm	14.6 mm	18 mm	20.2 mm
Actual		4.7	6.5	9.9
$0.5\,\Gamma_a$	34.9	5.9	7.8	11.5
$1.0\,\Gamma_a$	28.5	4.8	6.4	9.4
$1.5\,\Gamma_a$	27.1	4.5	6.1	8.9

16. Comparison of adhesive fracture energies and plastic dissipation

In comparing the 390 J/m^2 for the adhesive fracture energy that was obtained from the finite element model with the average values obtained from the pressure-central deflection and pressure-time responses (Table 3), it can be seen that the average adhesive fracture energy value from the pressure-central deflection was 16 percent higher while the adhesive fracture energy value from the linear portion of pressure-time responses was 20 percent lower. Including all the data in a linear fit of the pressure-time responses resulted in an adhesive fracture energy value which was in close agreement with that obtained from the finite element analysis. In order to better understand these differences, it is instructive to examine the finite element solution further.

The energy balance during quasi static debonding is given by

$$dW = dU_e + dU_a + dU_p, \tag{29}$$

where W, U_e, U_a and U_p represent the work done by fluid injection, the elastic strain energy stored in the film, the energy expenditure in debonding and the plastic dissipation outside the fracture process zone, respectively.

The interplay between the different energies alluded to in (28) is shown in Figure 16(a) for an elastoplastic finite element analysis of an initial crack length of 3.1 mm and an adhesive fracture energy of 390 J/m^2. For reference, the pressure and crack radius histories are plotted in Figure 16(b), where it can be seen that crack propagation initiated just prior to the maximum pressure. Initially (Figure 16(a)), all the work goes into stored elastic energy. This phase was followed by the onset (at $w_0 = 6$ mm) of plastic dissipation so that both it and the stored elastic energy contribute to the work. At about 8 mm, debonding began and all three quantities on the right-hand side of (28) consistently made up the external work. By the time the analysis was halted, more energy had been dissipated in fracturing the interface than in global plastic dissipation.

Rearranging (28) and determining the rates of energy change with respect to changes in crack surface area, A, we find

$$\Gamma_t = \Gamma_a + \Gamma_p, \tag{30}$$

where

$$\Gamma_t = -\frac{d}{dA}(U_e - W), \qquad \Gamma_a = \frac{dU_a}{dA} \quad \text{and} \quad \Gamma_p = \frac{dU_p}{dA}. \tag{31}$$

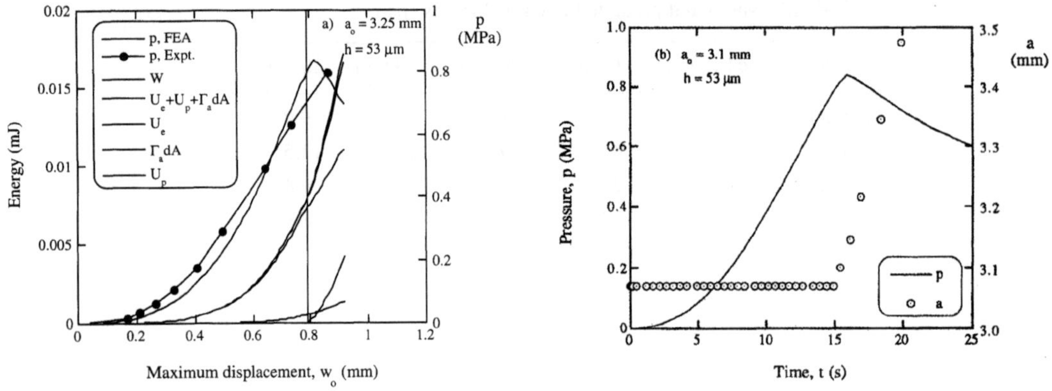

Figure 16. (a) The interplay between different energies for an initial crack length of 3.1 mm. (b) Pressure and crack radius histories.

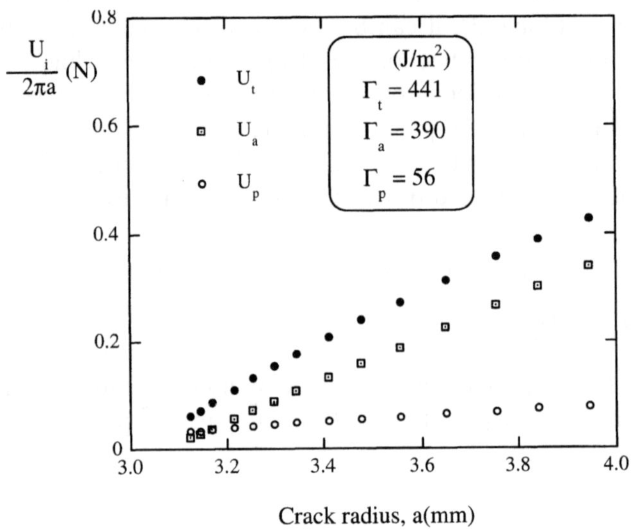

Figure 17. Extraction of plastic dissipation and the adhesive fracture energy.

Applying (30) and (31) to the data in Figure (16) yielded the results shown in Figure (17) where the work and energy terms divided by $2\pi a$ are plotted against crack length. As expected, the quantity $U_a/2\pi a$ was linear with crack length. There was some slight nonlinearity in $U_e - W/2\pi a$ and $U_p/2\pi a$ vs. a. However, taking the derivatives of each quantity did not give rise to the balance expected in (30). On the other hand, when linear fits were made to all quantities, the resulting energy release rate (Γ_t) and the dissipation rates (Γ_a and Γ_p), which are tabulated in the insert to Figure 17, did follow (30). These were then compared (Figure 18) with the adhesive fracture energy values extracted from (17) and (22). Again, it can be seen that the two dissipative terms Γ_a and Γ_p from the finite element analysis do indeed add up to the total energy release rate, Γ_t. However, the Γ_a values from (17) and (22) were not the same as the values from the finite element analysis.

Once it attained its maximum value, the adhesive fracture energy from (17) was close to Γ_t from the finite element analysis, but was about 16 percent higher than the finite element

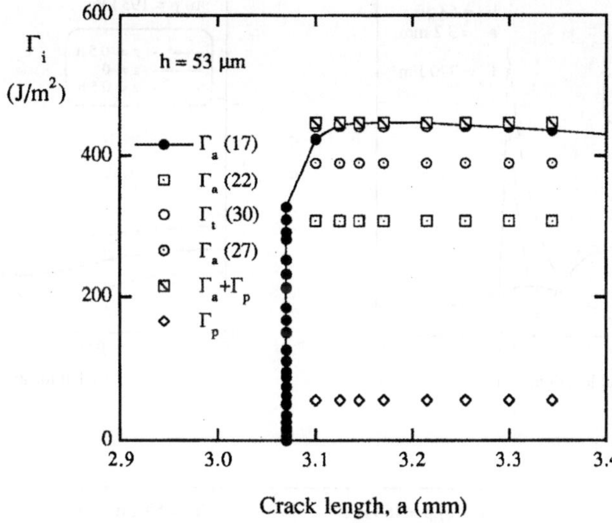

Figure 18. The comparison of adhesive fracture energy values extracted from various equations.

Table 5. Adhesive fracture energy values for various values of tangent and secant moduli.

Strain Level (%)	Tangent Modulus E_t (GPa)	Adhesive Fracture Energy (J/m²)	Secant Modulus E_s (GPa)	Adhesive Fracture Energy (J/m²)
1	1.508	374.3	2.025	352.9
2	0.703	436.0	1.523	373.6
5	0.214	553.1	0.811	423.7
10	0.077	678.6	0.476	471.4

value of Γ_a, as noted earlier. The plastic dissipation rate was 13 percent of Γ_a from the finite element analysis. The values of Γ_t from the finite element analysis and Γ_a from (17) were the same because both contained plastic components. In the former, the plastic disspation rate was added to the adhesive fracture energy. In the latter, even though the analysis was elastic, the measured values of central displacements that were used in (17) contained a plastic component.

The lower value of Γ_a from (22) may come about because of the fact that the initial elastic modulus was used whereas plastic deformation was occurring during steady state debonding, in which case the secant or tangent modulus would be more representative of material stiffness. Since these are both lower than the Young's modulus, the value of Γ_a from (22) would be increased. However, as can be seen in Table 5, strain levels greater than 2 percent soon produce unrealistically high adhesive fracture energy values using either the tangent or secant modulus. As can be seen from Table 2, global strain levels during delamination were closer to 10 percent. As a result, such a simple modification to the Chu et al. approach does not seem to be feasible.

The sources of plastic dissipation can be seen from the equivalent stress distributions shown in Figures 19 and 20 while the crack was stationary and growing, respectively. First yield occurred in a very small region near the crack front (Figure 19(a)) at approximately 36 kPa. Away from the crack front, the stress distribution had both membrane and bending components

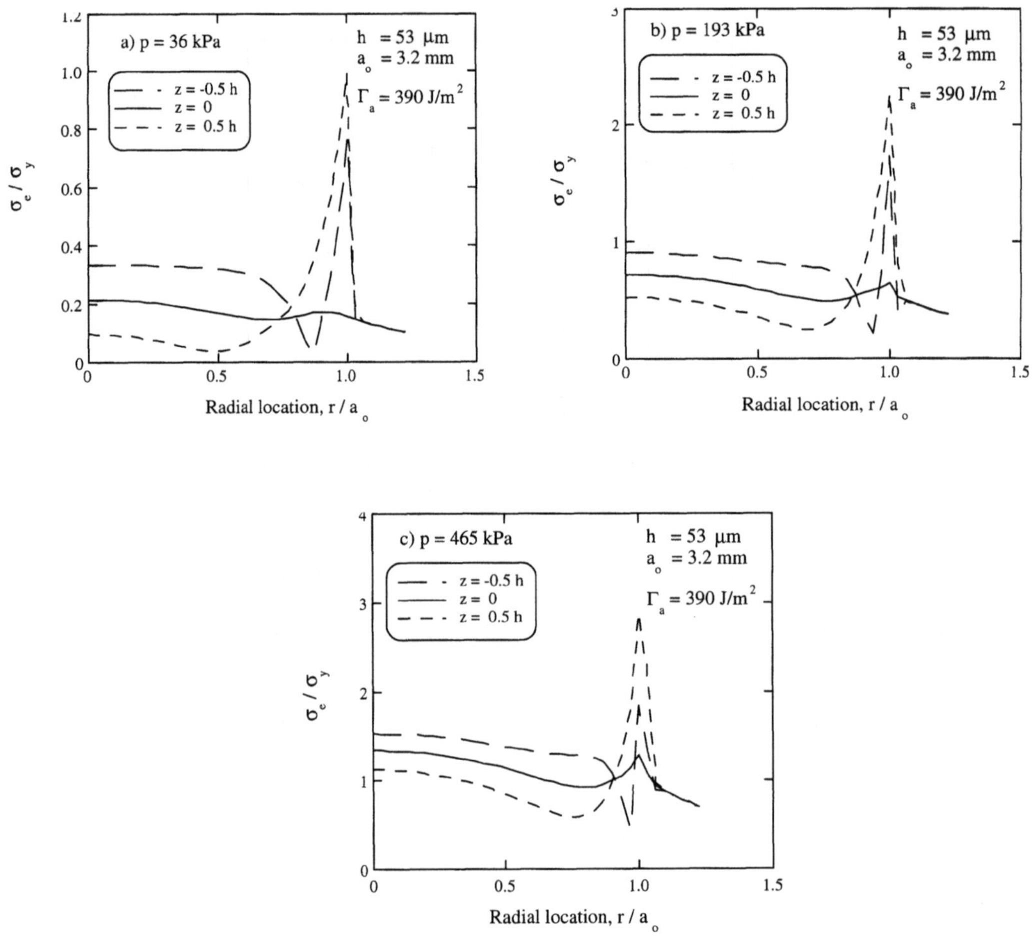

Figure 19. Von Mises effective stress distribution before intiation: (a) $p = 36\,\text{kPa}$, (b) $p = 193\,\text{kPa}$, and (c) $p = 465\,\text{kPa}$.

($a/h = 60$), all of which were below the yield strength of the polyimide. First yielding away from the crack front occurred (Figure 19(b)) at the center and top of the blister just above 193 kPa. By the time the pressure reached 465 kPa , yielding had nearly spread throughout the entire polyimide film. At the onset of delamination, the pressure was 740 kPa and the sharp peak at the mid surface near the crack front that had been seen while the crack was still stationary had disappeared (Figure 20(a)). By the time the crack had propagated about 0.4 mm, the pressure had dropped to 590 kPa and some unloading had occurred, especially at the bottom of the film ($z = 0.5\,h$) in the just delaminated region (Figure 20(b)) and to a lesser extent in the rest of the delaminated layer. This behavior leads to some residual strain which was included in the calculation of the elastic strain energy of the system.

X-ray Photoelectron Spectroscopy (XPS) and Infrared (RAIR) spectroscopic analyses were conducted of the fracture surfaces and the interface following imidization[2]. The interface probes identified the presence of carboxylate species in an interphase region between the bulk aluminum and polyimide. Fracture occurred in this interphase region which explains why the

[2] We are grateful to Professor F.J. Boerio and coworkers for obtaining this information.

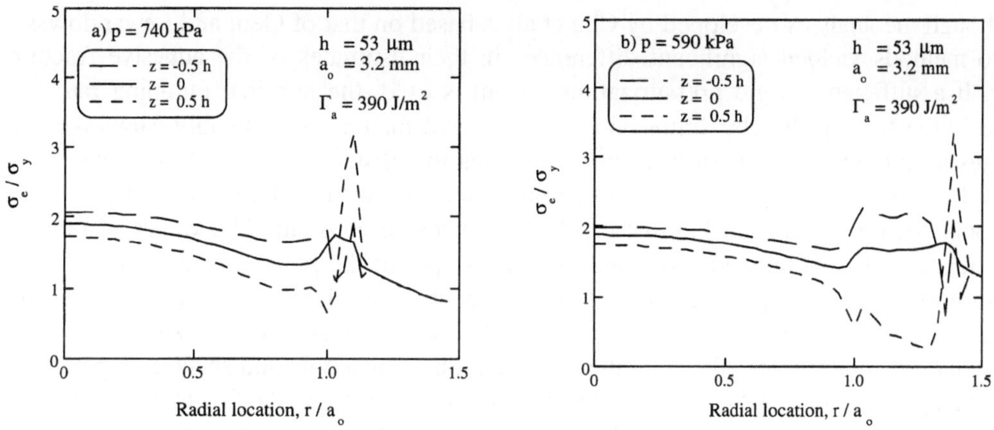

Figure 20. Von Mises effective stress distribution after initiation: (a) $p = 740\,\text{kPa}$, (b) $p = 590\,\text{kPa}$.

adhesive fracture energy $\Gamma_a = 390\,\text{J/m}^2$ was closer to the toughness of polyimide ($1\,\text{kJ/m}^2$) than the thermodynamic work of adhesion ($100\,\text{mJ/m}^2$). The fracture process zone modeling approach coupled with measurements of global and local response seems to have extracted a reasonable value of the adhesive fracture energy. A companion paper will examine its ability to detect differences in surface treatments and processing routes.

17. Conclusions

The extraction of the adhesive fracture energy of thin polyimide films on aluminium was considered in this study. The mechanical behavior of the polyimide was determined from bulge tests that were conducted prior to delamination. A series of delamination experiments was then conducted and methods for extracting the adhesive fracture energy from elastic and elastoplastic analyses were compared.

The circular blister was first essentially used as a bulge test for determining the mechanical properties of the thin polyimide films. The average value for elastic modulus E was determined to be 2.18 GPa and the yield strength and hardening exponent were 31 MPa and 6, respectively.

In reviewing elastic analyses for determining the adhesive fracture energy, the effect of radial displacement was added to the previous analyses of Hencky (1915), Gent and Lewandowski (1987) and Chu et al. (1992 a, b). This made notable differences to ((17) and (22)) that were used to extract the adhesive fracture energy. The Gent and Lewandowski formulation for extracting the adhesive fracture energy relies on measurements of pressure and central displacement (17), whereas the one developed by Chu et al. (22) makes use of the pressure history and the nominal volume flow rate of fluid. In order to determine the adhesive fracture energy of polyimide/metal interfaces in the presence of global yielding, a cohesive zone model based on the approaches of Needleman (1987, 1990) and Tvergaard and Hutchinson (1993, 1994) was also adopted for this work. This approach essentially involves attributing a traction-separation law to the interface. A set of nonlinear springs was chosen to model the interface. The calibration of the springs was made on the basis of measurements of the initiation pressure, p_c, the crack opening displacements and the pressure-central displacement responses during steady state debonding.

Although the analysis developed by Chu et al. is based on that of Gent and Lewandowski, the two methods yielded significant differences in their estimates of the adhesive fracture energy. If a sufficiently rigid pressurization system is used, the nominal pumping rate can be used in the former, thereby simplifying experimental matters considerably. However, the apparatus used here was noticeably compliant so that the blister volume had to be measured directly from its shape. In fact, it was determined that the volume flow rate and p^{-3} vs. t traces only became constant when the crack speed became constant. The values of the adhesive fracture energy that were extracted from the slopes of the p^{-3} vs. t traces were lower than those extracted from pressure-central displacement traces. This appears to be due to the fact that the method proposed by Gent and Lewandowski, although elastic in analysis, makes use of central displacements which, in this combination of materials and surface preparation, contained a plastic component. This conclusion was strengthened by the fact that the adhesive fracture energy derived from (17) was the same as the total fracture energy that was obtained from the elastoplastic finite element analysis. As can be seen from (30), the latter is made up of the adhesive fracture energy and the plastic dissipation rate.

The adhesive fracture energy values obtained from the pressure-time traces were 20 percent lower than the value that was used in the finite element analysis. Using secant or tangent moduli instead of the Young's modulus in (22) in order to account for the extensive yielding during delamination did not bring the adhesive fracture energies into agreement. Presumably plasticity effects would have to be more rigorously incorporated into a Chu et al. (1992 a, b) type of approach.

The 390 J/m^2 for the adhesive fracture energy is considerably higher than the work of adhesion of the polyimide/aluminum combination but lower than the 1 kJ/m^2 toughness of polyimide. This was because the locus of fracture was in a carboxylate-rich interphase region between the aluminum and the bulk polyimide and, presumably, the constraint applied by the aluminum on the thin interphase region. The global plastic dissipation rate was 13 percent of the adhesive fracture energy for this particular surface preparation.

Appendix A

We give the details of the analysis of the pressurized film, clamped along its circular boundary of radius a, containing residual stresses σ_R and subjected to a uniform pressure p on one side. Since the deflection surface in such a case is symmetrical with respect to the center of the blister, the displacement of a point in the middle plane of the plate can be resolved into two components; a component u in the radial direction and a component w perpendicular to the plane of the plate. In the deformed shape, the out-of-plane and axial displacements, w and u respectively are assumed to be of the form suggested by Timoshenko and Woinowsky–Kreiger (1987)

$$w = w_0\left(1 - \frac{r^2}{a^2}\right)^2 \tag{A.1}$$

and

$$u = r(a - r)(c_1 + c_2 r). \tag{A.2}$$

The boundary conditions are taken to be fully clamped, so that

$$\begin{aligned} &w(a) = 0, \quad w'(a) = 0, \quad w(0) = w_0, \quad \text{and} \quad w'(0) = 0, \\ &u(a) = 0, \quad \text{and} \quad u(0) = 0. \end{aligned} \tag{A.3}$$

Note that these boundary conditions are satisfied by (A.1) and (A.2).

The total strain energy U of the system can be divided into two parts.

$$U = U_b + U_s. \tag{A.4}$$

First, the strain energy due to bending is

$$U_b = \tfrac{1}{2}D \int_0^{2\pi} \int_0^a \left[\left(\frac{\partial^2 w}{\partial r^2}\right)^2 + \frac{1}{r^2}\left(\frac{\partial w}{\partial r}\right)^2 + \frac{2v}{r}\frac{\partial w}{\partial r}\frac{\partial^2 w}{\partial r^2} \right] r \, dr \, d\theta, \tag{A.5}$$

where

$$D = \frac{Eh^3}{12(1 - v^2)}$$

is the rigidity of the plate.

The strain energy due to stretching of the middle plane can be obtained by using the expression

$$U_s = \int_0^{2\pi} \int_0^a \tfrac{1}{2}(N_r \varepsilon_r + N_\theta \varepsilon_\theta) r \, dr \, d\theta, \tag{A.6}$$

where N_r, and N_θ are the membrane forces in the radial and circumferential directions, respectively. By applying Hook's law, their values are

$$\begin{aligned} N_r &= \frac{Eh}{1 - v^2}(\varepsilon_r + v\varepsilon_\theta) + \sigma_r h, \\ N_\theta &= \frac{Eh}{1 - v^2}(\varepsilon_\theta + v\varepsilon_r), \end{aligned} \tag{A.7}$$

where h is the thickness of the film, ε_r is the radial strain and ε_θ is the circumferential strain which are related to the displacement by

$$\begin{aligned} \varepsilon_r &= \frac{du}{dr} + \frac{1}{2}\left(\frac{dw}{dr}\right)^2, \\ \varepsilon_\theta &= \frac{u}{r}. \end{aligned} \tag{A.8}$$

The constants c_1 and c_2 in (A.2) are now determined from the condition that the total energy of the plate in a state of equilibrium is a minimum. Hence

$$\frac{\partial U}{\partial c_1} = 0 \quad \text{and} \quad \frac{\partial U}{\partial c_2} = 0. \tag{A.9}$$

Table A.1. The values of α', α'', β' and β''.

	α'	α''	β'	β''
Clamped circular plate	$1.4206 - 0.7063\nu$	$-1.881 + 0.3095\nu$	$\frac{25}{48}(1-\nu^2)$	$-\frac{5}{16}(1-\nu^2)$

As a result, c_1 and c_2 are

$$\begin{aligned} c_1 &= \alpha' \frac{w_0^2}{a^3} + \beta' \frac{\sigma_R}{Ea}, \\ c_2 &= \alpha'' \frac{w_0^2}{a^2} + \beta'' \frac{\sigma_R}{Ea^2}, \end{aligned} \tag{A.10}$$

where the values of α', α'', β', and β'' depend on the Poisson's ratio ν as shown in Table A.1.

The amount of work done by the external pressure is obtained by

$$W = \int_0^{2\pi} \int_0^a pwr \, \mathrm{d}r \, \mathrm{d}\theta. \tag{A.11}$$

The value of w_0 was then found by applying the principle of virtual displacements. From this principle it follows that

$$\frac{\mathrm{d}U}{\mathrm{d}w_0} \delta w_0 = \frac{\mathrm{d}W}{\mathrm{d}w_0} \delta w_0. \tag{A.12}$$

which yields the relationship between the maximum deflection and the applied pressure

$$\frac{pa^4}{Eh} = e(v)h^2 w_0 + f\left(v, \frac{\sigma_R}{E}\right) a^2 w_0 + g(v) w_0^3, \tag{A.13}$$

where functions e, f, and g are given by

$$\begin{aligned} e(v) &= \frac{5.333}{(1-\nu^2)}, \\ f\left(v, \frac{\sigma_R}{E}\right) &= \frac{0.552\sigma_0}{E(1-v^2)}[2.755 + v - 2.755v^2 - v^3], \\ g(v) &= \frac{0.552}{(1-v^2)}[4.114 + 2.329v - 1.529v^2]. \end{aligned} \tag{A.14}$$

The first term on the right-hand side of (A.13) represents the bending effect of the layer which can be ignored for very thin layers, while the second term reflects the effect of residual stresses and last term is due to membrane effects. For the films considered in this study, $200 < a/h < 600$, the third term is dominant. The quantity $g(v) = 2.834$ for $v = 0.3$ so that the central deflection was then given by

$$w_0 = C_2 \left(\frac{pa^4}{Eh}\right)^{1/3}, \tag{A.15}$$

where $C_2 = 0.706$ for $v = 0.3$. This is 6.6 percent larger than the value ($C_2 = 0.662$) determined by Hencky (1915) due to neglect of the radial displacement, u.

Appendix B

For linearly elastic debonding, we write the energy balance between the work W done by fluid injection, the elastic strain energy U_e stored in the film and the energy U_a expended in debonding as

$$W = U_e + U_a. \tag{B.1}$$

During quasi static debonding, for an infinitesimal increment in crack radius da, (B.1) can be rewritten as

$$\mathrm{d}W = \mathrm{d}U_e + \mathrm{d}U_a. \tag{B.2}$$

From variational calculus, the change in external work dW, is given by

$$\mathrm{d}W \equiv p\left(\frac{\partial(\Delta\nu)}{\partial a}\right)_p \mathrm{d}a. \tag{B.3}$$

Substituting (4) and (13) into (B.3) yields

$$\mathrm{d}W = \left(\frac{6e(v)h^2a^4 + 4f\left(v, \frac{\sigma_R}{E}\right)a^6 + \frac{10}{C_1^2\pi^2}g(v)(\Delta\nu)^2}{e(v)h^2a^5 + f\left(v, \frac{\sigma_R}{E}\right)a^7 + \frac{3}{C_1^2\pi^2}g(v)a(\Delta\nu)^2}\right) p(\Delta\nu)\,\mathrm{d}a. \tag{B.4}$$

Equations (4) and (13) also yield the amount of elastic energy stored in the inflated blister as

$$U_e = \frac{p(\Delta\nu)}{4} + \frac{Eh}{4C_1\pi a^6}\left[e(v)h^2 + f\left(v, \frac{\sigma_R}{E}\right)a^2\right](\Delta\nu)^2. \tag{B.5}$$

As a result, the change in elastic strain energy during an infinitesimal increase in the radius of the blister is

$$\mathrm{d}U_e \cong \tfrac{1}{4}p\left(\frac{\partial(\Delta\nu)}{\partial a}\right)_p \mathrm{d}a + (\Delta\nu)\zeta\left(\frac{\partial(\Delta\nu)}{\partial a}\right)_p \mathrm{d}a - 2(\Delta\nu)^2\frac{\zeta}{a}\,\mathrm{d}a, \tag{B.6}$$

where the values of $(\partial(\Delta\nu)/\partial a)_p$ and ζ are defined as

$$\left(\frac{\partial V}{\partial a}\right)_p = \left(\frac{6e(v)h^2a^4 + 4f\left(v, \frac{\sigma_R}{E}\right)a^6 + \frac{10}{C_1^2\pi^2}g(v)(\Delta\nu)^2}{e(v)h^2a^5 + f\left(v, \frac{\sigma_R}{E}\right)a^7 + \frac{3}{C_1^2\pi^2}g(v)a(\Delta\nu)^2}\right)(\Delta\nu) \tag{B.7}$$

and

$$\zeta = \frac{Eh}{2\pi C_1 a^4}\left[e(v)\frac{h^2}{a^2} + f\left(v, \frac{\sigma_0}{E}\right)\right]. \tag{B.8}$$

If the adhesive fracture energy Γ_a is the energy dissipated per unit area of debonding, then

$$dU_a = 2\pi a \Gamma_a \, da. \tag{B.9}$$

On substituting (B.4), (B.6), and (B.9) into (B.2), the adhesive fracture energy can be seen to be

$$\Gamma_a = \frac{1}{2\pi a}[\tfrac{3}{4}p - (\Delta\nu)\zeta]\left(\frac{\partial(\Delta\nu)}{\partial a}\right)_p - 2(\Delta\nu)^2\frac{\zeta}{a}. \tag{B.10}$$

When only membrane effects are considered, the external work term becomes

$$dW = \left(\frac{10p(\Delta\nu)}{3a}\right) da. \tag{B.11}$$

Using the same assumptions, the amount of energy stored in the inflated film is

$$U_e = \frac{p(\Delta\nu)}{4}. \tag{B.12}$$

The change in elastic strain energy is then

$$dU_e \cong \tfrac{1}{4}p\left(\frac{\partial(\Delta\nu)}{\partial a}\right)_p da. \tag{B.13}$$

The adhesive fracture energy becomes

$$\Gamma_a = \frac{1}{2\pi a}[\tfrac{3}{4}p]\left(\frac{\partial(\Delta\nu)}{\partial a}\right)_p. \tag{B.14}$$

From (4) and (5) we have

$$\left(\frac{\partial(\Delta\nu)}{\partial a}\right)_p = \frac{10(\Delta\nu)}{3a}, \tag{B.15}$$

so that

$$\Gamma_a = \frac{5p(\Delta\nu)}{4\pi a^2}. \tag{B.16}$$

In terms of the central displacement, we have, from (4)

$$\Gamma_a = C_3 p w_0, \tag{B.17}$$

where $C_3 = 0.619$, which is 4.6 percent lower than the value (0.649) from Gent and Lewandowski (1987).

Acknowledgement

The authors would like to acknowledge the financial support of the Office of Naval Research (Grant # N00014-95-1-0690) and the Natonal Science Foundation (Grant # cmS 9522964).

The cooperation and collaboration with Jim Boerio, Gilles Dillingham and graduate students at the University of Cincinnati, which was facilitated by Peter Schmidt at ONR, has been extremely rewarding.

References

Allen, M.G. and Senturia, S.D. (1989). Application of the island blister test for thin film adhesion measurement. *Journal of Adhesion* **29**, 219–231.

Beams, J.W. (1959). Mechanical properties of thin films of gold and silver. *Structure and Properties of Thin Films* (Edited by C.A. Neugebauer, J.B. Newkirk, and D.A. Vermilyea), Wiley Publishers, New York, 183–192.

Burton, J.D., Jones, W.B. and Williams, M.L. (1970). The theoretical and experimental treatment of fracture in an adhesive interlayer. *Transactions of the Society of Rheology* **15**, 39–50.

Chang, M.D., Devries, K.L. and Williams, M.L. (1972). The effects of plasticity in adhesive fracture. *Journal of Adhesion* **4**, 221–230.

Chu, Y.Z. and Durning, C.J. (1992). Application of the blister test to the study of polymer-polymer adhesion. *Journal of Applied Polymer Science* **45**, 1151–1164.

Chu, Y.Z., Jeong, H.S., White, R.C. and Durning, C.J. (1992). Characterization of adhesion in thin-film materials by the blister test. *Mat. Res. Symp. Proc.* **276**, 209–220.

Dillard, D.A. and Bao, Y. (1991). The peninsula blister test: a high and constant strain energy release rate fracture specimen for adhesives. *Journal of Adhesion* **33**, 253–272.

Gent, A.N. and Lewandowski, L.H. (1987). Blow-off pressure for adhering layers. *Journal of Applied Polymer Science.* **33**, 1567–1577.

Hencky, H. (1915). Uber den spannungszustand in kreisrunden platten mit verschwindender biegungssteifigkeit. *Zr. Math. Phys.* **63**, 311–317.

Hinkley, J.A. (1983). A blister test for adhesion of polymer films to SiO_2. *Journal of Adhesion* **16**, 115–125.

Jensen, H.M. (1991). On the blister test for interface toughness measurement. *Engineering Fracture Mechanics* **40**, 475–486.

Jensen, H.M. and Thouless, M.D. (1993). Effects of residual stresses in the blister test. *International Journal of Solids and Structures* **30**, 779–795.

Kim, K-S. and Aravas, N. (1988). Elastoplastic analysis of the peel test. *International Journal of Solids and Structures* **24**, 417–435.

Kinloch, A.J., Lau, C.C. and Williams, J.G. (1994). The peeling of flexible laminates. *International Journal of Fracture* **66**, 45–70.

Knauss, W.G. (1973). On the steady propagation of a crack in a viscoelastic sheet: experiments and analysis. *Deformation and Fracture of High Polymers* (Edited by H.H. Kausch, J.A. Hassel and R.I. Jaffee), Plenum Press, 501–541.

Liechti, K.M. and Hanson, E. (1988). Nonlinear effects in mixed mode delaminations *International Journal of Fracture* **36**, 199–217.

Liechti, K.M. and Liang, Y-M. (1992). The interfacial fracture characteristics of bimaterial and sandwich blister specimens. *International Journal of Fracture* **55**, 95–114.

Liechti, K.M. and Shirani, A. (1994). Large scale yielding in blister specimens. *International Journal of Fracture* **67**, 21–36.

Liechti, K.M. and Chai, Y-S. (1992). Asymmetric shielding in interfacial fracture under in-plane shear. *Journal of Applied Mechanics* **59**, 295–304.

Needleman A. (1987). A continuum model for void nucleation by inclusion debonding. *Journal of Applied Mechanics* **54**, 525–531.

Needleman A. (1990). An analysis of decohesion along an imperfect interface. *Journal of the Mechanics and Physics of Solids* **38**(3), 289–324.

Nix, W.D. (1996). Diamond films for electronic packaging. Final Technical Report (CRY93-C-0199) Stanford University, Ca.

Schapery, R.A. (1975a). A theory of crack initiation and growth in viscoelastic media: I. theoretical development. *International Journal of Fracture* **11**(1), 141–159.

Schapery, R.A. (1975b). A theory of crack initiation and growth in viscoelastic media: II. approximate methods of analysis. *International Journal of Fracture* **11**(3), 369–388.

Schapery, R.A. (1975c). A theory of crack initiation and growth in viscoelastic media: III. analysis of continuous growth. *International Journal of Fracture* **11**(4), 549–562.

Shirani, A. (1997). Determination of adhesive fracture energy from quasi static debonding of thin films. Engineering Mechanics Research Laboratory Report, EMRL 97-16. The University of Texas at Austin.

Swadener, J.G. and Liechti, K.M. (1996). Asymmetric shielding mechanisms in the mixed-mode fracture of a glass/epoxy interface. Engineering Mechanics Research Laboratory Report, EMRL 96-14. *Journal of Applied Mechanics*, accepted 1997.

Timoshenko, S. and Woinowsky-Kreiger, S. (1987). *Theory of Plates and Shells*, McGraw-Hill, New York, 2nd Edition.

Tvergaard V. and Hutchinson J.W. (1992). The relation between crack growth resistance and fracture process parameters in elastic-plastic solids. *Journal of the Mechanics and Physics of Solids* **40**, 1377–1397.

Tvergaard, V. and Hutchinson, J.W. (1993). The influence of plasticity on mixed mode interface toughness. *Journal of the Mechanics and Physics of Solids* **41**, 1119–1135.

Tvergaard, V. and Hutchinson, J.W. (1994). Effect of t-stress on mode I crack growth resistance in a ductile solid. *International Journal of Solids and Structures* **31**, 823–833.

Ungsuwarungsri, T. and Knauss, W.G. (1987). The role of damage-softened material behavior in the fracture of composites and adhesives. *International Journal of Fracture* **35**, 221–241.

Ungsuwarungsri, T. and Knauss, W.G. (1988a). A nonlinear analysis of an equilibrium craze: Part I – problem formulation and solution. *Journal of Applied Mechanics* **55**, 44–51.

Ungsuwarungsri, T. and Knauss, W.G. (1988b). A nonlinear analysis of an equilibrium craze: Part II – simulations of craze and crack growth. *Journal of Applied Mechanics* **55**, 52–58.

Wei, Y. and Hutchinson, J.W. (1996). Nonlinear delamination mechanics for thin films. Division of Engineering and Applied Sciences, Harvard University, Mech. 284.

Williams, M.L. (1969). The continuum interpretation for fracture and adhesion. *Journal of Applied Polymer Science* **13**, 29–40.

Williams, M.L. (1970). The fracture threshold for an adhesive interlayer. *Journal of Applied Polymer Science* **14**, 1121–1126.

International Journal of Fracture **93:** 315–333, 1998.

Interface strength, work of adhesion and plasticity in the peel test

YUEGUANG WEI and JOHN W. HUTCHINSON
Division of Engineering and Applied Sciences, Harvard University, Cambridge, MA 01238, USA
e-mail: hutchinson@husm.harvard.edu

Received 23 October 1997; accepted in revised form 3 September 1998

Abstract. A cohesive zone model is proposed and analyzed for steady-state peeling of a thin rate-independent, elastic-plastic film bonded to an elastic substrate. A traction-separation description of the interface is embedded within continuum characterizations of the film and substrate. The primary parameters characterizing the traction-separation relation are the work of adhesion and the peak separation stress, termed the interface strength. The objective of the study is the determination of the relationship of the peel force to the work of adhesion of the interface and its strength, with due regard for plastic deformation in the film. An example of an elastic film peeled from an elastic-plastic substrate is also presented.

Key words: Peel test, work of adhesion, interface strength, plastic yielding, adhesion of thin films.

1. Introduction

The study of the peel test conducted in this paper continues efforts by Kim and Kim (1988), Kim and Aravas (1988) and, most recently by Kinlock, Lau and Williams (1994) to characterize the mechanics of the test. Earlier work on this topic is cited in these papers. One of the primary aims of this general line of research has been to link the peel force or, equivalently, the macroscopic work of fracture, to the work of interface adhesion by accounting for plastic deformation accompanying the peeling process. In steady-state peeling, the peel force per unit width of film is simply related to the macroscopic work of fracture, which is effectively the sum of the work of adhesion and the plastic dissipation. Successful partitioning of these two contributions to the peel force would enable the work of adhesion to be inferred. Here, the research is carried one step beyond the previous studies. A cohesive zone model is introduced wherein a traction-separation relation is employed to model the interface as a condition linking continuum descriptions of the film and the substrate. The traction-separation relation is characterized by the work of adhesion (energy per unit area) Γ_0 and a maximum separation stress $\widehat{\sigma}$ which can be regarded as the interface strength under normal stressing. Under steady-state conditions, the model provides the peel force per unit width of the film P in terms of Γ_0, $\widehat{\sigma}$, the properties of the film and substrate, and the peel angle Φ. With one exception, the paper deals with elastic-plastic films bonded to elastic substrates.

When conditions are such that negligible plastic deformation occurs, the simple work balance $P(1-\cos\Phi)=\Gamma_0$ holds, independent of $\widehat{\sigma}$, assuming the stretch energy in the film can be neglected. Perhaps it is for this reason that prior work has not focused on the role of interface strength in the mechanics of the peel test. When plastic deformation does occur, the extent to which $P(1-\cos\Phi)$ exceeds Γ_0 depends strongly on $\widehat{\sigma}$. The two-parameter characterization $(\Gamma_0,\widehat{\sigma})$ of the interface in the present model has a parallel in the approach adopted by Kim and Kim (1988), Kim and Aravas (1988), and Kinloch et al. (1994) wherein

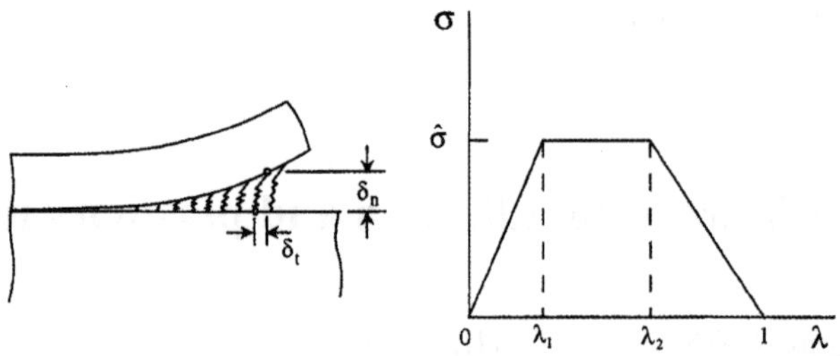

Figure 1. Traction-separation relation governing separation of the interface.

both the work of adhesion and an opening angle where the film leaves the substrate are used as parameters to specify the peeling process. Further discussion emphasizing contrast between the two approaches will be given later in the paper.

The traction-separation relation of the interface is postulated in Section 2, following earlier work on general interface fracture by Tvergaard and Hutchinson (1993). A complete specification of the model follows in Section 3. Details of the numerical method for steady-state peeling are given in Section 4, and numerical results are presented in Section 5. Conclusions, limitations of the model, and recommendations for further work are given in Section 6.

2. The interface traction-separation relation

The thickness of the interface in the unloaded state is taken to be zero. Following the notation for the interface traction-separation relation introduced in Tvergaard and Hutchinson (1993), let δ_n and δ_t be the normal and tangential components of the relative displacement of the respective faces across the interface in the zone where the separation process occurs, as indicated in Figure 1. Let δ_n^c and δ_t^c be critical values of these displacement components, and define a single dimensionless separation measure as

$$\lambda = \sqrt{(\delta_n/\delta_n^c)^2 + (\delta_t/\delta_t^c)^2}, \tag{2.1}$$

such that the tractions drop to zero when $\lambda = 1$. With $\sigma(\lambda)$ displayed in Figure 1, a potential from which the interface tractions in the separation zone are derived is defined as

$$\Pi(\delta_n, \delta_t) = \delta_n^c \int_0^\lambda \sigma(\lambda')\,\mathrm{d}\lambda'. \tag{2.2}$$

The normal and tangential components of the traction acting across the interface in the fracture process zone are given by

$$T_n = \frac{\partial \Pi}{\partial \delta_n} = \frac{\sigma(\lambda)}{\lambda}\frac{\delta_n}{\delta_n^c}, \qquad T_t = \frac{\partial \Pi}{\partial \delta_t} = \frac{\sigma(\lambda)}{\lambda}\frac{\delta_t}{\delta_t^c}\frac{\delta_n^c}{\delta_t^c}. \tag{2.3}$$

The traction law under a purely normal separation ($\delta_t = 0$) is $T_n = \sigma(\lambda)$ where $\lambda = \delta_n/\delta_n^c$. The peak normal traction under purely normal separation is $\widehat{\sigma}$, which will be termed the interface strength. Under a purely tangential displacement ($\delta_n = 0$), $T_t = (\delta_n^c/\delta_t^c)\sigma(\lambda)$ where $\lambda = \delta_t/\delta_t^c$. The peak shear traction is $(\delta_n^c/\delta_t^c)\widehat{\sigma}$ under a purely tangential displacement

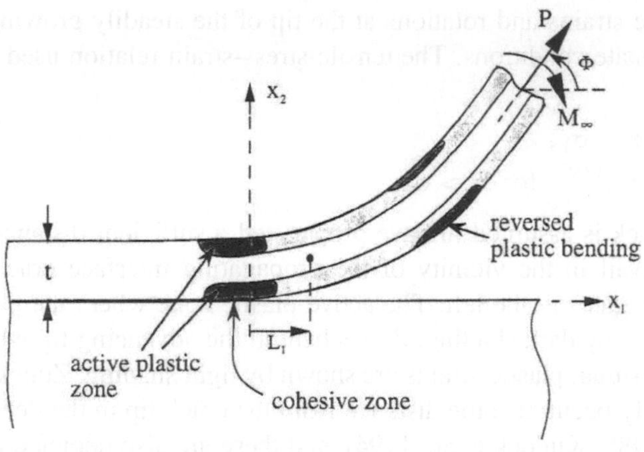

Figure 2. Model of steady-state peeling. A plane strain finite element formulation is used to produce the solution for the substrate and the portion of the film to the left of $x_1 = L_1$. An elastic–plastic 'elastica' solution is applied for the portion of the film to the right of $x_1 = L_1$.

of the faces. The work of separation per unit area of interface Γ_0 is given by (2.2) with $\lambda = 1$. For the separation function $\sigma(\lambda)$ specified in Figure 1

$$\Gamma_0 = \tfrac{1}{2}\hat{\sigma}\delta_n^c[1 - \lambda_1 + \lambda_2]. \tag{2.4}$$

The separation law is assumed to be independent of the time rate of deformation, as are the constitutive models characterizing the film and substrate. The parameters governing the separation law of the interface are the work of the fracture process Γ_0, the peak stress quantity $\hat{\sigma}$, and the critical displacement ratio δ_n^c/δ_t^c, together with the factors λ_1 and λ_2 governing the shape of the separation function. Use of the potential ensures that the work of separation is Γ_0 regardless of the combination of normal and tangential displacements taking place in the separation of the interface. Experience gained in the earlier studies suggests that details of the shape of the separation law are relatively unimportant. The two most important parameters characterizing interface separation are Γ_0 and $\hat{\sigma}$. The parameter δ_n^c/δ_t^c is the next most important, but the study of mixed mode interface toughness (Tvergaard and Hutchinson, 1993) indicates that predictions are relatively insensitive to this parameter as long as the interface process is dominantly normal separation. This is the case for the peel test under the range of peel angles considered here.

Attainment of $\lambda = 1$ at the end of the traction-separation zone is the condition for crack advance. In steady-state propagation, this condition must be imposed on the solution.

3. An embedded cohesive zone model for steady-state peeling

The geometry of the model is displayed in Figure 2 with a peel angle Φ. Plane strain conditions are assumed, as appropriate for a film whose width normal to the plane of deformation in Figure 2 is sufficiently large compared to its thickness t. Except for one example presented at the end of the paper, the film is taken to be elastic–plastic with Young's modulus E, Poisson's ratio ν, tensile yield stress σ_Y, and strain hardening exponent N. Rate-independent material behavior is assumed. The substrate is elastic with modulus E_s and Poisson's ratio ν_s. The standard J_2 flow theory of plasticity, based on the von Mises yield surface, is used to characterize plasticity in the film. The small strain version of the theory is employed, consistent with the fact that the strains and rotations at the tip of the steadily growing crack are indeed small under steady-state conditions. The tensile stress-strain relation used to characterize the film material is

$$\begin{aligned} \varepsilon &= \sigma/E \quad \text{for } \sigma \leqslant \sigma_Y, \\ &= (\sigma_Y/E)(\sigma/\sigma_Y)^{1/N} \quad \text{for } \sigma > \sigma_Y. \end{aligned} \tag{3.1}$$

The interface crack is assumed to have propagated a sufficient distance such that steady-state conditions prevail in the vicinity of the propagating interface crack. In Figure 2, the interface crack propagates to the left. The active plastic zone where the plastic strain rate $\dot{\varepsilon}_{ij}^{p}$ is nonzero is depicted by dark shading. Zones behind the advancing tip which have unloaded but which contain residual plastic strains are shown by light shading. Zones of reversed plastic loading will generally occur at some distance from the crack tip in the detached the film strip (Kim and Aravas, 1988; Kinlock et al., 1994), and these are also depicted by dark shading.

The peel angle Φ specifies the angle which the peel force per unit width of film P makes with the plane of the interface. Let $M(s)$ be the bending moment (per unit width) about the middle plane of the separated film at a distance s along the film measured from the crack tip. Take the origin of the coordinate system (x_1, x_2) at current location of the tip where $\lambda = 1$. For analysis purposes, the problem is sub-divided into two parts (cf. Figure 2): the substrate plus the film to the left of $s = x_1 = L_1$, and the separated infinite film segment to the right of L_1. Accurate results from the cohesive zone model require a full 2D plane strain, continuum analysis of the behavior in the vicinity of the interface crack. Representation of the film and substrate by a beam model of the film in the vicinity of the separation zone misses essential features of the phenomena. For example, the small scale yielding limit when the active plastic zone is confined to a region near the crack tip cannot be captured by a beam model. An Eulerian-based, finite element formulation designed to cope with steady-state conditions will be employed for this part of the problem. The point at $s = L_1$ is where the full 2D continuum analysis of the crack tip problem is matched to a 1D bending problem for the separated film strip. This point must lie to the right of the active plastic zone at the crack tip and to the left of any reversed plastic bending. Otherwise, the location of this matching point will be seen to have essentially no effect on the solution as long as the slope there θ_1 is small.

The film which emerges from the region to the left of L_1 has been subject to plastic deformation and has a residual curvature κ_0 prior to any reversed plastic deformation. The residual curvature κ_0 is computed as part of the solution to the crack tip problem. The film at the matching point $s = L_1$ has undergone elastic unloading sustaining a moment M_1 less than the maximum moment, which is attained at some point to the left of L_1. The moment-curvature relation of the film emerging into the region to the right of L_1 is displayed as the unloading branch in Figure 3. The initial portion of the curve of M versus κ is shown dashed since it is not used in the analysis. (This is the portion of the behavior computed using the full 2D representation of the film.) Reversed plastic deformation, if it occurs, takes place when M becomes sufficiently negative (Kim and Kim, 1988; Kim and Aravas, 1988; Kinlock et al., 1994). At distances s which are far from the tip the film becomes straight, corresponding to the state $M = -M_\infty$ with $\kappa = 0$. The final residual curvature of the *unloaded* film ($M = 0$) is labeled in Figure 3 as κ_f.

Plastic deformation due to reversed plastic bending contributes to the overall rate of plastic work in the system, and it reduces the final residual curvature of the film from the value, κ_0, inherited from the plastic deformation in the vicinity of the interface crack tip. Reversed plastic deformation depends on the Bauschinger behavior of the material, which is not well

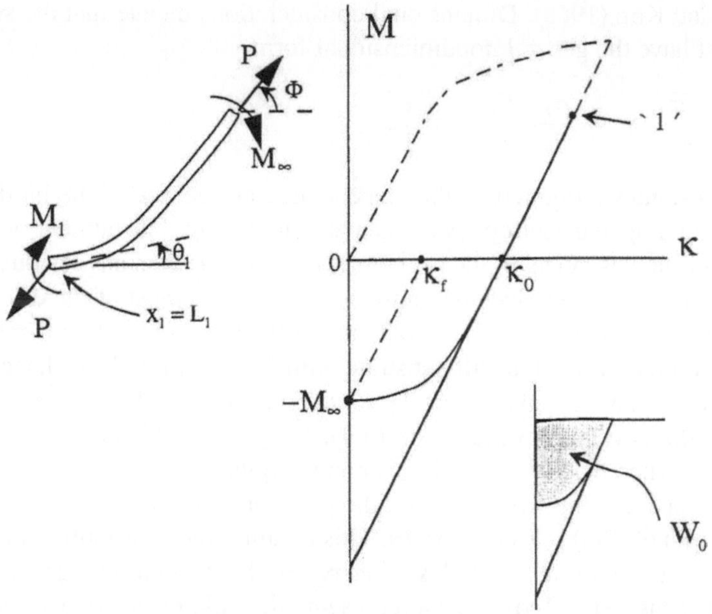

Figure 3. Moment-curvature relation governing behavior of the film to the right of $x_1 = L_1$.

quantified by conventional yield surface descriptions, especially not by an isotropic hardening theory. Because reversed plastic deformation occurs in the segment of the problem which is well characterized by 1D nonlinear bending theory, full details of the plastic constitutive behavior under reversed loading need not be considered. It will be shown that only the shaded area W_0 under the overall moment-curvature relation of the film in Figure 3 matters in the final results.

The study in this paper will consider the following parameters

$$E, \nu, \sigma_Y, N, t \text{ (film)}; \quad E_s, \nu_s \text{ (substrate)}; \quad \Gamma_0, \widehat{\sigma} \text{ (interface)}. \tag{3.2}$$

The peel angle Φ and the work per unit width of film W_0 characterizing reversed plastic bending complete the parameter set. An additional variable which is likely to have a major influence on the peel force is residual stress in the attached film acquired in the bonding process. When the residual stress is a nonnegligible fraction of the yield stress, it will alter the plastic dissipation. Residual prestress can be included in the model, but it will not be considered in this paper in the interest of reducing the number of variables.

The following material-based length quantity plays a fundamental role in the solution

$$R_0 = \frac{1}{3\pi(1-\nu^2)} \frac{E\Gamma_0}{\sigma_Y^2}. \tag{3.3}$$

The extent of the plastic zone scales with R_0. When P is only slightly larger than Γ_0, this length can be regarded as a estimate of the plastic zone height in the film at the crack tip (Tvergaard and Hutchinson, 1992, 1993). In general, however R_0 is less than the plastic zone height and should simply be regarded as a fundamental parameter with dimensions of length. The same length quantity (apart from a numerical constant) emerges as fundamental in the analysis of Kim and Kim (1988). Dimensional considerations dictate that the solution for the peel force P must have the general nondimensional form

$$\frac{P}{\Gamma_0} = F\left\{\frac{t}{R_0}, \frac{\widehat{\sigma}}{\sigma_Y}, N, \Phi, \frac{\sigma_Y}{E}, \frac{E_s}{E}, \nu, \nu_s\right\}, \tag{3.4}$$

where F is dimensionless. Implicit in this dependence are details of the hardening rule for reversed plastic straining, e.g. isotropic vs. kinematic hardening. The number of dimensionless variables in any mechanics model of the peel test is large, even when any residual stress arising during bonding is ignored. To reduce the number of variables, most of the calculations in the present study will be made with $\nu_s = \nu = 0.3$, and with either no elastic mismatch ($E_s = E$) or for a highly compliant film on a stiff substrate with $E = E_s/100$. The shape parameters in the traction separation law are taken to be $\lambda_1 = 0.15$ and $\lambda_2 = 0.5$, while δ_n^c/δ_t^c is fixed at 1. Of the remaining dimensionless variables in (3.4), t/R_0, $\widehat{\sigma}/\sigma_Y$, N and Φ all have significant influence on P/Γ_0 when plastic dissipation is nonnegligible.

In steady-state peeling, the work done by the peel force per unit advance of the interface crack (per unit width of film) is $P(1-\cos\Phi)$. This assumes the contribution of overall stretch energy of the film (including residual elastic energy in the bonded film) to the work balance can be neglected. Thus, $P(1-\cos\Phi)/\Gamma_0$ represents the ratio of the total work of fracture to the work of interface adhesion, while $P(1-\cos\Phi) - \Gamma_0$, represents the plastic dissipation in steady-state peeling.

4. Formulation and numerical solution

The two parts of the problem in Figure 2 are analyzed separately and coupled by requiring continuity of force, moment, displacement and rotation at $x_1 = L_1$. The first part to the left of L_1 is analyzed as a steady-state, plane strain continuum problem. The second part to the right of L_1 is treated as a finite rotation, bending problem (an elastic–plastic 'elastica') with a residual curvature κ_0 arising from plastic deformation in the first part. The analysis of each part is described, followed by a prescription of the coupling conditions.

4.1. Finite element solution for the first part of the problem

The stresses and strains well ahead of the advancing interface crack tip ($x_1 \to -\infty$) vanish because residual stresses in the film are assumed to be absent. In the small strain, small rotation steady-state problem, the rate at any point of a quantity such as stress or strain is related to the leftward velocity V of the crack tip and the gradient in the x_1 direction by

$$\dot{\varepsilon}_{ij} = V\partial\varepsilon_{ij}/\partial x_1. \tag{4.1}$$

It is this feature which makes it possible to directly solve the problem without having to consider transient behavior preceding attainment of steady-state. The Eulerian-based solution scheme was developed by Dean and Hutchinson (1980) and Parks, Lam and McMeeking (1981). It was applied to a related study of steady-state thin film delamination by Wei and Hutchinson (1997a). In outline, it is as follows.

Anticipating numerical implementation within a finite element framework, let E be the strain vector with components comprising the strains, and let Σ be the stress vector containing components of stress. Denote the plastic strain vector by E^p. The matrix of incremental moduli for plastic loading is denoted by $\mathbf{D}$ such that $\dot{\Sigma} = \mathbf{D}\dot{E}$; the corresponding elastic matrix of moduli is denoted by $\mathbf{D}^e$. The solution employs iteration. In each iteration, the core procedure is the finite element solution for the displacements, strains and stresses *assuming that the plastic strain distribution* E^p *is known.* In notation standard to the finite element method, let $\mathbf{U}$ be the vector of nodal displacements and let $\mathbf{B}$ be the strain matrix such that $E = \mathbf{BU}$. The finite element problem for $\mathbf{U}$ in terms of applied boundary forces $\mathbf{F}$ (distributions matching P and M are prescribed at $x_1 = L_1$) and any specified E^p is represented in the standard notation as

$$\mathbf{K}^e\mathbf{U} = \int_S \mathbf{N}^T\mathbf{F}\,\mathrm{d}S + \int_V \mathbf{B}^T\mathbf{D}^e\mathbf{E}^p\,\mathrm{d}V \quad \text{where } \mathbf{K}^e = \int_V \mathbf{B}^T\mathbf{D}^e\mathbf{B}\,\mathrm{d}V. \tag{4.2}$$

The iteration steps are as follows

(1) Use the distribution of E^p from the previous iteration in (4.2) to determine $\mathbf{U}$. In the first iteration take $E^p = \mathbf{0}$.
(2) Compute $\mathbf{E}$ from $\mathbf{U}$.
(3) Obtain a new estimate of the distribution of Σ. Use $\Sigma = \mathbf{D}^e\mathbf{E}$ in the region upstream of the current estimate of active plastic zone and use $\Sigma = \mathbf{D}^e(\mathbf{E} - E^p)$ downstream from the active zone. Where yield is currently met, make use the fact that for steady-state growth, $\dot{\Sigma} = \mathbf{D}\dot{E}$ can be replaced by $\partial\Sigma/\partial x_1 = \mathbf{D}\,\partial\mathbf{E}/\partial x_1$ such that for any point (x_1, x_2) within the active plastic zone

$$\Sigma(x_1, x_2) = \Sigma(x_1^*, x_2) - \int_{x_1}^{x_1^*} \mathbf{D}\,\partial\mathbf{E}/\partial x_1\,\mathrm{d}x_1, \tag{4.3}$$

where (x_1^*, x_2) is corresponding point on the leading edge of the active plastic zone (i.e., the left edge of the active zone). The integration in (4.3) is performed for fixed x_2 and applies to all points within the active plastic zone.
(4) Use $\mathbf{E}^p = \mathbf{E} - \mathbf{D}^{e^{-1}}\Sigma$ to compute the new estimate of $\mathbf{E}^p$ for the next iteration. Revise the boundaries of the active plastic zone using the new estimate of Σ. To the right of the active plastic zone in the downstream unloading region, $\mathbf{E}^p$ is a function only of x_2, corresponding to its value at the right edge of the active zone.
(5) If satisfactory convergence has not been achieved repeat steps (1) through (4).

Isoparametric elements with nine nodes with 2×2 Gauss integration are employed in the formulation. The traction-separation relation (2.3) is imposed on (4.2) by converting the tractions T_t and T_n to boundary forces linked to the displacement jumps across the crack face. These nonlinear equations are solved as part of the iterative process. The criterion for continuing crack propagation (i.e., $\lambda = 1$ at the end of the cohesive zone), is imposed on the solution.

4.2. SOLUTION IN THE SECOND PART OF THE FILM IN $s > L_1$

Within the first part, the film emerges from the active plastic zone with a plastic strain $\varepsilon_{ij}^p(x_2)$ which depends on x_2 but is independent of x_1. This is the state inherited by the film segment to the right of L_1. If the film were unloaded prior to any reversed plastic bending, it would have a residual curvature κ_0 given by

$$\kappa_0 = -\frac{12}{t^3}\int_0^t (x_2 - \tfrac{1}{2}t)\varepsilon_{11}^p(x_2)\,\mathrm{d}x_2 \tag{4.4}$$

which is computed from the solution to the first part. As noted earlier, reversed plastic deformation further to the right of L_1 will reduce the final residual curvature of the unloaded peeled film to κ_f.

The film segment to the right of L_1 is modeled as an inextensional elastic-plastic 'elastica', following earlier approaches. In this segment, the relation of the moment per unit width M and curvature κ is of the form shown in Figure 3. During unloading prior to reversed plastic bending, $M = B(\kappa - \kappa_0)$, where $B = Et^3/[12(1-\nu^2)]$ is the elastic bending stiffness per unit width of the film. With s as the distance along the film middle surface measured from the tip, and with θ as the rotation of the film middle surface relative to the x_1-axis, $\kappa = \mathrm{d}\theta/\mathrm{d}s$. Equilibrium requires $\mathrm{d}M/\mathrm{d}s = -P\sin(\Phi - \theta)$. By $\kappa = \mathrm{d}\theta/\mathrm{d}s$, the equilibrium equation can be rewritten as

$$\kappa\,\mathrm{d}M = -P\sin(\Phi - \theta)\,\mathrm{d}\theta. \tag{4.5}$$

Let M be any moment on the unloading curve in Figure 3 and let θ and κ be associated values. Integrate (4.5) from $M = -M_\infty$, where $\theta = \Phi$ and $\kappa = 0$, to M to obtain

$$W(M) = P[1 - \cos(\Phi - \theta)], \tag{4.6}$$

where $W(M)$ is the area to the right of the M-axis of the unloading moment-curvature curve between $-M_\infty$ and M

$$W(M) = \int_{-M_\infty}^{M} \kappa\,\mathrm{d}M. \tag{4.7}$$

Equation (4.6) provides the connection between M_1 and θ_1 at the point $s = L_1$, i.e.,

$$W(M_1) = P[1 - \cos(\Phi - \theta_1)]. \tag{4.8}$$

It proves to be revealing to bring into play the area $W_0 \equiv W(M = 0)$ of the unloading curve below the κ axis (cf. Figure 3). By (4.6),

$$W_0 = P[1 - \cos(\Phi - \theta_0)], \tag{4.9}$$

where θ_0 is the rotation at the point corresponding to $M = 0$. Subtract (4.9) from (4.8), noting that $W(M_1) - W_0$ is simply $M_1\kappa_0 + M_1^2/(2B)$. The resulting equation can manipulated and solved for M_1

$$M_1 = B\kappa_0\left[\sqrt{[1 - \cos(\Phi - \theta_1)][2P/(B\kappa_0^2)] + 1 - w_0} - 1\right], \tag{4.10}$$

where

$$w_0 = \frac{W_0}{\frac{1}{2}B\kappa_0^2}. \tag{4.11}$$

In this form, it can be seen that the relation between M_1 and θ_1 (which is an exact integration of the nonlinear elastic–plastic bending equation) depends on reversed plastic bending only through the area ratio w_0. As $B\kappa_0^2/2$ is the area below the κ-axis in Figure 3 in unloading

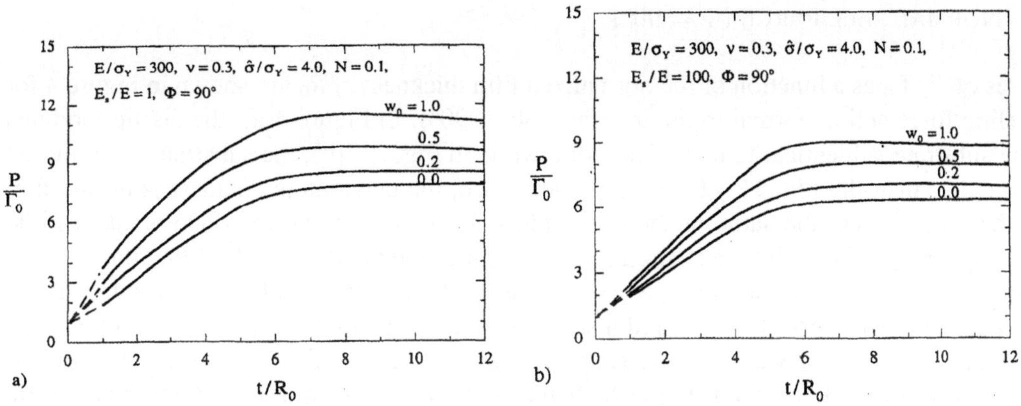

Figure 4. Normalized peeling force P/Γ_0 as a function of normalized film thickness t/R_0 for several values of the parameter characterizing reversed plastic bending w_0. The interface is moderately strong with $\widehat{\sigma}/\sigma_Y = 4$. (a) $E_S/E = 1$. (b) $E_S/E = 100$. Peel force is normal to the interface, $\Phi = 90°$. The other parameters are specified in the figure.

elastically to $\kappa = 0$, w_0 provides a measure of the effect of reversed plastic bending. It will be unity if no reversed plastic bending occurs. Smaller values of this ratio are pertinent when peeling involves significant plastic deformation. The quantity w_0 can be computed for any specific stress-strain description for reversed stressing. However, given the variety of plasticity descriptions for reversed stressing and the lack of agreement on which constitutes the best choice, we prefer to retain w_0 as an independent parameter in our analysis. It will be seen that its role is relatively minor.

4.3. Coupling of two parts of the solution at $s = L_1$

Continuity of displacement, rotation, force and moment are required at the point $s = x_1 = L_1$ where the two parts of the solution are matched. For the problem of the first part, a linear distribution of tractions is applied to the film along $x_1 = L_1$ with resultants chosen to coincide with the horizontal and vertical components of the force per unit width ($P \cos \Phi$, $P \sin \Phi$) and the moment per unit width M_1. The rotation θ_1 in the first part is computed as the rotation of the centerline of the film at $x_1 = L_1$. The continuity conditions are included in the set of equations to be satisfied in the iterative solution process. Thus, M_1 and θ_1 computed from the first part satisfy (4.10) at the end of the iteration process.

5. Numerical results for steady-state peeling

Calculations have been performed based on the formulation described above, and selected results will be presented in this section to bring out the dependence of the normalized peel force, P/Γ_0, on the dimensionless parameters identified in (3.4). The fraction w_0 measuring the contribution from reversed plastic bending is an additional parameter which will be considered. As emphasized in Section 3, $P(1-\cos \Phi)/\Gamma_0$ represents the ratio of the macroscopic work of fracture to the work of interface adhesion. The extent to which $P(1 - \cos \Phi)/\Gamma_0$ exceeds unity reflects the relative contribution of plastic dissipation in the film to the total work of fracture.

5.1. Normal peel force ($\Phi = 90°$)

Curves of P/Γ_0 as a function of the normalized film thickness t/R_0 are shown in Figure 4 for a peeling force acting normal to the interface ($\Phi = 90°$). In Figure 4(a), the elastic modulus of the substrate is identical to that of the film, while in Figure 4(b), the substrate is a hundred times stiffer than the film ($E_s/E = 100$). Otherwise, the parameters characterizing the film and the interface are the same in these two plots. In particular, the interface is taken to be relatively strong with an interface strength $\widehat{\sigma}$ four times the yield strength of the film σ_Y. The strain hardening index of the film is $N = 0.1$. The work of interface adhesion Γ_0 enters as the normalization of the peel force P and also in the length parameter R_0 defined in (3.3). The full range of reversed plastic bending is spanned by the curves in Figures 4(a,b). Recall that $w_0 = 1$ corresponds to no reversed plastic bending. At the other limit $w_0 = 0$ reversed plastic bending occurs as soon as M becomes negative. This latter limit would never be fully attained and would be approached only if the plastic deformation accompanying peeling were large. Suppression of reversed plastic bending has the effect of decreasing the moment carried by the film at the crack tip, thereby requiring a larger peel force to propagate the crack than when reversed plastic bending occurs. It is evident from Figure 4 that reversed plastic bending plays some role in determining the peel force, but not a dominant one. Similarly, a stiff substrate reduces the peel force relative to that for a more compliant substrate, but the effect is not large.

The influence of the location $x_1 = L_1$ where the two parts of the solution described in Section 4 are matched has been explored by repeating the calculations with different choices of L_1 to test for numerical sensitivity. Since the first part of the solution (the finite element solution) is based on a small rotation formulation, it is also essential that θ_1 be small. The calculations reported in this section were computed with $L_1/R_0 = 20$. This choice ensures that the matching point is well ahead of the active plastic zone at the crack tip and well to the left of the zone of reversed plastic bending. The angle θ_1 never exceeds $10°$. Repeating selected calculations such as those shown in Figure 4 for other choices of L_1/R_0, differing by as much as a factor of 2, produced at most, only a 2 or 3 percent change in the peel force.

The major trend brought out by Figure 4 is the dependence of the peel force on the film thickness t. Numerical values of P/Γ_0 in Figures 4(a,b) have been computed at all integer values of t/R_0 between 1 and 12. When t is small compared to R_0, plasticity occurs throughout the film, but plastic dissipation is nevertheless small compared to Γ_0 because the volume of film material is small. The peel test is an unusual illustration of a fracture phenomenon where large scale yielding is associated with lower toughness than small scale yielding. In the limit as t goes to zero, P approaches the interface adhesion Γ_0. The dashed sections of the curves between $t/R_0 = 0$ and 1 have been drawn to connect with $P/\Gamma_0 = 1$. At the other limit, when t/R_0 is sufficiently large, P/Γ_0 approaches an asymptote. This asymptote corresponds to a small scale yielding limit in the sense that the active plastic zone is confined to the crack tip, is small compared to the film thickness, and becomes independent of t/R_0. In this limit, there is no yielding on the top surface of the film above the tip and no reversed plastic bending such that $w_0 = 1$. Steady-state toughness in this limit is the same as that for an interface crack in small scale yielding at the same mode mixity (Tvergaard and Hutchinson, 1993).

The relatively strong interface ($\widehat{\sigma}/\sigma_Y = 4$) gives rise to considerable plastic dissipation such that the total work of fracture is approximately an order of magnitude greater than the work of interface adhesion when the film is sufficiently thick. For normalized interface strengths $\widehat{\sigma}/\sigma_Y$ above 4 in Figure 5, the peel force attains a maximum when the normalized thickness is approximately $t/R_0 = 6$ for the case $w_0 = 1$. The peak becomes somewhat

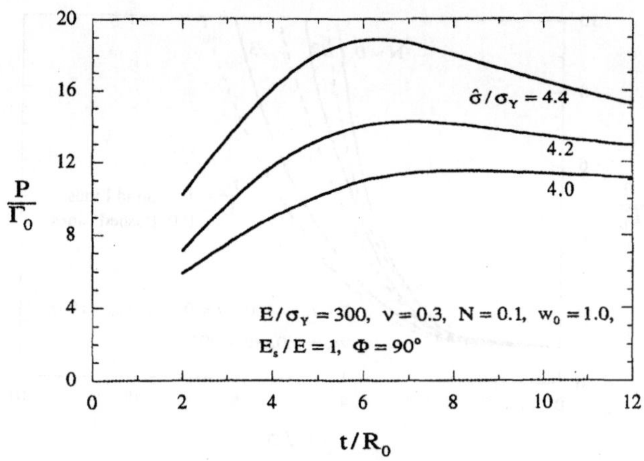

Figure 5. Normalized peeling force P/Γ_0 as a function of normalized film thickness t/R_0 for several values of the normalized interface strength, $\widehat{\sigma}/\sigma_Y$. The peel force is normal to the interface.

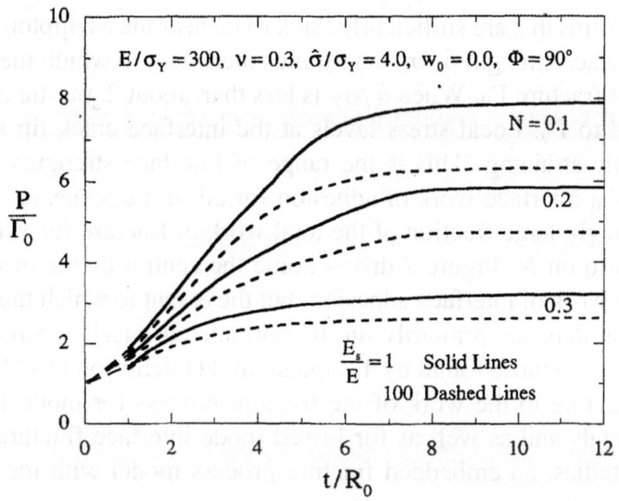

Figure 6. Influence of strain hardening index N on normalized peel force for a force acting perpendicular to the interface.

more prominent at smaller peel angles, as will be seen below. Peel tests on metal films with very strong interfaces and values of P/Γ_0 as large as 100 (Kim and Kim, 1989) display a pronounced peak at intermediate film thicknesses. The beam model of Kim and Kim (1989) predicts a peak peel force at roughly comparable values of normalized thickness.

The effect of the strain hardening index N of the film is displayed in Figure 6 for the case $w_0 = 0$. Otherwise, the parameters specifying the film/substrate system are the same as those in Figure 4. Strain hardening elevates crack tip stresses and, accordingly, makes it possible to attain a given interface strength $\widehat{\sigma}$ at a reduced peel force. It also increases the moment experienced near the tip at a given peel force. The relative effect is largest for thicker films with the highest levels of plastic dissipation.

Of all the parameters, the interface strength $\widehat{\sigma}$ has the most influence on ratio P/Γ_0, as seen in Figure 7. Curves are shown for two ratios of film to substrate moduli and three strain hardening indices. The curves are all for the case $w_0 = 0$, and they are computed with $t/R_0 =$

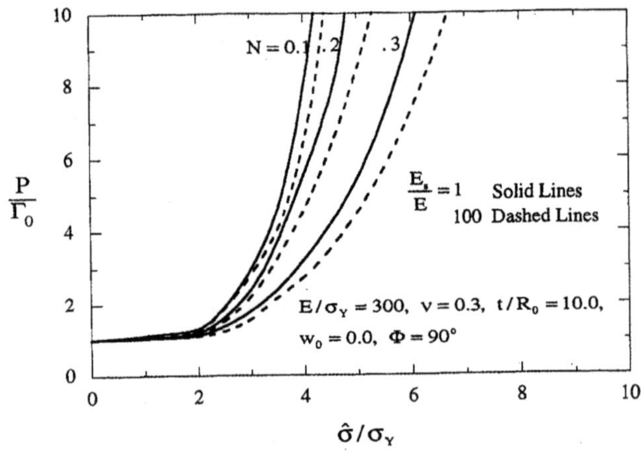

Figure 7. Dependence of the peel force on the interface strength for various levels of strain hardening for $t/R_0 = 10$. The force acts perpendicular to the interface.

10 corresponding to films that are sufficiently thick to lie near the asymptote in Figures 4 and 5. The normalized interface strength $\widehat{\sigma}/\sigma_Y$ determines the extent to which the peel force exceeds the interface work of fracture Γ_0. When $\widehat{\sigma}/\sigma_Y$ is less than about 2, plastic dissipation is nearly negligible compared to Γ_0. Local stress levels at the interface crack tip are low and induce relatively little plastic straining. This is the range of interface strengths for which the peel force is essentially the interface work of adhesion for all film thicknesses. Plastic dissipation becomes an increasingly large fraction of the total work of fracture for values of $\widehat{\sigma}/\sigma_Y$ larger than 2, depending also on N. Figure 7 drives home the central theme of this paper: the peel force scales with the work of interface adhesion, but the extent to which the peel force exceeds the work of adhesion depends primarily on the normalized interface strength. Qualitatively, these trends are similar to those found by Tvergaard and Hutchinson (1992, 1993) for the ratio thc total work of fracture to the work of the fracture process for mode I crack propagation in homogeneous metals and as well as for mixed mode interface fracture under small scale yielding. In those studies, an embedded fracture process model with the traction-separation law (2.3) was also employed to represent the fracture process on the extended crack plane or on the interface.

To complete the set of results for peeling under a normal force, the residual curvature κ_0 (4.4) emerging from the crack tip region (and prior to reversed plastic bending) and the opening angle α under load measured at the interface crack tip are presented. Curves of $Et\kappa_0/[\sigma_Y(1-\nu^2)]$ as a function of t/R_0 are plotted in Figure 8 for the same set of film/substrate parameters used in plotting Figures 4 and 5. The companion plots for the crack opening angle are given in Figure 9. The opening profile of a crack near its tip in steady-state propagation can be closely approximated by the angle the separated film surface makes with substrate surface, with due account for the fact that the opening displacement at the tip itself is at the critical separation value $\delta_n(0)$ given by the traction-separation law (2.3). An effective definition can be made in terms the crack face opening $\delta_n(r)$ a small distance r behind the tip according to

$$\tan\alpha = \frac{\delta_n(r) - \delta_n(0)}{r}. \tag{5.1}$$

In the present study we have taken $r = t$. Further discussion of α is deferred to the next section.

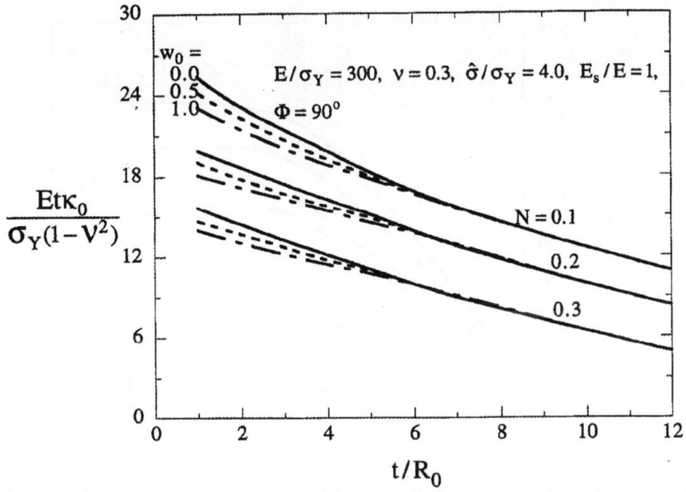

Figure 8. Variation of residual curvature κ_0 emerging from the active plastic zone at the interface crack tip as a function of film thickness for a moderately strong interface, $\widehat{\sigma}/\sigma_Y = 4$.

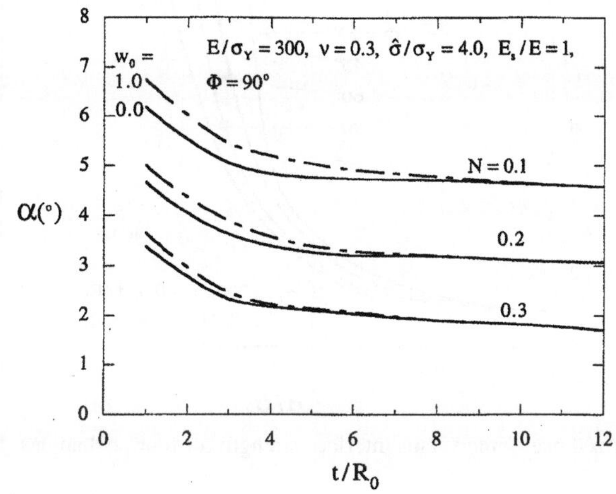

Figure 9. Variation of crack tip opening angle α at the interface crack tip as a function of film thickness for a moderately strong interface $\widehat{\sigma}/\sigma_Y = 4$.

5.2. The role of the peel angle Φ

Curves of the normalized peel force as a function of film thickness are shown in Figure 10 for four peel angles. The parameters characterizing the system are the same as those in Figure 4(a). If plasticity makes a significant contribution to the total work of fracture, then decreasing the peel angle decreases $P(1 - \cos\Phi)$ but increases the peel force magnitude P. For peel angles less than about 60°, a peak in the peel force exists at an intermediate film thickness (about $t/R_0 = 6$ for the case shown). The peak is even more prominent for larger values of w_0, corresponding to less relative reversed plastic bending.

The dependence of $P(1 - \cos\Phi)/\Gamma_0$ on the normalized interface strength $\widehat{\sigma}/\sigma_Y$ is displayed in Figure 11 for four peel angles ranging from 20° to 90°. These results were computed for relatively thick films ($t/R_0 = 10$), corresponding fairly closely to the asymptote for large t/R_0. The trend with interface strength is similar to that discussed for the normal peel force.

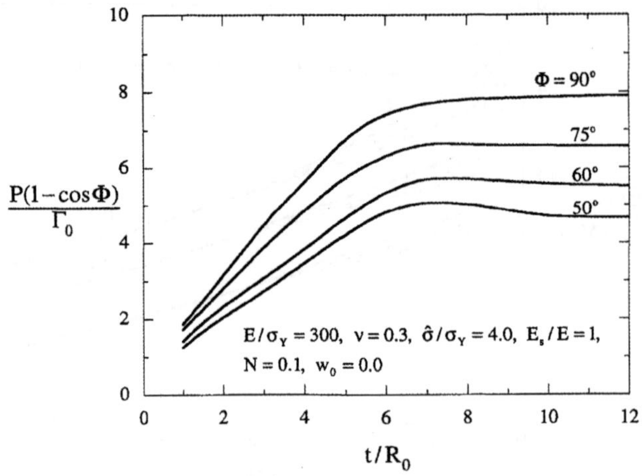

Figure 10. Normalized peel force versus film thickness for four peel angles Φ for a moderately strong interface with $\widehat{\sigma}/\sigma_Y = 4$.

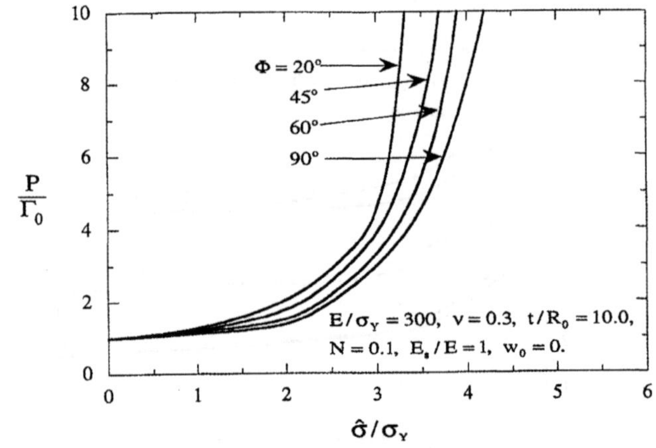

Figure 11. Normalized peel force versus interface strength for four peel angles Φ for $t/R_0 = 10$.

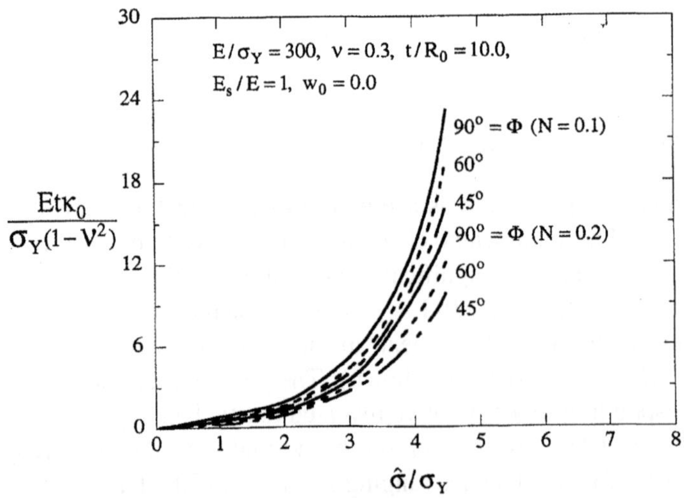

Figure 12. Residual curvature κ_0 emerging from the active plastic zone at the interface crack tip as a function of interface strength for various peel angles and strain hardening levels for $r/R_0 = 10$.

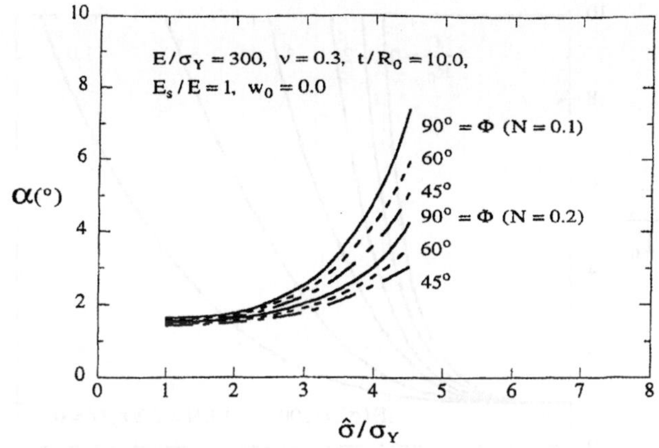

Figure 13. Crack tip opening angle α at the interface crack tip as a function of interface strength for various peel angles and strain hardening levels for $t/R_0 = 10$.

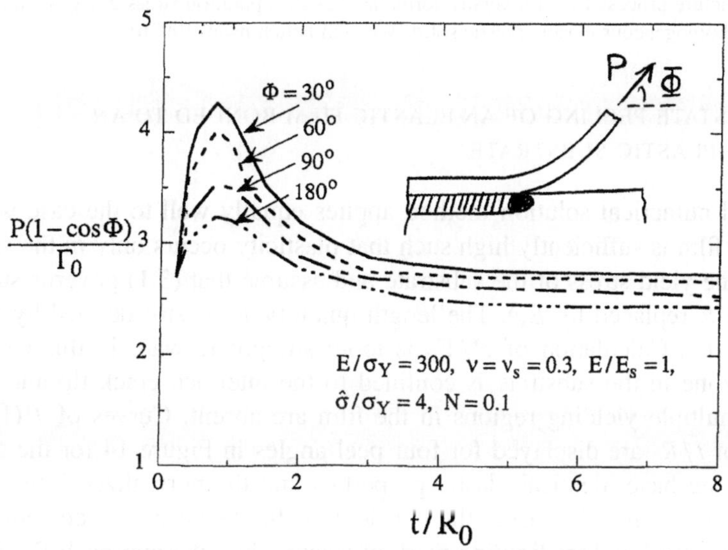

Figure 14. An elastic film peeled from an elastic–plastic substrate with yield stress σ_Y and strain hardening exponent $N = 0.1$. Normalized peel force versus film thickness for four peel angles Φ for a moderately strong interface with $\hat{\sigma}/\sigma_Y = 4$.

It can be seen, however, that the macroscopic work of fracture, $P(1 - \cos\Phi)$, decreases with decreasing peel angle at a given interface strength, with plasticity making less of a contribution. The peel force itself P is larger at a given interface strength, the smaller is the peel angle. The residual curvature emerging from the tip region κ_0 and the opening angle α are plotted as functions of the normalized interface strength in Figures 12 and 13 for three peel angles and two levels of strain hardening, again for normalized film thickness, $t/R_0 = 10$. The trend of these quantities with interface strength is similar to that of the peel force itself. At low interface strength, both quantities are independent of $\hat{\sigma}$, Φ and N. As $\hat{\sigma}/\sigma_Y$ increases, κ_0 and α increase sharply and become dependent on Φ and N.

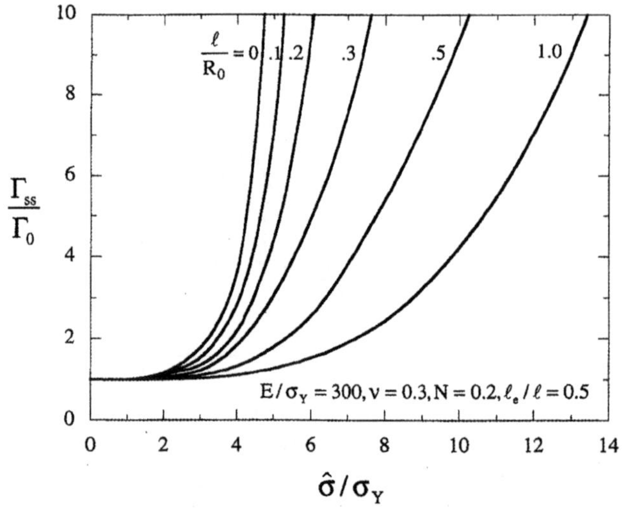

Figure 15. Influence of length parameter ℓ in strain gradient plasticity on the ratio of the total work of fracture to the work of the fracture process Γ_{ss}/Γ_0 versus normalized peak separation stress $\widehat{\sigma}/\sigma_Y$ for mode I, plane strain crack growth in a homogeneous elastic–plastic solid (Wei and Hutchinson, 1997b).

5.3. STEADY-STATE PEELING OF AN ELASTIC FILM BONDED TO AN ELASTIC–PLASTIC SUBSTRATE

The model and numerical solution method applies equally well to the case where the yield strength of the film is sufficiently high such that plasticity occurs only in the substrate. Now, let σ_Y denote the yield stress of the substrate and assume that (3.1) governs substrate tensile behavior (with E replaced by E_s). The length quantity R_0 is still defined by (3.3), but with E replaced by E_s. Calculation of P/Γ_0 is more straight forward in this case because the active plastic zone in the substrate is confined to the interface crack tip and complications arising from multiple yielding regions in the film are absent. Curves of $P(1 - \cos\Phi)/\Gamma_0$ as a function of t/R_0 are displayed for four peel angles in Figure 14 for the case where the film and substrate have identical elastic properties and the normalized interface strength is $\widehat{\sigma}/\sigma_Y = 4$. When the film is elastic, the plastic zone in the substrate does not shrink to zero when t becomes small and the limiting work of fracure does not approach Γ_0. A distinct peak in the peel force occurs at $t/R_0 \cong 1$; small scale yielding conditions are rapidly approached as t/R_0 increases. The ratio $P(1 - \cos\Phi)/\Gamma_0$ depends only weakly on the peel angle in the small scale yielding limit implying that there is relatively little variation in mode mixity on the peel angle.

6. Implications and limitations of the model

6.1. IMPLICATIONS

Aside from its apparent simplicity and popularity, the peel test does not provide an easy means of determining the interface adhesion energy Γ_0, except under conditions where plastic dissipation is negligible. The cohesive zone model brings out the combined roles of the two parameters characterizing the interface, Γ_0 and interface strength $\widehat{\sigma}$, in determining the peel force and the contribution of plastic deformation to the total work of fracture. The coupling is highly nonlinear. At high interface strength, as measured by $\widehat{\sigma}/\sigma_Y$, plastic dissipation can constitute a large fraction of the total work of fracture, such that the peel force is many times Γ_0. The model provides a computational means of partitioning the total work of fracture into the interface adhesion energy and plastic dissipation. As a practical matter, however, the task of backing out the adhesion energy from peel test data remains a major challenge.

The peel force necessarily scales with the interface adhesion energy, but a dependence on Γ_0 also enters through R_0 in (3.3) via the normalized film thickness parameter t/R_0. The film yield strength comes into play through R_0 and, most significantly, through the normalized interface strength, $\widehat{\sigma}/\sigma_Y$. In prior modeling efforts, other less fundamental quantities have been surrogate to the interface strength itself, such as the crack tip opening angle α. Nevertheless, it is clear from the numerical results presented here that the opening angle may be a useful alternative indicator of interface strength, in the sense that the larger is $\widehat{\sigma}/\sigma_Y$, the larger is α. From an observational perspective, the crack tip opening angle is far more assessable than the interface strength. Nevertheless, the two quantities are not equivalent. This is highlighted by the fact that the peel force *decreases* with increasing substrate to film modulus E_s/E at fixed $\widehat{\sigma}$, while the peel force *increases* with increasing E_s/E at fixed α (Kim et al., 1989).

The present model reveals that $P(1 - \cos\Phi)/\Gamma_0$ asymptotes to a constant value for films of sufficiently large thickness. It has been noted that this is a small scale yielding limit in the sense that the active plastic zone has a height that is small compared to the film thickness. Moreover, the active plastic zone is confined to the interface crack tip and reversed plastic bending does not occur in this limit such that the macroscopic work of fracture is identical to that for an interface crack under small scale yielding at the same mode mixity. The elevation of the peel force over the adhesion energy depends on the other parameters of system, and in particular on the interface strength. Models which represent the entire film by a beam, such as those of Kim and Kim (1988) and Kim et al. (1989), are unable to capture the elevation of the peel force due to plasticity in the limit of thick films. They predict that $P(1 - \cos\Phi)/\Gamma_0$ approaches unity as t/R_0 becomes large (cf. Kim and Kim, 1988). A full 2D analysis of the crack tip region such as that employed in the present approach is required to delineate role of plastic dissipation in this important limit.

6.2. LIMITATIONS

As it stands, the model is limited to systems whose normalized interface strength $\widehat{\sigma}/\sigma_Y$ is not larger than about 4 to 6, depending on strain hardening (cf. Figure 7). Larger values of $\widehat{\sigma}/\sigma_Y$ would be expected for metal films which bond strongly to ceramic substrates (cf. Kim et al., 1989), and possibly even for certain classes of polymer films bonded to metal or ceramic substrates. The conventional plasticity theory employed in the model (J_2 flow theory) does not give rise to sufficient stress elevation at the crack tip to produce separation when the normalized interface strength is larger than about 4 to 6, corresponding to values of $P(1 - \cos\Phi)/\Gamma_0$ larger than about 10 to 20. As it stands, the model predicts that peeling will not take place when the interface is very strong. For metals, there are compelling reasons to believe that conventional plasticity theory significantly underestimates hardening and stresses when the gradients of plastic strain are large as, for example, at a crack tip (Fleck and Hutchinson, 1997). More and more experimental evidence is accruing indicating that, when nonuniform deformation occurs on the micron scale, elevations in tractions occur which are several times those observed at larger scales at equivalent strain levels. Stress elevation fosters interface separation at the tip of a crack on a strong interface.

A quantitative study of the effect of stress elevation on plane strain, Mode I crack growth in homogeneous metals was carried out by Wei and Hutchinson (1997b) using the same type of cohesive zone model employed here. However, in place of J_2 flow theory, these authors used a strain gradient theory of plasticity (Fleck and Hutchinson, 1997) which incorporates a length parameter ℓ characterizing the scale at which strain gradient hardening becomes important. Figure 15, from Wei and Hutchinson (1997), displays a set of results for the ratio of the total work of fracture Γ_{ss} to the work of the fracture process Γ_0 as a function of normalized interface strength for various levels of the new nondimensional length parameter ℓ/R_0. These results apply to steady-state propagation in small scale yielding. The reference length R_0 is the same as defined here in (3.3). The limiting curve for $\ell/R_0 = 0$ corresponds to the behavior predicted using the conventional plasticity theory J_2 flow theory. Values of ℓ/R_0 in the range from 0.1. to 1 produce crack tip stress elevation and give rise to crack propagation at much larger normalized separation strengths $\widehat{\sigma}/\sigma_Y$ than can be encompassed by the conventional plasticity theory.

The implications of these results for strong interfaces become apparent when one considers values of the reference length R_0. For relatively strong metal/ceramic interfaces failing by atomic decohesion, representative values of R_0 in (3.3) typically fall in the range from about one tenth to several microns. From available experimental data, the length parameter ℓ appears to be on the order of a micron for many metals (Fleck and Hutchinson, 1997). Thus, it is evident from Figure 15 that strain gradient effects should have a profound influence on the relationship between the total work of fracture and the work of adhesion for strong metal/ceramic interfaces. For less strong interfaces and for systems for which ℓ/R_0 is small, the present model based on conventional plasticity theory should be adequate.

Acknowledgment

This work was supported in part by the ONR through grant N00014-96-10059, by the NSF through grant NSF-CMS-96-34632, and by the Division of Engineering and Applied Sciences, Harvard University.

References

Dean R.H. and Hutchinson, J.W. (1980). Quasi-static steady crack growth in small scale yielding. In *Fracture Mechanics*, ASTM STP 700, American Society for Testing Materials, 383–405.

Fleck, N.A. and Hutchinson, J.W. (1997). Strain gradient plasticity. *Advances in Applied Mechanics* (Edited by J.W. Hutchinson and T.Y. Wu) **33**, 295–361.

Kim, J., Kim, K.-S. and Kim, Y.H. (1989). Mechanical effects of peel adhesion test. *Journal of Adhesion Science and Technology* **3**, 175–187.

Kim, K.-S. and Aravas, N. (1988). Elasto–plastic analysis of the peel test. *International Journal of Solids and Structures* **24**, 417–435.

Kim, K.-S. and Kim, J. (1988). Elasto–plastic analysis of the peel test for thin film adhesion. *Journal of Engineering Materials and Technology* **110**, 266–273.

Kinloch, A.J., Lau, C.C. and Williams, J.G. (1994). The peeling of flexible laminates. *International Journal of Fracture* **66**, 45–70.

Parks, D.M., Lam, P.S. and McMeeking, R.M. (1981). Some effects of inelastic constitutive models on crack tip fields in steady crack growth. *Advances in Fracture Research* (Edited by D. Francois), Pergamon Press, **5**, 2607–2614.

Tvergaard, V. and Hutchinson, J.W. (1992). The relation between crack growth resistance and fracture process parameters in elastic–plastic solids. *Journal of the Mechanics and Physics of Solids* **40**, 1377–1397.

Tvergaard, V. and Hutchinson, J.W. (1993). The influence of plasticity on mixed mode interface toughness. *Journal of the Mechanics and Physics of Solids* **41**, 1119–1135.

Wei, Y. and Hutchinson, J.W. (1997a). Nonlinear delamination mechanics for thin films. *Journal of the Mechanics and Physics of Solids* **45**, 1137–1159.

Wei, Y. and Hutchinson, J.W. (1997b). Steady-state crack growth and work of fracture for solids characterized by strain gradient plasticity. *Journal of the Mechanics and Physics of Solids* **45**, 1253–1273.

International Journal of Fracture **93:** 335–349, 1998.

Predicting crack initiation in composite material systems due to a thermal expansion mismatch

E.S. FOLIAS[1], M. HOHN[1] and T. NICHOLAS[2]
[1]*Department of Mathematics, University of Utah, Salt Lake City UT 84112, USA; e-mail: folias@math.utah.edu*
[2]*Wright Laboratory Materials Directorate, Wright – Paterson Air Force Base, Ohio 45433-7817, USA*

Received 14 January 1998; accepted in revised form 26 August 1998

Abstract. Residual stresses due to curing and thermal stresses due to differences between the thermal expansion coefficients of the matrix and fiber may have a major effect on the micro-stresses within a composite material system and must be added to the stresses induced by the external mechanical loads. Such microstresses are often sufficient to produce micro-cracking even in the absence of external loads, example during the cooling process.

In this investigation, a micro-mechnics approach is used in which the fibers of a composite material system are modeled as cylindrical inclusions that are embedded into a matrix plate. The model is then used to predict, analytically, the residual stresses due to a thermal expansion mismatch, e.g. during a cooling process. Additionally, some critical effects due to a load transverse to the direction of the fibers are examined. The analysis provides a better understanding of how residual stresses are developed and how they may be controlled in material systems where small strains are present.

Moreover, the results are used to identify locations of possible crack failure and to derive a fracture criterion for crack initiation at the local level. Comparison with experimental evidence for matrix cracking in intermetallic composites caused by thermal expansion mismatch shows a good agreement.

Key words: Residual stresses, thermal fracture, composites, micromechanics.

1. Introduction

In order to achieve revolutionary advances in turbopropulsion-powered systems for the next century, advanced high-temperature ceramic matrix composites will be required. Yet, fabrication of these advanced materials is still being accomplished by traditional trial-and-error approaches. Rarely such an approach, however, leads to an optimized material process that yields components of a high quality. Perhaps material modeling coupled with numerical simulation may be an intelligent approach to material processing which ultimately may lead to lower manufacturing costs.

Thus, if rational designs of ceramic composites are to be made, their performance under static, dynamic, thermally fatigued and environmental loads need to be predictable. The first step towards this goal is the realization that the ultimate failure, as well as many other aspects of the composite behavior, is the result of the growth and accumulation of microdamage to the fibers, matrix and their interfaces. Thus, it appears that any generally successful model of performance and failure must incorporate the effects of this damage at the microlevel. Coupling, therefore, between the micro-mechanical with that of the macro-mechanical behavior will be very helpful in developing rational methods for the prediction of the structural life span of such materials. This certainly represents a challenge.

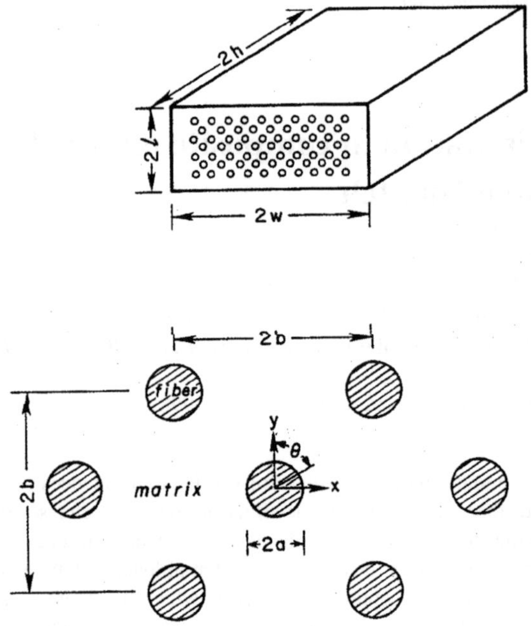

Figure 1.

For example, the residual stresses due to curing and thermal stresses due to differences between the thermal expansion coefficients of the matrix and fiber may have a major effect on the microstresses within a composite material system and must be added to the stresses induced by the external mechanical loads. Such microstresses are often sufficient to produce microcracking even in the absence of external loads, e.g. during the cooling process. Furthermore, if the material system is thermally fatigued, these residual stresses may cause some of the existing micro-cracks to grow and coalesce and thus form the presence of larger cracks.

In this investigation, a systematic 3D, micromechanics approach is used in which the fibers of a composite material system are modeled as cylindrical inclusions that are embedded into a matrix plate. The analytical model is then solved and the results are used to predict, the residual stresses due to a thermal expansion mismatch. The model provides a better understanding of how the residual stresses are being developed and how they may be controlled particularly in relation to ceramic materials where there is no ductility present to accommodate plastic deformation.

The analysis reveals the dependence of the residual stress field on the fiber volume fraction ratio, identifies the critical locations where a crack is most likely to initiate and subsequently propagate, recovers the interface shear stress profile and provides important information and guidance to material designers for the pre-selection of fiber and matrix materials in order to alleviate some of the residual stresses.

2. Mathematical model

Consider, for example, an infinite plate matrix which consists of material Beta21 (i.e. a titanium alloy), see Figure 1. The matrix plate is assumed to extend to infinity both in the x- and y-directions. In order to capture any possible 3D effects that may be present, the matrix plate is assumed to have a finite dimension $2h$, along the z-direction. A uniform and square periodic arrangement of cylindrical fibers, for example SCS-6 (i.e. carbon-coated silicon carbide), is embedded into the matrix plate in the directions of the x- and y-axes. Although in the present study we are primarily interested in the residual stresses due to the thermal expansion mismatch, the material system is also vulnerable, in certain regions, to loadings that are transverse to the fibers. For this reason two different types of loads are being considered:

(i) a uniform temperature load ΔT (cooling) that is applied throughout the material system and

(ii) a uniform transverse load σ_0 perpendicular to the direction of the fibers and along the y-direction.

Such consideration will identify some of the critical locations of the material system that material designers should be aware of. Both fiber and matrix materials are assumed to be homogeneous and linearly elastic. While it is true that for this particular composite system the matrix is rather elastoplastic, in the present study we restrict the analysis to small strains. Moreover, this particular material system is used only as a vehicle to show some of the details of the model analysis.

The governing equations, are the well known Navier's equations coupled with the Energy Balance equation. More specifically,

$$\frac{1}{1-2\nu}\frac{\partial u}{\partial x} + \nabla^2 u - \frac{2(1+\nu)}{1-2\nu}\frac{\partial T}{\partial x} = 0, \tag{1}$$

$$\frac{1}{1-2\nu}\frac{\partial v}{\partial y} + \nabla^2 v - \frac{2(1+\nu)}{1-2\nu}\frac{\partial T}{\partial y} = 0, \tag{2}$$

$$\frac{1}{1-2\nu}\frac{\partial w}{\partial z} + \nabla^2 w - \frac{2(1+\nu)}{1-2\nu}\frac{\partial T}{\partial z} = 0, \tag{3}$$

$$\nabla^2 T = 0. \tag{4}$$

Where ∇^2 represents the Laplacian operator, T the temperature and ν Poison's ratio. The stress field can be obtained from the stress strain equations:

$$\sigma_{ij} = \lambda\delta_{ij}\varepsilon_{kk} + 2G\varepsilon_{ij} - \alpha(3\lambda + 2G)(T - T_o)\delta_{ij}. \tag{5}$$

In this paper, ΔT = constant, which implies that all temperature derivatives with respect to space are zero.

As to boundary conditions, we require that:

(i) the appropriate stresses do vanish at the free edges, i.e.

$$\text{at } |z| = h, \quad \sigma_{zz}^{(j)} = 0, \quad \tau_{rz}^{(j)} = 0, \quad \tau_{\theta z}^{(j)} = 0; \quad j = f, m \tag{6–8}$$

(ii) perfect bonding is assumed to prevail at the fiber/matrix interface, i.e.

$$\text{at } r = a, \quad u_r^{(f)} = u_r^{(m)}, \quad v_\theta^{(f)} = v_\theta^{(m)}, \quad w^{(f)} = w^{(m)}, \tag{9–11}$$

$$\sigma_{rr}^{(f)} = \sigma_{rr}^{(m)}, \quad \tau_{rz}^{(f)} = \tau_{rz}^{(m)}, \quad \tau_{\theta r}^{(f)} = \tau_{\theta r}^{(m)}, \tag{12–14}$$

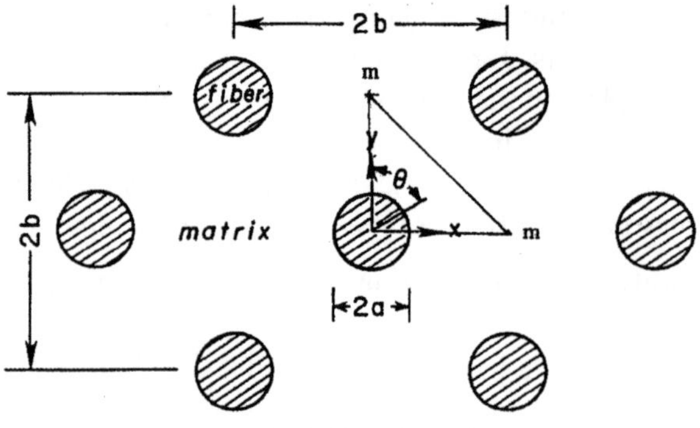

Figure 2. Cell configuration.

(iii) finally the boundary conditions within the cell configuration[1]. (See Figure 2), i.e. on the plane mm, $u_n^{(m)} = 0$, (15)

$$\tau_{mn}^{(m)} = 0, \qquad \tau_{mz}^{(m)} = 0, \qquad \int_{mm} F_x \, ds / A_{mm} = \sigma_0, \qquad \int_{mm} F_y \, ds = 0, \qquad (16\text{–}17)$$

where F_x, F_y are the resultant force components in the x- and y-direction respectively and A_{mm} the cross-sectional area of the plane mm. Moreover, the planes $x = 0$ and $y = 0$ are symmetry planes and, therefore, the solution must satisfy the symmetry boundary conditions.

3. The 3D displacement field

Without going into the mathematical details, the general 3D complementary displacement field satisfying boundary conditions (6–8) has already been constructed by Folias (1976) and the results may be expressed as:

$$\begin{aligned} u^{(c)(j)} &= \frac{1}{m_j - 2} \sum_{\nu=1}^{\infty} \frac{\partial H_\nu^{(j)}}{\partial x} \{2(m_j - 1) f_1(\beta_\nu z) + m_j f_2(\beta_\nu z)\} \\ &+ \sum_{n=1}^{\infty} \frac{\partial H_\nu^{(j)}}{\partial y} \cos(\alpha_n h) \cos(\alpha_n z) + l_1^{(j)} - y \frac{\partial l_3^{(j)}}{\partial x} + \frac{1}{m_j + 1} z^2 \frac{\partial^2 l_3^{(j)}}{\partial x \partial y}, \end{aligned} \qquad (18)$$

$$\begin{aligned} v^{(c)(j)} &= \frac{1}{m_j - 2} \sum_{\nu=1}^{\infty} \frac{\partial H_\nu^{(j)}}{\partial y} \{2(m_j - 1) f_1(\beta_\nu z) + m_j f_2(\beta_\nu z)\} \\ &- \sum_{n=1}^{\infty} \frac{\partial H_n^{(j)}}{\partial x} \cos(\alpha_n h) \cos(\alpha_n z) + \frac{3m_j - 1}{m_j + 1} l_3^{(j)} + l_2^{(j)} \\ &- y \frac{\partial l_3^{(j)}}{\partial y} - \frac{1}{m_j + 1} z^2 \frac{\partial^2 l_3^{(j)}}{\partial x^2}, \end{aligned} \qquad (19)$$

[1] The reader should note that displacements are unique up to an additive constant.

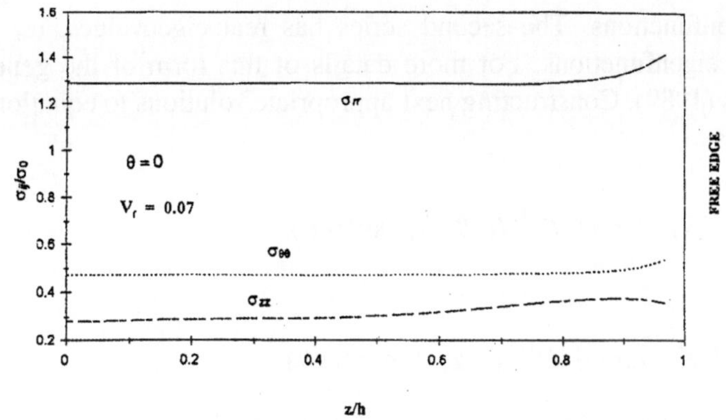

Figure 3a. Interface matrix stresses as a function of z/h and $a/h = 0.10$.

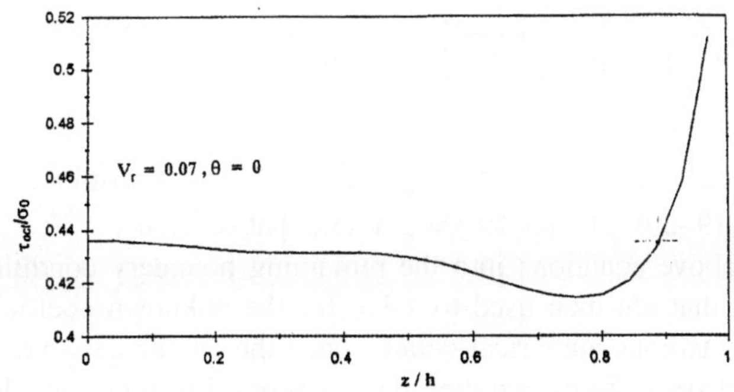

Figure 3b. Interface matrix octahedral shear stress as a function of z/h and $a/h = 0.10$.

$$w^{(c)(j)} = \frac{1}{m_j - 2}\sum_{\nu=1}^{\infty}\frac{\partial H_\nu^{(j)}}{\partial z}\{-2(m_j-1)f_1(\beta_\nu z) + m_j f_2(\beta_\nu z)\}$$

$$-\frac{1}{m_j+1}z\frac{\partial l_3^{(j)}}{\partial y}, \tag{20}$$

where $m = 1/\nu$, and

$$f_1(\beta_\nu z) = \cos(\beta_\nu h)\cos(\beta_\nu z), \tag{21}$$

$$f_2(\beta_\nu z) = (\beta_\nu h)\sin(\beta_\nu h)\cos(\beta_\nu z) - (\beta_\nu z)\cos(\beta_\nu h)\sin(\beta_\nu z), \tag{22}$$

$$\left(\frac{\partial^2}{\partial x^2} + \frac{\partial^2}{\partial y^2} - \beta_\nu^2\right)H_\nu^{(j)} = 0, \tag{23}$$

$$\left(\frac{\partial^2}{\partial x^2} + \frac{\partial^2}{\partial y^2} - \alpha_n^2\right)H_n^{(j)} = 0, \tag{24}$$

and where l_1, l_2, l_3 are 2D harmonic functions. Perhaps it is appropriate here to note that the displacement fields are expressed in terms of two infinite series the first of which has complex eigenvalues β_ν, which are the roots of the equation

$$\sin(2\beta_\nu h) = -(2\beta_\nu h), \tag{25}$$

and complex eigenfunctions. The second series has real eigenvalues, $\alpha_n = n\pi/h$, $n = 1, 2\ldots$, and real eigenfunctions. For more details of this form of the general solution see Penado and Folias (1989). Constructing next appropriate solutions to equations (23–24) in the polar form one has:

$$H_\nu^{(j)} = \sum_{k=0}^{\infty}\{A_k^{(j)}K_k(\beta_\nu r) + B_k^{(j)}I_k(\beta_\nu r)\}\exp(ik\theta) \tag{26}$$

$$H_n^{(j)} = \sum_{k=0}^{\infty}\{C_k^{(j)}K_k(\alpha_n r) + D_k^{(j)}I_k(\alpha_n r)\}\exp(ik\theta) \tag{27}$$

and for the 2D harmonic functions

$$l_l^{(j)} = \sum_{n=-\infty}^{\infty}\left\{c_{nl}^{(j)}\left(\frac{r}{a}\right)^n + d_{nl}^{(j)}\left(\frac{r}{a}\right)^{-n}\right\}\exp(in\theta); \quad l = 1, 2, 3, \tag{28}$$

where the constants $A_k^{(j)}$, $B_k^{(j)}$, $C_k^{(j)}$, $D_k^{(j)}$, $c_{nl}^{(j)}$, $d_{nl}^{(j)}$ are to be determined from the remaining boundary conditions (9–20). The reader should note that constants $A_k^{(j)}$, $B_k^{(j)}$ are complex.

Substituting the above equations into the remaining boundary conditions we arrive at a system of equations that are then used to solve for the unknown coefficients. Without going into the long and tedious numerical details, once the coefficients have been determined, the displacement and stress fields can then be recovered. For more details of the numerical approach, see Penado and Folias (1989).

4. Uniform loading transverse to the fibers ($\Delta T = 0$)

Although in the present study we are primarily interested in the residual stresses due to the thermal expansion mismatch, the material system is also vulnerable, in certain regions, to loadings that are transverse to the fibers in which case the effects may be even more pronounced. For this reason, we give some of the important results due to a transverse loading

(i) *Interior stress field*

For a constant transverse loading only along the direction of the x-axis (i.e. $\Delta T = 0$), the 3D stress field at the interface, for a volume fraction ratio of $V_f = 0.07$, was found to be constant all along the interior (see Figure 3(a)) and that as one approaches the free surface a boundary layer was observed to prevail, where the stresses increase rather rapidly. This suggests, therefore, that a stress singularity may be present in this region. It may also be noted that for fiber volume fractions of $V_f = 0.07$ or less, all interactions between fibers have for all practical purposes subsided.

There are four important 3D characteristics that one can draw from this figure. First, the width of this boundary layer is, approximately, two fiber diameters away from the free edge. Second, the amplitude of the stresses at the center of the fiber length is, in general, a function of the ratio of fiber diameter/fiber length. If, however, that ratio happens to be less than or equal to 1/10, then the magnitude of the stresses is precisely that of plane strain. Third, for ratios between 1/10 and 10 a state of 'pseudo plane strain' condition prevails whereby the amplitude

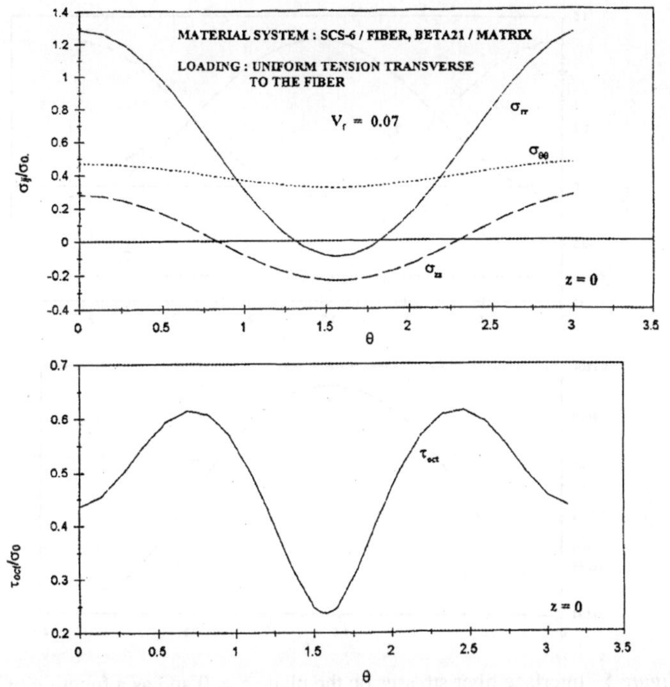

Figure 4. Interface matrix stresses on the plane $z = 0$ and as a function of θ.

of the stresses now increases, even on the plane $z = 0$, by as much as 10–12 percent. The term pseudo is used because away from the boundary layer, the condition

$$\sigma_{zz} = \nu(\sigma_{rr} + \sigma_{\theta\theta}) \tag{29}$$

is still satisfied even though the stresses are functions of the variable z. Fourth, for ratios greater than 10 the magnitude of the stresses is precisely that of plane stress. Finally, similar stress profiles exist for V_f ratios greater than 0.07, where the magnitude now is a function of V_f.

Figures 3(a,b) depict the profile of the interface stresses along the line $\theta = 0$ as a function of the ratio z/h. The numerical results are specialized for the material system: SCS-6 / fibers, Beta21 / matrix. Figures 4 and 5, show typical interface stress profiles of the matrix and fiber on the plane $z = 0$, and as functions of the angle θ.

(ii) *Edge stress field*

As it was previously noted, in the neighborhood of the free surface e.g. the edge of the plate or in the vicinity of crack bridging (see Figure 6), there may very well be present a stress singularity. By utilizing a local 3D asymptotic analysis one can substantiate the presence of a weak stress singularity. Complete details of this analysis can be found in the work of Folias (1989). Without going into the mathematical details, a summary of the results, for room temperature and for the material system discussed, is given below.

The local stress field is given by

$$\sigma_{ik}^{(j)} = \rho^{-\alpha} F_{ik}^{(j)}(\theta, \phi), \tag{30}$$

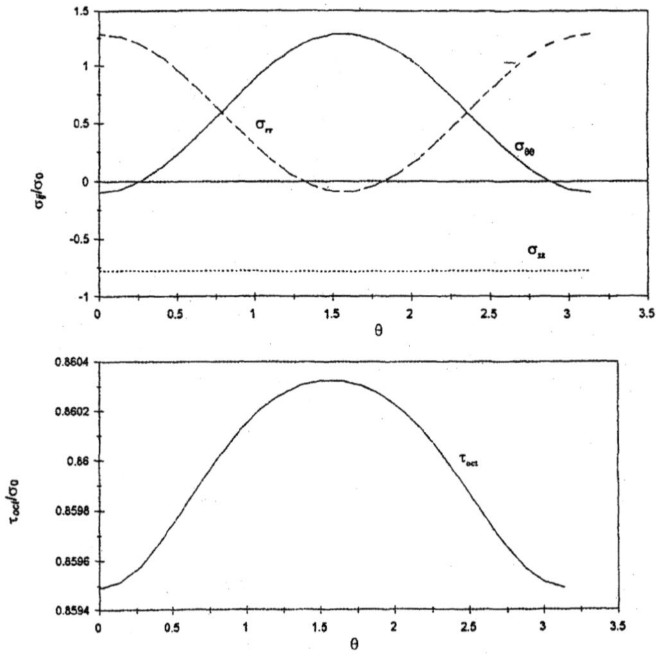

Figure 5. Interface fiber stresses on the plane $z = 0$ and as a function of θ.

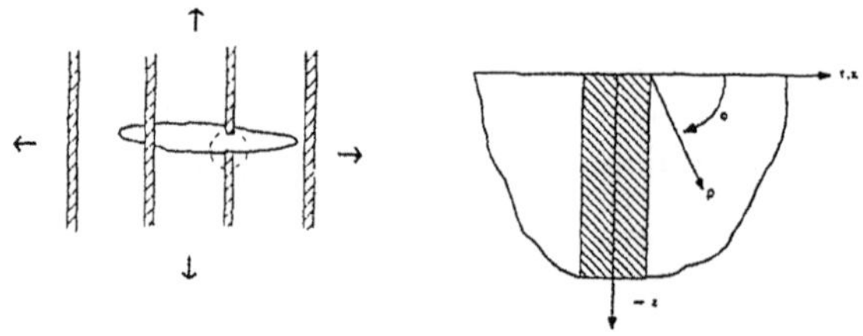

Figure 6. Fiber meeting a free surface.

where the $F_{ik}^{(j)}$ are explicit functions of the angles θ and ϕ and the stress singularity α is a function of the material constants. For a titanium matrix and SCS fibers it is found that

$$\alpha = 0.110 \quad \text{at room temperature} \tag{31a}$$

$$\alpha = 0.190 \quad \text{at } 900°\text{C}. \tag{31b}$$

While, in order to determine the residual stresses at room temperature it is only necessary to use the material constants of the system at room temperature, the material constants for 900°C were also used in order to see the difference of the stress singularity exponent α. The reader however, should be cautioned that this calculation is based on linear elasticity and on the presence of a constant temperature field.

Moreover, at $\phi = \frac{1}{2}\pi$ and for $G_f/G_m = 3.608$ it is found that

$$\sigma_{rr}^{(mm)} = -11.219\rho^{-\alpha}B^{(f)}, \tag{32a}$$

$$\sigma_{\theta\theta}^{(mm)} = -4.823\rho^{-\alpha}B^{(f)}. \tag{32b}$$

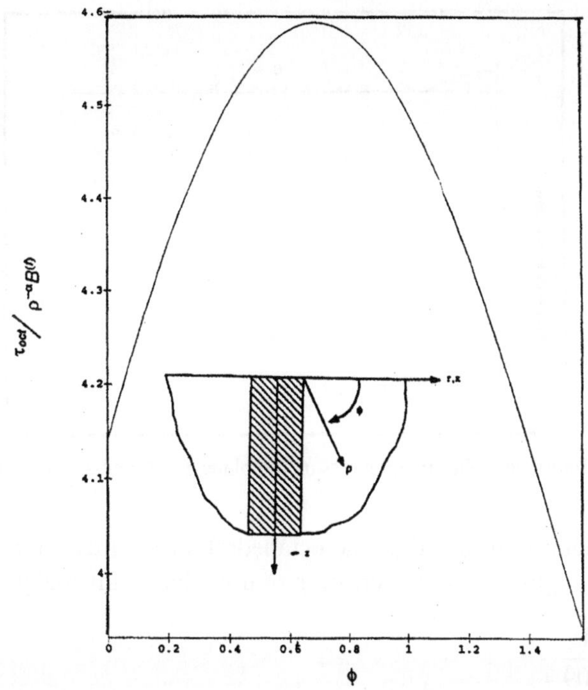

Figure 7. The octahedral shear stress in the matrix and close to the edge where the fiber meets the free surface.

Similarly, at $\phi = 0$ it is found that

$$\sigma_{rr}^{(m)} = -9.970\rho^{-\alpha} B^{(f)}, \tag{33a}$$

$$\sigma_{\theta\theta}^{(m)} = -3.391\rho^{-\alpha} B^{(f)}, \tag{33b}$$

where $B^{(f)}$ stands for an arbitrary constant. The octahedral shear stress is plotted in Figure 7 where it is noted that its maximum occurs at $\theta = 40°$.

The following observations are also worthy of note. First, as the ratio of the shear moduli increases, so is the stress singularity. This is compatible with our physical expectations and within the assumptions of our theory. Second, all things being equal, at the edge the controlling stress for failure is the radial stress particularly at the location $\phi = 0$ and $\theta = 0$. Perhaps it is also appropriate here to note that the carbon coatings on the SCS-6 fibers produce a particularly weak interface. However, the authors would like to emphasize that this material system is only used as a vehicle to show how the details of this analysis work.

	Room Temp.	900°C
at $\phi = 0$:	$\dfrac{\sigma_{rr}^{(m)}}{\sigma_{\theta\theta}^{(m)}} = 2.94,$	$\dfrac{\sigma_{rr}^{(m)}}{\sigma_{\theta\theta}^{(m)}} = 2.94,$
at $\phi = \frac{1}{2}\pi$:	$\dfrac{\sigma_{rr}^{(m)}}{\sigma_{\theta\theta}^{(m)}} = 2.33,$	$\dfrac{\sigma_{rr}^{(m)}}{\sigma_{\theta\theta}^{(m)}} = 2.18.$

The reader may notice that there is a very small difference in the ratios.

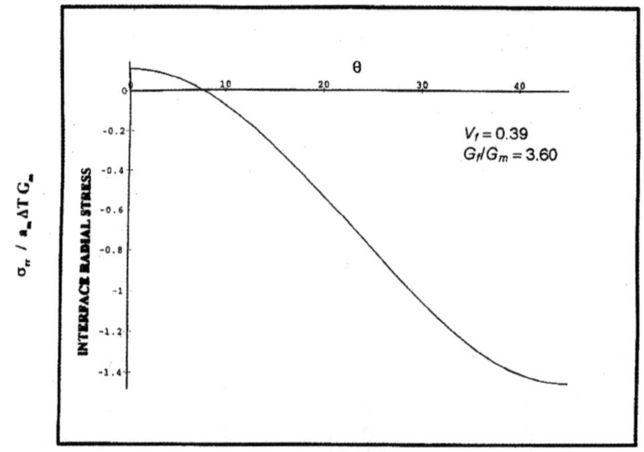

Figure 8. Interface radial matrix stress on the plane $z = 0$ and as a function of θ.

Similarly, in the vicinity of the edge, the octahedral shear stress attains a maximum at an angle of $\phi = 40°$ (see Figure 7). It is interesting to note that in this neighborhood, the ratio of

$$\frac{\tau_{\text{octmax}\,\text{roomtemp}}}{\tau_{\text{octmax}\,900\text{C}}} = \frac{1}{2.09}$$

which suggests that, at high temperatures, the material will undergo substantial plastic deformation in this region if subjected to a load transverse to the fibers.

Thus, a mode I and III crack failure may initiate at the edge and along the interface, particularly when the material system is subjected to a transverse (to the fibers) load in which case it becomes more pronounced.

5. Residual stresses due to ΔT ($\sigma_0 = 0$)

Residual stresses due to curing and thermal stresses due to differences between the thermal expansion coefficients of the matrix and fiber may have a major effect on the microstresses within a composite material system and must be added to the stresses induced by the external mechanical loads. Such microstresses are often sufficient to produce microcracking even in the absence of external mechanical loads, for example during a cooling process.

In this section we let the applied load $\sigma_0 = 0$ and furthermore assume ΔT to be a constant. The analytical model is then used to predict, the residual stresses due to the thermal expansion mismatch between the fiber and matrix.

Without going into the mathematical details, we consider a composite material system consisting of SCS-6 fibers which are embedded into a beta21 matrix plate and the entire system is then exposed to an environment of a uniform cooling temperature ΔT. While it is true that the material constants do change continuously as a function of the temperature, the thermal coefficients appear in the solution as a ratio, which ratio changes very little. On the other hand, the ratio of the shear moduli changes considerably as the temperature varies. Consequently, the results are very much dependent on the material properties which one uses. Thus, if one bases the analysis on the shear moduli ratio at room temperature, the following stress profiles are recovered at the fiber/matrix interface. Figure 8 depicts, for a $V_f = 0.39$, the radial matrix stress on the plane $z = 0$ and as function of the angle θ. It is noted that the radial

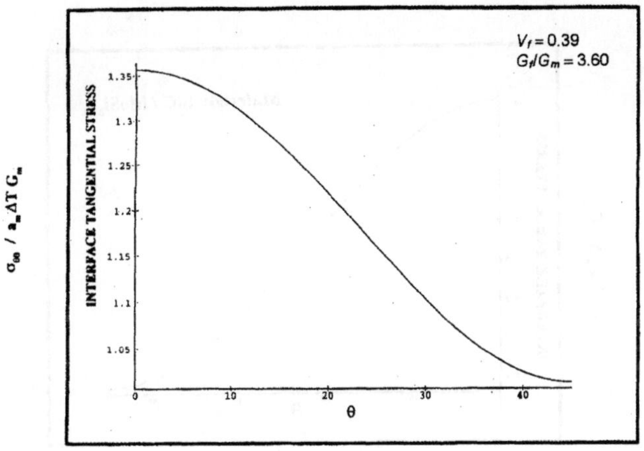

Figure 9. Interface $\sigma_{\theta\theta}$ matrix stress on the plane $z = 0$ and as a function of θ.

stress is compressive. Similarly, the tangential stress is tensile in nature and its maximum occurs at the location $\theta = 0$ (see Figure 9). In general, the location of this maximum is a function of the material properties and particularly of the shear moduli ratio. Moreover, in this analysis perfect bonding was assumed to prevail at the fiber/matrix interface. If, however, we were to relax the conditions at the interface whereby we allow slippage to occur, then the maximum will occur elsewhere. More specifically in this case it occurs at $\theta = 45°$.

Examining next the possibility of matrix cracking, it becomes evident from the above that no cracking will occur in the matrix for $\Delta T = 900°$C. This matrix material is too strong for preexisting microcracks to grow. Examination of the σ_{zz} stress also shows that no cracks will develop in that direction either. The results are in line with those obtained by Kroupa (1994) based on a finite element analysis.

Finally, it should be noted that the effect of the shear moduli ratio on the interface stresses can be substantial. This can be seen by the following comparison of the tangential interface stress when using three different shear moduli ratios that reflect three different and discrete temperature levels

V_f	$\dfrac{\sigma_{\theta\theta}}{\alpha_m G_m \Delta T}$	$\dfrac{\sigma_{\theta\theta}}{\alpha_m G_m \Delta T}$	$\dfrac{\sigma_{\theta\theta}}{\alpha_m G_m \Delta T}$
	Room Temp.	*Mid Temp.*	*High Temp.*
0.39	1.36	1.59	1.64

Although it would be desirable to have a program in which the material properties can vary continuously with temperature, one can compensate by taking the results corresponding to the high shear moduli ratio. The thermal expansion coefficients on the other hand appear as a ratio which ratio does not vary appreciably to make any significant differences.

The variation of the normalized tangential interface stress as a function of the fiber volume fraction for this material, is almost linear and may be approximated with the equation

$$\frac{\sigma_{\theta\theta}}{\alpha_m G_m \Delta T} = 0.92 + 1.02V_f + 0.28V_f^2. \tag{34}$$

In view of the above, this matrix will not exhibit any cracking as a result of the residual stresses which are developed during the cooling process. Experimental evidence showed no such cracking either.

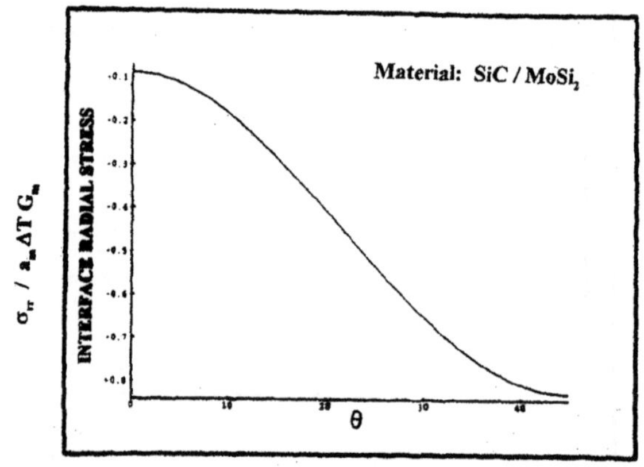

Figure 10. Interface radial matrix stress on the plane $z = 0$ and as a function of θ.

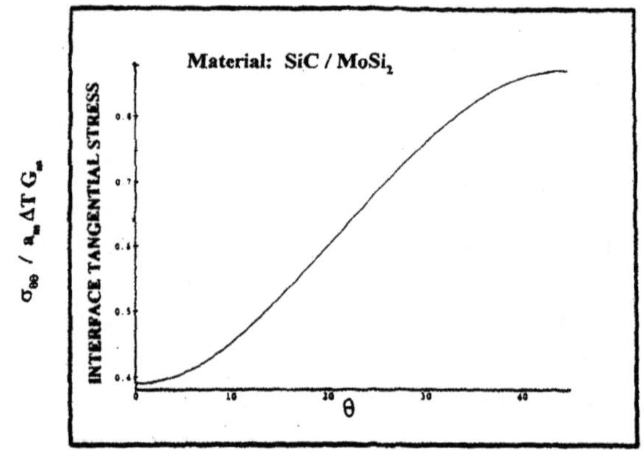

Figure 11. Interface $\sigma_{\theta\theta}$ matrix stress on the plane $z = 0$ and as a function of θ.

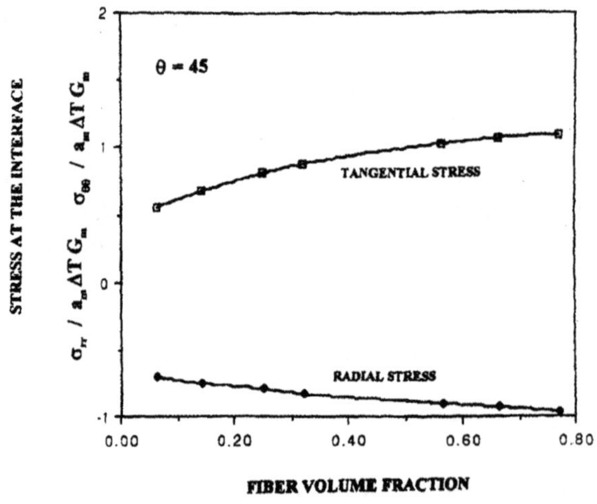

Figure 12. Normalized interface stresses on the plane $z = 0$ and as a function of V_f.

As a second physical example, we consider the material system consisting of SiC fibers embedded into an $MoSi_2$ matrix. The material system is characterized by the material properties $\nu_f = \nu_m = 0.25$, and $G_f/G_m = 1.05$, $\alpha_f/\alpha_m = 0.56$. Again omitting the mathematical details for a $V_f = 0.49$, the interface tangential, and radial stresses, on the plane $z = 0$ and as functions of the angle θ, are given in Figures 10–11, respectively. The reader may notice here that the maximum value of the tangential stress occurs at $\theta = 45°$, while the radial stress is compressive with a maximum absolute at $\theta = 0$. Finally, in Figure 12, we plot the variation of the maximum value of the tangential and radial stresses as a function of V_f. The following remarks are worthy of note

(i) for a decrease in temperature (e.g. during cooling), $\sigma_{\theta\theta}$ is the controlling stress,
(ii) for an increase in temperature (e.g. operating temperature), σ_{rr} is the controlling stress.

In general, the location of the maximum interface stress depends heavily on the material properties and not on the fiber spacing. This suggests, therefore, that all things being equal, during the cooling process a crack is most likely to develop at the interface and at $\theta = 45°$ and then advance into the matrix until it reaches the adjacent fiber. Alternatively, during the operating temperature the crack is most likely to develop along the interface and at $\theta = 0°$ and then advance along the interface towards the position $\theta = 45°$. Moreover, as the volume fraction ratio increases, the stresses increase in magnitude in the absolute value sense.

6. Fracture criterion for crack initiation

In view of our previous discussion, it is clear that the model may be used

(i) to identify the location where cracks are most likely to initiate and propagate, and
(ii) to identify the controlling stress that governs the failure.

Thus, if one assumes the presence of a small crack, of length c, in the matrix and adjacent to the interface, and along the direction of $\theta = 45°$, it is now possible to derive an approximate fracture criterion for crack initiation with which one can estimate the magnitude of the critical tangential stress that may cause the matrix to crack along this direction. More specifically (Hellan, 1984)

$$\{\sigma_{\theta\theta}^{(m)}\}_{\text{critical}} 1.12\sqrt{\pi c}\left\{1 - 0.15\frac{c}{b\left(1 - \sqrt{2}\frac{a}{b}\right)}\right\} = K_c, \tag{35}$$

where K_c is the material fracture toughness and where

$$\{\sigma_{\theta\theta}^{(m)}\}_{\text{critical}} = \{\sigma_{\theta\theta}^{(m)}\}_{\theta=45°,\, r=a}\{\alpha_m \Delta T G_m\}, \tag{36}$$

of our previous analysis. Specializing the fracture criterion for the material system MoSi2/SIC one finds the critical cooling temperature T to be 1,356 Kelvin which corresponds to 1,083°C. In this case we assume a crack size of $c/a = 0.1$. The rational for this assumption is based not on a specific material but on previous knowledge of the behavior of 3D stress fields in regions where a boundary layer may be present as a result of an interface or a free boundary. Comparison with experimental observations carried out at Santa Barbara by Lu et al. (1992) on the same material system shows good agreement. The processing temperature reported was 1330°C, which is way beyond the critical value and for this reason cracks were expected

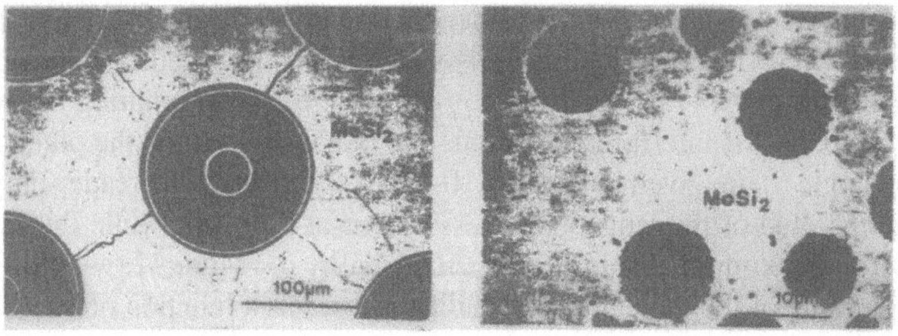

Figure 13. Experimental evidence of cracking in intermetallic composites.

to form, and they did (see Figure 13(a)). The model furthermore shows that, as the ratio V_f decreases, the magnitude of the residual stress also decreases and as a result the 'safe' cooling temperature increases to 1,197°C for a $V_f = 0.41$. Thus, the cracks in this case should be a little less visible, which they are (see Figure 13(b), where we have accounted for the different scale). Notice also that for this material system the cracks are at $\theta = 45°$, as predicted by the model. A similar criterion can also be developed for the radial, as well as the z-direction.

In conclusion, the model predicts an estimate of a safe processing temperature which will suppress the growth of microcracks. It may also be worthy to note that the theoretical model is applicable to ceramic, metal/matrix, as well as organic composites provided that the respective plastic deformations are relatively small. For moderately large plastic deformations, however, a correction factor to account for the nonelastic behavior of the matrix should be introduced.

7. Conclusions

An analytical model has been developed, in order to estimate the residual stresses for matrix cracking in ceramics and intermetalic composite material systems. The model may be used

- to identify the location where cracks are most likely to initiate and propagate, and
- to identify the controlling stress that governs the primary failure,
- to identify the controlling stress that governs the secondary failure,
- to provide information to material designers for the pre-selection of fiber and matrix materials.

Moreover, an approximate fracture criterion for crack initiation has been developed that may be used to predict the critical temperature, e.g. during the operating and cooling, beyond which micro-cracks are most likely to initiate and propagate within the material system. The criterion may also be used to search for possible optimization features between fiber spacing and material properties in order to achieve maximum strength.

Acknowledgments

This work was supported in part by the Air Force Office of Scientific Research Grant No. AFOSR-F49620-93-1-0074P00002. The authors wish to thank Dr. Walter Jones for this support.

References

Folias, E.S. (1975). On the three-dimensional theory of cracked plates. *Journal of Applied Mechanics* **42**, 663–674.

Folias, E.S. (1989). On the stress singularities at the intersection of a cylindrical inclusion with the free surface of a plate. *International Journal of Fracture* **39**, 25–34.

Hellan, K. (1984). *Introduction to Fracture Mechanics*, McGraw-Hill.

Isida, M. et al. (1991). Analysis of zig-zag array of circular inclusions in solid under uniaxial tension, *International Journal of Solids and Structures* **27**, 1515–1535.

Kroupa, J. (1994). Private communication.

Lu, T.C., Yang, J., Suo, Z., Evans, A.G., Hecht, R. and Mehrabian, R. (1992). Matrix cracking in intermetallic composites caused by thermal expansion mismatch.

Penado, E. and Folias, E. S. (1989). The three-dimensional stress field around a cylindrical inclusion in a plate of arbitrary thickness, *International Journal of Fracture* **39**, 129–146.

International Journal of Fracture **93:** 351–371, 1998.

Role of controlled debonding along fiber/matrix interfaces in the strength and toughness of metal matrix composites

Dedicated to Max and Melba Williams on the occasion of their retirement from the editorship of the International Journal of Fracture.

A.S. ARGON[1], M.L. SELEZNEV[1], C.F. SHIH[2*] and X.H. LIU[2**]
[1]*Massachusetts Institute of Technology, Cambridge, MA 02139, USA; e-mail: oona@mit.edu*
[2]*Brown University Providence, R.I. 02912, USA*

Received 11 November 1997; accepted in revised form 15 June 1998

Abstract. In metal matrix composites toughness is derived primarily from the plastic work of rupture of ductile matrix ligaments between the fractured fibers and from the plastic work of simple shear separation along steps connecting major fracture terraces. In the optimization of tensile strength in the longitudinal and transverse directions together with the respective works of fracture the most important factor is the control of the extent of debonding along interfaces between the fibers and the matrix, which develops locally in the course of deformation in a continuously changing mix of modes. In Al alloy matrix composites reinforced with Al_2O_3 fibers an effective means of controlling the key interface fracture toughness is through coarsening of Al_2Cu intermetallic interface precipitates which prescribe a ductile fracture separation layer. A combined experimental approach and micromechanical modeling, utilizing a specially tailored novel tension/shear: traction/separation law provides the means for further optimization of overall behavior.

Key words: Interface fracture, debonding, toughness of composites, traction/separation model.

1. Introduction

While metal matrix composites offer remarkable potential for applications in load bearing structural elements at intermediate to high homologous temperatures, short of long term creep conditions in the metal matrix, barring a few outstanding examples such as diesel engine piston heads, this potential has rarely been fully realized. This is because the development of these composites has largely been fragmented into pre-conceived areas of importance such as fiber development, matrix/fiber compatibility, processing and a limited set of performance controlling factors, and the like, rather than being based on coherent considerations of the major factors and their interactions that govern composite performance. Particularly in the area of achieving combined high axial and transverse strength and high levels of work of fracture, very substantial improvements are possible through a newly developed process of interface structure control and its analysis. This process is based on placement of a pedigreed population of intermetallic compound precipitates on the fiber/matrix interface that provide a wide range of control of the interface traction separation relationship, and thereby achieve desired levels of interface debonding that governs subsequent matrix ligament rupture and resistance to axial shearing off.

* Present address: National University of Singapore, Singapore.
** Present address: Mechanical and Aerospace Engineering, Princeton University, NJ 08544, USA.

Here we present the details of our recent findings and the means of analysis of related composite behavior through some specific micromechanics models specially developed for this purpose.

2. Mechanisms of interface controlled fractures in composites

2.1. Basic strategies and mechanisms

In ceramic-matrix composites consisting of strong fibers and brittle matrixes fiber pull-out during the final stages of fracture often provides the only means of affecting the overall work of fracture. To nurture such toughness in continuous fiber composites with ceramic matrixes, a certain variability of fiber strength is needed to arrange for a degree of randomness of primary fractures in the fibers by balancing the tendency of development of planar fracture cascades arising from strong mechanical coupling of fiber fractures against variability in fiber strength. While this accomplishes the desired goal of producing non-planar primary fractures and a range of broken fiber engagement lengths ready for pull-out, a substantial potential for composite strength becomes compromised by the required variability in fiber strength (Argon, 1974). When aligned discontinuous fibers are used instead, the length of these fibers is chosen to maximize the pull-out work while just preventing additional fiber fractures – again all at the expense of the ultimate strength of the composite.

In metal-matrix composites with strong fibers and ductile matrixes, under ideal conditions this compromise is not necessary because the post fiber-fracture work of separation is not derived from fiber pull out but rather from the plastic work of fracture of the metal-matrix itself. Thus, fibers of higher strength and lower variability can be used resulting in composites of higher axial strength. While this produces planar fracture cascades in parallel fibers with little energy absorption, substantial overall fracture work is, nevertheless, supplied by the work of rupture of the remaining metal ligaments, as shown in the micrograph of Figure 1 of a transverse section of a fractured metal-matrix composite that is the subject of the presentation which follows.

The sketches of Figure 1(a) show two components of the axial fracture process in a continuous-fiber/metal-matrix composite. Primary fiber fracture cascades initiated at different levels produce terraces of planar fractures which we label as mode A. In these terraces the dissipative processes affecting fracture work are the ligament ruptures in the metal matrix over a debonding length L_D. The primary mode A fractures are bridged by mode B type simple shear fractures joining up the terraces. In the fracture of strong composites the fraction of mode B fracture is generally small in comparison with the mode A fracture. Therefore, in what follows we consider only the mode A type separations. However, the methodology that will be presented is equally capable of dealing with the mode B contribution to the overall fracture work.

A first order, elementary analysis of the mode A fracture work, ignoring all other aspects except the rupture of the metal matrix ligaments as depicted in Figure 1(a), results in a simple expression for the specific fracture work U_A of,

$$U_A = \frac{1}{\sqrt{3}}(1 - v_f)\sigma_0 L_D, \tag{1}$$

where v_f is the volume fraction of fibers, σ_0 the matrix tensile plastic resistance in a non-hardening model, and L_D is the debonding length (Friler et al., 1993). Clearly, this debond

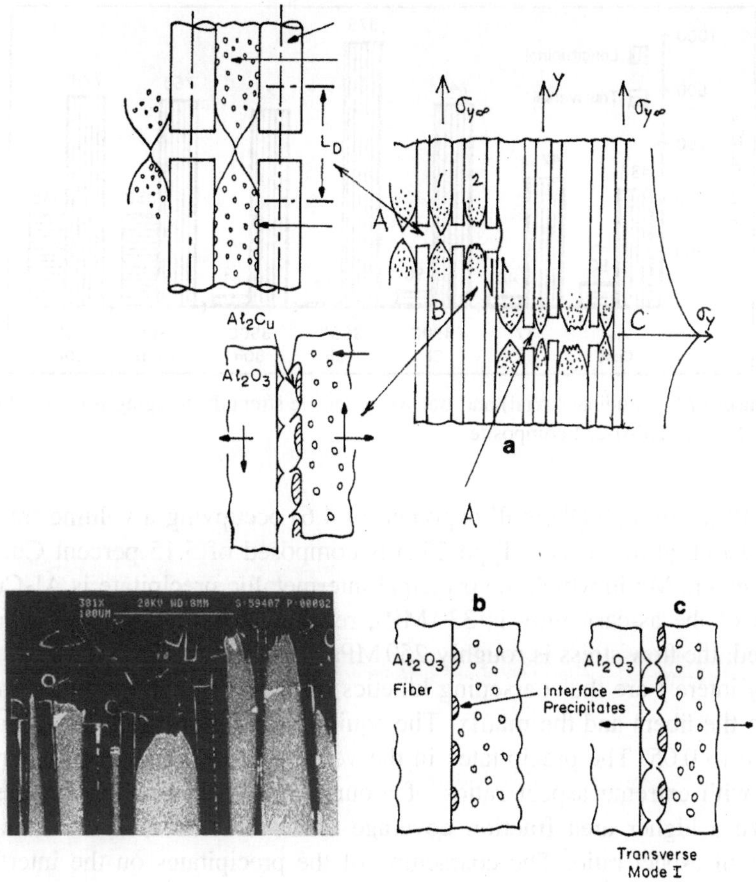

Figure 1. Forms of interface fracture in a metal matrix composite: (a) modes A of tensile, and B of simple shear fracture in a composite, with insets depicting the mixed mode of interface debonding fractures ocurring in modes A and B; (b) depiction of interface fracture of the composite, in the transverse direction.

length is an important element in this fracture process. Very tough interfaces result in negligible debonding and in very low energy absorption in the axial fracture of the composite (albeit at an enhanced stress due to the strong local plastic constraint effects), while excessive debonding beyond the natural necking length of the remaining metal ligaments adversely affects the transverse strength of the composite. The actual development of debonding in response to a deformation-governed mix of modes of interface fracture is the subject of the micromechanical model. In the particular metal-matrix systems composed of sol-gel derived, polycrystalline Al_2O_3 fibers and aluminum alloy matrixes, fracture in the particular mix of modes is governed by the interface precipitates, their volume fraction and actual size. The separation toughness by an arbitrary mix of modes is controlled by the Al_2Cu precipitates on the interfaces, the control of which is the basis of the present approach in governing the overall properties of the composite.

2.2. The composite

The specific metal-matrix system of interest is made up of sol-gel derived polycrystalline Al_2O_3 (Nextel 610) fibers of 10 μm diameter with an average strength of 1.78 GPa and a

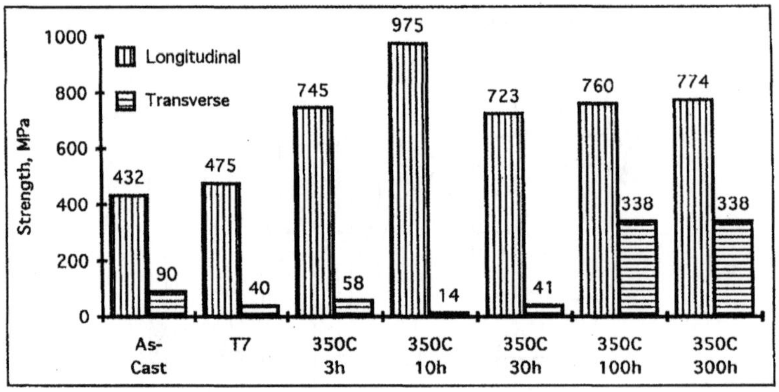

Figure 2. Responses of longitudinal (axial) and transverse tensile strengths to aging time, producing precipitate coarsening in the Nextel 610/Al-224 composite.

variance of 0.18 in strength (Weibull exponent of 4.6) occupying a volume fraction of $v_f = 0.5$. The Al casting alloy matrix (Type 224) is composed of 5.15 percent Cu, 0.12 percent Fe, and 0.34 percent Mn in which the principal intermetallic precipitate is Al_2Cu. The tensile yield strength of the as-cast alloy is 330 MPa, reaching 530 MPa at peak aging condition. When overaged, the flow stress is roughly 250 MPa.

Of primary interest are the coarsening kinetics of the precipitate particles and their states of adhesion to the fibers and the matrix. The equilibrium volume fraction of particles in the matrix is close to 0.05. The precipitates in the volume have shapes ranging from equiaxed to lenticular, with extreme aspect ratios of around 5. The precipitates on the fiber/matrix interfaces have a higher area fraction coverage close to 0.2 (Seleznev et al., 1993) with similar ranges of aspect ratio. The coarsening of the precipitates on the interfaces obeys a classical $t^{1/4}$ law, indicating interface diffusion control. Over the range of time observed (100 hrs) at 350–400C the average interface particle size increases from 0.4 μm to about 1.0 μm. Contact angle measurements lead to estimates of works of adhesion (decohesion) between the Al_2Cu particles and Al_2O_3 fibers and the Al matrix of 0.3 J/m^2 and 0.85 J/m^2 respectively. Parenthetically, it is well known that Al_2O_3 (the self oxide) adheres to Al tenaciously.

2.3. Phenomenology of Composite Fracture

The results of fracture experiments on initially intact (free of macrocracks) composites in the axial and transverse direction are summarized in Figure 2. The figure shows that with aging at 350C the composite tensile strength increases monotonically by more than a factor of 2 and reaches a level of 975 MPa at peak aging condition of the composite in about 10 hrs. This is qualitatively what should be expected from a rule of mixtures response of the composite partly due to a systematic increase of the plastic resistance of the matrix with possible coarsening. For longer periods of aging the strength decreases and levels off at roughly 750 MPa as the matrix overages. Meanwhile, the transverse strength of the composite remains roughly flat at about 40 MPa over aging times up to 30 hrs, but increases sharply to a level of 338 MPa after 100 hrs. The dramatic rise of transverse strength at about 50 hrs of aging coincides roughly with the full coarsening of the interface precipitates. This must indicate, at least in part, an increase of the resistance to crack growth parallel to the fiber/matrix interfaces, where the increased resistance is derived from the well known increase of fracture toughness with increase in the microstructural scale. A more quantitative evaluation of these results discussed

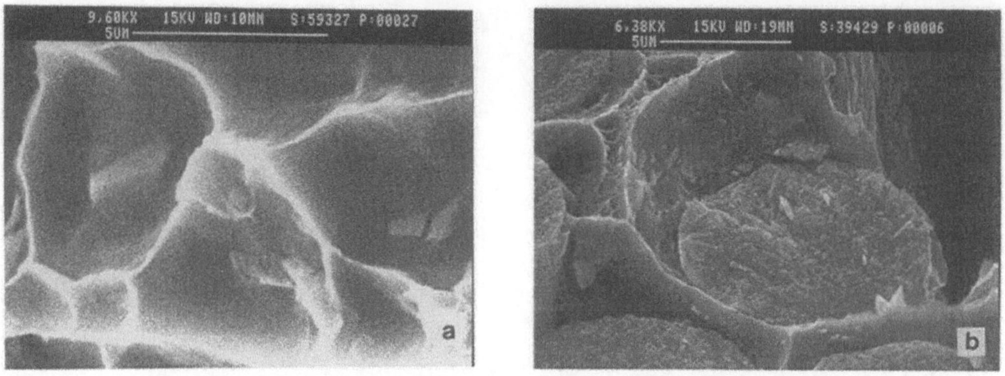

Figure 3. (a) Transverse fracture surface with dimples on the matrix side showing precipitate particles that continue to adhere to the Al matrix, (b) fully developed rupture in remaining metal ligament, following fiber fracture.

below will indicate that additional effects, possibly involving flaw healing, must also be at play.

2.4. Fracture along fiber/matrix interfaces

Detailed examinations of the fracture surfaces of the composites have verified the mechanisms of separation sketched out in Figure 1. These examinations showed that along the interface debonding must occur under a deformation induced mix of modes I and II, prior to matrix ligament rupture. Such debonding initiates preferentially between the Al_2Cu precipitates and the Al_2O_3 fibers where the work of adhesion is the lowest. The precipitate particles remain well adhered to the Al matrix which itself remains well adhered to the Al_2O_3 fibers. The separation at the interface then occurs by a series of ductile dimple ruptures in a combination of normal separation and shear modes, depicted in the lower left inset of Figure 1(a). Transverse composite fracture similarly occurs, by a direct mode I type separation by ductile dimple ruptures as depicted in Figure 1(b). Figures 3(a) and 3(b) show direct evidence for this form of separation.

Clearly, the key of the fiber/matrix debonding process, establishing the length L_D resides in the effective traction/separation characteristic of the matrix at the interface by ductile cavitation under the local mix of mode I and II. A model of this behavior is discussed in the following section.

3. Micromechanics models of interface fracture in metal-matrix composites

3.1. Forms of interface fracture

In Sections 1 and 2 the types of interface fracture were introduced in idealized form and shown how they manifest themselves through fractographic evidence in the actual composite. A key element of the specific system of Al alloy matrixes containing Al_2Cu precipitate particles both in the matrix and on fiber/matrix interfaces is the ductile cavity growth mode of separation occurring at interfaces as a precursor process, setting the stage of the eventual fracture of the composite through the ductile rupture of matrix ligaments. The most important element of this interface fracture is its response to a deformation induced changing mix of tension

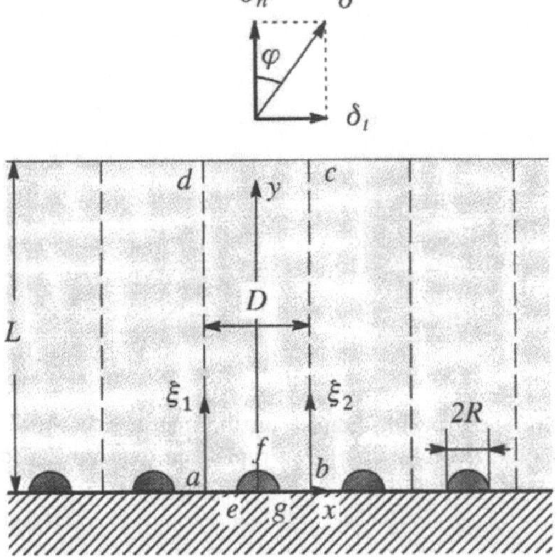

Figure 4. Interface with a row of uniformly spaced semi-circular precipitates.

and shear. In a more formal sense it is attractive to model this behavior by a tension/shear: traction/separation law as has been done by a number of investigators (Nagpal et al., 1973; Needleman 1987; Tvergaard and Hutchinson, 1993). While we will adopt some form of this representation in our modelling of the complete fracturing process, we will start out first in Section 3.2 below to model the tension/shear: traction/separation process by a specific micromechanical model of void nucleation at interface particles followed by their growth by plastic flow in various combinations of imposed tension and shear to demonstrate the complexity of the actual process. In Section 3.3 we consider a simplified representation of this tension/shear: traction/separation response to explore the overall response of the composite in the mode A form of fracture to explore some of the sensitivities of the model to possible variations in material response.

3.2. The tension/shear: traction/separation response model of interfaces

In both the interface response and the overall composite fracture models the constitutive law for the metal matrix is taken as elastic-plastic-power-law hardening, which under uniaxial tension is stated as

$$\frac{\varepsilon^e}{\varepsilon_0} = \frac{\sigma}{\sigma_0} \qquad \frac{\varepsilon^p}{\varepsilon_0} = \left(\frac{\sigma}{\sigma_0}\right)^{1/N} - 1 \quad (\sigma \geqslant \sigma_0), \tag{2a,b}$$

where σ_0 is the yield strength, $\varepsilon_0 = \sigma_0/E$, $1/N$ is the reciprocal of the strain hardening exponent, and E is the Young's modulus (which for Al is 70 GPa). The plasticity is taken to follow the J_2 flow rule.

In the interface debonding model the fiber/matrix material system is treated as a two dimensional plane strain problem. The precipitate particle area fractions on the interface is taken to be 0.2, as mentioned above, and is translated into the plane strain geometry as the ratio of $2R/D$ illustrated in Figure 4. While the actual shapes of the interface precipitate particles are somewhat different, the hemi-cylindrical shapes were chosen to facilitate the computation of the elastic–plastic cavity growth following the decohesion of particles from the fiber. The tensile bond strength T_{ip} of the particle with the fiber is taken to be a definite fraction of the Young's modulus of the matrix which for a chosen value of 1 GPa is $1.42 \times 10^{-2}E$ and represents a realistic level encountered in the particular alloy systems. Additional material parameters chosen for the interface debonding model were $\sigma_0/E = 5 \times 10^{-3}$, Poisson's ratio $\nu = 0.3$ and $N = 0.05$. While matrix ligaments between particles can separate only by rupture in the actual system, in the model a terminal strain to fracture $\varepsilon_f^p = 1.86$ has been used to facilitate the final separation.

In the interface debonding model a combined proportional displacement loading condition is prescribed at the distant boundary $y = L$ where the latter length was chosen to be large enough relative to the precipitate spacing D to make the deformation essentially a uniform field at this distance. The combined loading is applied through the normal and tangential displacements δ_n and δ_t as fixed ratios of δ_t/δ_n during specific radial loading problems characterized by specific phase angles $\varphi = \tan^{-1}(\delta_t/\delta_n)$. The specific values chosen for φ were 0°, 30°, 60°, and 85° ranging from pure tension to nearly simple shear (since the actual interface debonding condition was taken to be a critical level of interface tension, developing this condition in simple shear, i.e. $\varphi = 90°$ becomes computationally too demanding). Throughout the deformation, debonding along the matrix/fiber interface between the particles is considered to be impossible because of the tenacious adhesion between the two media. Therefore, the following boundary conditions were specified.

$$u_1 = u_2 = 0 \text{ at } y = 0 \text{ along } -D/2 \leqslant x \leqslant -R$$
$$\text{and at } y = 0 \text{ along } R \leqslant x \leqslant D/2 \tag{3}$$

and

$$u_1 = \delta \sin\varphi, \qquad u_2 = \delta \cos\varphi \quad \text{at } y = L, \tag{4}$$

where δ is the total displacement imposed at $y = L$ as depicted in Figure 4. Due to the periodicity of the precipitates the solution is periodic along the interface, and all cells in Figure 4 have identical fields, making it necessary to analyse only the region *abcd*.

Before precipitate decohesion from the fiber the displacements along the precipitate/matrix interface, *efg* are zero. Debonding along the entire interface is taken to begin when the average normal stress at the interface, *efg* between precipitate and fiber reaches the critical value T_{ip}. Once this occurs, the precipitate is considered separated along the hemi-cylindrical interface *efg* by equating all tractions along this interface to zero. Since nucleation of the initial cavity, while critical, is only a small portion of the entire debonding response, this critical assumption of convenience, departing from reality, is of little consequence. After nucleation the cavity grows at the matrix side by plastic deformation and finally complete debonding along the entire interface occurs by the necking down and rupture of the matrix ligaments between adjacent cavities. As stated above, the actual ligament rupture process is terminated when the local ligament strain reaches ε_f^p. More detailed descriptions of the implementations of the finite element model can be found elsewhere (Liu 1998; Shih and Liu, 1998).

Figures 5(a,b) give the numerical results of the debonding simulation in pure tension, i.e., for $\varphi = 0°$. The results are presented not in normalized form but for actual material values of $E = 70\,\text{GPa}$, $\sigma_0 = 350\,\text{MPa}$, $T_{ip} = 1\,\text{GPa}$, $D = 5 \times 10^{-6}\,\text{m}$.

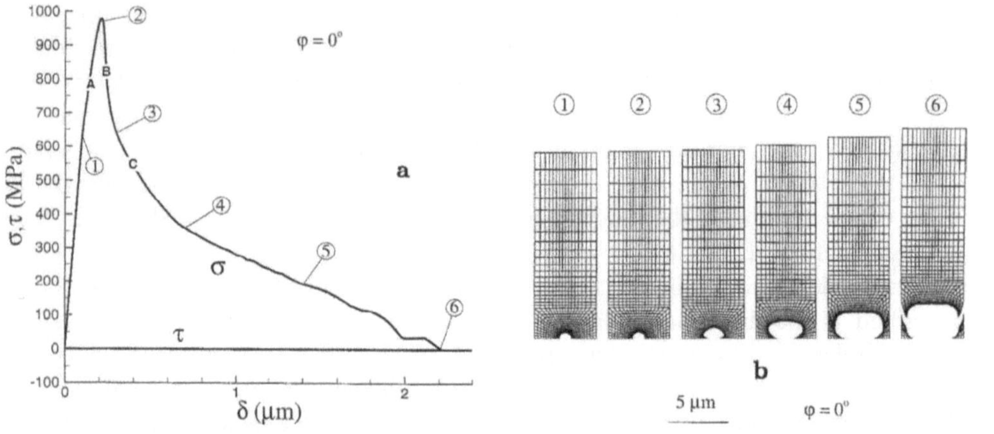

Figure 5. (a) Calculated normal stress and shear stress versus total loading displacement under pure tension; (b) sequence of deformed configurations under pure tension ($\varphi = 0°$).

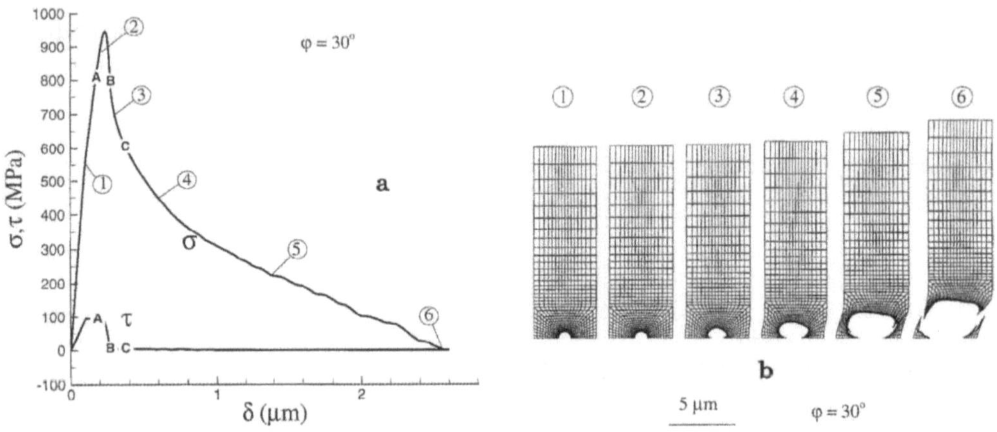

Figure 6. (a) Calculated normal stress and shear stress versus total loading displacement under combined tension and shear with phase angle $\varphi = 30°$; (b) sequence of deformed configurations in mixed tension and shear loading.

In Figure 5(a) the letters A, B, and C indicate the levels at which the precipitate decoheres from the fiber, where cavity nucleation is complete and where matrix rupture sets in by intense shear in matrix ligaments. The numbers ① to ⑥, in turn, relate to the various states of debonding by cavity growth illustrated in Figure 5(b). The curves for σ and τ represent the evolving tensile and shear separation resistances. Since $\varphi = 0$, τ remains zero throughout the debonding process. At point ① yielding occurs in the matrix while the precipitate is still attached to the fiber – a response necessitated by the fact that $\sigma_0 < T_{ip}$. Peak resistance occurs at ②, after precipitate decohesion, when the interface traction is released and concentrated plasticity spreads to the cell borders. At position ③ the cavity begins to dilate quickly under decreasing tensile resistance; and at positions ④ and ⑤ matrix shearing in the ligaments intensifies as the neighboring cavity interactions become ever more intense. Finally at ⑥ the remaining ligament reaches the critical plastic strain to fracture ε_f^p in lieu of full rupture, and the ligament is permitted to shear off as depicted.

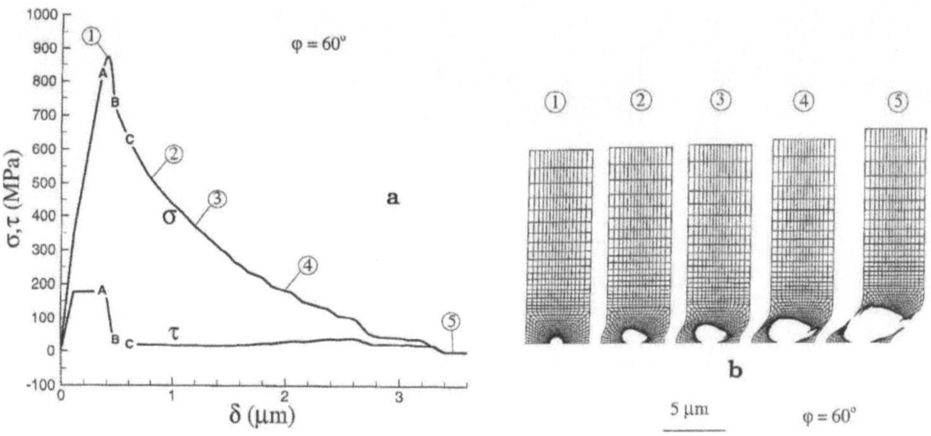

Figure 7. (a) Calculated normal stress and shear stress versus total loading displacement under combined tension and shear with phase angle $\varphi = 60°$; (b) sequence of deformed configurations in mixed tension and shear loading.

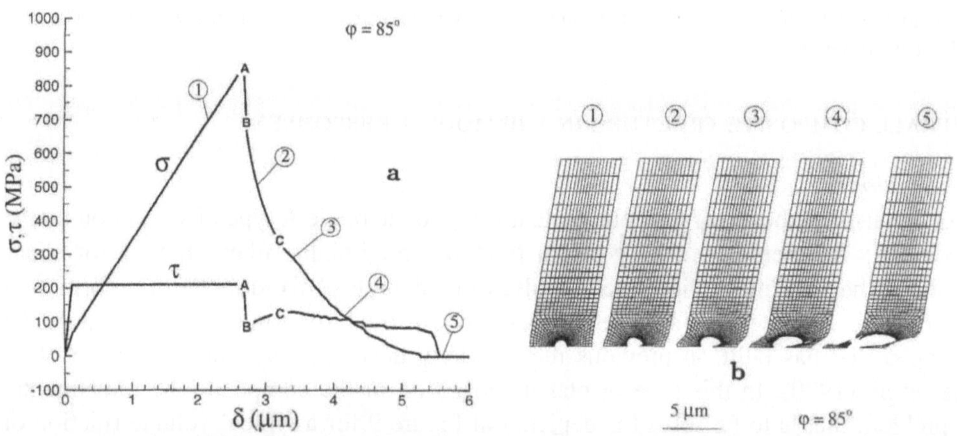

Figure 8. (a) Calculated normal stress and shear stress versus total loading displacement under combined tension and shear with phase angle $\varphi = 85°$; (b) sequence of deformed configurations in mixed tension and shear loading.

The debonding simulations for $\varphi = 30°$, $60°$ and $85°$ are depicted in Figures 6(a,b), 7(a,b) and 8(a,b) respectively, giving both the σ and τ resistance behaviors and the developing distortions of the growing and shearing cavities. The letters and numbers on the resistance curves have the same meanings given above for the pure tensile separation.

The figures show clearly that the presence of an imposed shear displacement promotes enhanced plastic response which delays the precipitate decohesion step to larger overall combined displacements δ but also, predictably, decreases the peak levels of tensile resistance σ. Prior to decohesion of the precipitates from the fiber the principal source of the shear resistance τ is derived from the precipitates. This resistance increases with increasing phase angle φ but is sharply reduced after decohesion of the precipitate occurs. Nevertheless, with increasing φ the peak shear resistance levels increase gradually as the precipitates support some of the shear load even though the principal driving force for cavity expansion is provided by the tensile distortions. Figures 6(b), 7(b) and 8(b) show the increasing shear distortions of the cavity growth process with increasing φ.

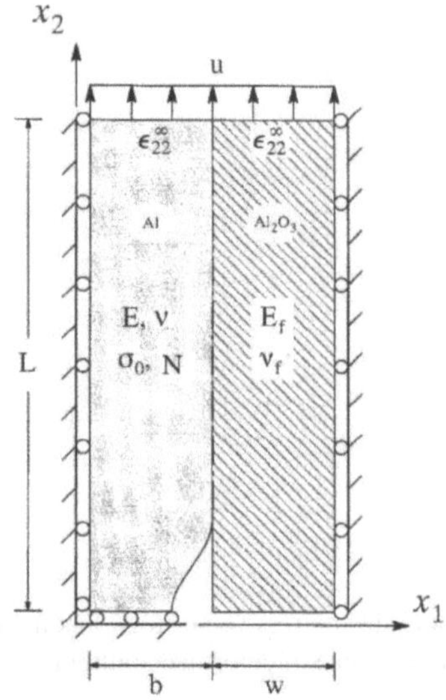

Figure 9. Computational cell for longitudinal tension in a plane strain idealization of composite fracture. The fiber is assumed to have fractured.

3.3. Overall composite fracture in the mode A response

Model description

In the overall tensile response of the composite leading to the mode A type of separation by the rupture of matrix ligaments, following planar fractures in adjoining fibers, the tension/shear response along the fiber/matrix interfaces evolves in a complex mixed mode of deformation depicted in Figure 1. To develop a model of such mode A fractures a micromechanical model was developed that has built on previous methodology developed by Suo and Shih (1993) and Varias et al. (1990). In this type of planar separation of the composite the fundamental unit cell problem that is to be solved is depicted in Figure 9 for a typical volume fraction of fibers of $v_f = 0.5$, again idealized as a plane strain problem. The Al_2O_3 sheet (fiber) is taken as initially cracked and responding only elastically, while the constitutive law for the metal matrix is taken as elastic–plastic, power-law hardening as given in equations (2a,b) above.

The mixed modes I and II of interface loading in the present model are induced by the inhomogenous plastic deformation of the matrix ligaments. To deal with this complexity in the present micromechanics model the detailed and complex tension/shear: traction/separation model, presented in Section 3.2 above, must be idealized into a more tractable form which was taken to be in the form of a special traction potential Φ used by us earlier (Argon et al., 1997, Liu, 1998). The specific potential formulated earlier by Tvergaard and Hutchinson (1993) was in the form of

$$\Phi(\delta_n, \delta_t) = \delta_n^c \int_o^{\lambda} \sigma(\lambda')\, d\lambda'. \tag{5}$$

It defines the normal T_n and tangential T_t, components of the tractions offered by the separating interface as follows

$$T_n = \frac{\partial \Phi}{\partial \delta_n} = \frac{\sigma(\lambda)}{\lambda} \frac{\delta_n}{\delta_n^c}, \qquad T_t = \frac{\partial \Phi}{\partial \delta_t} = \left(\frac{\delta_n^c}{\delta_t^c}\right) \frac{\sigma(\lambda)}{\lambda} \frac{\delta_t}{\delta_t^c}, \tag{6a,b}$$

where δ_n and δ_t are the normal and tangential components of separation across the interface, and δ_n^c and δ_t^c, their critical terminal values in the pure response modes of tension and shear. The combined separation measure is defined by λ, as an operationally convenient mode of response that can be taken as a geometrical mean of the two pure response modes as

$$\lambda = \sqrt{\left(\frac{\delta_n}{\delta_n^c}\right)^2 + \left(\frac{\delta_t}{\delta_t^c}\right)^2}. \tag{7}$$

Finally, the form of the effective traction response $\sigma(\lambda)$ along the interface is taken to be, (Needleman, 1987)

$$\sigma(\lambda) = \sigma_{\max}\frac{\lambda}{\lambda_0}\left(\frac{1-\lambda}{1-\lambda_0}\right)^{(1/\lambda_0)-1} \qquad (0 < \lambda_0 < 0.5), \tag{8}$$

similar to the well-known, so-called universal Binding Energy Relationship of Rose et al., (1984), where $\sigma_{\max}$ is the peak stress and λ_0 is the value of λ where the peak stress is reached, and represents the principal shape factor of the traction/separation (t-s) law, while $\sigma_{\max}$ scales its height. In the present case, of course, the t-s relation is intended to model the ductile cavitation and dimple fracture process across the interface where $\sigma_{\max}$ should represent the interface stress that is reached during decohesion of the precipitates from the fibers (as it appears in Figures 5(b)–8(b)) and where the effective range of action of λ respresents a characteristic rupture separation distance of the order of interparticle spacing. Alternatively, the interfacial separation energy, Γ_0 and $\sigma_{\max}$ by plastic dimpling can be taken as the two characterizing parameters where,

$$\Gamma_0 = \frac{\lambda_0(1-\lambda_0)^{1-(1/\lambda_0)}}{(1+\lambda_0)}\sigma_{\max}\delta_n^c. \tag{9}$$

As a first approximation δ_n^c/δ_t^c was taken as unity (closer to 2 in the model of Section 3.2) and the value of λ_0 as 0.3. Parametric studies considered $\sigma_{\max}$ and Γ_0 as the two principal scaling factors. These were introduced in normalized form as $(\Gamma_0/\sigma_0 b)$ and $(\sigma_{\max}/\sigma_0)$ in sensitivity studies of the model. For composites with long fibers, $L \gg b, w$, and L_D (the length of debonded interface), the dependence of the axial load should be independent of the cell length L. Indeed, the inelastic separation displacement u^*, defined as

$$u* = u - \varepsilon_{22}^{\infty}L \tag{10}$$

is independent of L, where $\varepsilon_{22}^{\infty}$ is the longitudinal strain at the end, with u being the imposed axial displacement. Therefore, the dependence of axial load P on imposed inelastic displacement can be written as

$$\frac{P}{\sigma_0 b} = f\left(\frac{u^*}{\varepsilon_0 b}; v_f; \frac{\sigma_0}{E}, N, \nu; \frac{E_f}{E}, \nu_f; \frac{\Gamma_0}{\sigma_0 b}, \frac{\sigma_{\max}}{\sigma_0}, \frac{\delta_n^c}{\delta_t^c}, \lambda_0\right), \tag{11}$$

together with parameters of geometry, materials and interface. During the progressive failure of the composite, the work done by the applied load is dissipated in plastic flow of the matrix ligaments and in the surface creation due to interfacial decohesion (followed by the elastic–plastic cavity growth mode of debonding discussed in Section 3.2 above) except in the component, U_e, stored in the material as elastic strain energy. This gives

$$\int_0^u P\,\mathrm{d}u = \int_V \int_0^{\varepsilon^p} \boldsymbol{\sigma} : \mathrm{d}\boldsymbol{\varepsilon}^p\,\mathrm{d}V + \Gamma_0 L_D(u^*) + U^e, \tag{12}$$

where L_D is the length of the debonded interface. Both the work input and elastic strain energy U^e depend on the cell length L. However, their difference

$$U^* = \int_0^u P\,\mathrm{d}u - U^e = U^p + U^s \tag{13}$$

is independent of L, representing the total irreversible energy dissipation through matrix plastic flow and the concurrent interfacial debonding. This energy dissipation can be expressed as a function of applied displacement, with geometry, material and interface parameters, as,

$$\frac{U^*}{\sigma_0 \varepsilon_0 b^2} = g\left(\frac{u^*}{\varepsilon_0 b}; v_f; \frac{\sigma_0}{E}, N, \nu; \frac{E_f}{E}, \nu_f; \frac{\Gamma_0}{\sigma_0 b}, \frac{\sigma_{\max}}{\sigma_0}, \frac{\delta_n^c}{\delta_t^c}, \lambda_0\right). \tag{14}$$

For parametric studies, the effects of interface and matrix properties on composite strength, ductility and energy dissipation have been investigated. Parameters of secondary importance were fixed at appropriate levels. The most significant parameters are the bond strength $\sigma_{\max}$, interfacial separation energy Γ_0 and exponent N of matrix strain hardening. The values of other parameters for typical Al_2O_3 fiber and Al alloy matrix were taken as

$$v_f = w/(w+b) = 0.5, \qquad \nu = \nu_f = 0.3,$$

$$\frac{\sigma_0}{E} = 1.5 \times 10^{-3}, \qquad \frac{E_f}{E} = 6.$$

With the fixed parameters, the load-displacement (11) and the energy dissipation (work of fracture) (14) relations can be simplified as

$$\frac{P}{\sigma_0 b} = f\left(\frac{u^*}{\varepsilon_0 b}; N; \frac{\Gamma_0}{\sigma_0 b}, \frac{\sigma_{\max}}{\sigma_0}\right), \tag{15}$$

$$\frac{U^*}{\sigma_0 \varepsilon_0 b^2} = g\left(\frac{u^*}{\varepsilon_0 b}; N; \frac{\Gamma_0}{\sigma_0 b}, \frac{\sigma_{\max}}{\sigma_0}\right). \tag{16}$$

Main findings

Figure 10 shows a series of deformed configurations of evolving states of matrix ligament rupture with increasing inelastic tensile displacement for a realistic case of 10 μm diameter fibers where this diameter can be taken as a characteristic normalization length. Contours of effective plastic strain are displayed in every configuration. The plastic deformation originates in the matrix at the corner of the broken fiber due to the local strain concentration. As the load increases to the peak load, debonding begins at the fiber/matrix interface governed by the t-s law. Beyond the peak load, more and more interfacial debonding occurs as the imposed displacement increases further (see figures for u = 0.2, 0.3, 0.4 μm etc.). This process,

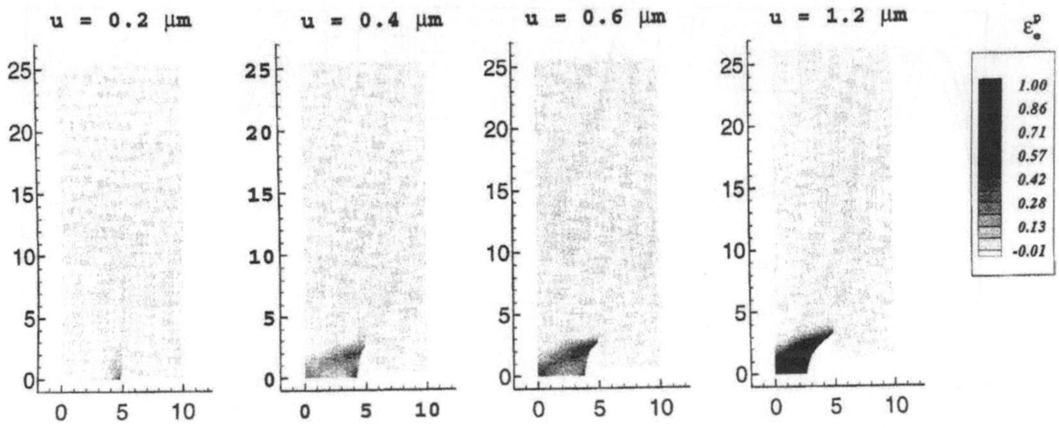

Figure 10. Deformed configurations of the matrix ligament at a fiber break, for four different levels of axial displacement. Shading shows different levels of effective plastic strain.

accompanied with significant decrease in the axial load, is controlled by the fiber/matrix debonding by the plastic cavitation processes of Section 3.2.

Interface debonding gradually slows down with increasing imposed axial displacement as the traction on the lateral fiber surfaces decreases. The axial load on the fiber is gradually transfered to the matrix and the plastic strains build up in the necking and rupturing matrix ligament between the two adjacent broken fibers, providing ductility to the composite. This process, dominated by the rupturing matrix has a slower load decrease than the regime controlled by interface debonding. The distinctive stages, controlled by interface debonding, followed by one dominated by necking and interface ligament rupture, are illustrated clearly in the load-displacement and energy dissipation curves of Figures 11(a)–11(c).

Figure 11(a) displays the effect of interfacial separation energy Γ_0 on the load-displacement curves and the energy dissipation during the progression of failure for $\sigma_{max}/\sigma_0 = 3$ and $N = 0.1$. The peak load remains roughly at a constant level for various Γ_0's. Actually, for a matrix with low hardening the peak load is chiefly determined by the interfacial bond strength between precipitates and fiber. The energy dissipation (fracture energy) increases with increasing Γ_0, because more work is required to separate the fiber/matrix interface.

Curves of load-displacement and energy dissipation are shown in Figure 11(b) for various normalized bond strengths σ_{max}/σ_0 (with $\Gamma_0/\sigma_0 b = 6.0 \times 10^{-2}$ and $N = 0.1$). The peak load increases with the interfacial bond strength. Under our assumed conditions of earlier fiber fracture high axial strength can be achieved with high bond strength at fiber/matrix interface. However, the bond strength has comparatively little effect on the energy dissipation since most of it is governed by the elastic–plastic cavity growth process. The small increase in the energy dissipation comes from slight increases of the plastic strains in the matrix.

The most significant factor for the toughness of the composite is matrix strain hardening. Figure 11(c) displays the load-displacement curves and energy dissipations for various strain hardening exponents N with fixed interfacial property ($\Gamma_0/\sigma_0 b = 9.0 \times 10^{-2}$ and $\sigma_{max}/\sigma_0 = 3$). Both matrix strain hardening and interfacial bond strength contribute to the axial strength of the composite. The increases in peak load with matrix hardening is shown in the figure. The fracture energy increases significantly with the strain hardening exponent. This is due to the larger plastic dissipation in the matrix and a little more debonding along the interface. In contrast to the interfacial bond strength and separation energy, the matrix strain hardening can

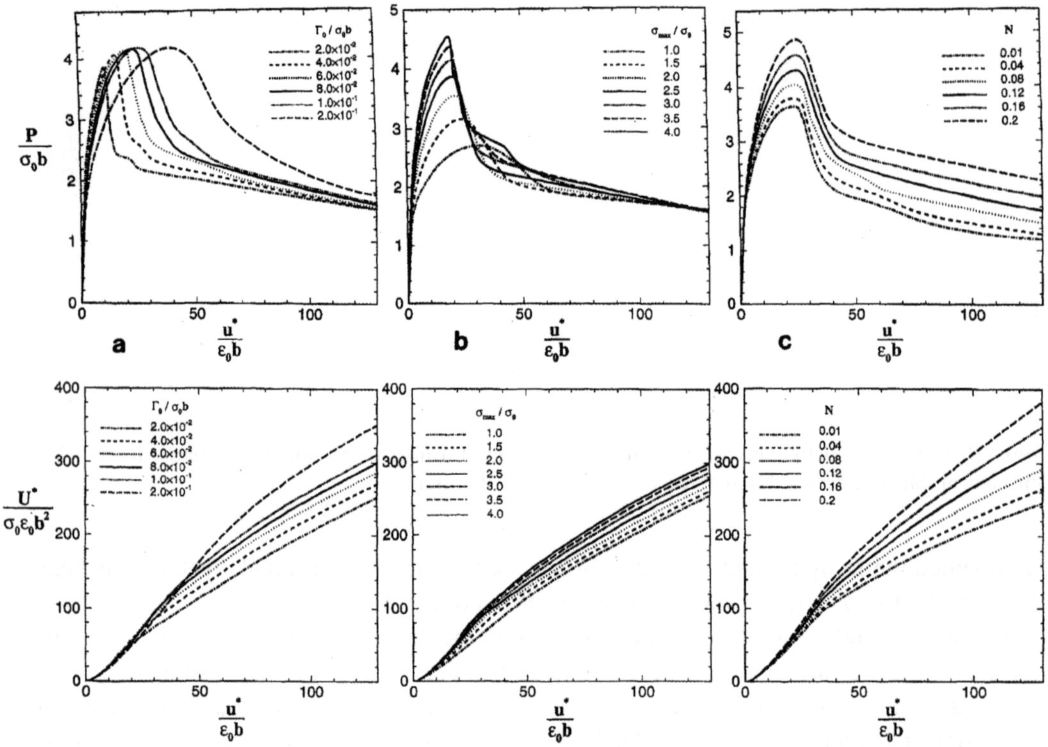

Figure 11. Load displacement and associated overall energy dissipation curves as a function of normalized inelastic axial displacement: (a) effect of debonding energy Γ_o on behavior, (for $\sigma_{max}/\sigma_0 = 3$ and $N = 0.1$); (b) effect of peak stress σ_{max} on behavior, (for $\Gamma_0/\sigma_0 b = 6 \times 10^{-2}$ and $N = 0.1$); (c) effect of hardening exponent N on behavior (for $\Gamma_0/\sigma_0 b = 6 \times 10^{-2}$ and $\sigma_{max}/\sigma_0 = 3$).

provide both higher peak loads and larger energy dissipation. Unfortunately, in the alloys of interest where precipitation strengthening is the major source of increase of plastic resistance strain hardening is weak.

The model for mode A behavior has confirmed that the fiber/matrix interface plays a very important role in the toughness of the composite system. If the interface strength is low, Γ_0 as defined in (9) will be very low. Interfacial cracks will propagate long distances upon fiber fracture, decoupling the matrix from the fiber. Matrix ligaments will still neck and rupture resulting in acceptable levels of axial work of fracture, but the transverse strength of the composite will be compromised substantially. This appears to be what happens for aging periods less than 100 hrs at 350C in Figure 2. In the other extreme case, an excessively well bonded fiber/matrix interface will assure a high transverse strength but will minimize the development of matrix ligament rupture by promoting premature dilatant plastic behavior and ductile dimple fracture from matrix particles (type '2' behavior shown in Figure 1(a)). This appears to be what happens for aging periods much above 100 hrs at 350C, resulting in significantly reduced work of fracture in the longitudinal direction due to inadequate extent of debonding. Only concurrent interface debonding, accompanying the neck formation and rupture of matrix ligaments will maximize the work of fracture in the mode A process.

Table 1. Results of the mixed mode traction/separation simulation ($c = 0.2$; $D = 5 \times 10^{-6}$m; $N = 0.05$; $\sigma_0 = 350$ MPa; $E_m = 70$ GPa)

φ degrees	σ MPa	σ/σ_0	τ MPa	τ/σ_0	Γ_0 J/m^2	$\Gamma_0/\sigma_0 D$	$\alpha(\varphi)$
0	980	2.80	0	0	649	0.371	0.500
30	950	2.71	100	0.286	727	0.415	0.559
60	875	2.50	175	0.500	1079	0.617	0.831
85	850	2.43	179	0.511	1733	0.990	1.342

4. Discussion

4.1. Evaluation of Simulation Results

Since the traction/separation simulation, in mixed tension and shear, is central to the understanding of the deformation induced debonding of the rupturing matrix, as it separates from the broken fiber, we evaluate first the results of Section 3.2. Our simulations, albeit in great detail, represent only a special case in the broader slip line field investigation of Nagpal et al. (1973). We find this model very convenient to relate our findings to a more general framework.

The basic results of our simulation are summarized in Table 1, for the special case of a precipitate areal coverage of $c = 0.2$ at the interface, for material parameters listed in the heading of the table. The tractions σ and τ listed are the peak values at the start of the ductile cavity expansion process. The debond energy Γ_0 is the area under the tensile separation curve, the level of which incorporates the effect of the imposed shear. In Figure 12 the normalized values of the peak tensile and shear tractions are plotted as a generalized flow process locus. The portion of the locus for displacement vector angles φ larger than 85° is shown dotted, descending down to an asymptotic value of $\tau/\sigma_0 = 1/\sqrt{3}$, and not $(1-c)/\sqrt{3}$ since decohesion of the precipitates that sets the stage of cavity growth is not expected to occur if the tensile tractions across the interface fall below the level that accomplishes this (in the tension case for $\varphi = 0$ such decohesion occurs at a local tensile stress of about 780 MPa for a decohesion strength of 1 GPa, indicating the presence of a deformation induced stress concentration of about 1.28). This difficulty in particle decohesion is already apparent in the case for $\varphi = 85°$ of Figures 8(a,b) where plastic flow begins early in the uncavitated material in response to the overwhelming shear character of the deformation, and the tensile tractions required for precipitate decohesion build up very slowly. Upon decohesion, the shear tractions drop to a lower level as the tensile tractions still perform the principal function of ductile debonding. Parenthetically, Figure 12 shows that the associated flow rule is obeyed, demonstrated by the normality of the displacement vectors to the generalized yield surface.

The simulation of Section 3.2 was performed for a precipitate area fraction of $c = 0.2$ and an inter-precipitate spacing $D = 5 \times 10^{-6}$m. Since our approach of property control is based on precipitate coarsening, it is necessary to generalize our findings to obtain a more universal expression. Here the study of Nagpal et al (1973), who have performed an incremental slip line analysis of the plastic expansion of circular cavities arranged in a plane, supplies the missing information. According to Nagpal et al. the debond work by ductile cavitation for an initial porosity c should be

$$\Gamma_0 = \Gamma_{00} + \alpha(\varphi)\sigma_0 D(1 - \beta c), \tag{17}$$

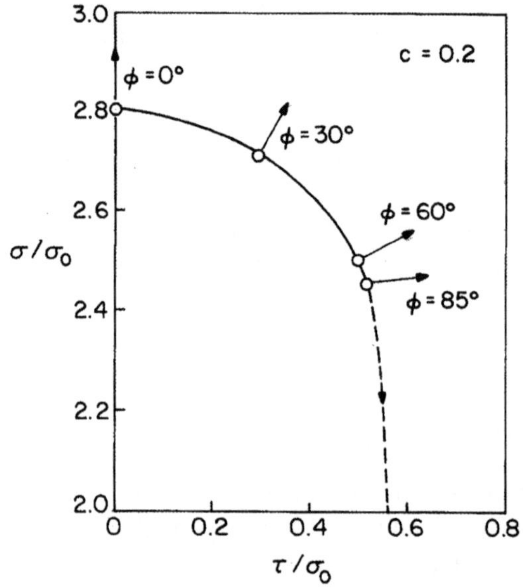

Figure 12. Generalized yield locus representing the inception of the debonding process along the fiber/matrix interface by ductile cavity expansion, as modeled in Section 3.2 for an initial interface porosity level of $c = 0.2$.

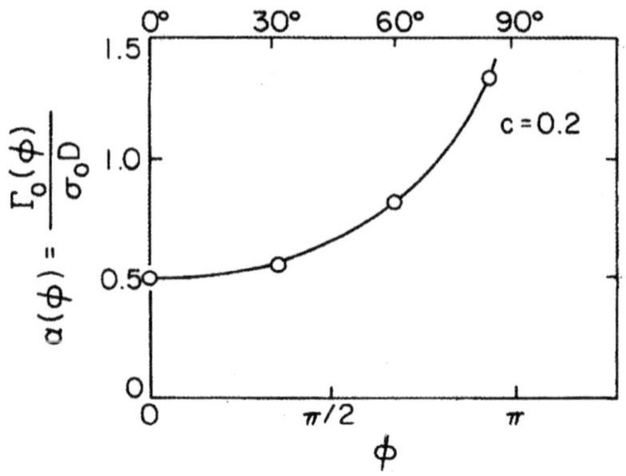

Figure 13. Dependence of the specific debond work by ductile cavitation on the mix of tension and shear, represented by the angle φ that the imposed displacement vector of separation makes with the interface normal.

where $\beta = 1.67(= 1/0.6)$. Considering our debond energy for $\varphi = 0$, listed in Table 1, we obtain from (17) $\alpha(\varphi = 0) = 0.5$ which agrees exactly with the results of Nagpal et al. For completeness we have added an intrinsic decohension energy Γ_{00} to the debond work in (17). Since this is expected to be only of the order of 1–$2\,\mathrm{J/m^2}$, it is generally negligible for most cases of interest – provided it is recognized that decohesion of the precipitates is essential since without this there would be no cavities to expand – as we already indicated above. We take (17) as the central finding that guides our particle coarsening strategy for property control. The results of the simulation, summarized in Table 1 show that the debonding work by ductile cavitation increases substantially with the increasing shear component of the imposed separation vector. This increase in $\alpha(\varphi)$ is shown in Figure 13.

The principal result of the simulation of the overall composite fracture by the rupture of matrix ligaments, strongly affected by the deformation induced changing mixture of the interface ductile debonding process, is represented in Figure 11(a) which gives the development of the overall axial fracture work U^* as a function of the debond work Γ_0 along the interfaces. Considering the terminal levels of $U^*/\sigma_0\varepsilon_0 b^2$ at $u^*/\varepsilon_0 b = 130$ shown in the figure (which amounts to a peak ligament strain of roughly 1.5) as representative, we have plotted the dependence of $U^*(u^*/\varepsilon_0 b = 130)/\sigma_0\varepsilon_0 b^2$ on $\Gamma_0/\sigma_0 b$ in Figure 14 (which, of course, corresponds to the performance of a composite with a fiber volume fraction of $v_f = 0.5$). We note that the axial work of fracture $U^*/\sigma_0\varepsilon_0 b^2$ increases monotonically with the level of the interface debond work but that this increase appears to level off beyond the range of the simulation at about $\Gamma_0/\sigma_0 b$ of roughly 0.25 where it appears to reach a level of $U^*/\sigma_0\varepsilon_0 b^2 \cong 360$. Determination of the actual levels of U^* for $\sigma_0 = 350\,\text{MPa}$, $\varepsilon_0 = 5 \times 10^{-3}$, $b = 5 \times 10^{-6}\text{m}$, and relating the findings to a specific level by dividing with $b/(1 - v_f) = 2b$ gives the values shown on the right hand side of the figure. Figure 14 is quite instructive. It shows that if $\Gamma_0/\sigma_0 b \rightarrow 0$ the axial work of fracture (i.e. energy release rate G_{IC}, provided that a mode A type of planar fracture continues to prevail) $U^*/\sigma_0\varepsilon_0 b^2 \rightarrow 230$ or on the absolute scale $U^*/2b \rightarrow 1000\,\text{J/m}^2$. This would represent the fracture work given by the idealized representation of (1) which considers the debond work to be negligible in the overall energy balance. Accordingly, from (1) for the above work of fracture, for $\sigma_0 = 350\,\text{MPa}$, and $v_f = 0.5$, a debond length of $L_D = 9.90 \times 10^{-6}\text{m}$ is obtained. This is nearly equal to the fiber diameter, or the initial matrix ligament thickness in the plane strain model. This should be a good representation of the perceived natural neck length. In this limiting form of behavior, while the axial work of fracture is at a respectable level, the transverse strength should be negligibly low. Figure 14 demonstrates that the ductile debonding work along the interfaces makes a substantial contribution to the axial fracture work as it increases to levels of $\Gamma_0/\sigma_0 b = 0.25$ (or $\Gamma_0 = 437\,\text{J/m}^2$). Based on the interpretations of the trend shown in Figure 14, beyond this level, increases in Γ_0 into the ranges that should actually be present in the experimental study, the axial fracture work should begin to decrease with increasing difficulty in debonding. Clearly, an ideal level of ductile debond work would appear to be the above value of $\Gamma_0 = 437\,\text{J/m}^2$ which maximizes the axial work of fracture at a level of $U^*/\sigma_0\varepsilon_0 b^2 = 360$ (or $U^*/2b = 1550\,\text{J/m}^2$).

4.2. Relevance of the simulation to the experimental findings

Axial properties

In the aligned fiber composite under consideration we expect the axial strength to be governed by a simple rule of mixtures, in the absence of cracks and discontinuities. Thus

$$\sigma_a = v_f\overline{\sigma}_f^* + (1 - v_f)\sigma_m^*(\varepsilon_f^*), \tag{18}$$

where $\overline{\sigma}_f^*$ is the average tensile fracture strength of the fibers (Argon, 1974) which we take to be 1.78 GPa as mentioned in Section 2.2 above, and $\sigma_m^*(\varepsilon_f^*) = \overline{\sigma}_f^* E_m/E_f$, provided this value is less than the tensile flow stress of the matrix which we take to be between 350 MPa and 530 MPa (when in its peak aged condition). Since $E_m/E_f = 1/6$ we have for $\sigma_m^* = 297\,\text{MPa}$ which is indeed less than the matrix flow stress. The axial strength calculated from (18) is 1039 MPa while the measured peak axial strength was 975 MPa which is well within the calculated value. We now note that experiments show a gradual decrease of axial strength with aging, down to about 750–770 MPa. Since the tensile plastic resistance of the overaged Al alloy does not fall below 330 MPa, which continues to be larger than σ_m^* as interpreted above, we must conclude that the decline in axial strength is not likely to come from a decrease

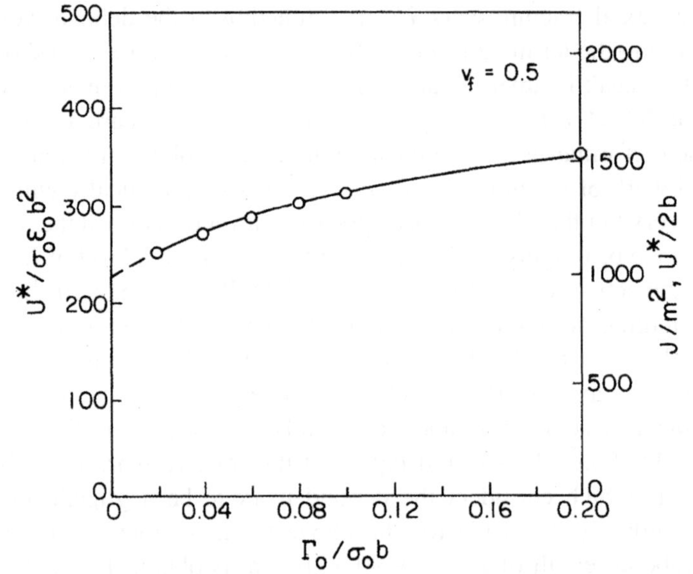

Figure 14. Summary of the dependence of axial fracture energy on the interface debond energy at the fiber matrix interface, at a stage of local matrix ligament strain of $\epsilon \cong 1.5$ which is taken as the termination condition as represented in Figure 11(a).

in matrix plastic resistance due to overaging but could only be a result of loss of fiber strength. This loss is readily ascertainable if the interpretations of σ_m^* above hold, since for that case

$$\sigma_a = v_f\sigma_f + (1 - v_f)\sigma_f E_m/E_f \to \tfrac{7}{12}\sigma_f, \tag{18a}$$

from which we determine $\sigma_f = 1.29$ GPa. This is a substantial loss of fiber strength compared with $\overline{\sigma}_f^* = 1.78$ GPa. For this we have no ready explanation. An alternative explanation for the loss of tensile strength of the composite is that flaws develop in certain regions of fiber bundles which then lower the strength. If this were the case the tensile strength should be governed by

$$\sigma_a = \sqrt{\frac{U^* E_{ca}}{(1 - \nu^2)\pi a_a}}. \tag{19}$$

If U^* is taken to be close to the peak value of 1500 J/m^2 determined in Section 4.1 above, for an upper bound estimate of axial modulus $E_{ca} = v_f E_f + (1 - v_f)E_m$ or 245 GPa we determine an effective crack half length a_a to be 2.28×10^{-4}m. While applying continuum fracture mechanics notions to a composite at this microstructure scale is risky, it is unlikely that cracks of this size can appear as a result of a relatively benign process of overaging. Thus, if the decline in composite strength is neither due to a loss of fiber strength nor due to appearance of substantial size cracks we would have to conclude that the matrix ligament toughness is significantly reduced due to an aging induced embrittlement.

Transverse properties

To start out, we estimate the level of transverse resistance of the composite based on its being governed by plastic flow. We obtain an estimate of the composite plastic resistance from an appropriate model of Huang (1971) based on considering the cylindrical fibers to be acting as rigid filler. The expression of Huang for the transverse plastic resistance σ_{tp} is

$$\frac{\sigma_{tp}}{\sigma_0} = \frac{(1 - v_f)}{(1 - Rv_f)} \frac{1}{(1 - v_f)^N}, \tag{20}$$

where $R \to 4/3$ for $1/N$ exceeding 5 (which for our case is 20). Taking $\sigma_0 = 350\,\text{MPa}$ for the overaged alloy and $N = 0.05$, we calculate σ_{tp} to be 543 MPa which exceeds the largest measured transverse tensile strength of 338 MPa. Thus, we must conclude that the transverse strength is flaw governed where now the flaws under consideration must be imperfectly formed fiber/matrix interfaces. Thus, using an expression similar to (19), i.e.,

$$\sigma_t = \sqrt{\frac{\Gamma_0 E_{ct}}{(1 - \nu^2)\pi a_t}}, \tag{21}$$

and taking $\Gamma_0 = 649\,\text{J/m}^2$ for $\varphi = 0$, as determined by the traction/separation simulation and E_{ct} as the lower bound estimate to the composite modulus as 120 GPa, we estimate a transverse flaw half size, a_t to be 235 μm. Interface flaws of this magnitude along fiber/matrix interfaces are not unlikely and can result from any number of processing imperfections, but being most likely due to inadequate fiber preparation governing nonuniform wetting of the fiber by the molten matrix. Finally, we explore whether or not the low transverse tensile strengths in the manufactured product are consistent with this interpretation. Prior to attainment of good transverse strength for aging periods less than 100 hrs the transverse tensile strength is around 50 MPa. If the same transverse processing flaws are present, the low transverse strength would have to be a result of a low debond strength Γ_0. Thus, considering the same transverse interface flaw size of 235 μm estimated above and the transverse modulus of 120 GPa, we estimate Γ_0 to be 14 J/m^2. From (17) this implies an interface precipitate spacing $D = 0.120\,\mu\text{m}$ and an interface precipitate size of 24 nm. An interface precipitate size of this magnitude in the as cast alloy, prior to coarsening is quite likely. This indicates that, if flaw healing as a result of aging can be discounted, the coarsening of particles would indeed be the principal cause of the improvement of the properties of the present metal matrix composite.

General observations

The traction/separation model of Section 3.2 can be adapted to model the mode B types of simple shear separation processes that accompany the mode A types of primary fractures. This, and its projection to experiments will be the subject of a future communication.

The traction/separation law at interfaces, as well as those of Section 3.3 for the total mode A fracture of the composite, have the features of a classical tension/shear: traction/separation response of solids on the atomic level, as it has been fashioned by Rice et al. (1992) for dislocation emission at crack tips and further generalized by Xu et al., (1995, 1997) for the same purpose. In the present context, however, the traction/separation response of Section 3.2 by elastic–plastic cavity growth and the tensile fracture response of Section 3.3 are on scales larger than the atomistic processes by 3–5 orders of magnitude. Moreover, we note that the ductile debonding at a precipitate covered interface under a mixture of modes simulated in Section 3.2 gives a considerably different and less smoothed response than the traction/separation law under combined tension and shear used in Section 3.3 via (8). When presently active experimental measurements of the actual separation under various mixtures of loading are complete a more realistic functional form to replace that of (8) should emerge. At present the form used in (8) has definite operational advantages.

While only preliminary in their exploration of the complex fracture response of metal matrix composites and being in need of much further development in both micro-mechanical models and in definitive experiments, we view the present results as important in presenting methodologies for further development.

We note with interest that ideas similar to ours and on very similar Al/Al_2O_3 systems have also been explored by Shang and co-workers (Liu and Shang, 1996) but with considerably different emphasis.

Acknowledgement

The early experimental research was supported by the ONR through Contract N00014-85-K-0645. The present and expanded program is supported by the AFOSR under Grant F49620-97-0347. The associated micro-mechanical model studies performed at Brown University were supported by ONR Grant N00014-95-1-0399, with additional support coming from NSF through the Materials Research Science and Engineering Center at Brown University under Grant DMR-9632524. We acknowledge the collaboration of Dr. J.A. Cornie of MMCC Inc. in the earlier ONR supported experimental studies.

References

Argon, A.S. (1974). *Statistical Aspects of Fracture*. In: Composite Materials: Fracture and Fatigue (Edited by L.J. Broutman), vol. 5, chap. 4, pp. 153–190. Academic Press, New York.

Argon, A.S., Seleznev, M.L., Shih, C.F., and Liu, X.H. (1997). *Mechanisms and Mechanics of Debond Fractures along Metal-Matrix Composite Interfaces* In: Advances in Fracture Research, (Edited by Karihaloo, B.L., et al.), vol. 5, pp. 2377–2389, Pergamon Press, Oxford.

Friler, J.B., Argon, A.S. and Cornie, J.A. (1993). Strength and toughness of carbon fiber reinforced aluminum matrix composites. *Material Science Engineering* **A162**, 143–152.

Huang, W.-C. (1971). *Journal of Composite Materials* **5**, 320–338.

Liu, G., and Shang, J.K. (1996). *Fracture Behavior of Interphase – Modified Metal/Ceramic Interfaces*, In: Fracture Mechanics of Ceramics (Edited by Bradt, R.C., et al.), pp. 451–462, Plenum Press, New York.

Liu, X.H. (1998). *Plastic Kinking in Polymer Matrix Composites and Interface Debonding in Metal-Matrix Composites*, PhD Thesis, Brown University, Providence RI.

Nagpal, V., McClintock, F.A., Berg, C.A., and Subudhi, M. (1973). *Traction-displacement Boundary Conditions for Plastic Fracture by Hole Growth*, In: Foundations of Plasticity (Edited by Sawczuk, A.), pp. 365–385, Noordhoff. Intern. Publ., Leyden, Netherlands.

Needleman, A., (1987). A continuum model for void nucleation and inclusion debonding. *Journal of Applied Mechanics* **54**, 523–531.

Rice, J.R., Beltz, G.E., and Sun, Y. (1992). *Peierls Framework for Dislocation Nucleation from a Crack Tip*, In: Topics in Fracture and Fatigue, (Edited by Argon, A.S.), pp. 1–58, Springer Verlag, New York.

Rose, J.H., Smith, J.R., Guinea, F., and Ferrante, J. (1984). Universal features of the equation of state of metals. *Physical Review* **B29**, 2963–2969.

Seleznev, M.L., Cornie, J.A., and Armantis Jr., F.A. (1993). Improving mechanical properties of nextel 610 reinforced Al-224 alloy through θ phase precipitation at fiber/matrix interface: kinetics of the precipitation process *Journal of Material Engineering Performance* **2**, 347–352.

Shih, C.F., and Liu, X.H. (1998). To be published.

Suo, Z., and C.F. Shih, (1993). *Models for Metal/Ceramic Interface Fracture*. In: Metal Matrix Composites (Edited by S. Suresh et al.) chap. 12, pp. 217–232. Butterworth-Heineman, London.

Tvergaard, V., and Hutchinson, J.W. (1993). The influence of plasticity on mixed-mode interface toughness, *Journal of Mechancial and Physical of Solids* **41**, 1119–1135.

Varias, A.G., O'Dowd, N.P., Asaro, R.J., and Shih, C.F. (1990) Failure of bimaterial interfaces. *Materials Science Engineering*. **A126**, 65–93.

Xu, G., Argon, A.S., and Ortix, M. (1995). Nucleation of dislocations from crack tips under mixed modes of loading: implications for brittle against ductile behavior of crystals. *Philosophical Magazine* **A72**, pp. 415–451; (1997) Critical configurations for dislocation nucleation from crack tips. *Philosophical Magazine* **75**, 341–367.

International Journal of Fracture **93:** 373–386, 1998.

Analysis of sharp angular notches in anisotropic materials

C. BLANCO[1], J.M. MARTÍNEZ-ESNAOLA[1] and C. ATKINSON[2]

[1]*Centro de Estudios e Investigaciones Técnicas de Guipúzcoa (CEIT), P°. Manuel Lardizábal 15, 20009 San Sebastián, Spain; e-mail: jmesnaola@ceit.es and Escuela Superior de Ingenieros, Universidad de Navarra, P°. Manuel Lardizábal 13, 20009 San Sebastián, Spain*

[2]*Department of Mathematics, Imperial College of Science, Technology and Medicine, Huxley Building, 180 Queen's Gate, London SW7 2BZ, U.K.*

Received 11 July 1997; accepted in revised form 18 March 1998

Abstract. The singular stresses at the tip of a sharp angular notch are analysed for the most general case of elastic anisotropy. The problem is stated in relation with the kinked crack and is modelled by means of continuous distributions of dislocations which are assumed to be singular at the notch vertex, the kind of the main singularity λ being unknown and weaker than at the crack tip. The Mellin transform is applied to obtain a system of simultaneous functional equations that enables one to find the parameter λ. The reciprocal theorem is used to compute the generalised stress intensity factor which characterises the singular stresses in a neighbourhood of the notch tip.

Key words: Stress singularities, general elastic anisotropy, kinked cracks, angular notches, dislocation models, reciprocal theorem, generalised stress intensity factors.

1. Introduction

Problems related to stress singularities have received much attention in the theory of Elasticity, see, for example, Williams (1952), Muskhelishvili (1977), Cherepanov (1979), Atkinson (1979). In particular, near the tip of a sharp angular notch, the elastic stress field is singular with singularity weaker than that of a crack, that is, the stresses behave like $\sigma \approx r^{-\lambda}$, with $0 < \lambda < 1/2$, where r is measured from the notch tip. In isotropic materials, the stress exponent λ only depends on the notch angle, see, for example, Williams (1952), Atkinson et al. (1988). However, in the case of anisotropic materials, the severity of the stress singularity is also dependent upon the material properties. For example, Bogy (1972), Heinzelmann et al. (1994) and Busch et al. (1994) have developed asymptotic analyses for angular notches in orthotropic materials, which show such a dependence on the elastic constants.

In this paper we analyse the notch problem in the most general case of elastic anisotropy. In Section 2 an analytical-numerical method is developed to calculate the dominant stress singularity near the notch tip. The more general problem we have in mind (which is discussed in an accompanying paper) is the analysis of the effect of the stress singularity at the apex of a kinked crack on the stress intensity factors at the kinked and main crack tips. For this application it is convenient to model the whole kinked crack as a continuous distribution of dislocations which is also singular at the kink. Applying the Mellin transform, we obtain a system of functional equations that enables one to find the notch singularity λ. In Section 3 an equation is obtained for the calculation of the corresponding generalised stress intensity factor (the coefficient of the notch singularity) by applying a reciprocal work contour integral which has been proved to be useful for problems related to stress singularities (Stern et al.,

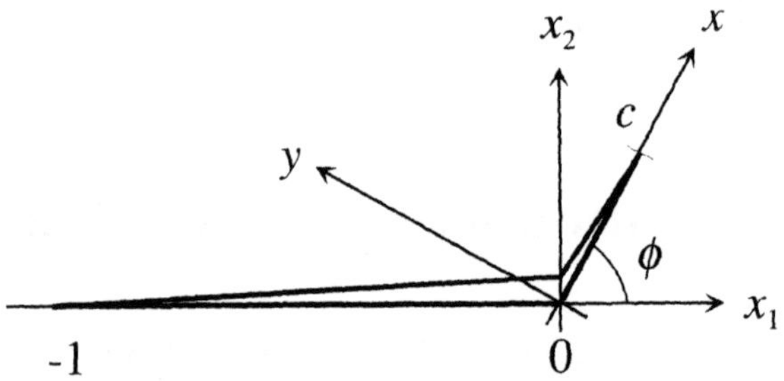

Figure 1. Kinked crack: coordinate system.

1976; Atkinson et al., 1988; Bastero et al., 1989). Using the reciprocal theorem, the singular stress field near the notch tip can be determined through the known remote stress field. The present formulation is checked against solutions reported in the literature for particular cases of anisotropy. A noteworthy result is that the dominant singularity at the notch tip is arbitrary up to a single constant in the most general cases where there is little or no symmetry.

2. Analysis of the singularity at the notch tip

To model a kinked crack consider a coordinate system (x_1, x_2, x_3) such that the main crack lies on the x_1-axis, the origin coincides with the vertex and the crack tip with the point $(-1, 0)$, see Figure 1. The branch of length c forms an angle ϕ with the main crack. A new coordinate system (x, y, z), $z = x_3$, is introduced so that the branch line corresponds to the x-axis.

The kinked crack is modelled by means of continuous distributions of dislocations that are designated as f_j and g_j, if they correspond to the main crack or to the branch, respectively.

The solution for the stress field produced by an isolated dislocation located at point (a_1, a_2) with Burgers' vector b_i in an infinite anisotropic medium is (Stroh, 1958)

$$\sigma_{i1}(x_1, x_2) = \frac{-1}{4\pi} \sum_\alpha L_{i\alpha} M_{\alpha j} p_\alpha \frac{d_j}{z_\alpha - \zeta_\alpha} + C.C., \tag{2.1a}$$

$$\sigma_{i2}(x_1, x_2) = \frac{1}{4\pi} \sum_\alpha L_{i\alpha} M_{\alpha j} \frac{d_j}{z_\alpha - \zeta_\alpha} + C.C., \tag{2.1b}$$

where C.C. denotes the complex conjugate of the preceding expression and the convention of summing over repeated Latin indices is used. In the above equations, $\zeta_\alpha = a_1 + p_\alpha a_2$ and $z_\alpha = x_1 + p_\alpha x_2$. The magnitudes p_α, $L_{i\alpha}$, $M_{\alpha j}$ and d_j are designated with the same name in Stroh's work, and are given in Appendix 1 for clarity.

Integrating equations (2.1) along the whole crack, the tractions produced by the dislocation densities f_j and g_j at a point on the main crack $(x_1, x_2 = 0)$ can be expressed as

$$\begin{aligned} \sigma_{k2}(x_1) = {} & \frac{1}{4\pi} \sum_\alpha L_{k\alpha} M_{\alpha j} \\ & \times \left\{ \int_{-1}^{0} \frac{f_j(\zeta)}{x_1 - \zeta} \, d\zeta + \int_0^c \frac{g_j(\eta)}{x_1 - \eta\tau_\alpha} \, d\eta \right\} + C.C. \quad (-1 < x_1 < 0) \end{aligned} \tag{2.2a}$$

and the tractions at a point $(x, y = 0)$ of the branch are given by

$$\sigma'_{k2}(x) = \frac{l_{ki}}{4\pi} \sum_{\alpha} L_{i\alpha} M_{\alpha j} \tau_\alpha \times \left\{ \int_{-1}^{0} \frac{f_j(\zeta)}{x\tau_\alpha - \zeta} \, d\zeta + \int_{0}^{c} \frac{g_j(\eta)}{x\tau_\alpha - \eta\tau_\alpha} \, d\eta \right\} + C.C. \quad (0 < x < c) \tag{2.2b}$$

where l_{ij} are the direction cosines of the axes (x, y) with respect to (x_1, x_2), and

$$\tau_\alpha = \cos\phi + p_\alpha \sin\phi. \tag{2.3}$$

To determine the functions f_j and g_j use is made of the condition that the crack is traction free, therefore,

$$-t_k(x_1) = \frac{1}{4\pi} \sum_{\alpha} L_{k\alpha} M_{\alpha j} \times \left\{ \int_{-1}^{0} \frac{f_j(\zeta)}{x_1 - \zeta} \, d\zeta + \int_{0}^{c} \frac{g_j(\eta)}{x_1 - \eta\tau_\alpha} \, d\eta \right\} + C.C. \quad (-1 < x_1 < 0), \tag{2.4a}$$

$$-t'_k(x) = \frac{l_{ki}}{4\pi} \sum_{\alpha} L_{i\alpha} M_{\alpha j} \tau_\alpha \times \left\{ \int_{-1}^{0} \frac{f_j(\zeta)}{x\tau_\alpha - \zeta} \, d\zeta + \int_{0}^{c} \frac{g_j(\eta)}{x\tau_\alpha - \eta\tau_\alpha} \, d\eta \right\} + C.C. \quad (0 < x < c), \tag{2.4b}$$

where t_k and t'_k are the tractions on the main crack and on the branch crack respectively, induced by the external load in the medium without the crack.

Equations (2.4a) and (2.4b) can be extended to the intervals $(-\infty, 0)$ and $(0, \infty)$, respectively, by defining the unknown functions $s_k(x_1)$ and $s'_k(x)$ as follows

$$\frac{1}{4\pi} \sum_{\alpha} L_{k\alpha} M_{\alpha j} \left\{ \int_{-1}^{0} \frac{f_j(\zeta)}{x_1 - \zeta} \, d\zeta + \int_{0}^{c} \frac{g_j(\eta)}{x_1 - \eta\tau_\alpha} \, d\eta \right\} + C.C. = \begin{cases} -t_k(x_1) & (-1 < x_1 < 0) \\ s_k(x_1) & (x_1 < -1) \end{cases} \tag{2.5a}$$

$$\frac{l_{ki}}{4\pi} \sum_{\alpha} L_{i\alpha} M_{\alpha j} \tau_\alpha \left\{ \int_{-1}^{0} \frac{f_j(\zeta)}{x\tau_\alpha - \zeta} \, d\zeta + \int_{0}^{c} \frac{g_j(\eta)}{x\tau_\alpha - \eta\tau_\alpha} \, d\eta \right\} + C.C. = \begin{cases} -t'_k(x) & (0 < x < c) \\ s'_k(x) & (c < x). \end{cases} \tag{2.5b}$$

Applying the change of variable

$$\begin{cases} x_1 = -x' \\ \zeta = -\zeta_1 \end{cases} \tag{2.6}$$

we obtain

$$\frac{1}{4\pi}\sum_{\alpha} L_{k\alpha}M_{\alpha j}$$

$$\times\left\{-\int_0^1 \frac{f_j(-\zeta_1)}{x'-\zeta_1}\,\mathrm{d}\zeta_1 - \int_0^c \frac{g_j(\eta)}{x'+\eta\tau_\alpha}\,\mathrm{d}\eta\right\} + C.C. = t_k^*(x') \quad (x'>0) \tag{2.7a}$$

$$\frac{l_{ki}}{4\pi}\sum_{\alpha} L_{i\alpha}M_{\alpha j}\tau_\alpha$$

$$\times\left\{\int_0^1 \frac{f_j(-\zeta_1)}{x\tau_\alpha+\zeta_1}\,\mathrm{d}\zeta_1 + \int_0^c \frac{g_j(\eta)}{x\tau_\alpha-\eta\tau_\alpha}\,\mathrm{d}\eta\right\} + C.C. = t_k'^*(x) \quad (x>0), \tag{2.7b}$$

where

$$\begin{aligned} t_k^*(x') &= \begin{cases} -t_k(-x') & (0<x'<1) \\ s_1(-x') & (1<x') \end{cases} \\ t_k'^*(x) &= \begin{cases} -t_k'(x) & (0<x<c) \\ s_k'(x) & (c<x). \end{cases} \end{aligned} \tag{2.8}$$

Applying the Mellin transform and doing the algebra, equations (2.7) turn into

$$\begin{aligned} \tilde{t}_k^*(s) &= \tfrac{1}{2}\cot(\pi s)\int_0^1 f_k(-\zeta_1)\zeta_1^{s-1}\,\mathrm{d}\zeta_1 - \tfrac{1}{4}\csc(\pi s) \\ &\quad\times\left\{\sum_\alpha L_{k\alpha}M_{\alpha j}\tau_\alpha^{s-1} + \sum_\alpha \bar{L}_{k\alpha}\bar{M}_{\alpha j}\bar{\tau}_\alpha^{s-1}\right\} \\ &\quad\times\int_0^c g_j(\eta)\eta^{s-1}\,\mathrm{d}\eta \quad (0<\mathrm{Re}(s)<1), \end{aligned} \tag{2.9a}$$

$$\begin{aligned} \tilde{t}_k'^*(s) &= \frac{l_{ki}}{4}\csc(\pi s) \\ &\quad\times\left\{\sum_\alpha L_{i\alpha}M_{\alpha j}\tau_\alpha^{1-s} + \sum_\alpha \bar{L}_{i\alpha}\bar{M}_{\alpha j}\bar{\tau}_\alpha^{1-s}\right\}\int_0^1 f_j(-\zeta_1)\zeta_1^{s-1}\,\mathrm{d}\zeta_1 \\ &\quad-\frac{l_{kj}}{2}\cot(\pi s)\int_0^c g_j(\eta)\eta^{s-1}\,\mathrm{d}\eta \quad (0<\mathrm{Re}(s)<1), \end{aligned} \tag{2.9b}$$

where $\tilde{f}(s)$ denotes the Mellin transform of the function $f(x)$, that is,

$$\tilde{t}_k^*(s) = \mathcal{M}(t_k^*(x'), s) = \int_0^\infty x'^{s-1}t_k^*(x')\,\mathrm{d}x' \tag{2.10a}$$

and

$$\tilde{t}_k'^*(s) = \mathcal{M}(t_k'^*(x), s) = \int_0^\infty t_k'^*(x)x^{s-1}\,\mathrm{d}x. \tag{2.10b}$$

Equations (2.9) constitute a system of functional equations defined in the domain $0 < \mathrm{Re}(s) < 1$, with unknowns

$$\hat{f}_j(s) = \int_0^1 f_j(-\zeta_1)\zeta_1^{s-1}\,\mathrm{d}\zeta_1 \quad j = 1, 2, 3 \tag{2.11a}$$

$$\hat{g}_j(s) = \int_0^c g_j(\eta)\eta^{s-1}\,\mathrm{d}\eta \quad j = 1, 2, 3 \tag{2.11b}$$

which can be written in matrix form as

$$\mathbf{M}(s)\mathbf{F}(s) = \mathbf{T}(s), \tag{2.12}$$

where

$$\mathbf{M}(s) = \begin{pmatrix} \frac{1}{2}\cot(\pi s)\delta_{kj} & -\frac{1}{2}\csc(\pi s)H_{kj}(s-1) \\ \cdots & \cdots \\ \frac{1}{2}l_{ki}\csc(\pi s)H_{ij}(1-s) & -\frac{1}{2}l_{kj}\cot(\pi s) \end{pmatrix} \tag{2.13}$$

$$\mathbf{F}(s) = \begin{pmatrix} \hat{f}_j(s) \\ \cdots \\ \hat{g}_j(s) \end{pmatrix}, \tag{2.14}$$

$$\mathbf{T}(s) = \begin{pmatrix} \tilde{t}_k^*(s) \\ \cdots \\ \tilde{t}_k^{\prime *}(s) \end{pmatrix}, \tag{2.15}$$

with

$$H_{ij}(z) = \frac{1}{2}\left\{\sum_\alpha L_{i\alpha}M_{\alpha j}\tau_\alpha^z + \sum_\alpha \bar{L}_{i\alpha}\bar{M}_{\alpha j}\bar{\tau}_\alpha^z\right\}. \tag{2.16}$$

Since $H_{ij}(\zeta)$ are entire functions, the elements of the system matrix, $\mathbf{M}(s)$, are analytic within the strip $0 < \mathrm{Re}(s) < 1$. Likewise, the elements of the matrix of independent terms, $\mathbf{T}(s)$, are analytic in the same domain.

The dislocation densities f_j and g_j are expected to be singular at the origin and at both crack tips, with singularities $r^{-\lambda}$ and $r^{-1/2}$, respectively, with $0 < \lambda < \frac{1}{2}$ (note that in order not to disallow a complex singularity in the near notch tip behaviour we denote $r^{-\lambda}$ as the dominant part of the singular behaviour); therefore, we can express these functions as follows

$$f_j(\zeta) = \frac{D_j(\zeta)}{(1+\zeta)^{1/2}(c-\zeta)^{1/2}} + \frac{F_j(\zeta)}{(-\zeta)^\lambda} \quad (-1 < \zeta < 0), \tag{2.17a}$$

$$g_j(\eta) = \frac{D_j(\eta)}{(1+\eta)^{1/2}(c-\eta)^{1/2}} + \frac{G_j(\eta)}{\eta^\lambda} \quad (0 < \eta < c), \tag{2.17b}$$

where F_j, G_j and D_j are bounded functions in $[-1, 0]$, $[0, c]$ and $[-1, c]$, respectively. Using equations (2.17), the unknowns of the system (2.9) can be written as

$$\hat{f}_j(s) = \int_0^1 \frac{D_j(-\zeta_1)\zeta_1^{s-1}}{(1-\zeta_1)^{1/2}(c+\zeta_1)^{1/2}}\,\mathrm{d}\zeta_1 + \int_0^1 F_j(-\zeta_1)\zeta_1^{s-1-\lambda}\,\mathrm{d}\zeta_1 \tag{2.18a}$$

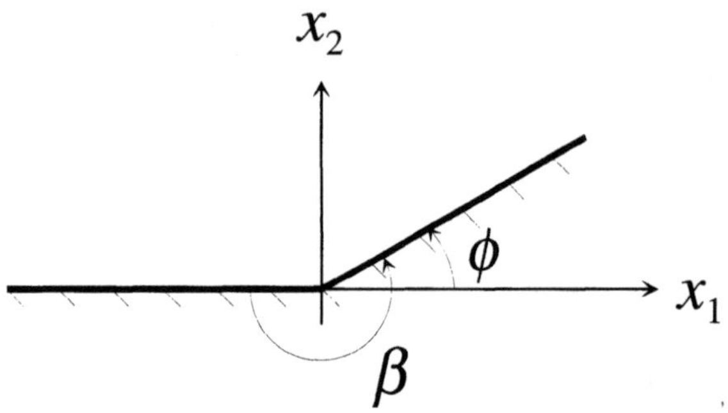

Figure 2. Sharp angular notch.

$$\hat{g}_j(s) = \int_0^c \frac{D_j(\eta)\eta^{s-1}}{(1+\eta)^{1/2}(c-\eta)^{1/2}}\,\mathrm{d}\eta + \int_0^c G_j(\eta)\eta^{s-1-\lambda}\,\mathrm{d}\eta \tag{2.18b}$$

and therefore, they are analytic in the domain $\lambda < \mathrm{Re}(s) < 1$.

From these observations, it can be concluded that the inverse of the system matrix, $\mathbf{M}^{-1}(s)$, must be singular at a point s_0 with $\mathrm{Re}(s_0) = \lambda$. In particular, we are interested in the parameter λ that characterises the severity of the dominant stress singularity in a neighbourhood of the vertex, which will be given by the root of greatest real part of the characteristic equation

$$\det[\mathbf{M}(s)] = 0. \tag{2.19}$$

For the solution to have physical sense, the root s_0 should be real. Although this property has not been proved analytically, its validity has been assessed by numerical calculations for different notch angles and material constants.

Note that the matrix $\mathbf{M}(s)$ depends both upon the elastic constants – through the Stroh's constants $L_{i\alpha}$ and $M_{\alpha j}$ – and the notch angle ϕ. Therefore, in anisotropic materials, the stress singularity depends on both sets of parameters, while in the isotropic case it only depends on the notch angle (see, for example, Atkinson et al., 1988; Sih and Ho, 1991; Williams, 1952). Moreover, $M(s)$ is independent of the branch length, see Equations (2.13) and (2.16), and then, as could be expected, λ does not depend on the dimensions of the kinked crack taken to develop the formulation.

3. The generalised stress intensity factor

In this section we develop a method to analyse the singular stress field near the tip of a sharp angular notch. By similarity with the preceding formulation, the notch is described by the material angle $\beta = \pi + \phi$, with $0 < \phi < \pi$, and we assume that the faces of the notch are traction free (Figure 2).

3.1. Statement of the Problem

Numerical calculations have been used to check that the rank of matrix $\mathbf{M}(\lambda)$ is five for a number of notch angles and material constants (Blanco, 1997). Therefore the eigenvector $\mathbf{V} = v_j (j = 1, \ldots, 6)$ which is solution of the equation

$$\mathbf{M}(\lambda) \cdot \mathbf{V} = \mathbf{0} \tag{3.1}$$

is determined up to a multiplicative constant. Note that in the case of a crack the rank of $\mathbf{M}(\lambda)$ will be 3 with $\lambda = 1/2$.

The asymptotic stress field near the notch tip is modelled as a continuous distribution of dislocations on the notch faces with density functions

$$f_j(\zeta) = Av_j(-\zeta)^{-\lambda} \quad (-1 < \zeta < 0) \tag{3.2a}$$

$$g_j(\eta) = Av_{j+3}\eta^{-\lambda} \quad (0 < \eta < c) \tag{3.2b}$$

where A is an unknown constant. The stresses induced by this distribution of dislocations can be expressed as

$$\sigma_{k1}(z_\alpha) = -\frac{A}{2\pi} \times \mathrm{Re}\left[\sum_\alpha L_{k\alpha}M_{\alpha j}p_\alpha \left\{v_j \int_{-1}^{0} \frac{(-\zeta)^{-\lambda}}{z_\alpha - \zeta}\,\mathrm{d}\zeta + v_{j+3}\int_0^c \frac{\eta^{-\lambda}}{z_\alpha - \eta\tau_\alpha}\,\mathrm{d}\eta\right\}\right] \tag{3.3a}$$

$$\sigma_{k2}(z_\alpha) = \frac{A}{2\pi} \times \mathrm{Re}\left[\sum_\alpha L_{k\alpha}M_{\alpha j} \left\{v_j \int_{-1}^{0} \frac{(-\zeta)^{-\lambda}}{z_\alpha - \zeta}\,\mathrm{d}\zeta + v_{j+3}\int_0^c \frac{\eta^{-\lambda}}{z_\alpha - \eta\tau_\alpha}\,\mathrm{d}\eta\right\}\right]. \tag{3.3b}$$

Taking $z_\alpha = \varepsilon(\cos\theta + p_\alpha \sin\theta)$ with $\varepsilon \to 0$ in Equations (3.3), the asymptotic stresses can be obtained as

$$\sigma_{k1}(\varepsilon \to 0) = -A\frac{\varepsilon^{-\lambda}}{2}\csc(\pi\lambda) \times \mathrm{Re}\left[\sum_\alpha L_{k\alpha}M_{\alpha j}p_\alpha\tau_\alpha(\theta)^{-\lambda}\left\{v_j - v_{j+3}\frac{(-1)^{-\lambda}}{\tau_\alpha^{1-\lambda}}\right\}\right] \tag{3.4a}$$

$$\sigma_{k2}(\varepsilon \to 0) = A\frac{\varepsilon^{-\lambda}}{2}\csc(\pi\lambda) \times \mathrm{Re}\left[\sum_\alpha L_{k\alpha}M_{\alpha j}\tau_\alpha(\theta)^{-\lambda}\left\{v_j - v_{j+3}\frac{(-1)^{-\lambda}}{\tau_\alpha^{1-\lambda}}\right\}\right], \tag{3.4b}$$

where

$$\tau_\alpha(\theta) = \cos\theta + p_\alpha \sin\theta. \tag{3.5}$$

Equations (3.4) show that the stresses σ_{ij} have a singular behaviour in the neighbourhood of the notch vertex, that is, $\sigma_{ij} \approx A\varepsilon^{-\lambda} f_{ij}(\theta)$, where $f_{ij}(\theta)$ are known functions and A is an unknown constant. Note also that, as expected, the asymptotic stresses at the notch tip are independent of the segments $[-1, 0]$ and $[0, c]$ along which the distributions of dislocations are applied in the present approach.

In what follows, use is made of the reciprocal theorem to develop a method for the numerical calculation of the generalised stress intensity factor A.

3.2. Determination of the generalised stress intensity factor

3.2.1. *The reciprocal theorem*

The reciprocal theorem for elastostatics, assuming no body forces are present, is

$$\int_S (\sigma_{ij}\hat{u}_i - \hat{\sigma}_{ij}u_i)n_j \,\mathrm{d}S = 0, \tag{3.6}$$

where n_j is the outward pointing normal to surface S, and $\hat{\sigma}_{kj}$ and $\hat{u}_k$ are two auxiliary (dual) fields of stress and displacement which satisfy the same equilibrium and constitutive equations as the actual fields σ_{kj} and u_k,

$$\sigma_{ij,j} = 0 = \hat{\sigma}_{ij,j} \tag{3.7}$$

$$\sigma_{ij} = c_{ijkl}u_{k,l} \quad \text{and} \quad \hat{\sigma}_{ij} = c_{ijkl}\hat{u}_{k,l}, \tag{3.8}$$

and S is a closed surface containing no singularities.

As our problem is two dimensional, Equation (3.6) becomes

$$\int_C (\sigma_k\hat{u}_k - \hat{\sigma}_k u_k)\,\mathrm{d}l = 0, \quad \text{with} \quad \sigma_k = \sigma_{kj}n_j \quad \text{and} \quad \hat{\sigma}_k = \hat{\sigma}_{kj}n_j \tag{3.9}$$

with C a simple closed contour.

3.2.2. *The dual fields*

In order to apply the reciprocal theorem to the notch problem, the suitable dual fields of stress and displacement must be chosen so that the integrals along an infinitesimal path that shrinks to the notch tip are bounded. As we concluded before, the parameter λ that characterises the stress singularity is a root of equation (2.19). It can be easily proved that $s = 2 - \lambda$ also verifies this equation.

As dual fields we take the fields induced by the distribution of dislocations $\hat{d}'_j(r)$ along both faces of the notch, with density functions defined by

$$\hat{d}'_j(r) = \begin{cases} \dfrac{w_j}{(-r)^{2-\lambda}} & (-\infty < r < 0) \\ \dfrac{w_{j+3}}{r^{2-\lambda}} & (0 < r < \infty) \end{cases} \tag{3.10}$$

where $\mathbf{W} = w_i (i = 1, \ldots, 6)$ is the eigenvector of matrix $\mathbf{M}$ corresponding to the eigenvalue $s = 2 - \lambda$.

Using again Stroh's solution for a single dislocation, we obtain for the dual fields

$$\begin{aligned} \hat{\sigma}_{i1}(z_\alpha) &= -\frac{1}{4\pi}\sum_\alpha L_{i\alpha}M_{\alpha j}p_\alpha \\ &\times \left[w_j \int_{-\infty}^{0} \frac{(-r)^{\lambda-2}}{z_\alpha - r}\,\mathrm{d}r + w_{j+3}\int_0^\infty \frac{r^{\lambda-2}}{z_\alpha - \tau_\alpha r}\,\mathrm{d}r\right] + C.C. \end{aligned} \tag{3.11a}$$

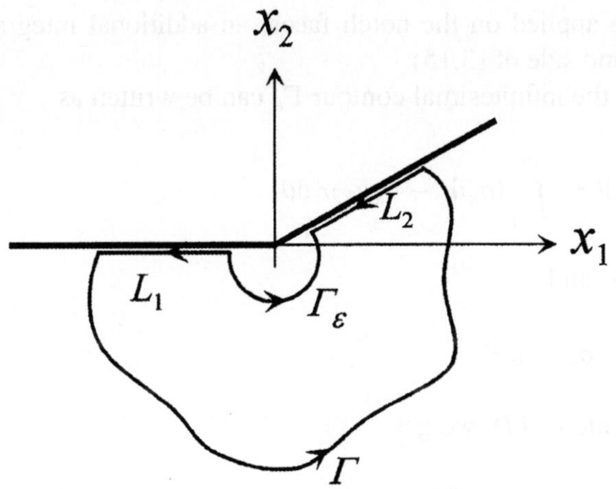

Figure 3. Integration contour to apply the reciprocal theorem.

$$\hat{\sigma}_{i2}(z_\alpha) = \frac{1}{4\pi}\sum_\alpha L_{i\alpha}M_{\alpha j} \times \left[w_j \int_{-\infty}^{0} \frac{(-r)^{\lambda-2}}{z_\alpha - r}\,\mathrm{d}r + w_{j+3}\int_0^{\infty} \frac{r^{\lambda-2}}{z_\alpha - \tau_\alpha r}\,\mathrm{d}r\right] + C.C. \tag{3.11b}$$

which can be expressed as

$$\hat{\sigma}_{i1}(z_\alpha) = -\sum_\alpha L_{i\alpha}p_\alpha \hat{f}'_\alpha(z_\alpha) + C.C. \tag{3.12a}$$

$$\hat{\sigma}_{i2}(z_\alpha) = \sum_\alpha L_{i\alpha}\hat{f}'_\alpha(z_\alpha) + C.C. \tag{3.12b}$$

with

$$\hat{f}'_\alpha(z_\alpha) = \frac{1}{4\pi}M_{\alpha j}\left[w_j \int_{-\infty}^{0} \frac{(-r)^{\lambda-2}}{z_\alpha - r}\,\mathrm{d}r + w_{j+3}\int_0^{\infty} \frac{r^{\lambda-2}}{z_\alpha - \tau_\alpha r}\,\mathrm{d}r\right]. \tag{3.13}$$

With this notation, the dual displacement field can be written as (Eshelby et al., 1953)

$$\hat{u}_k(z_\alpha) = \sum_\alpha A_{k\alpha}\hat{f}_\alpha(z_\alpha) + C.C. \tag{3.14}$$

3.2.3. *Application of the reciprocal theorem*

Consider a closed contour as shown in Figure 3 with Γ_ε a circular path of radius $\varepsilon(\varepsilon \to 0)$. It is easy to prove that the dual tractions $\hat{\sigma}_k$ vanish on L_1 and L_2. Then assuming that the notch faces are traction free, the reciprocal theorem can be expressed

$$\int_{\Gamma_\varepsilon} (\sigma_k\hat{u}_k - \hat{\sigma}_k u_k)\,\mathrm{d}l = \int_\Gamma (\sigma_k\hat{u}_k - \hat{\sigma}_k u_k)\,\mathrm{d}l. \tag{3.15}$$

Note that if loads are applied on the notch faces, an additional integral $\int_{L_1+L_2} \sigma_k \hat{u}_k \, dl$ will appear on the right hand side of (3.15).

The integral along the infinitesimal contour Γ_ε can be written as

$$\int_{\Gamma_\varepsilon} (\sigma_k \hat{u}_k - \hat{\sigma}_k u_k) \, dl = \int_{-\pi}^{\phi} (\sigma_k \hat{u}_k - \hat{\sigma}_k u_k) r \, d\theta \tag{3.16}$$

On Γ_ε, $z_\alpha = \varepsilon \tau_\alpha(\theta)$ and

$$\sigma_k = -\sigma_{k1} \cos \theta - \sigma_{k2} \sin \theta. \tag{3.17}$$

Substituting (3.4) into (3.17), we get

$$\begin{aligned} \sigma_k(\varepsilon\tau_\alpha(\theta)) &= A\frac{\varepsilon^{-\lambda}}{4} \csc(\pi\lambda) \sum_\alpha L_{k\alpha} M_{\alpha j} (-\sin\theta + p_\alpha \cos\theta) \tau_\alpha(\theta)^{-\lambda} \\ &\quad \times \left[v_j - v_{j+3} \frac{(-1)^{-\lambda}}{\tau_\alpha^{1-\lambda}} \right] + C.C. \end{aligned} \tag{3.18}$$

and, therefore

$$\begin{aligned} u_k(\varepsilon\tau_\alpha(\theta)) &= A\frac{\varepsilon^{1-\lambda}}{4(1-\lambda)} \csc(\pi\lambda) \sum_\alpha A_{k\alpha} M_{\alpha j} \tau_\alpha(\theta)^{1-\lambda} \\ &\quad \times \left[v_j - v_{j+3} \frac{(-1)^{-\lambda}}{\tau_\alpha^{1-\lambda}} \right] + C.C. \end{aligned} \tag{3.19}$$

Likewise, from (3.12) and (3.14) the dual fields can be expressed as

$$\begin{aligned} \hat{\sigma}_k(\varepsilon\tau_\alpha(\theta)) &= \frac{\varepsilon^{\lambda-2}}{2} \csc(\pi\lambda) \\ &\quad \times \mathrm{Re}\Bigg\{ \sum_\alpha L_{k\alpha} M_{\alpha j} (-\sin\theta + p_\alpha \cos\theta) \tau_\alpha(\theta)^{\lambda-2} \\ &\qquad \times \left[-w_j + w_{j+3} \frac{(-1)^{\lambda}}{\tau_\alpha^{\lambda-1}} \right] \Bigg\}. \end{aligned} \tag{3.20}$$

$$\begin{aligned} \hat{u}_k(\varepsilon\tau_\alpha(\theta)) &= \frac{\varepsilon^{\lambda-1}}{2(\lambda-1)} \csc(\pi\lambda) \\ &\quad \times \mathrm{Re} \left\{ \sum_\alpha A_{k\alpha} M_{\alpha j} \tau_\alpha(\theta)^{\lambda-1} \left[-w_j + w_{j+3} \frac{(-1)^{\lambda}}{\tau_\alpha^{\lambda-1}} \right] \right\}. \end{aligned} \tag{3.21}$$

Substituting (3.18) to (3.21) into (3.16), we obtain

$$\int_{\Gamma_\varepsilon} (\sigma_k \hat{u}_k - \hat{\sigma}_k u_k) \, dl = A(C_1 - C_2), \tag{3.22}$$

where

$$C_1 = \frac{1}{4(\lambda-1)}\csc^2(\pi\lambda)$$

$$\times \int_{-\pi}^{\phi}\left(\mathrm{Re}\left\{\sum_{\alpha} L_{k\alpha}M_{\alpha j}(-\sin\theta + p_\alpha\cos\theta)\tau_\alpha(\theta)^{-\lambda}\left[v_j - v_{j+3}\frac{(-1)^{-\lambda}}{\tau_\alpha^{1-\lambda}}\right]\right\}\right.$$

$$\left.\times \mathrm{Re}\left\{\sum_{\beta} A_{k\beta}M_{\beta q}\tau_\beta(\theta)^{\lambda-1}\left[-w_q + w_{q+3}\frac{(-1)^{\lambda}}{\tau_\beta^{\lambda-1}}\right]\right\}\right)\,\mathrm{d}\theta \qquad (3.23)$$

and

$$C_2 = \frac{1}{4(1-\lambda)}\csc^2(\pi\lambda)$$

$$\times \int_{-\pi}^{\phi}\left(\mathrm{Re}\left\{\sum_{\alpha} A_{k\alpha}M_{\alpha j}\tau_\alpha(\theta)^{1-\lambda}\left[v_j - v_{j+3}\frac{(-1)^{-\lambda}}{\tau_\alpha^{1-\lambda}}\right]\right\}\right.$$

$$\times \mathrm{Re}\left\{\sum_{\beta} L_{k\beta}M_{\beta q}(-\sin\theta + p_\beta\cos\theta)\tau_\beta(\theta)^{\lambda-2}\right.$$

$$\left.\left.\times\left[-w_q + w_{q+3}\frac{(-1)^{\lambda}}{\tau_\beta^{\lambda-1}}\right]\right\}\right)\,\mathrm{d}\theta. \qquad (3.24)$$

Finally, from (3.15) the generalised stress intensity factor A can be computed as

$$A = \frac{1}{C_1 - C_2}\int_{\Gamma}(\sigma_k\hat{u}_k - \hat{\sigma}_k u_k)\,\mathrm{d}l. \qquad (3.25)$$

This allows the calculation of A using numerical estimates of the remote stress and displacement fields obtained via standard numerical techniques (for example, finite elements) without a special care for the near tip singular fields.

4. Results

The stress and displacement fields in a neighbourhood of the tip of an angular notch for isotropic materials are known (Atkinson et al., 1988; Sih and Ho, 1991). Table 1 compares the results obtained by Atkinson et al. (1988) for different angles with the values obtained with the present analysis, approximating the isotropic material as a limit case of slight anisotropy.

For anisotropic materials, the stress singularity depends both upon the notch angle and the material elastic constants. The three last columns of Table 1 show the variation of λ for different anisotropic materials. *MS* and *EDT* denote monocrystalline silicon – orthotropic – and ethylene diamine tartrate – monoclinic – (Huntington, 1958), respectively. *AN* denotes an anisotropic material which is not real: a positive-definite, symmetric matrix with arbitrary elements was taken to define the elastic constants. The elasticity tensors of the three materials are given in Appendix 2.

Table 1. Values of λ for anisotropic materials.

	Material				
ϕ(o)	Isotropic (Atkinson et al., 1988)	Slight anisotropy	*EDT*	*MS*	*AN*
0	0.0000	0.0000	0.0000	0.0000	0.0000
20	0.1813	0.1813	0.2393	0.1652	0.2400
40	0.3028	0.3028	0.3238	0.2938	0.3688
60	0.3842	0.3842	0.3725	0.3824	0.4272
80	0.4371	0.4371	0.4220	0.4407	0.4545
100	0.4696	0.4696	0.4639	0.4712	0.4709
120	0.4877	0.4877	0.4861	0.4884	0.4829
140	0.4965	0.4965	0.4954	0.4968	0.4924
160	0.4995	0.4995	0.4992	0.4996	0.4986
180	0.5000	0.5000	0.5000	0.5000	0.5000

Note that for $\phi = 0°$, both notch faces lie on the same plane; the value $\lambda = 0$ is then consistent with the absence of singularity as there is no tip in this case. The other limit case is when $\phi = 180°$, for which the value $\lambda = 0.5$ agrees with the known stress singularity at the crack tip.

Another check has been done for the particular case of an orthotropic material. For a notch of monocrystalline silicon with an opening angle of $70.52°$ ($\phi = 109.48°$ in our formulation), Heinzelmann et al. (1994) give as the maximum singular stress exponent $\lambda = 0.4814$, while the value obtained through our formulation is $\lambda = 0.4806$.

5. Conclusions

An analytical-numerical method has been developed to obtain the stress singularity in the neighbourhood of the tip of an angular notch in materials with general anisotropy. The validity of this method has been checked against the solution for angular notches in orthotropic and isotropic materials, the latter being analysed as a limit case of our formulation. Results have been presented that show the influence of the material parameters on the kind of singularity when the case of general anisotropy is considered.

It has been shown that the application of the reciprocal theorem allows the calculation of the generalised stress intensity factor at the notch tip through the remote stress and displacement fields which can be efficiently approximated using standard numerical techniques.

Appendix 1

With regard to Equations (2.1), the $p_\alpha (\alpha = 1, 2, 3)$ are the roots with imaginary positive part of the characteristic equation

$$\det[c_{i1k1} + p(c_{i1k2} + c_{i2k1}) + p^2 c_{i2k2}] = 0, \tag{A1.1}$$

where c_{ijkl} is the tensor of elastic constants, i.e., $\sigma_{ij} = c_{ijkl}u_{k,l}$, which satisfies the symmetry conditions

$$c_{ijkl} = c_{ijlk} = c_{jikl} = c_{klij}. \tag{A1.2}$$

The matrix $L_{i\alpha}$ is given by

$$L_{i\alpha} = A_{k\alpha}(c_{i2k1} + p_\alpha c_{i2k2}), \tag{A1.3}$$

where $A_{k\alpha}$ denotes the eigenvector corresponding to the eigenvalue p_α above ($\alpha = 1, 2, 3$). The matrix $M_{\alpha j}$ is defined as the inverse of $L_{i\alpha}$,

$$M_{\alpha j}L_{j\beta} = \delta_{\alpha\beta} \tag{A1.4}$$

Finally, the vector d_j is related to the Burgers' vector b_k through the equation

$$b_k = B_{kj}d_j \tag{A1.5}$$

with

$$B_{kj} = \frac{i}{2}\sum_\alpha(A_{k\alpha}M_{\alpha j} - \bar{A}_{k\alpha}\bar{M}_{\alpha j}) = -\mathrm{Im}\left(\sum_\alpha A_{k\alpha}M_{\alpha j}\right), \tag{A1.6}$$

where the overbar denotes the complex conjugate.

Appendix 2

EDT: Ethylene diamine tartrate ($C_2H_6N_2{\cdot}H_2C_4H_4O_6$)

$$c^{-1}_{ijkl} = \begin{bmatrix} 3.34 & -3.0 & -3.0 & 0.0 & -1.7 & 0.0 \\ & 3.65 & -1.8 & 0.0 & -1.5 & 0.0 \\ & & 10.0 & 0.0 & -2.65 & 0.0 \\ & & & 19.2 & 0.0 & 0.38 \\ & \text{sym} & & & 11.7 & 0.0 \\ & & & & & 19.1 \end{bmatrix} 10^{-12}\ \mathrm{cm^2/dyne}$$

MS: Monocrystalline silicon

$$c_{ijkl} = \begin{bmatrix} 1.16564 & 0.6394 & 0.6394 & 0.0 & 0.0 & 0.0 \\ & 1.943 & 0.3528 & 0.0 & 0.0 & 0.0 \\ & & 1.943 & 0.0 & 0.0 & 0.0 \\ & & & 0.5085 & 0.0 & 0.0 \\ & \text{sym} & & & 0.7951 & 0.0 \\ & & & & & 0.7951 \end{bmatrix} 10^{11}\ \mathrm{Pa}$$

AN: Anisotropic material, not real

$$c_{ijkl} = \begin{bmatrix} 20.3 & -3.1 & 1.6 & -1.0 & 4.0 & 0.8 \\ & 4.0 & 20.0 & 1.5 & 2.0 & -0.1 \\ & & 150.5 & 2.0 & 10.0 & 6.2 \\ & & & 1.75 & 1.7 & 5.7 \\ & \text{sym} & & & 9.5 & 4.1 \\ & & & & & 222.2 \end{bmatrix}$$

Acknowledgements

One of the authors (C.B.) wishes to thank the Asociación de Amigos of the University of Navarra for the grant received to carry out her Ph.D.

References

Atkinson, C. (1979). Stress singularities and fracture mechanics. *Applied Mechanics Reviews* **32**, 123–135.

Atkinson, C., Bastero, J.M. and Martínez-Esnaola, J.M. (1988). Stress analysis in sharp angular notches using auxiliary fields. *Engineering Fracture Mechanics* **31**, 637–646.

Bastero, J.M., Atkinson, C. and Martínez-Esnaola, J.M. (1989). Use of path-independent integral A for mixed-mode fast crack propagation in uncoupled thermoelastic media. *Engineering Fracture Mechanics* **34**, 325–335.

Blanco, C. (1997). *Modelos de dislocaciones para grietas en recubrimientos y para grietas acodadas en el caso general de anisotropía elástica*, Tesis Doctoral, Escuela Superior de Ingenieros de San Sebastián, Universidad de Navarra.

Bogy, D.B. (1972). The plane solution for anisotropic elastic wedges under normal and shear loading. *Journal of Applied Mechanics* **39**, 1103–1109.

Busch, M., Heinzelmann, M. and Maschke, H.G. (1994). A cohesive zone model for the failure assessment of V-notches in micromechanical components. *International Journal of Fracture* **69**, R15–R21.

Cherepanov, G.P. (1979). *Mechanics of Brittle Fracture*, McGraw-Hill, New-York.

Eshelby, J.D., Read, W.T. and Shockley, W. (1953). Anisotropic elasticity with applications to dislocations theory. *Acta Metallurgica* **1**, 251–259.

Heinzelmann, M., Kuna, M. and Busch, M. (1994). FEM and BEM analyses of notch stress intensity factors in anisotropic materials, *ECF 10 – Structural Integrity: Experiments, Models and Applications*, Vol. **1** (Edited by K-H. Schwalbe and C. Berger), EMAS, Warley (UK), 329–334.

Huntington, H.B. (1958). *Solid State Physics*, Vol. **7** (Edited by F. Seitz and Turubull), Academic Press, New York, 214–351.

Muskhelishvili, N.I. (1977). *Some Basic Problems of the Mathematical Theory of Elasticity*, Noordhoff Int. Pub., Leyden, The Netherlands.

Sih, G.C. and Ho, J.W. (1991). Sharp notch fracture strength characterised by critical energy density. *Theoretical and Applied Fracture Mechanics* **16**, 179–214.

Stern, M., Becker, E.B. and Dunham, R.S. (1976). A contour integral computation of mixed-mode stress intensity factors. *International Journal of Fracture* **12**, 359–368.

Stroh, A.N. (1958). Dislocations and cracks in anisotropic elasticity, *Philosophical Magazine* **3**, 625–646.

Williams, M.L. (1952). Stress singularities resulting from various boundary conditions in angular corners of plates in extension. *Journal of Applied Mechanics* **19**, 526–528.

International Journal of Fracture **93:** 387–407, 1998.

Kinked cracks in anisotropic elastic materials

C. BLANCO[1], J.M. MARTÍNEZ-ESNAOLA[1] and C. ATKINSON[2]
[1]*Centro de Estudios e Investigaciones Técnicas de Guipúzcoa (CEIT). P°. Manuel Lardizábal 15, 20009 San Sebastián, Spain; and Escuela Superior de Ingenieros, Universidad de Navarra, P°. Manuel Lardizábal 13, 20009 San Sebastián, Spain. e-mail: jmesnaola@ceit.es*
[2]*Department of Mathematics, Imperial College of Science, Technology and Medicine, Huxley Building, 180 Queen's Gate, London SW7 2BZ, U.K.*

Received 11 July 1997; accepted in revised form 18 March 1998

Abstract. The problem of a kinked crack is analysed for the most general case of elastic anisotropy. The kinked crack is modelled by means of continuous distributions of dislocations which are assumed to be singular both at the crack tips and at the kink vertex. The resulting system of singular integral equations is solved numerically using Chebyshev polynomials and the reciprocal theorem. The stress intensity factors for modes I, II and III and the generalised stress intensity factor at the vertex are obtained directly from the dislocation densities.

Key words: Stress singularities, general elastic anisotropy, kinked cracks, dislocation models, singular integral equations, reciprocal theorem, stress intensity factors, generalised stress intensity factors.

1. Introduction

The problem of the kinked crack is relevant to situations where cracks deviate from their original planes. Anisotropic materials are particularly prone to this behaviour because of the variation of crack resistance with orientation and thus crack extension often occurs along directions different from the plane containing the crack.

Most attempts to solve this problem have been for isotropic materials and based on Muskhelishvili's theory of complex potentials (Muskhelishvili, 1977), combined both with conformal mapping (Chatterjee, 1975; Leblond, 1989; Amestoy and Leblond, 1992) and with techniques based on continuous distributions of dislocations (Vitek, 1977; Lo, 1978; Hayashi and Nemat-Nasser, 1981; Melin, 1986). Theocaris and Makrakis (1986, 1987) solved the problem using the Mellin transform, and Cotterell and Rice (1980) used perturbation techniques. The solution of Khrapkov (1971) is based on Mellin transform and Bilby and Cardew (1975) using a solution of Khrapkov, solved the problem of a semi-infinite kinked crack.

Few general solutions to the problem of the kinked crack in an anisotropic solid have been attempted despite its interest. Obata et al. (1989) studied the orthotropic plane case and Gao and Chiu (1992) solved the problem via a perturbation analysis.

It should be noted that in a neighbourhood of the kinked crack vertex, the stresses are singular like r^{-s} with $s < \frac{1}{2}$, r being the radial distance to the vertex (Bogy, 1971). This has not always been taken into account. Melin (1986) and Vitek (1977), for example, did not consider this singularity in their analyses. Karihaloo et al. (1980) point out that the singularity near the vertex is weaker than at the crack tip, but their analysis proceeds with a singularity $r^{-1/2}$ at the vertex.

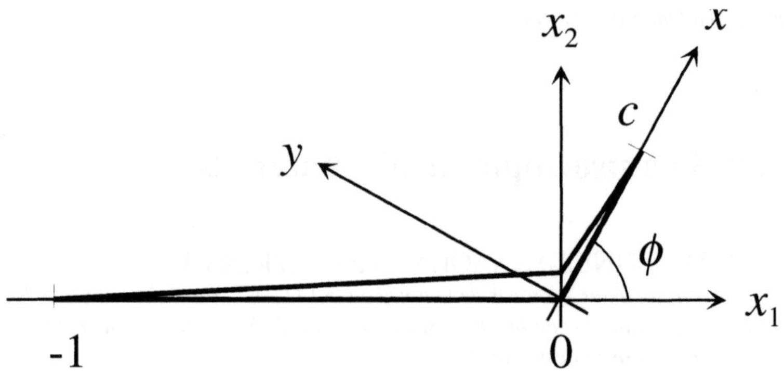

Figure 1. Kinked crack: coordinate system.

In this paper a method is presented for the analysis of a curved crack in an anisotropic solid, which consists in replacing the crack by a continuous distribution of dislocations along the crack path. Recently, Blanco et al. (1998) have developed an analytical-numerical method to obtain the stress singularity (and the corresponding stress field) in the neighbourhood of the vertex of a kinked crack in materials with general anisotropy, as well as a method to compute the generalised stress intensity factor at the vertex of an angular notch. Based on that preceding work, the general method is applied to model a kinked crack in an anisotropic material taking a dislocation density which is singular at both crack tips and at the vertex of the kinked crack. The integral equations which are obtained are solved numerically by the Chebyshev polynomials technique (Erdogan, 1969; Atkinson, 1972), and applying the reciprocal theorem. The stress intensity factors at the crack tips and the generalised stress intensity factor at the kink vertex can be obtained directly from the resulting dislocation densities.

2. Statement of the problem

Consider a coordinate system (x_1, x_2, x_3) such that the main crack lies on the x_1-axis, the origin coincides with the vertex and the crack tip with the point $(-1, 0)$, see Figure 1. The branch of length c forms an angle ϕ with the main crack. In this coordinate system, the equations which describe the crack geometry are

$$X_1(r) = \begin{cases} r & -1 \leqslant r \leqslant 0 \\ r\cos\phi & 0 < r \leqslant c \end{cases},$$
$$X_2(r) = \begin{cases} 0 & -1 \leqslant r \leqslant 0 \\ r\sin\phi & 0 < r \leqslant c \end{cases}. \tag{2.1}$$

The kinked crack is modelled by means of continuous distributions of dislocations that are designated as $d'_j(r)$. In a preceding paper (Blanco et al., 1998), the authors have shown that the stress singularity λ near the vertex of the kinked crack is an eigenvalue of the characteristic equation

$$\det[\mathbf{M}(\lambda)] = 0. \tag{2.2}$$

The form of matrix $\mathbf{M}(\lambda)$ is given in equation (2.13) of Blanco et al. (1998). Since the dislocation densities $d'_j(\rho)$ are expected to have square-root singularities at the ends of the interval $[-1, c]$ and a weaker singularity at $\rho = 0$, we can express these functions as follows

$$d'_j(\rho) = \begin{cases} (1+\rho)^{-1/2}(c-\rho)^{-1/2}D_j(\rho) + Av_j(-\rho)^{-\lambda} & (-1 < \rho < 0), \\ (1+\rho)^{-1/2}(c-\rho)^{-1/2}D_j(\rho) + Av_{j+3}\rho^{-\lambda} & (0 < \rho < c), \end{cases} \tag{2.3}$$

where $D_j(\rho)$ are unknown bounded functions in $[-1, c]$, A is an unknown constant and $\mathbf{V} = v_j$ $(j = 1, \ldots, 6)$ is the eigenvector corresponding to the eigenvalue λ.

The solution for the stress field produced by an isolated dislocation located at point (a_1, a_2) with Burgers' vector b_i in an infinite anisotropic medium is (Stroh, 1958)

$$\sigma_{i1}(x_1, x_2) = \frac{-1}{4\pi}\sum_\alpha L_{i\alpha}M_{\alpha j}p_\alpha \frac{d_j}{z_\alpha - \zeta_\alpha} + C.C. \tag{2.4a}$$

$$\sigma_{i2}(x_1, x_2) = \frac{1}{4\pi}\sum_\alpha L_{i\alpha}M_{\alpha j} \frac{d_j}{z_\alpha - \zeta_\alpha} + C.C. \tag{2.4b}$$

where C.C. denotes the complex conjugate of the preceding expression and the convention of summing over repeated Latin indices is used. In the above equations, $\zeta_\alpha = a_1 + p_\alpha a_2$ and $z_\alpha = x_1 + p_\alpha x_2$. The magnitudes p_α, $L_{i\alpha}$, $M_{\alpha j}$ and d_j are designated with the same names in Stroh's work, and are given in Appendix 1 for clarity.

Integrating (2.4) along the whole crack, the stress field produced by the dislocation densities $d'_j(\rho)$ – equation (2.3) – at a point (x_1, x_2) can be expressed as

$$\begin{aligned}\sigma_{i1}(x_1, x_2) &= \frac{-1}{4\pi}\sum_\alpha L_{i\alpha}M_{\alpha j}p_\alpha \int_{-1}^{c} \frac{(1+\rho)^{-1/2}(c-\rho)^{-1/2}D_j(\rho)}{z_\alpha - (X_1(\rho) + p_\alpha X_2(\rho))}\,\mathrm{d}\rho \\ &\quad - \frac{A}{4\pi}\sum_\alpha L_{i\alpha}M_{\alpha j}p_\alpha \left\{ v_j \int_{-1}^{0} \frac{(-\rho)^{-\lambda}}{z_\alpha - \rho}\,\mathrm{d}\rho + v_{j+3}\int_0^c \frac{\rho^{-\lambda}}{z_\alpha - \rho\tau_\alpha}\,\mathrm{d}\rho \right\} + C.C.,\end{aligned} \tag{2.5a}$$

$$\begin{aligned}\sigma_{i2}(x_1, x_2) &= \frac{1}{4\pi}\sum_\alpha L_{i\alpha}M_{\alpha j} \int_{-1}^{c} \frac{(1+\rho)^{-1/2}(c-\rho)^{-1/2}D_j(\rho)}{z_\alpha - (X_1(\rho) + p_\alpha X_2(\rho))}\,\mathrm{d}\rho \\ &\quad + \frac{A}{4\pi}\sum_\alpha L_{i\alpha}M_{\alpha j} \left\{ v_j \int_{-1}^{0} \frac{(-\rho)^{-\lambda}}{z_\alpha - \rho}\,\mathrm{d}\rho + v_{j+3}\int_0^c \frac{\rho^{-\lambda}}{z_\alpha - \rho\tau_\alpha}\,\mathrm{d}\rho \right\} + C.C.,\end{aligned} \tag{2.5b}$$

where

$$\tau_\alpha = \cos\phi + p_\alpha \sin\phi. \tag{2.6}$$

To determine the dislocation density use is made of the condition that the crack is traction free. Using (2.5) evaluated at the crack surface and denoting $T_i(r)$ the tractions induced on the crack surface by the external load in the medium without the crack, this condition gives

$$-T_i(r) = \frac{1}{4\pi}\sum_{\alpha} L_{i\alpha}M_{\alpha j}\Omega_\alpha(r)\int_{-1}^{c}\frac{(1+\rho)^{-1/2}(c-\rho)^{-1/2}D_j(\rho)}{X_1(r)-X_1(\rho)+p_\alpha(X_2(r)-X_2(\rho))}\,\mathrm{d}\rho$$

$$+\frac{A}{4\pi}\sum_{\alpha} L_{i\alpha}M_{\alpha j}\Omega_\alpha(r)\left\{v_j\int_{-1}^{0}\frac{(-\rho)^{-\lambda}}{X_1(r)+p_\alpha X_2(r)-\rho}\,\mathrm{d}\rho\right.$$

$$\left.+v_{j+3}\int_{0}^{c}\frac{\rho^{-\lambda}}{X_1(r)+p_\alpha X_2(r)-\rho\tau_\alpha}\,\mathrm{d}\rho\right\} + C.C. \quad (-1<r<c) \tag{2.7}$$

where

$$\Omega_\alpha(r) = \frac{\mathrm{d}X_1}{\mathrm{d}r}(r) + p_\alpha\frac{\mathrm{d}X_2}{\mathrm{d}r}(r) \tag{2.8}$$

represents the different crack plane orientations in the intervals $-1 \leqslant r \leqslant 0$ and $0 < r \leqslant c$.

Furthermore the condition that the crack is closed at both ends gives

$$\int_{-1}^{c}(1+\rho)^{-1/2}(c-\rho)^{-1/2}D_j(\rho)\,\mathrm{d}\rho$$

$$+A\left\{v_j\int_{-1}^{0}(-\rho)^{-\lambda}\,\mathrm{d}\rho + v_{j+3}\int_{0}^{c}\rho^{-\lambda}\,\mathrm{d}\rho\right\} = 0. \tag{2.9}$$

Equations (2.7) and (2.9) constitute a system of singular integral equations with unknowns $D_j(\rho)$ and A, which will be solved numerically in the next section.

3. Numerical solution of the system of integral equations

3.1. THE SINGULAR INTEGRAL EQUATIONS

We define the normalised parameters t and ζ as

$$r = \frac{1+c}{2}(1+t) - 1 \tag{3.1}$$

$$\rho = \frac{1+c}{2}(1+\zeta) - 1. \tag{3.2}$$

Then, making in (2.7) and (2.9) the substitutions

$$x_i(t) = X_i(r) \tag{3.3}$$

$$\omega_\alpha(t) = \Omega_\alpha(r)\frac{1+c}{2} \tag{3.4}$$

$$t_i(t) = T_i(r)\frac{1+c}{2} \tag{3.5}$$

$$B_j(\zeta) = D_j(\rho)\frac{2}{1+c} \tag{3.6}$$

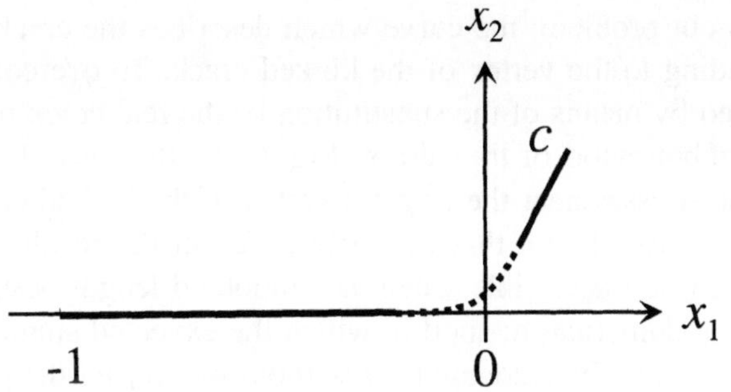

Figure 2. Modified (smooth) kinked crack.

we obtain

$$\frac{-4\pi}{1+c}t_i(t) = \int_{-1}^{1}\frac{(1-\zeta^2)^{-1/2}B_i(\zeta)}{t-\zeta}\,\mathrm{d}\zeta + \int_{-1}^{1}k_{ij}(t,\zeta)(1-\zeta^2)^{-1/2}B_j(\zeta)\,\mathrm{d}\zeta + AJ_i(t) \quad (-1<t<1) \tag{3.7a}$$

$$\int_{-1}^{1}(1-\zeta^2)^{-1/2}B_j(\zeta)\,\mathrm{d}\zeta + \frac{2A}{(c+1)(1-\lambda)}[v_j + v_{j+3}c^{1-\lambda}] = 0, \tag{3.7b}$$

where

$$k_{ij}(t,\zeta) = \mathrm{Re}\left\{\sum_{\alpha} L_{i\alpha}M_{\alpha j}\omega_\alpha(t)K_\alpha(t,\zeta)\right\} \tag{3.8}$$

with

$$K_\alpha(t,\zeta) = \frac{1}{t-\zeta}\left\{\frac{1}{\dfrac{x_1(t)-x_1(\zeta)}{t-\zeta} + p_\alpha\dfrac{x_2(t)-x_2(\zeta)}{t-\zeta}} - \frac{1}{\omega_\alpha(t)}\right\} \tag{3.9}$$

and

$$J_i(t) = \frac{2}{(1+c)}\mathrm{Re}\left\{\sum_{\alpha} L_{i\alpha}M_{\alpha j}\omega_\alpha(t)\left[v_j\int_{-1}^{0}\frac{(-\rho)^{-\lambda}}{x_1(t)+p_\alpha x_2(t)-\rho}\,\mathrm{d}\rho + v_{j+3}\int_{0}^{c}\frac{\rho^{-\lambda}}{x_1(t)+p_\alpha x_2(t)-\rho\tau_\alpha}\,\mathrm{d}\rho\right]\right\}. \tag{3.10}$$

It is easy to prove that

$$\lim_{\zeta\to t}K_\alpha(t,\zeta) = \frac{1}{2}\left[\frac{\mathrm{d}^2x_1}{\mathrm{d}t^2}(t) + p_\alpha\frac{\mathrm{d}^2x_2}{\mathrm{d}t^2}(t)\right][\omega_\alpha(t)]^{-2} \tag{3.11}$$

(provided the derivatives exist), and therefore if the crack path is smooth, the functions $k_{ij}(t,\zeta)$ can be considered as bounded functions within the square $[-1,1]\times[-1,1]$. However, this property is not true in our problem: the curve which describes the crack is not differentiable at the point corresponding to the vertex of the kinked crack. To overcome this difficulty the curve may be smoothed by means of the substitution of the real crack path by a polynomial, for example, in a neighbourhood of the vertex (Figure 2). Because of the singularity of the dislocation density, the stresses near the original vertex of the kinked crack are still singular. Numerical calculations have shown that the differences in the results obtained using both paths (real and modified) are negligible when the smoothed length is sufficiently small. It is also possible to derive a numerical method in which the expected singular behaviour at both vertex and crack tips is correctly accounted for without modifying the geometry of the kink, but we have not used such a method here.

3.2. Solution by Chebyshev polynomials

Using the property that the fundamental function $(1-\zeta^2)^{-1/2}$ of equation (3.7a) is the weight function of the Chebyshev polynomials of the first kind, the approximate solution may be expressed as

$$B_j(\zeta) = \sum_{n=0}^{N} B_{jn} T_n(\zeta), \tag{3.12}$$

where B_{jn} are unknown constants and $T_n(\zeta)$ are the Chebyshev polynomials of the first kind.

Substituting (3.12) into (3.7), we get the system

$$\begin{aligned}\frac{-4\pi}{1+c} t_i(t) &= \sum_{n=0}^{N} B_{in} \int_{-1}^{1} \frac{T_n(\zeta)}{(1-\zeta^2)^{1/2}(t-\zeta)} \mathrm{d}\zeta \\ &+ \sum_{n=0}^{N} B_{jn} \int_{-1}^{1} k_{ij}(t,\zeta) \frac{T_n(\zeta)}{(1-\zeta^2)^{1/2}} \mathrm{d}\zeta + AJ_i(t) \quad (-1 < t < 1)\end{aligned} \tag{3.13a}$$

$$\sum_{n=0}^{N} B_{jn} \int_{-1}^{1} \frac{T_n(\zeta)}{(1-\zeta^2)^{1/2}} \mathrm{d}\zeta + \frac{2A}{(c+1)(1-\lambda)} [v_j + v_{j+3} c^{1-\lambda}] = 0 \tag{3.13b}$$

with unknowns A and B_{jn} ($j = 1, 2, 3$ and $n = 0, 1, \ldots, N$).

Applying the change of variable

$$\zeta = \cos\theta \tag{3.14}$$

and the relations (Abramowitz and Stegun, 1970)

$$T_n(\cos\theta) = \cos(n\theta) \tag{3.15}$$

$$\int_{-1}^{1} \frac{T_n(\zeta)}{(1-\zeta^2)^{1/2}(t-\zeta)} \mathrm{d}\zeta = \begin{cases} 0 & n = 0, \\ -\pi U_{n-1}(t) & n > 0, \end{cases} \tag{3.16}$$

where $U_n(t)$ are the Chebyshev polynomials of second kind, the system (3.13) becomes

$$\frac{-4\pi}{1+c}t_i(t) = -\pi\sum_{n=1}^{N} B_{in}U_{n-1}(t)$$

$$+\sum_{n=0}^{N} B_{jn}\int_{-1}^{1} k_{ij}(t,\zeta)\frac{T_n(\zeta)}{(1-\zeta^2)^{1/2}}\,\mathrm{d}\zeta + AJ_i(t) \quad (-1 < t < 1) \tag{3.17a}$$

$$\pi B_{j0} + \frac{2A}{(c+1)(1-\lambda)}[v_j + v_{j+3}c^{1-\lambda}] = 0. \tag{3.17b}$$

The integral on the right-hand side of (3.17a) can be computed numerically by means of the formula (Erdogan and Gupta, 1972)

$$\int_{-1}^{1} f(t)(1-t^2)^{-1/2}\,\mathrm{d}t = \frac{\pi}{Q}\sum_{q=1}^{Q} f(\cos\theta_q) \tag{3.18}$$

where

$$\theta_q = \frac{(2q-1)\pi}{2Q}. \tag{3.19}$$

Using (3.18) in (3.17a) we get

$$\frac{-4}{1+c}t_i(t) = -\sum_{n=1}^{N} B_{in}U_{n-1}(t)$$

$$+\frac{1}{Q}\sum_{n=0}^{N} B_{jn}\sum_{q=1}^{Q} k_{ij}(t,\cos\theta_q)T_n(\cos\theta_q) + \frac{A}{\pi}J_i(t) \quad (-1 < t < 1). \tag{3.20}$$

Multiplying both sides of (3.20) by $U_m(t)(1-t^2)^{1/2}$ $(m = 0, 1, \ldots, N-1)$ and integrating we obtain

$$\frac{-4}{1+c}\int_{-1}^{1} U_m(t)(1-t^2)^{1/2}t_i(t)\,\mathrm{d}t$$

$$= -\sum_{n=1}^{N} B_{in}\int_{-1}^{1} U_m(t)(1-t^2)^{1/2}U_{n-1}(t)\,\mathrm{d}t$$

$$+\frac{1}{Q}\sum_{n=0}^{N} B_{jn}\sum_{q=1}^{Q} T_n(\cos\theta_q)\int_{-1}^{1} U_m(t)(1-t^2)^{1/2}k_{ij}(t,\cos\theta_q)\,\mathrm{d}t$$

$$+\frac{A}{\pi}\int_{-1}^{1} U_m(t)(1-t^2)^{1/2}J_i(t)\,\mathrm{d}t \quad m = 0, 1, \ldots, N-1 \tag{3.21}$$

Now, the suitable integration formula for the integrals in (3.21) is (Erdogan and Gupta, 1972)

$$\int_{-1}^{1} f(t)(1-t^2)^{1/2}\,\mathrm{d}t = \frac{\pi}{R+1}\sum_{r=1}^{R} f(\cos\varphi_r)\sin^2\varphi_r, \tag{3.22}$$

where

$$\varphi_r = \frac{r\pi}{R+1}. \tag{3.23}$$

Using (3.15) and the property of the Chebyshev polynomials of the second kind (Abramowitz and Stegun, 1970)

$$U_m(\cos\varphi) = \frac{\sin[(m+1)\varphi]}{\sin\varphi}, \tag{3.24}$$

equation (3.22) can be expressed as

$$\begin{aligned}
&\frac{-4}{1+c}\sum_{r=1}^{R}\sin[(m+1)\varphi_r]t_i(\cos\varphi_r)\sin\varphi_r \\
&\quad = -\sum_{n=1}^{N} B_{in}\sum_{r=1}^{R}\sin[(m+1)\varphi_r]\sin(n\varphi_r) \\
&\qquad +\frac{1}{Q}\sum_{n=0}^{N} B_{jn}\sum_{q=1}^{Q}\cos(n\theta_q)\sum_{r=1}^{R}\sin[(m+1)\varphi_r]k_{ij}(\cos\varphi_r,\cos\theta_q)\sin\varphi_r \\
&\qquad +\frac{A}{\pi}\sum_{r=1}^{R}\sin[(m+1)\varphi_r]J_i(\cos\varphi_r)\sin\varphi_r \quad (m=0,1,\ldots,N-1).
\end{aligned} \tag{3.25}$$

Equations (3.25) and (3.17b) constitute a system of $3N+3$ equations and $3N+4$ unknowns: the constant A and the coefficients B_{in} ($i = 1,2,3$; $n = 0,1,\ldots,N$). Then an additional equation is needed to solve this system. This is obtained in the next section by application of the reciprocal theorem.

3.3. A SUPPLEMENTARY EQUATION

Blanco et al. (1998) have developed a method to calculate the generalised stress intensity factor at the tip of a sharp angular notch applying the reciprocal theorem. Similarly, the reciprocal theorem can be used to obtain a supplementary equation for the present problem that allows us to solve the above system of equations.

The reciprocal theorem for two dimensional elastostatics, assuming no body forces are present, is

$$\int_C (\sigma_k\hat{u}_k - \hat{\sigma}_k u_k)\,\mathrm{d}l = 0, \quad \text{with } \sigma_k = \sigma_{kj}n_j \text{ and } \hat{\sigma}_k = \hat{\sigma}_{kj}n_j, \tag{3.26}$$

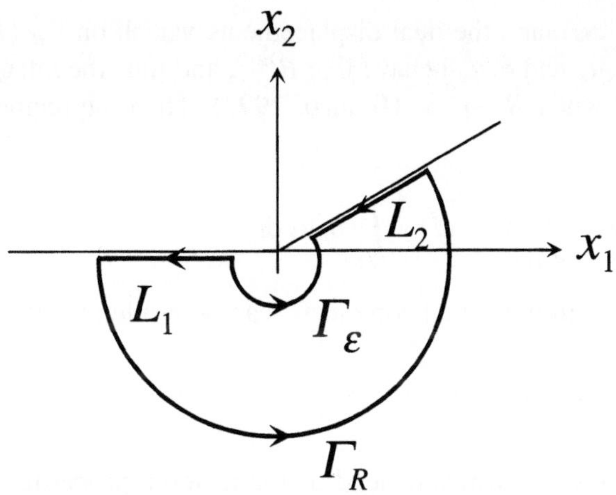

Figure 3. Integration contour for the application of the reciprocal theorem.

where n_j is the outward pointing normal to the simple closed contour C containing no singularities, and $\hat{\sigma}_{kj}$ and $\hat{u}_k$ are two auxiliary (dual) fields of stress and displacement which satisfy the same equilibrium and constitutive equations as the actual fields σ_{kj} and u_k.

As dual fields we take the fields derived in the preceding work (Blanco et al., 1998), namely, those induced by the distribution of dislocations $\hat{d}'_j(r)$ along both the crack and the branch planes, with density functions defined by

$$\hat{d}'_j(r) = \begin{cases} \dfrac{w_j}{(-r)^{2-\lambda}} & (-\infty < r < 0), \\ \dfrac{w_{j+3}}{r^{2-\lambda}} & (0 < r < \infty), \end{cases} \tag{3.27}$$

where $\mathbf{W} = w_i$ $(i = 1, \dots, 6)$ is the eigenvector of matrix $\mathbf{M}$ corresponding to the eigenvalue $s = 2 - \lambda$ (Blanco et al., 1998).

The dual fields can be expressed as (Blanco et al., 1998)

$$\hat{\sigma}_{i1}(z_\alpha) = -\sum_\alpha L_{i\alpha} p_\alpha \hat{f}'_\alpha(z_\alpha) + C.C. \tag{3.28a}$$

$$\hat{\sigma}_{i2}(z_\alpha) = \sum_\alpha L_{i\alpha} \hat{f}'_\alpha(z_\alpha) + C.C. \tag{3.28b}$$

$$\hat{u}_k(z_\alpha) = \sum_\alpha A_{k\alpha} \hat{f}_\alpha(z_\alpha) + C.C. \tag{3.29}$$

with

$$\hat{f}'_\alpha(z_\alpha) = \frac{1}{4\pi} M_{\alpha j} \left[w_j \int_{-\infty}^{0} \frac{(-\rho)^{\lambda-2}}{z_\alpha - \rho} \mathrm{d}\rho + w_{j+3} \int_0^\infty \frac{\rho^{\lambda-2}}{z_\alpha - \tau_\alpha \rho} \mathrm{d}\rho \right]. \tag{3.30}$$

Consider a closed contour as shown in Figure 3 with Γ_ε and Γ_R circular paths of radius ε ($\varepsilon \to 0$) and R ($R \to \infty$), respectively. It is easy to prove that the dual tractions $\hat{\sigma}_k$ vanish on

L_1, L_2 and Γ_R, $(R \to \infty)$, and the dual displacements vanish on Γ_R $(R \to \infty)$. Moreover, on Γ_R both products $\sigma_k \hat{u}_k$ and $\hat{\sigma}_k u_k$ behave like $R^{\lambda-2}$, and thus the integrals of the reciprocal theorem on Γ_R vanish when $R \to \infty$ (Blanco, 1997). Then the reciprocal theorem can be expressed in the form

$$\int_{\Gamma_\varepsilon} (\sigma_k \hat{u}_k - \hat{\sigma}_k u_k)\,\mathrm{d}l = \int_{L_1} \sigma_k \hat{u}_k\,\mathrm{d}l + \int_{L_2} \sigma_k \hat{u}_k\,\mathrm{d}l. \tag{3.31}$$

The integral along the infinitesimal contour Γ_ε can be written as (Blanco et al., 1998)

$$\int_{\Gamma_\varepsilon} (\sigma_k \hat{u}_k - \hat{\sigma}_k u_k)\,\mathrm{d}l = A(C_1 - C_2), \tag{3.32}$$

where C_1 and C_2 are constants which depend on the material properties and the kink angle ϕ and are designated with the same name in Blanco et al. (1998).

On the kinked crack the tractions σ_k are opposite to the tractions T_k induced by the external load in the medium without the crack, so the integrals along the contours L_1 and L_2 can be rewritten as

$$\int_{L_1} \sigma_k \hat{u}_k\,\mathrm{d}l = \int_\varepsilon^1 T_k(-r)\hat{u}_k(-r)\,\mathrm{d}r - \int_1^\infty \sigma_k(-r)\hat{u}_k(-r)\,\mathrm{d}r \tag{3.33}$$

and

$$\int_{L_2} \sigma_k \hat{u}_k\,\mathrm{d}l = \int_\varepsilon^c T_k(r)\hat{u}_k(r)\,\mathrm{d}r - \int_c^\infty \sigma_k(r)\hat{u}_k(r)\,\mathrm{d}r. \tag{3.34}$$

Analysing both line integrals we obtain

$$\int_{L_1} \sigma_k \hat{u}_k\,\mathrm{d}l = \sum_{n=0}^{N} (B_{jn} Z_k^{(1)} U_{jkn}^{(1)}) + A Z_k^{(1)} V_k^{(1)} + Z_k^{(1)} W_k^{(1)} \tag{3.35}$$

$$\int_{L_2} \sigma_k \hat{u}_k\,\mathrm{d}l = \sum_{n=0}^{N} (B_{jn} Z_k^{(2)} U_{jkn}^{(2)}) + A Z_k^{(2)} V_k^{(2)} + Z_k^{(2)} W_k^{(2)}, \tag{3.36}$$

where $U_{jkn}^{(i)}$, $V_k^{(i)}$, $W_k^{(i)}$ and $Z_k^{(i)}$ $(i = 1, 2)$ are constants which depend on the material properties and the geometry of the kinked crack. Details of the analysis and the resulting constants are given in Appendix 2.

Substituting (3.32), (3.35) and (3.36) into (3.31) we obtain the additional equation which, together with equations (3.25) and (3.17b), constitute a system of $3N+4$ equations and $3N+4$ unknowns.

4. The stress intensity factors

4.1. The stress intensity factors (SIFs) at both crack tips

A new coordinate system (x, y, z), $z = x_3$, is introduced so that the branch line corresponds to the x-axis (Figure 1).

The SIFs can be directly obtained from the resulting dislocation densities. As the stresses are singular near the crack tip, they are governed by the first singular part on the right-hand side of (2.5). Hence, the asymptotic behaviour can be obtained by neglecting the nonsingular terms. In particular, at the branch tip

$$\sigma'_{k2}(c+\varepsilon) = \frac{l_{ki}}{4\pi}\sum_{\alpha} L_{i\alpha}M_{\alpha j}\tau_{\alpha}\int_0^c \frac{(1+\rho)^{-1/2}(c-\rho)^{-1/2}D_j(\rho)}{(c+\varepsilon)\tau_{\alpha}-\rho\tau_{\alpha}}\,\mathrm{d}\rho + C.C. \tag{4.1}$$

where σ'_{kj} is the stress tensor referred to the coordinate system (x, y, z) and l_{ki} are the direction cosines of the new base with respect to the old one.

Applying the change of variable

$$\rho = \frac{1+c}{2}(\cos\varphi + 1) - 1 \tag{4.2}$$

we get

$$\sigma'_{k2}(c+\varepsilon) \approx \frac{1+c}{4\pi} l_{ki}\int_0^{\varphi_0} \frac{B_i(\cos\varphi)}{c+\varepsilon - \frac{1+c}{2}(\cos\varphi+1)+1}\,\mathrm{d}\varphi \tag{4.3}$$

with

$$\varphi_0 = \cos^{-1}\left(\frac{1-c}{1+c}\right). \tag{4.4}$$

Using the series expansion of $B_i(\cos\varphi)$ around $\cos\varphi = 1$ and neglecting again the nonsingular terms, it then follows that

$$\sigma'_{k2}(c+\varepsilon) \approx \frac{l_{ki}}{4}\sqrt{\frac{1+c}{\varepsilon}}B_i(1). \tag{4.5}$$

Now, the SIFs K_i $(i = 1, 2, 3)$ are defined by

$$K_{\eta(i)} = \lim_{\varepsilon\to 0}\sqrt{2\pi\varepsilon}\,\sigma'_{i2}(c+\varepsilon) \tag{4.6}$$

with $\eta(1) = 2$, $\eta(2) = 1$ and $\eta(3) = 3$. Substituting (4.5) into (4.6), we obtain for the branch tip

$$K_{\eta(k)} = \frac{l_{ki}}{2}\sqrt{\pi\frac{1+c}{2}}B_i(1) = \frac{l_{ki}}{2}\sqrt{\pi\frac{1+c}{2}}\sum_{n=0}^{N} B_{in}. \tag{4.7}$$

A completely parallel analysis for the main crack tip yields

$$K_{\eta(k)} = -\frac{1}{2}\sqrt{\pi\frac{1+c}{2}}B_k(-1) = -\frac{1}{2}\sqrt{\pi\frac{1+c}{2}}\sum_{n=0}^{N}(-1)^n B_{kn}. \tag{4.8}$$

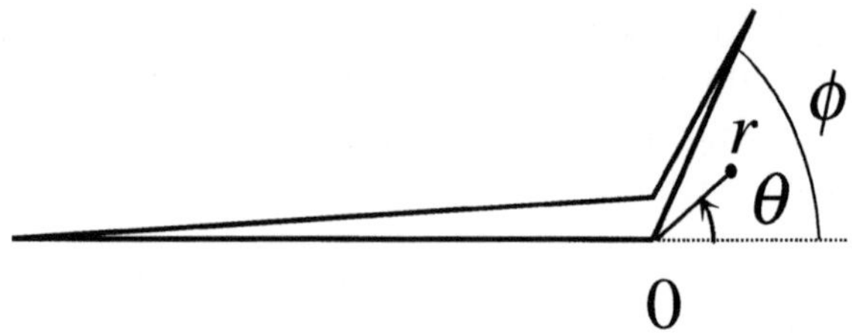

Figure 4. Polar coordinate system with origin at the kink vertex.

4.2. THE GENERALISED STRESS INTENSITY FACTOR (GSIF) AT THE KINK

Consider the polar coordinate system shown in Figure 4 with origin at the kink vertex. Now, the stresses near the vertex of the kinked crack are governed by the second singular term on the right-hand side of (2.5). Neglecting the first term, at $z_\alpha = \varepsilon(\cos\theta + p_\alpha \sin\phi)$ with $\varepsilon \to 0$ we obtain

$$\sigma_{k1}(\varepsilon \to 0)$$
$$= -A\frac{\varepsilon^{-\lambda}}{2}\csc(\pi\lambda)\mathrm{Re}\left[\sum_\alpha L_{k\alpha}M_{\alpha j}p_\alpha\tau_\alpha(\theta)^{-\lambda}\left\{v_j - v_{j+3}\frac{(-1)^{-\lambda}}{\tau_\alpha^{1-\lambda}}\right\}\right], \tag{4.9a}$$

$$\sigma_{k2}(\varepsilon \to 0)$$
$$= A\frac{\varepsilon^{-\lambda}}{2}\csc(\pi\lambda)\mathrm{Re}\left[\sum_\alpha L_{k\alpha}M_{\alpha j}\tau_\alpha(\theta)^{-\lambda}\left\{v_j - v_{j+3}\frac{(-1)^{-\lambda}}{\tau_\alpha^{1-\lambda}}\right\}\right], \tag{4.9b}$$

where

$$\tau_\alpha(\theta) = \cos\theta + p_\alpha\sin\theta. \tag{4.10}$$

Equations (4.9) show that the stresses σ_{ij} have a singular behaviour in the neighbourhood of the kinked crack vertex, that is, $\sigma_{ij} \approx A\varepsilon^{-\lambda}f_{ij}(\theta)$, where $f_{ij}(\theta)$ are known functions and the constant A represents a generalised stress intensity factor which characterises the singular stress fields at the kink, which is obtained from the system of equations derived in the preceding section.

5. Results

Using the analysis described in the previous sections, we present here SIF and GSIF results for various angles and branch lengths of a kinked crack lying in an anisotropic material. Comparisons of the present analysis with some existing solutions for the isotropic case are also given.

Table 1. SIF in the main crack tip for uniform tension perpendicular to the main crack ($\phi = -45°$, $N = 30$, $Q = R = 150$).

	Kitagawa et al. (1975)		Present work	
c/b	$F_{\mathrm{I}}(\mathrm{B})$	$F_{\mathrm{II}}(\mathrm{B})$	$F_{\mathrm{I}}(\mathrm{B})$	$F_{\mathrm{II}}(\mathrm{B})$
0.01	1.000	0.003	0.9996	0.0025
0.10	0.998	0.019	0.9966	0.0200
0.20	0.995	0.028	0.9954	0.0273
0.50	0.988	0.033	0.9859	0.0342
0.80	0.985	0.024	0.9840	0.0242
1.00	0.985	0.016	0.9841	0.0173
1.50	0.988	−0.004	0.9876	−0.0036
2.00	0.993	−0.022	0.9908	−0.0164

Table 2. SIF in the branch tip for uniform tension perpendicular to the main crack ($N = 30$, $Q = R = 150$).

	$\phi = 15°$				$\phi = 45°$			
c/b	Kitagawa et al. (1975)		Present work		Kitagawa et al. (1975)		Present work	
	$F_{\mathrm{I}}(\mathrm{C})$	$F_{\mathrm{II}}(\mathrm{C})$	$F_{\mathrm{I}}(\mathrm{C})$	$F_{\mathrm{II}}(\mathrm{C})$	$F_{\mathrm{I}}(\mathrm{C})$	$F_{\mathrm{II}}(\mathrm{C})$	$F_{\mathrm{I}}(\mathrm{C})$	$F_{\mathrm{II}}(\mathrm{C})$
0.1	0.9540	0.2120	0.9538	0.2214	0.6339	0.5053	0.6311	0.5222
0.2	0.9496	0.2346	0.9496	0.2394	0.5983	0.5578	0.5971	0.5697
0.4	0.9466	0.2555	0.9457	0.2562	0.5741	0.6072	0.5745	0.6021
0.6	0.9457	0.2648	0.9451	0.2652	0.5679	0.6283	0.5678	0.6273
0.8	0.9456	0.2694	0.9455	0.2709	0.5678	0.6376	0.5678	0.6396
1.0	0.9457	0.2718	0.9447	0.2754	0.5694	0.6413	0.5673	0.6423
1.5	0.9463	0.2737	0.9453	0.2779	0.5744	0.6414	0.5718	0.6446
2.0	0.9468	0.2733	0.9466	0.2737	0.5785	0.6377	0.5791	0.6375

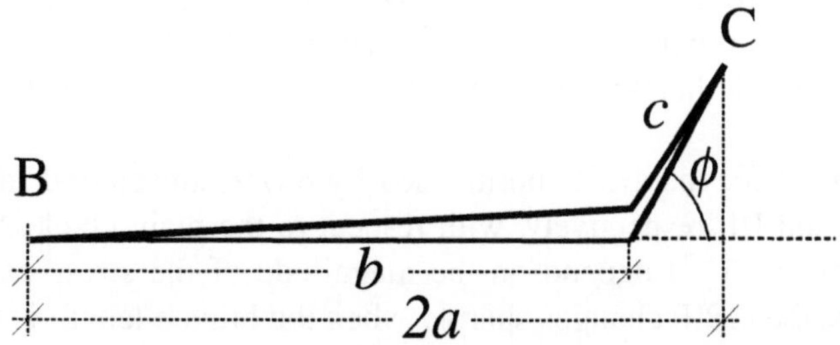

Figure 5. Kinked crack in an anisotropic medium.

Kitagawa et al. (1975) studied the two-dimensional isotropic case. In Tables 1 and 2 we compare the results reported by these authors with the values obtained with the present analysis, approximating the isotropic material as a limit case of slight anisotropy. The SIFs are normalised with respect to the SIF for a straight crack of length $2a$, that is

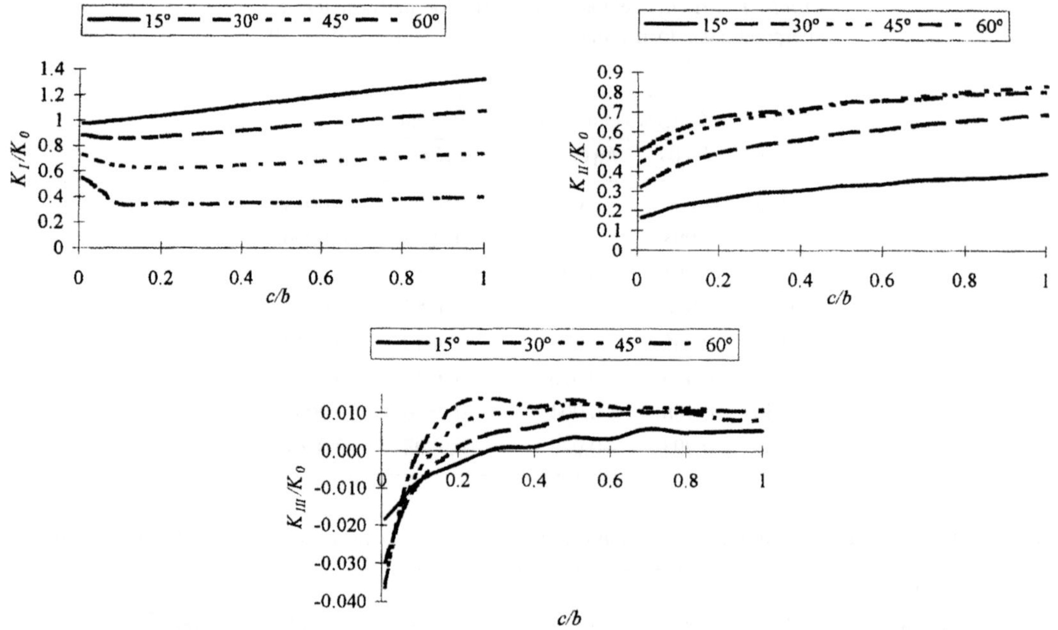

Figure 6. SIF at the branch tip as a function of branch length and branch angle for uniform tension perpendicular to the main crack ($N = 30$, $Q = R = 200$).

$$F_i = \frac{K_i}{\sigma\sqrt{\pi a}} \tag{5.1}$$

with $2a = b + c\cos\phi$ (Figure 5).

Figures 6 and 7 show the variation of the SIFs with the branch length at various branching angles for a crack lying in a monoclinic material (ethylene diamine tartrate: *EDT*). The elastic constants of this material are included in Appendix 3. In this case, the SIFs are normalised with respect to the SIF of the straight crack before branching takes place, that is $K_0 = \sigma\sqrt{\pi\frac{1}{2}b}$ (Figure 5). As expected, for small values of c/b, K_{I} at the main crack tip tends to K_0, see Figure 7. We can observe for K_{II} and K_{III} an oscillatory behaviour with the length of the kinked crack that has never been reported for isotropic materials. We have no physical explanation for this behaviour and think that the effect (which is small) may be an artifact of the numerical procedure.

The results obtained for the GSIF, normalised by $\sigma\sqrt{\pi}$, are plotted in Figures 8–10 for loading modes I, II and III, respectively, with respect to the main crack. Note that the units of this parameter ($\mathrm{ML}^{1+\lambda}\mathrm{T}^{-2}$) depend on the magnitude of the stress exponent λ. For the three loading modes, the GSIF changes sharply when the branch length is small and tends to stabilise towards the stationary value corresponding to an infinite notch as the branch length increases.

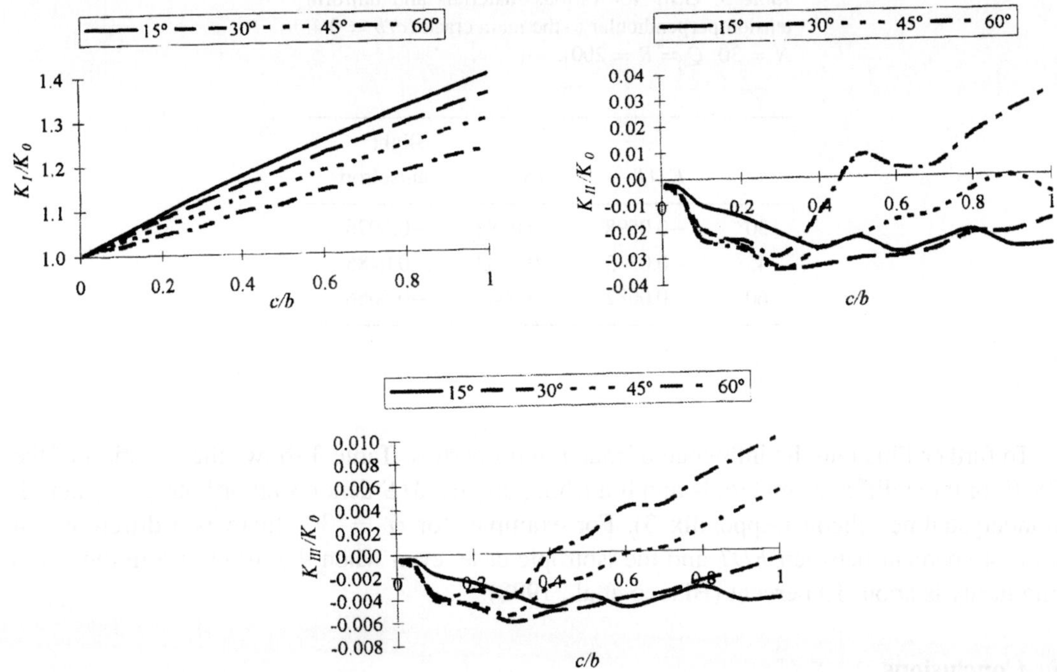

Figure 7. SIF at the main crack tip as a function of branch length and branch angle for uniform tension perpendicular to the main crack ($N = 30$, $Q = R = 200$).

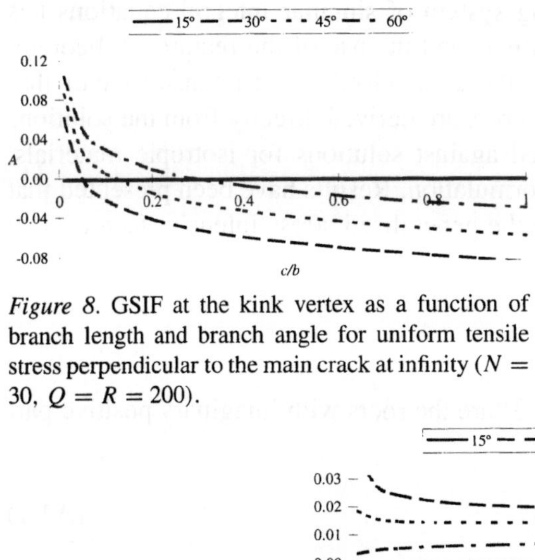

Figure 8. GSIF at the kink vertex as a function of branch length and branch angle for uniform tensile stress perpendicular to the main crack at infinity ($N = 30$, $Q = R = 200$).

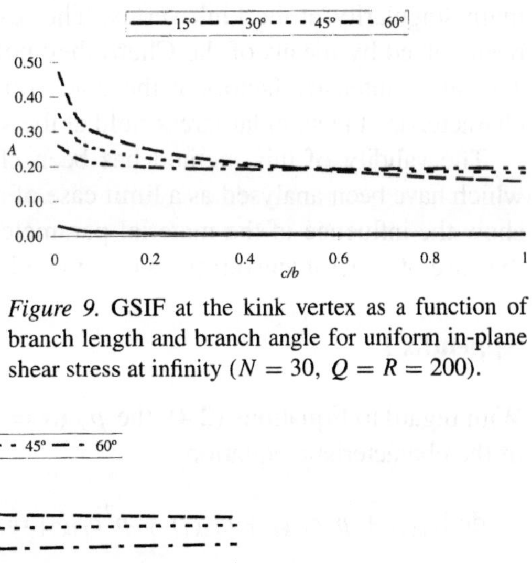

Figure 9. GSIF at the kink vertex as a function of branch length and branch angle for uniform in-plane shear stress at infinity ($N = 30$, $Q = R = 200$).

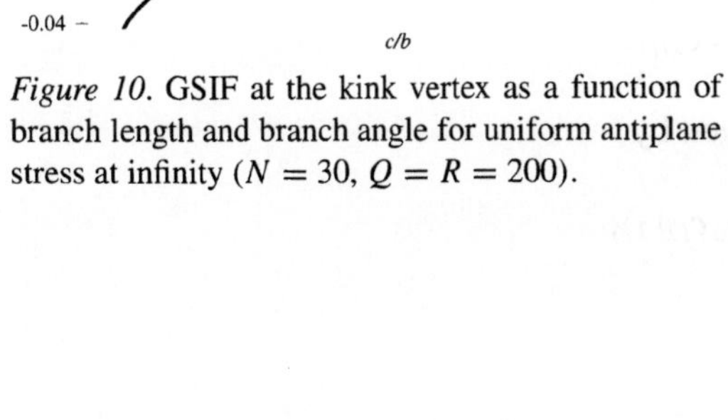

Figure 10. GSIF at the kink vertex as a function of branch length and branch angle for uniform antiplane stress at infinity ($N = 30$, $Q = R = 200$).

Table 3. GSIF for various materials and uniform tension perpendicular to the main crack ($c/b = 0.4$, $N = 30$, $Q = R = 200$).

ϕ	EDT	MS	Slight anisotropy
30°	−0.0598	−0.0782	−0.0976
45°	−0.0333	−0.0382	−0.0485
60°	−0.0082	−0.0042	−0.0066

To further illustrate the influence of material properties, Table 3 shows the variation of the GSIF for three different materials and branching angles. MS denotes an orthotropic material: monocrystalline silicon (Appendix 3). For example, for $\phi = 30°$ there is a difference of about 40 percent between *EDT* and the isotropic case, even when the difference in the stress singularity is about 10 percent (Blanco et al., 1998).

6. Conclusions

The problem of a kinked crack in an elastic solid with general anisotropy has been modelled using continuous distributions of dislocations which incorporate the accurate analysis of the main singularity at the kink vertex. The resulting system of singular integral equations has been solved by means of the Chebyshev polynomials and the use of the reciprocal theorem. The stress intensity factors at the crack tips and the generalised stress intensity factor that characterises the singular stress field at the kink vertex are derived directly from the solution.

The validity of this method has been checked against solutions for isotropic materials, which have been analysed as a limit case of our formulation. Results have been presented that show the influence of the material parameters in the generalised stress intensity factor when the case of general anisotropy is considered.

Appendix 1

With regard to Equations (2.4), the p_α ($\alpha = 1, 2, 3$) are the roots with imaginary positive part of the characteristic equation

$$\det[c_{i1k1} + p(c_{i1k2} + c_{i2k1}) + p^2 c_{i2k2}] = 0, \tag{A1.1}$$

where c_{ijkl} is the tensor of elastic constants, i.e., $\sigma_{ij} = c_{ijkl}u_{k,l}$, which satisfies the symmetry conditions

$$c_{ijkl} = c_{ijlk} = c_{jikl} = c_{klij}. \tag{A1.2}$$

The matrix $L_{i\alpha}$ is given by

$$L_{i\alpha} = A_{k\alpha}(c_{i2k1} + p_\alpha c_{i2k2}), \tag{A1.3}$$

where $A_{k\alpha}$ denotes the eigenvector corresponding to the eigenvalue p_α above ($\alpha = 1, 2, 3$).

The matrix $M_{\alpha j}$ is defined as the inverse of $L_{i\alpha}$,

$$M_{\alpha j} L_{j\beta} = \delta_{\alpha\beta}. \tag{A1.4}$$

Finally, the vector d_j is related to the Burgers' vector b_k through the equation

$$b_k = B_{kj} d_j \tag{A1.5}$$

with

$$B_{kj} = \frac{i}{2} \sum_\alpha (A_{k\alpha} M_{\alpha j} - \overline{A}_{k\alpha} \overline{M}_{\alpha j}) = -\mathrm{Im}\left(\sum_\alpha A_{k\alpha} M_{\alpha j} \right), \tag{A1.6}$$

where the overbar denotes the complex conjugate.

Appendix 2

The integral $\int_{L_1} \hat{u}_k \sigma_k \, \mathrm{d}l$

Substituting $z_\alpha = -r$ into (3.29) and (3.30), we get

$$\hat{u}_k(-r) = -\frac{r^{\lambda-1}}{2(\lambda-1)} \csc(\pi\lambda) \mathrm{Re} \left\{ \sum_\alpha A_{k\alpha} M_{\alpha j} [-\cos(\pi\lambda) w_j + \tau_\alpha^{1-\lambda} w_{j+3}] \right\}. \tag{A2.1}$$

The real tractions $\sigma_k = \sigma_{k2}$ can be obtained from (2.5b) with $z_\alpha = -r$. Using the change of variable (3.2) and the Chebyshev expansion for the dislocation density, we obtain

$$\begin{aligned} &\sigma_{k2}(-r) \\ &= \frac{1+c}{8\pi} \sum_{n=0}^{N} B_{jn} \left\{ \sum_\alpha L_{k\alpha} M_{\alpha j} \int_{-1}^{1} (1-\zeta^2)^{-1/2} T_n(\zeta) [\delta(r,\zeta) + \chi_\alpha(r,\zeta)] \, \mathrm{d}\zeta \right\} \\ &\quad + \frac{A}{4\pi} \sum_\alpha L_{k\alpha} M_{\alpha j} \left\{ v_j \int_{-1}^{0} \frac{(-\rho)^{-\lambda}}{-r-\rho} \mathrm{d}\rho + v_{j+3} \int_0^c \frac{\rho^{-\lambda}}{-r-\rho\tau_\alpha} \mathrm{d}\rho \right\} + C.C. \end{aligned} \tag{A2.2}$$

with

$$\delta(r,\zeta) = \begin{cases} \dfrac{1}{-r - \frac{1+c}{2}(1+\zeta) + 1} & \left(-1 < \zeta \leqslant \dfrac{1-c}{1+c}\right) \\ 0 & \left(\dfrac{1-c}{1+c} < \zeta < 1\right) \end{cases} \tag{A2.3}$$

and

$$\chi_\alpha(r,\zeta) = \begin{cases} 0 & \left(-1 < \zeta \leqslant \dfrac{1-c}{1+c}\right) \\ \dfrac{1}{-r - \left(\frac{1+c}{2}(1+\zeta) - 1\right)\tau_\alpha} & \left(\dfrac{1-c}{1+c} < \zeta < 1\right) \end{cases}. \tag{A2.4}$$

Numerical integration of the first integral in (A2.2), see (3.18), gives

$$\sigma_{k2}(-r) = \frac{1+c}{4Q}\sum_{n=0}^{N} B_{jn}\sum_{q=1}^{Q}\cos(n\theta_q)\left\{\delta_{kj}\delta(r,\cos\theta_q) + \mathrm{Re}\left[\sum_{\alpha} L_{k\alpha}M_{\alpha j}\chi_\alpha(r,\cos\theta_q)\right]\right\}$$

$$+\frac{A}{2\pi}\mathrm{Re}\left[\sum_{\alpha} L_{k\alpha}M_{\alpha j}\left\{v_j\int_{-1}^{0}\frac{(-\rho)^{-\lambda}}{-r-\rho}\,\mathrm{d}\rho + v_{j+3}\int_0^c \frac{\rho^{-\lambda}}{-r-\rho\tau_\alpha}\,\mathrm{d}\rho\right\}\right]. \qquad \text{(A2.5)}$$

Finally, substituting (A2.1) and (A2.5) into (3.32), the integral along L_1 can be written as

$$\int_{L_1}\hat{u}_k\sigma_k\,\mathrm{d}l = \sum_{n=0}^{N}(B_{jn}Z_k^{(1)}U_{jkn}^{(1)}) + AZ_k^{(1)}V_k^{(1)} + Z_k^{(1)}W_k^{(1)} \qquad \text{(A2.6)}$$

with

$$Z_k^{(1)} = \frac{\csc(\pi\lambda)}{2(\lambda-1)}\mathrm{Re}\left\{\sum_{\alpha} A_{k\alpha}M_{\alpha j}[-\cos(\pi\lambda)w_j + \tau_\alpha^{1-\lambda}w_{j+3}]\right\} \qquad \text{(A2.7)}$$

$$V_k^{(1)} = \frac{1}{2\pi}\left\{v_k\int_1^\infty r^{\lambda-1}\int_{-1}^{0}\frac{(-\rho)^{-\lambda}}{-r-\rho}\,\mathrm{d}\rho\,\mathrm{d}r + \mathrm{Re}\left[\sum_{\alpha} L_{k\alpha}M_{\alpha j}v_{j+3}\int_1^\infty r^{\lambda-1}\int_0^c\frac{\rho^{-\lambda}}{-r-\rho\tau_\alpha}\,\mathrm{d}\rho\,\mathrm{d}r\right]\right\}, \qquad \text{(A2.8)}$$

$$W_k^{(1)} = -\int_0^1 T_{k2}(-r)r^{\lambda-1}\,\mathrm{d}r, \qquad \text{(A2.9)}$$

$$U_{jkn}^{(1)} = \frac{1+c}{4Q}\sum_{q=1}^{Q}\cos(n\theta_q) \times\left\{\delta_{kj}\int_1^\infty r^{\lambda-1}\delta(r,\cos\theta_q)\,\mathrm{d}r + \mathrm{Re}\left[\sum_{\alpha} L_{k\alpha}M_{\alpha j}\int_1^\infty r^{\lambda-1}\chi_\alpha(r,\cos\theta_q)\,\mathrm{d}r\right]\right\}. \qquad \text{(A2.10)}$$

The integral $\int_{L_2}\hat{u}_k\sigma_k\,\mathrm{d}l$

Using the same procedure as above, a similar expression is obtained for the integral along L_2

$$\int_{L_2} \hat{u}_k \sigma_k \, dl = \sum_{n=0}^{N} (B_{jn} Z_k^{(2)} U_{jkn}^{(2)}) + A Z_k^{(2)} V_k^{(2)} + Z_k^{(2)} W_k^{(2)} \tag{A2.11}$$

with

$$Z_k^{(2)} = -\frac{\csc(\pi\lambda)}{2(\lambda-1)} \mathrm{Re}\left\{ \sum_\alpha A_{k\alpha} M_{\alpha j} \tau_\alpha^{\lambda-1} \left[-w_j + w_{j+3} \frac{\cos(\pi\lambda)}{\tau_\alpha^{\lambda-1}} \right] \right\}, \tag{A2.12}$$

$$V_k^{(2)} = \frac{1}{2\pi} \left\{ v_j \mathrm{Re}\left[\sum_\alpha L_{k\alpha} M_{\alpha j} \tau_\alpha \int_c^\infty r^{\lambda-1} \int_{-1}^{0} \frac{(-\rho)^{-\lambda}}{r\tau_\alpha - \rho} \, d\rho \, dr \right] \right.$$

$$\left. + v_{k+3} \int_c^\infty r^{\lambda-1} \int_0^c \frac{\rho^{-\lambda}}{r-\rho} \, d\rho \, dr \right\}, \tag{A2.13}$$

$$W_k^{(2)} = -\int_0^c [-\sin\phi T_{k1}(r) + \cos\phi T_{k2}(r)] r^{\lambda-1} \, dr, \tag{A2.14}$$

$$U_{jkn}^{(2)} = \frac{1+c}{4Q} \sum_{q=1}^{Q} \cos(n\theta_q)$$

$$\times \left\{ \delta_{kj} \int_1^\infty r^{\lambda-1} \mu(r, \cos\theta_q) \, dr \right.$$

$$\left. + \mathrm{Re}\left[\sum_\alpha L_{k\alpha} M_{\alpha j} \tau_\alpha \int_1^\infty r^{\lambda-1} v_\alpha(r, \cos\theta_q) \, dr \right] \right\}, \tag{A2.15}$$

where

$$\mu(r,\zeta) = \begin{cases} 0 & \left(-1 < \zeta \leqslant \dfrac{1-c}{1+c}\right) \\ \dfrac{1}{r - \frac{1+c}{2}(1+\zeta) + 1} & \left(\dfrac{1-c}{1+c} < \zeta < 1\right) \end{cases} \tag{A2.16}$$

$$v_\alpha(r,\zeta) = \begin{cases} \dfrac{1}{r\tau_\alpha - \frac{1+c}{2}(1+\zeta) + 1} & \left(-1 < \zeta \leqslant \dfrac{1-c}{1+c}\right) \\ 0 & \left(\dfrac{1-c}{1+c} < \zeta < 1\right). \end{cases} \tag{A2.17}$$

Appendix 3

EDT: Ethylene diamine tartrate ($C_2H_6N_2H_2C_4H_4O_6$)

$$c^{-1}_{ijkl} = \begin{bmatrix} 3.34 & -3.0 & -3.0 & 0.0 & -1.7 & 0.0 \\ & 3.65 & -1.8 & 0.0 & -1.5 & 0.0 \\ & & 10.0 & 0.0 & -2.65 & 0.0 \\ & & & 19.2 & 0.0 & 0.38 \\ & \text{sym} & & & 11.7 & 0.0 \\ & & & & & 19.1 \end{bmatrix} 10^{-12}\,\text{cm}^2/\text{dyne}$$

MS: Monocrystalline silicon

$$c_{ijkl} = \begin{bmatrix} 1.16564 & 0.6394 & 0.6394 & 0.0 & 0.0 & 0.0 \\ & 1.943 & 0.3528 & 0.0 & 0.0 & 0.0 \\ & & 1.943 & 0.0 & 0.0 & 0.0 \\ & & & 0.5085 & 0.0 & 0.0 \\ & \text{sym} & & & 0.7951 & 0.0 \\ & & & & & 0.7951 \end{bmatrix} 10^{11}\,\text{Pa}$$

Acknowledgements

One of the authors (C.B.) wishes to thank the Asociación de Amigos of the University of Navarra for the grant received to carry out her Ph.D.

References

Abramowitz, M. and Stegun, I.A. (1970). *Handbook of Mathematical Functions*, Dover Publications, Inc., New York.

Amestoy, M. and Leblond, J.B. (1992). Crack paths in plane situations-II. Detailed form of the expansion of the stress intensity factors, *International Journal of Solids and Structures* **29**, 465–501.

Atkinson, C. (1972). On dislocation densities and stress singularities associated with cracks and pile-ups in inhomogeneous media, *International Journal of Engineering Science* **10**, 45–71.

Bilby, B.A. and Cardew, G.E. (1975). The crack with a kinked tip. *International Journal of Fracture Mechanics* **11**, 708–712.

Blanco, C. (1997). *Modelos de dislocaciones para grietas en recubrimientos y para grietas acodadas en el caso general de anisotropía elástica.* Tesis Doctoral, Escuela Superior de Ingenieros de San Sebastián, Universidad de Navarra.

Blanco, C., Martínez-Esnaola, J.M. and Atkinson, C. (1998). Analysis of sharp angular notches in anisotropic materials. *International Journal of Fracture*, this issue, 373–386.

Bogy, D.B. (1971). Two edge-bonded elastic wedges of different materials and wedge angles under surface tractions, *Journal of Applied Mechanics* **38**, 377–386.

Chatterjee, S.N. (1975). The stress field in the neighbourhood of a branched crack in an infinite elastic sheet. *International Journal of Solids and Structures* **11**, 521–538.

Cotterell, B. and Rice, J.R. (1980). Slightly curved or kinked cracks. *International Journal of Fracture* **16**, 155–159.

Erdogan, F. (1969). Approximate solution of systems of singular integral equations. SIAM *Journal of Applied Mathematics* **17**, 1041–1059.

Erdogan, F. and Gupta, G. (1972). On the numerical solution of singular integral equations. *Quarterly of Applied Mathematics* **30**, 525–534.

Gao, H. and Chiu, C.-H. (1992). Slightly curved kinked cracks in anisotropic elastic solids. *International Journal of Solids and Structures* **29**, 947–972.

Hayashi, K. and Nemat-Nasser, S. (1981). Energy-release rate and crack kinking under combined loading. *Journal of Applied Mechanics* **48**, 520–524.

Karihaloo, B.L., Keer, L.M. and Nemat-Nasser, S. (1980). Crack kinking under nonsymmetric loading. *Engineering Fracture Mechanics* **13**, 879–888.

Khrapkov, A.A. (1971). The first problem for a notch at the apex of an infinite wedge. *International Journal of Fracture Mechanics* **7**, 373–382.

Kitagawa, H., Yuuki, R. and Ohira, T. (1975). Crack-morphological aspects in fracture mechanics. *Engineering Fracture Mechanics* **7**, 515–529.

Leblond, J.B. (1989). Crack paths in plane situations-I. General form of the expansion of the stress intensity factors. *International Journal of Solids and Structures* **25**, 1311–1325.

Lo, K.K. (1978). Analysis of branched cracks. *Journal of Applied Mechanics* **45**, 797–802.

Melin, S. (1986). On singular integral equations for kinked cracks. *International Journal of Fracture* **30**, 57–65.

Muskhelishvili, N.I. (1977). *Some Basic Problems of the Mathematical Theory of Elasticity*. Noordhoff Int. Pub., Leyden, The Netherlands.

Obata, M., Nemat-Nasser, S. and Goto Y. (1989). Branched cracks in anisotropic elastic solids. *Journal of Applied Mechanics* **56**, 858–864.

Stroh, A.N. (1958). Dislocations and cracks in anisotropic elasticity. *Philosophical Magazine* **3**, 625–646.

Theocaris, P.S. and Makrakis, G.N. (1986). The kinked crack solved by Mellin transform. *Journal of Elasticity* **16**, 393–411.

Theocaris. P.S. and Makrakis, G.N. (1987). Crack kinking in anti-plane shear solved by the Mellin transform. *International Journal of Fracture* **34**, 251–262.

Vitek, V. (1977). Plane strain stress intensity factors for branched cracks. *International Journal of Fracture* **13**, 481–501.

International Journal of Fracture **93:** 409–429, 1998.

Williams meets von Karman: Mode coupling and nonlinearity in the fracture of thin plates

CHUNG-YUEN HUI, ALAN T. ZEHNDER and YOGESH K. POTDAR
Department of Theoretical and Applied Mechanics, Cornell University, 313-4, Kimball Hall, Ithaca, NY 14853-1503 USA; e-mail: ch45@cornell.edu

Received 28 January 1998; accepted in revised form 11 August 1998

Abstract. The stress field near the tip of a crack in a plate subjected to membrane and bending loads and undergoing large deflections, is investigated by performing an asymptotic analysis in the context of von Karman plate theory. It is demonstrated that the character of the near tip fields is identical to those of the linear plate theory. However, the determination of the crack tip stress intensity factors requires the solution of a large deflection, and hence nonlinear, problem due to the coupling of the membrane and bending modes. This effect is illustrated through the solution of three fracture problems involving plates of simple geometries loaded by pressure, tension and shearing. In two of these problems, the energy release rate is obtained exactly. Nonlinear finite element computations are performed to obtain the stress intensity factors and energy release rate associated with tension, bending and shearing. These results are compared to the theoretical results for energy release rate and stress intensity factors.

Key words: Stress intensity factor, plate theory, energy release rate, finite element method, large deflection.

1. Introduction

Recent interest in aircraft structural integrity has led to renewed research in the problem of fracture of cracked, elastic thin plates (Williams, 1962; Sih et al. 1962; Knowles and Wang, 1960; Wang, 1970; Delale and Erdogan, 1979; Murthy et al. 1981; Boduroglu and Erdogan, 1983; Simmonds and Duva, 1981; Young and Sun, 1993ab; Alwar and Thiagarajan, 1989; Hui and Zehnder, 1993, 1994; Viz et al. 1995ab; Viz and Zehnder, 1994; Viz, 1996) under combined bending, twisting and in-plane loading conditions. Research in this field, however, is not new; and has its origins in William's 1961 paper on crack tip fields in the small deflection, Kirchhoff plate theory (Kirchhoff, 1950). Theories of fracture based on those results had considerable success, but, in principle, are limited in application to structures in which the deflections are small. In practice, this limitation to small deflections has stopped no one. It does, however, lead to the question: Do the Kirchhoff theory fields still describe the crack tip stresses when large deflections occur? This question is answered here by performing an asymptotic analysis in the context of the large deflection plate theory developed by von Karman (1910). The assumption of small deflections used in the existing analyses of crack tip fields enables the membrane stresses due to in-plane loading to be decoupled from the stresses due to bending, twisting, and shearing loading. In Kirchhoff theory, the crack tip stress and displacement fields, with respect to a polar coordinate system (r, θ) located at the mid-plane

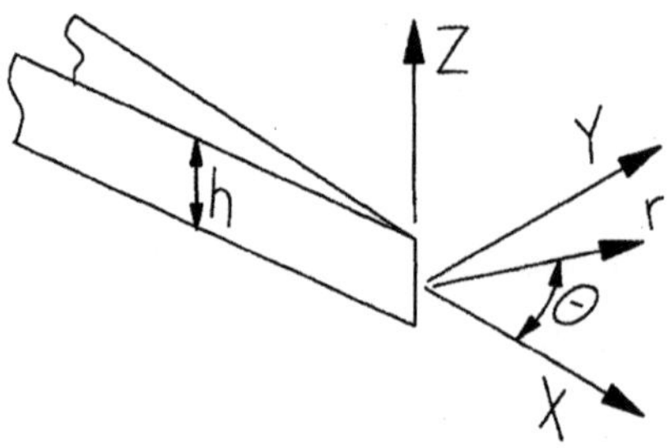

Figure 1. Geometry illustrating the crack tip coordinate system.

of the plate, $z = 0$, (see Figure 1), have the form (as $r \to 0$) (Williams, 1961; Sih et al. 1962; Murakami, 1987)

$$\boldsymbol{\sigma} = \frac{K_{\rm I}}{\sqrt{2\pi r}}\mathbf{f}^{\rm I}(\theta) + \frac{K_{\rm II}}{\sqrt{2\pi r}}\mathbf{f}^{\rm II}(\theta) + \frac{k_1}{\sqrt{2r}}\frac{2z}{h}\mathbf{k}^{\rm I}(\theta) + \frac{k_2}{\sqrt{2r}}\frac{2z}{h}\mathbf{k}^{\rm II}(\theta), \tag{1a}$$

$$\mathbf{u} = \frac{(1+\nu)K_{\rm I}\sqrt{r}}{E\sqrt{2\pi}}\hat{\mathbf{u}}^{\rm I}(\theta) + \frac{(1+\nu)K_{\rm II}\sqrt{r}}{E\sqrt{2\pi}}\hat{\mathbf{u}}^{\rm II}(\theta), \tag{1b}$$

$$w = \frac{k_1(2r)^{3/2}}{2Eh}\hat{w}^{\rm I}(\theta) + \frac{k_2(2r)^{3/2}}{2Eh}\hat{w}^{\rm II}(\theta), \tag{1c}$$

where E, ν are the Young's modulus and the Poisson's ratio respectively and h is the plate thickness. The in-plane stress tensor and displacement vector are denoted by $\boldsymbol{\sigma}$ and $\mathbf{u}$ respectively. The out of plane plate deflection is denoted by w. All tensor and vector functions are denoted by boldface letters. The classical Mode I and Mode II stress intensity factors are denoted by $K_{\rm I}$, $K_{\rm II}$ respectively. The Kirchhoff stress intensity factors are denoted by k_1, k_2 which represent, respectively, a symmetric bending mode and an antisymmetric twisting/transverse shearing mode. These modes are illustrated schematically in Figure 2. The dimensionless functions $\mathbf{f}^{\rm I}(\theta)$, $\mathbf{f}^{\rm II}(\theta)$, $\mathbf{k}^{\rm I}(\theta)$, $\mathbf{k}^{\rm II}(\theta)$ and $\hat{\mathbf{u}}^{\rm I}(\theta)$, $\hat{\mathbf{u}}^{\rm II}(\theta)$, $\hat{w}^{\rm I}(\theta)$, $\hat{w}^{\rm II}(\theta)$ describe the angular variations of the crack tip fields and can be found in (Williams, 1961; Sih et al. 1962; Hui and Zehnder, 1993; Viz, 1996). The bending and membrane modes are decoupled since, for small deflections, $K_{\rm I}$, $K_{\rm II}$ and k_1, k_2 can be obtained independently by solving two uncoupled, linear elastic crack problems. A consequence is that these stress intensity factors are directly proportional to the external loading parameters.

The crack tip fields resulting from Kirchhoff plate theory differ from the usual three dimensional elasticity solution. Indeed, the transverse shear stresses $(\sigma_{xz}, \sigma_{yz})$ vary asymptotically as $r^{-3/2}$ instead of $r^{-1/2}$ as $r \to 0$. Furthermore, the angular distributions of the Kirchhoff theory stress field differ from the linear elastic planar fracture mechanics results. For these reasons, the asymptotic stress field has also been calculated by many researchers using Reissner plate theory (Knowless and Wang, 1960; Wang, 1970; Dalale and Erdogan, 1979; Murthy et al. 1981). The asymptotic crack tip fields in Reissner plate theory all vary as $r^{-1/2}$ as $r \to 0$. Furthermore, they have identical angular distributions as those from planar elasticity. However, the use of Kirchhoff plate theory is justified by an asymptotic argument similar to that of small scale yielding in elastic–plastic fracture mechanics. Specifically, the Kirchhoff crack tip fields may be interpreted as the far field boundary conditions for the near tip Reissner crack tip fields. Indeed, Hui and Zehnder (1993) as well as Young and Sun (1993a) have shown independently

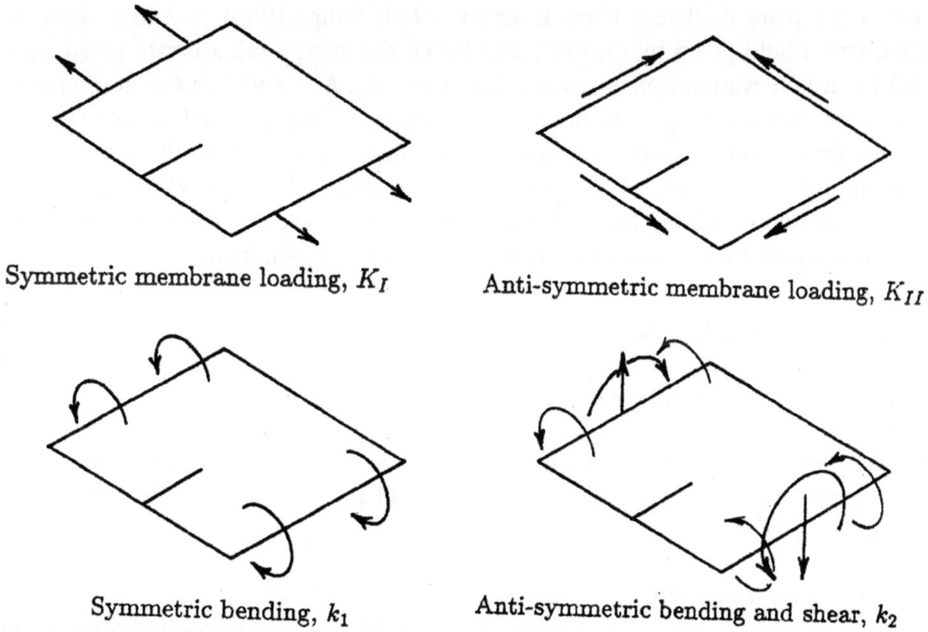

Figure 2. Fracture modes for a Kirchhoff cracked plate and their associated stress intensity factors, K_{I}, K_{II} represent the plane stress modes and k_1, k_2 represent the Kirchhoff plate theory modes.

that there is a unique relation between the Kirchhoff stress intensity factors and the Reissner stress intensity factors in the limit of very thin plates. This unique relation is universal in the sense that it is independent of the specimen geometry and loading configurations.

For self-similar crack growth, the energy release rate G is related to the stress intensity factors by (Hui and Zehnder, 1993)

$$\begin{aligned} G &= \frac{K_{\mathrm{I}}^2}{E} + \frac{K_{\mathrm{II}}^2}{E} + \frac{\pi(1+\nu)k_1^2}{3(3+\nu)E} + \frac{\pi(1+\nu)k_2^2}{3(3+\nu)E}, \\ &= G_{\mathrm{I}} + G_{\mathrm{II}} + g_1 + g_2, \end{aligned} \tag{2}$$

where the G_{I}, G_{II}, g_1, g_2 are the energy release rate components associated with the different fracture modes. It is important to note that since (2) follows from (1a–c) through a virtual crack extension argument, it is valid as long as the crack tip fields have the form given by (1); even if K_{I}, K_{II} and k_1, k_2 are interdependent and are nonlinear functions of the external loads.

A serious limitation of the linearized Kirchhoff theory is that it is only valid when the plate deflection is much smaller than the plate thickness. For cracks in aircraft structures, and in the experiments performed by Viz et al. (1995b), the plate deflections can be many times the plate thickness. In this regime, the coupling of the membrane and bending stresses is extremely important and the simple approximation of independently computing (K_{I}, K_{II}) from the membrane loads and (k_1, k_2) from the bending loads leads to large errors, as we will demonstrate below. Furthermore, it is not clear whether the near tip fields are still given by (1a–1c) if a large deflection plate theory is used in the analysis.

The goal of this work is to characterize the crack tip fields and to investigate the nonlinear coupling of the membrane and bending modes of fracture using von Karman's nonlinear, large deflection theory of elastic plates. This theory allows for the plate deflection to be on the order of the plate thickness (Von Karman, 1910; Fung, 1965). We first show that the asymptotic stress field given by (1a–1c), and hence the energy release rate given by (2) are still valid for a von Karman plate except that now $(K_{\mathrm{I}}, K_{\mathrm{II}})$ and (k_1, k_2) are generally not independent of each other, but are nonlinear functions of the external loads. The coupling between membrane and bending/shearing modes will be illustrated by three crack problems involving thin, cracked elastic plates of simple geometries. These problems will be studied analytically and numerically. In two of these problems, the energy release rate can be obtained in closed form. In the third problem, approximate results are obtained.

2. Crack tip fields in a Von Karman plate

2.1. Summary of von Karman plate equations

The governing equations of a von Karman plate in the absence of in plane body forces, expressed in terms of the plate deflection, $w(x, y)$, and the Airy stress function, $F(x, y)$ are (Fung, 1965)

$$\nabla^4 w = D^{-1}[p + B(F, w)], \tag{3a}$$

$$\nabla^4 F = EhA(w), \tag{3b}$$

where $D = Eh^3/12(1 - \nu^2)$, p is the pressure acting on the plate, and B and A are nonlinear differential operators defined by

$$B(F, w) \equiv F_{,yy}w_{,xx} + 2F_{,xy}w_{,xy} + F_{,xx}w_{,yy}, \tag{3c}$$

$$A(w) \equiv (w_{,xy})^2 + w_{,xx}w_{,yy}. \tag{3d}$$

The comma denotes partial derivative, e.g. $w_{,xy} = \partial^2 w/\partial x \partial y$.

The membrane forces (force/unit length) $\mathbf{N}$ are related to F by

$$N_{xx} = F_{,yy}, N_{yy} = F_{,xx}, N_{xy} = -F_{,xy}. \tag{4a}$$

The membrane forces and the bending and twisting moments $\mathbf{M}$ are related to the displacements by

$$\begin{aligned}
N_{xx} &= \frac{Eh}{1-\nu^2}\left[u_{x,x} + \nu u_{y,y} + \tfrac{1}{2}(w_{,x})^2 + \frac{\nu(w_{,y})^2}{2}\right],\\
N_{yy} &= \frac{Eh}{1-\nu^2}\left[u_{y,y} + \nu u_{x,x} + \tfrac{1}{2}(w_{,y})^2 + \frac{\nu(w_{,x})^2}{2}\right],\\
N_{xy} &= \frac{Eh}{2(1+\nu)}(u_{x,y} + u_{y,x} + w_{,y}w_{,x}),\\
M_{xx} &= -D(w_{,xx} + \nu w_{,yy}), \qquad M_{yy} = -D(w_{,yy} + \nu w_{,xx}),\\
M_{xy} &= -D(1-\nu)w_{,xy}.
\end{aligned} \tag{4b}$$

2.2. Asymptotic analysis

Since we are interested in crack tip fields, the crack can be considered as semi-infinite. Assume that the crack lies on the negative x axis with its tip at the origin. The traction free boundary conditions consistent with plate theory are

$$N_{yy}(r, \theta = \pm\pi) = N_{xy}(r, \theta = \pm\pi) = 0, \tag{5a}$$

$$M_{yy}(r, \theta = \pm\pi) = [Q_y(r, \theta = \pm\pi) + M_{xy},_x (r, \theta = \pm\pi)] = 0, \tag{5b}$$

where Q_y is the y component of the transverse shear resultant Q which is given by

$$\begin{aligned} Q_x &= -D\nabla^2 w,_x + N_{xx} w,_x + N_{xy} w,_y\,, \\ Q_y &= -D\nabla^2 w,_y + N_{xy} w,_x + N_{yy} w,_y\,. \end{aligned} \tag{5c}$$

Note that the nonlinear terms appear both in the differential equations, i.e., $B(F, w)$, $A(w)$, as well as in the boundary conditions. The linear theory neglects these nonlinear terms resulting in the uncoupling of the differential equations and of the boundary conditions. Specifically, the asymptotic results of the linear theory (1a–c) satisfy equations (3,4,5) with all the nonlinear terms removed.

We now show that the asymptotic results given by (1a–c) are still valid for a von Karman plate. The simplest way to see this is to note that the nonlinear terms, $B(F, w)$ and $A(w)$, in the differential equations, as well as in the boundary conditions vanish in comparison to the linear terms as $r \to 0$. In the language of asymptotic analysis, these terms are subdominant. Indeed, using (1a–1c), one can easily verify that

$$B(F, w) = O(r^{-1}), \qquad A(F, w) = O(r^{-1}), \qquad r \to 0, \tag{6}$$

whereas the asymptotic behavior of the LHS of (3a–b) is

$$\nabla^4 w = O(r^{-5/2}), \qquad \nabla^4 F = O(r^{-5/2}), \qquad r \to 0 \tag{7}$$

Likewise, in the boundary conditions, e.g. $N_{yy}(r, \theta = \pm\pi) = 0$, we have, as $r \to 0$

$$\begin{aligned} N_{yy} &= \frac{Eh}{1-\nu^2}\left[u_y,_y + \nu u_x,_x + \frac{(w,_y)^2}{2} + \frac{\nu (w,_x)^2}{2}\right] \\ &\approx \frac{Eh}{1-\nu^2}[u_y,_y + \nu u_x,_x], \end{aligned}$$

since the term

$$\frac{(w,_y)^2}{2} + \frac{\nu(w,_x)^2}{2} = O(r), \qquad r \to 0,$$

whereas the term

$$u_y,_y + \nu u_x,_x = O(r^{-1/2}), \qquad r \to 0,$$

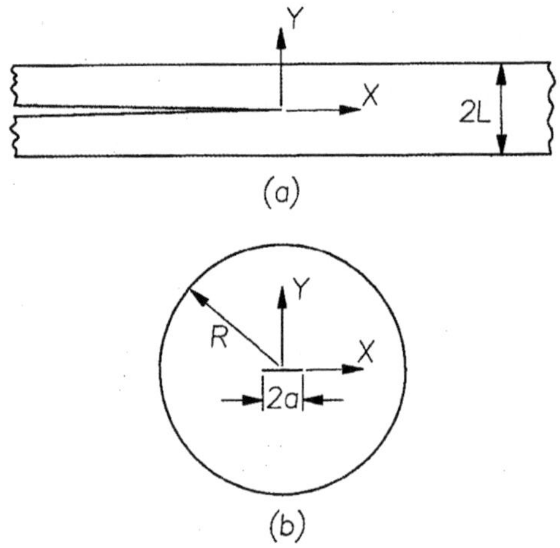

Figure 3. Example problems analyzed: (a) Infinite strip with a semi-infinite crack. Top and bottom edges clamped. Plate loaded either by stretching combined with uniform pressure, or by out of plane shearing. (b) Clamped circular plate of radius R, under uniform pressure, containing a small crack of length $2a$.

so that the term $[(w,_y)^2 + \nu(w,_x)^2]/2$ becomes negligible with respect to $u_y,_y + \nu u_x,_x$. In other words, the first order asymptotic behavior of the stress and deformation fields is completely determined by the linear terms of the governing equations and the boundary conditions. Therefore, the asymptotic results of (1a–1c) still apply for the case of a von Karman plate. This, however, does not imply that the stress intensity factors can be determined by the linear theory, since they can only be determined by solving the full set of nonlinear, coupled equations.

Of course, one can start with the formal approach assuming F and w must have the form

$$F = r^{\lambda}\hat{F}(\theta), \qquad w = r^{\lambda+1}\hat{w}(\theta)$$

and proceed with the asymptotic method of William's. As before, this procedure will show that the nonlinear terms are subdominant so that the results will be identical to that of (1a–c).

3. Nonlinear solutions illustrating mode coupling: three examples

In most real-world problems involving cracked plates or shells the out of plane deflections are relatively large. Depending on the boundary conditions these large deflections can result in membrane stresses as well as the bending stresses associated with the linear plate theories. As seen in equation (3), when the membrane stresses must be included there results a set of nonlinear, coupled equations. For complex geometries these equations must be solved numerically using geometrically nonlinear finite element methods. There are however a number of simple geometries for which analytical solutions can be obtained. These solutions demonstrate the coupling between the membrane and bending fracture modes, show how the different stress intensity factor and energy release rate components scale with the applied loads or displacements, and provide nonlinear benchmark solutions against which numerical analysts can test their numerical procedures.

The three examples given in this paper are illustrated in Figure 3. The first two are problems involving an infinite strip with an semi-infinite crack. The edges of the strip, located at $y =$

$\pm L$ are clamped and the strip is loaded either by a uniform pressure combined with in-plane stretching or by out-of-plane shearing. The third problem is for a small crack in a clamped, circular plate under uniform pressure. The infinite strip problems are solved exactly for G using an energy approach. The circular plate solution is approximate, but quite accurate for small cracks. It is a stress based method that uses a simple interpolation of the membrane stress and bending moment to determine K_{I} and k_1. Finite element solutions of all three problems are given since in some cases the individual components of energy release rate, and hence the stress intensity factors cannot be determined from the analytical results, although the total energy release rate may be known exactly.

3.1. Finite element procedures

The energy release rates defined in (2) were computed using the modified crack closure integral (MCCI) (Viz et al. 1995a) method in conjunction with geometrically nonlinear finite element analyses. Both four noded (S4R in ABAQUS) and eight noded (S8R), six degrees of freedom per node, elements were used. Using previous convergence studies as a guideline (Viz et al. 1995a and Viz, 1996), the crack tip region was meshed using square elements with a nodal spacing of $0.02L$, where L is the crack length or other relevant dimension. Special crack tip, or singular elements are not used. The computations were geometrically nonlinear, but materially linear.

To minimize the transverse shear stiffness of the elements, and hence simulate the von-Karman theory, the shell thickness, h, was taken to be small. The finite element model for the problem of a semi-infinite crack in an infinite strip is shown in Figure 4. The model dimensions are $L = 1, h = 0.001$, and a length of 20. Symmetry (or anti-symmetry) about $y = 0$ is used to reduce the problem size and to simplify the modeling. The boundary conditions and mesh are given in Figure 4. A simple rectangular mesh of four noded elements is used throughout. A refined mesh of 20×10 elements, each of length $0.02L$ is used in the region surrounding the crack tip.

The finite element model (FEM) for the problem of a small crack in a circular plate is shown in Figure 5. For the analytical approximation given in Section 3.4 to be valid, the crack must be small compared with typical dimensions. Thus the dimensions are $R = 1, a = 0.1$ and $h = 0.01$. Past experience shows that $R/a = 10$ is sufficient to approximate a finite crack in an infinite plate. Quarter symmetry is used along $x = 0$ and $y = 0$ to reduce the problem size. Eight node elements are used, and as in the case of infinite strip, the crack tip mesh consists of square elements with a nodal spacing of $0.02a$, element length of $0.04a$.

Let (f_x^n, f_y^n, f_z^n) be the forces, $(M_{xx}^n, M_{yy}^n, M_{zz}^n)$, the moments, (u_x^n, u_y^n, u_z^n) the displacements, and $(\phi_x^n, \phi_y^n, \phi_z^n)$ the rotations at node 'n'. For four node elements, the components of the energy release rate are then given by (Viz et al. 1995a)

$$G_{\mathrm{I}} = \frac{f_y^i \Delta u_y^{i-1} + M_{zz}^i \Delta\phi_z^{i-1}}{2h\Delta a}, \qquad G_{\mathrm{II}} = \frac{f_x^i \Delta u_x^{i-1}}{2h\Delta a},$$

$$g_1 = \frac{M_{xx}^i \Delta\phi_x^{i-1}}{2h\Delta a}, \qquad g_2 = \frac{f_z^i \Delta u_z^{i-1} + M_{yy}^i \Delta\phi_y^{i-1}}{2h\Delta a}, \tag{8a}$$

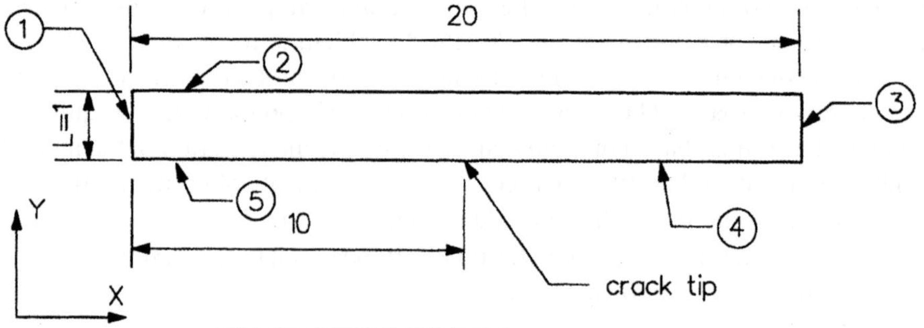

Surface	Boundary Conditions
2	$(u_x=\phi_x=\phi_y=\phi_z=0)$ $(u_z=0,\ u_y=\varepsilon)$, pressure load case $(u_z=w_L,\ u_y=0)$, shear load case
1, 3	$(u_x=\phi_y=\phi_z=0)$
4	$(u_y=\phi_x=\phi_z=0)$, pressure load case $(u_y=u_z=\phi_y=\phi_z=0)$, shear load case
5	traction free

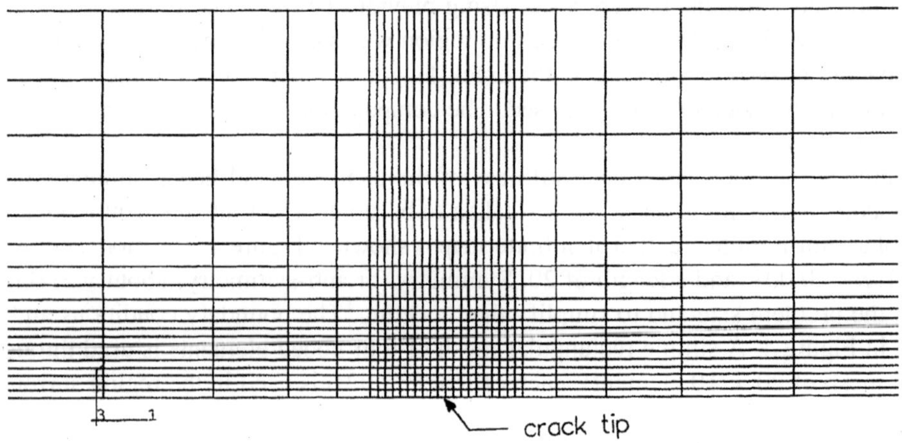

Figure 4. Half plate finite element model used for strip problems (not to scale) and crack tip mesh.

where Δa is the crack tip element length, the $\mathbf{f}^i$ and $\mathbf{M}^i$ are the nodal forces and moments at the crack tip, and $\Delta\mathbf{u}^{i-1}$, $\Delta\phi^{i-1}$ are the jumps across the crack faces in displacements and rotations one node behind the crack tip. For eight node elements the computation is similar

$$
\begin{aligned}
G_{\mathrm{I}} &= \frac{f_y^i \Delta u_y^{i-2} + f_y^{i+1} \Delta u_y^{i-1} + M_{zz}^i \Delta\phi_z^{i-2} + M_{zz}^{i+1} \Delta\phi_z^{i-1}}{2h\Delta a},\\
G_{\mathrm{II}} &= \frac{f_x^i \Delta u_x^{i-2} + f_x^{i+1} \Delta u_x^{i-1}}{2h\Delta a},\\
g_1 &= \frac{M_{xx}^i \Delta\phi_x^{i-2} + M_{xx}^{i+1} \Delta\phi_x^{i-1}}{2h\Delta a},\\
g_2 &= \frac{f_z^i \Delta u_z^{i-2} + f_z^{i+1} \Delta u_z^{i-1} + M_{yy}^i \Delta\phi_y^{i-2} + M_{yy}^{i+1} \Delta\phi_y^{i-1}}{2h\Delta a},
\end{aligned}
\tag{8b}
$$

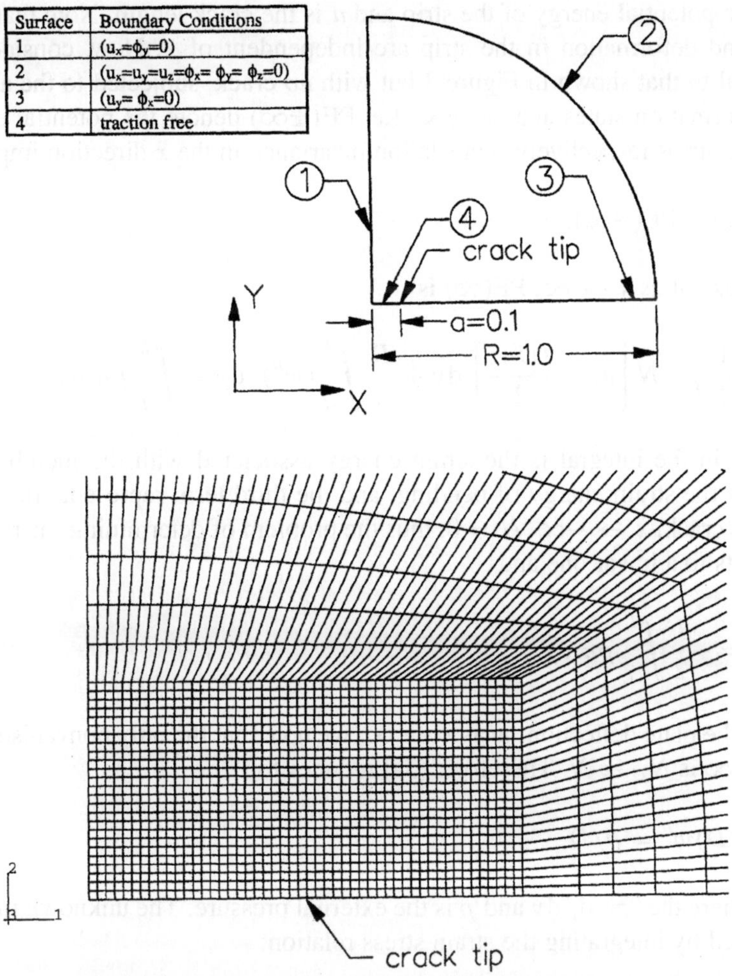

Figure 5. Quarter-plate finite element model used for circular plate problem.

where i is the crack tip node, $i+1$ is one node ahead of the crack, $i-1$ is one node behind the crack, and $i-2$ is two nodes behind the crack. In the $G_{\rm I}$ calculation, the $M_{zz}\Delta\phi_z$ term is negligible while the $f_z\Delta u_z$ and $M_{yy}\Delta\phi_y$ terms are of equal magnitude in the g_2 calculation. To verify the accuracy of the numerical computations, linear analyses were performed of each case and the computed value of G compared to exact analytical solutions. In all cases the numerical results differed from the exact results by less than 1.5 percent.

3.2. Infinite strip under pressure and in-plane stretching (energy method)

3.2.1. *Problem formulation*

The procedure to calculate the energy release rate for the infinite strip problems illustrated in Figure 3 is outlined here. The total energy release rate is calculated based on the definition of G, i.e.

$$G = -\frac{\partial \Pi}{\partial a},$$

where Π is the potential energy of the strip and a is the crack length. Note that as $x \to \pm\infty$, the stresses and deformation in the strip are independent of x. Now, consider two infinite strips, identical to that shown in Figure 3 but with *no* crack, subjected to the aforementioned stress and deformation states at $x = \pm\infty$. Let PE($\pm\infty$) denote the potential energy per unit length of these strips respectively. Translation invariance in the x direction implies that

$$G = \mathrm{PE}(\infty) - \mathrm{PE}(-\infty). \tag{9}$$

Since N is constant as $x \to \infty$, PE(∞) is

$$\mathrm{PE}(\infty) = \frac{1}{2}\int_{-L}^{L} N\left[u_y' + \frac{(w')^2}{2}\right]\mathrm{d}y + \frac{D}{2}\int_{-L}^{L}(w'')^2\,\mathrm{d}y - \int_{-L}^{L} pw\,\mathrm{d}y. \tag{10a}$$

The first term in the integral is the strain energy associated with the membrane stress, the second term is the strain energy of bending, and the third term represents the external work. Far behind the crack, $x \to -\infty$, $N = 0$; thus the problem of determining the relation between p and w reverts to a linear one and

$$\mathrm{PE}(-\infty) = -\frac{1}{2}\int_{-L}^{L} pw\,\mathrm{d}y. \tag{10b}$$

To determine the plate deflection, w as $x \to \infty$, we note that the only nonvanishing membrane force as $x \to \infty$ is $N_{22} \equiv N$. Therefore, (3a) reduces to

$$w'''' - (N/D)w'' = p/D, \tag{11}$$

as $x \to \infty$, where the $' \equiv \mathrm{d}/\mathrm{d}y$ and p is the external pressure. The unknown membrane force N is determined by integrating the strain stress relation,

$$\varepsilon_{yy} = u_y' + \frac{(w')^2}{2} = \frac{(1-\nu^2)N}{Eh}, \tag{12}$$

from $y = -L$ to L.

The boundary conditions are $w(\pm L) = w'(\pm L) = 0$, and $u_y(\pm L) = \pm\varepsilon L$, (the prestrain). The plate deflection as $x \to \infty$ is obtained by solving (11) with $p =$ constant,

$$\hat{w} = \frac{12(1-\nu^2)P}{\alpha^3}\left[\frac{\cosh\alpha\eta - \cosh\alpha}{\sinh\alpha} + \tfrac{1}{2}\alpha(1-\eta^2)\right], \tag{13}$$

where the following normalisations are used

$$\alpha = L\sqrt{N/D}, \qquad \eta = y/L, \qquad \hat{w} = w/h, \qquad P = pL^4/(Eh^4). \tag{14}$$

Integration of (11) using $u_y(\pm L) = \pm\varepsilon L$, yields an equation for the normalized membrane force α as a function of P,

$$2(\alpha/\delta)^2\varepsilon + [12(1-\nu^2)P]^2\Psi_1(\alpha) - \alpha^4/6 = 0, \tag{15a}$$

where $\delta = h/L$, ε is the prestrain and

$$\Psi_1(\alpha) = \alpha^{-2}\left[-\frac{3\coth\alpha}{2\alpha} + 2\alpha^{-2} + \frac{1}{3} - \frac{\mathrm{cosech}^2\alpha}{2}\right]. \tag{15b}$$

Given P and ε (15) can be solved for the normalised membrane force α. Once α is known, the energy release rate is found by substituting (13) into (10a) to obtain PE(∞). Then, using (10a,b) in (9) gives

$$\hat{G} \equiv \frac{\mathrm{G}L^3}{Eh^4} = (1-\nu^2)P^2\left[\frac{3}{5} - 6\Psi_1(\alpha) - 12\Psi_2(\alpha)\right] + \frac{\varepsilon(\alpha/\delta)^2}{12(1-\nu^2)}, \tag{16a}$$

where $\Psi_2(\alpha)$ is

$$\Psi_2(\alpha) = -\alpha^{-3}\coth\alpha + \alpha^{-4} + \alpha^{-2}/3, \tag{16b}$$

and $\hat{G}$ is the normalised energy release rate.

3.2.2. *Solution in the case of no prestrain, $\varepsilon = 0$*

Consider the special case where the strip has no prestrain. The behavior of the normalised membrane force α for small and large pressure can be obtained from the asymptotic behavior of $\Psi_1(\alpha)$, i.e.,

$$\Psi_1(\alpha) \approx \begin{cases} 2\alpha^2/945 & \alpha \ll 1, \\ \alpha^{-2}/3 & \alpha \gg 1, \end{cases} \tag{17}$$

resulting in

$$\alpha^2 \cong \frac{4[12(1-\nu^2)P]^2}{315} \quad \text{or} \quad N \cong \frac{48(1-\nu^2)p^2L^6}{315Eh^5}, \qquad P \ll 1, \tag{18a}$$

$$\alpha \cong (2[12(1-\nu^2)P]^2)^{1/6} \quad \text{or} \quad N \cong \left[\frac{Eh^5L^2}{6(1-\nu^2)}\right]^{1/3} p^{2/3}, \qquad P \gg 1. \tag{18b}$$

Note that the membrane force is proportional to p^2 and $p^{2/3}$ at low and high loads respectively.

The behavior of the energy release rate at low and high pressures can be obtained using (16) and the following asymptotic behavior of $\Psi_2(\alpha)$:

$$\Psi_2(\alpha) \approx \begin{cases} 1/45 & \alpha \ll 1, \\ \alpha^{-2}/3 & \alpha \gg 1, \end{cases} \tag{19}$$

resulting in

$$\hat{G} = (1-\nu^2)P^2/3, \qquad P \ll 1, \tag{20a}$$

$$\hat{G} = 3(1-\nu^2)P^2/5, \qquad P \gg 1. \tag{20b}$$

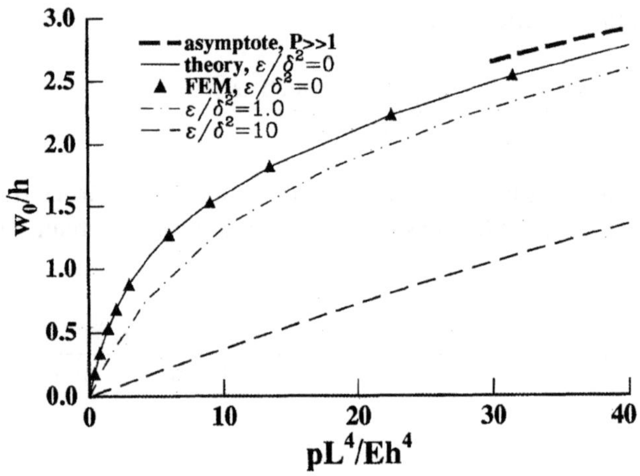

Figure 6. Deflection at the center of strip ($y = 0, x = \infty$) under uniform pressure for various amounts of prestrain.

When the applied pressure is small, we expect that membrane effects can be neglected. The membrane force contribution to the energy release rate is given by the $\Psi_1(\alpha)$ term in (16a), which, at low pressures, according to (17), vanishes as α^2. The energy of the bending term at $x = \infty$ is given by $\Psi_2(\alpha)$, which is a constant 1/45 at low pressures. Note also that, a comparison of (20a) and (20b) shows that, at sufficiently high pressures, G is underestimated by the Kirchhoff plate theory by about 44 percent.

The deflection at the middle of the plate ($x \to \infty, y = 0$), denoted by w_0, is plotted in Figure 6 as a function of P. From (13), one can show that

$$w_0 \to \frac{[6(1-\nu^2)^{1/3}P^{1/3}h}{2} \tag{21}$$

in the limit $P \gg 1$. This limit is also plotted in Figure 6 which shows that the membrane limit is approached for large plate deflections. The normalised energy release rate $\hat{G}$ versus the normalised pressure P is plotted in Figure 7. The low and high pressure limits given by (20a–b) are also plotted to indicate the range of validity of these asymptotic expressions. This figure shows that over the entire range the energy release rate is approximately proportional to P^2. The FEM results for $\hat{G}$ are also shown in Figure 7. For $P < 20$, the FEM results agree with the analytical solution to within 2 percent. For $P > 20$, the results begin to diverge slightly since the analytical solution does not take into account very large (elastica like) deflection behind the crack. The close agreement provides confidence in using the finite element program to determine the stress intensity factors, which cannot be obtained in closed form.

In this problem the modes of fracture are the symmetric membrane and bending so that $k_2 = K_{\mathrm{II}} = 0$, i.e.,

$$G = G_{\mathrm{I}} + g_1.$$

In the limit of low loads, the contribution of the membrane force to the energy release rate is small compared with those due to bending, so that

$$g_1 \equiv \frac{\pi(1+\nu)k_1^2}{3(3+\nu)E} \cong G \cong \frac{Eh^4(1-\nu^2)P^2}{3L^3}, \quad \text{or} \tag{22a}$$

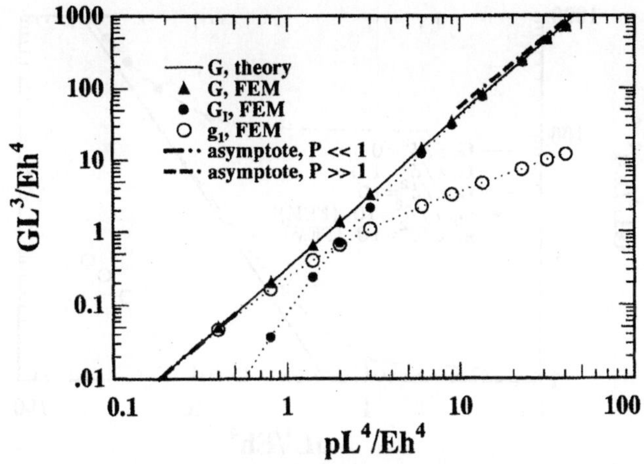

Figure 7. Energy release rate and its components versus pressure for strip under pressure loading, with no prestrain.

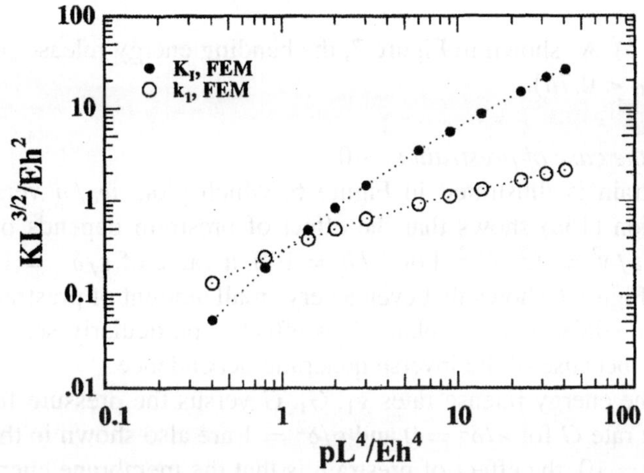

Figure 8. Stress Intensity factors versus pressure for strip under pressure loading with no prestrain.

$$k_{\mathrm{I}} \cong \frac{Eh^2 P}{3L^{3/2}} \sqrt{(3+\nu)(1-\nu)/\pi}, \quad K_{\mathrm{I}}/k_{\mathrm{I}} \to 0, \quad P \ll 1. \tag{22b}$$

Conversely, in the membrane limit, the bending effects are negligible, so that

$$G_{\mathrm{I}} \equiv \frac{K_{\mathrm{I}}^2}{E} \cong G \cong \frac{3Eh^4(1-\nu^2)P^2}{5L^3}, \quad \text{or} \tag{23a}$$

$$K_{\mathrm{I}} \cong \frac{Eh^2 P}{L^{3/2}} \sqrt{\frac{3}{5}(1-\nu^2)}, \quad k_{\mathrm{I}}/K_{\mathrm{I}} \to 0, \quad P \gg 1. \tag{23b}$$

With the exception of these limiting cases, we were unable to find a simple way to decompose G into its components. Thus, FEM was used to determine g_{I} and G_{I} and hence k_{I}, K_{I}. The FEM results for the energy release rates g_{I}, G_{I}, G and stress intensities k_{I}, K_{I} are plotted against the normalised pressure P, in Figures 7 and 8 respectively. These figures show that the membrane limit (20a,b) is approached for $P > 5$ corresponding to plate deflection in excess

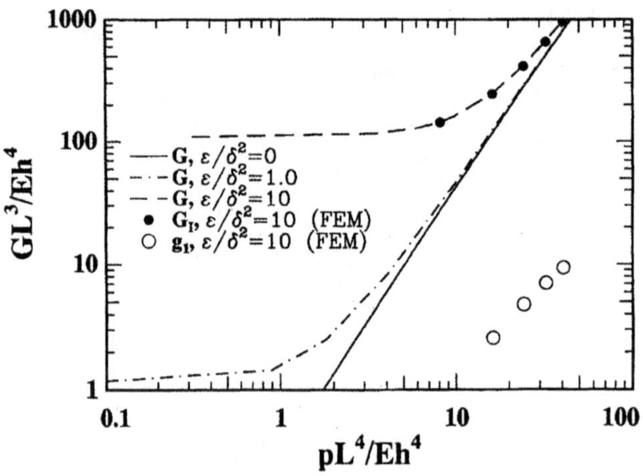

Figure 9. Energy release rate versus pressure for strip under pressure loading, with and without prestrain. Also shown are the energy release rate components for $\varepsilon/\delta^2 = 10$.

of $1.3h$ (see Figure 5). As shown in Figure 7, the bending energy release rate g_{I} is significant only for $P < 2$, ($w_0 < 0.7h$).

3.2.3. *Solution in the case of prestrain* $\varepsilon > 0$

The effect of prestrain is illustrated in Figure 6, which plots w_0/h versus P for $\varepsilon/\delta^2 = 0, 1.0, 10.0$. Equation (15a) shows that the effect of prestrain depends only on the dimensionless parameter $\varepsilon/\delta^2 = L^2\varepsilon/h^2$. For $L/h = 100$, a value of $\varepsilon/\delta^2 = 10$ corresponds to a prestrain of 0.001. Figure 6 shows that even a very small amount of prestrain can significantly affect the transverse stiffness of the plate. This effect is particularly sensitive to the relative plate thickness h/L because of the inverse quadratic dependence.

Figure 9 plots the energy release rates $g_{\mathrm{I}}, G_{\mathrm{I}}, G$ versus the pressure for $\varepsilon/\delta^2 = 10$. The total energy release rate G for $\varepsilon/\delta^2 = 0$ and $\varepsilon/\delta^2 = 1$ are also shown in the same figure. It is clear that, for $\varepsilon/\delta^2 = 10$, the effect of prestrain is that the membrane energy release rate G_{I}, completely dominates. The energy release rate G, for small P is given approximately by its plane stress value of

$$G = \frac{EL\varepsilon^2}{(1-\nu^2)}, \quad \text{or} \quad \hat{G} = \frac{GL^3}{Eh^4} = \frac{\varepsilon^2}{\delta^4(1-\nu^2)}. \tag{24}$$

3.3. Infinite strip with applied shear displacements (energy method)

In this problem no pressure loading or prestrain are applied. The plate is loaded by out-of -plane shearing, which induces a combination of K_{I} and k_2 fracture modes. We follow the same normalisation of equations and procedure as in 3.2.1. Let the applied shear displacements on the edges $y = \pm L$ be $\pm w_L$ respectively. The boundary conditions are $w(\pm L) = \pm w_L$, $w'(\pm L) = 0$. The plate deflection as $x \to \infty$, found by solving (11), is

$$\hat{w} = \hat{w}_L \frac{\sinh \alpha\eta - \alpha\eta \cosh \alpha}{\sinh \alpha - \alpha \cosh \alpha}, \tag{25}$$

where $\hat{w}_L = w_L/h$ is the normalised applied displacement. The normalised membrane stress α is determined by integrating (12), to yield

$$6\hat{w}_L^2\left(1+\frac{\cosh 2\alpha}{2}-\frac{3\sinh 2\alpha}{4\alpha}\right)(\sinh\alpha-\alpha\cosh\alpha)^{-2}-1=0. \tag{26}$$

The normalised energy release rate is

$$\hat{G}=\frac{\alpha^4}{144(1-\nu^2)}\left[1+\left[1+\frac{\cosh 2\alpha}{2}-\frac{3\sinh 2\alpha}{4\alpha}\right]^{-1}\left(\frac{\sinh 2\alpha}{2\alpha}-1\right)\right]. \tag{27}$$

The behavior of the normalised membrane force α for small and large applied displacements is

$$\alpha^2 \cong 36\hat{w}_L^2/5 \quad \text{or} \quad N \cong \frac{3Ew_L^2h}{[5L^2(1-\nu^2)]}, \quad w_L \ll h, \tag{28a}$$

$$\alpha^2 \cong 6\hat{w}_L^2 \quad \text{or} \quad N \cong \frac{Ew_L^2h}{[4L^2(1-\nu^2)]}, \quad w_L \ll h, \tag{28b}$$

Note that the membrane force is proportional to w_L^2 at both low and high loads.

Using (27) and (28), the energy release rates for small and large applied displacements are

$$\hat{G} \approx \frac{\hat{w}_L^2}{[4(1-\nu^2)]}, \quad w_L \ll h, \tag{29a}$$

$$\hat{G} \approx \frac{\hat{w}_L^4}{[4(1-\nu^2)]}, \quad w_L \gg h, \tag{29b}$$

Figure 10 plots the normalised energy release rate $\hat{G}$ as a function of the normalised displacement $\hat{w}_L$. The low and high load limits given by (29a,b) are also plotted to indicate the range of validity of these asymptotic expressions. The FEM results for the total energy release rate agree with the analytic solution to within 2 percent.

Symmetry implies that $k_1 = K_{\mathrm{II}} = 0$. In the limit of low loads, the membrane effects are negligible, so that by (29a),

$$g_2 \equiv \frac{\pi(1+\nu)k_2^2}{3(3+\nu)E} \cong G \cong \frac{Eh^4\hat{w}_L^2}{4L^3(1-\nu^2)}, \quad \text{and} \tag{30a}$$

$$k_2 \cong \frac{Eh^2\hat{w}_L}{L^{3/2}}\sqrt{\frac{3(3+\nu)}{4\pi(1-\nu^2)(1+\nu)}}, \quad w_L \ll h. \tag{30b}$$

In the membrane limit, the bending effects are negligible, so that by (29b),

$$G_{\mathrm{I}} \equiv \frac{K_{\mathrm{I}}^2}{E} \cong G \cong \frac{Eh^4\hat{w}_L^4}{4(1-\nu^2)L^3}, \tag{31a}$$

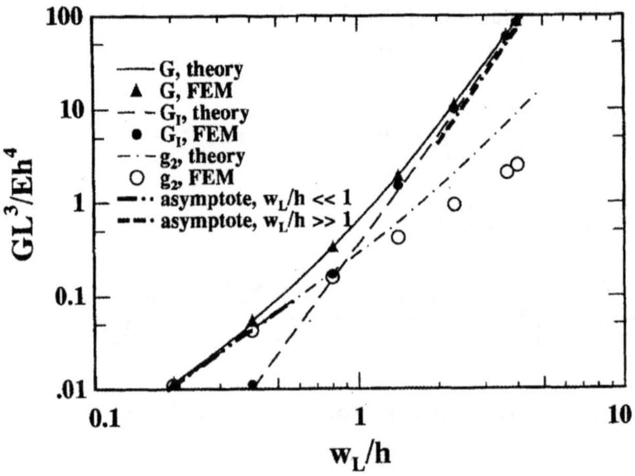

Figure 10. Energy release rate and its components versus out of plane edge shearing deflection for strip under shear loading.

$$K_{\mathrm{I}} \cong \frac{Eh^2\hat{w}_L^2}{2L^{3/2}\sqrt{1-\nu^2}}, \quad w_L \gg h. \tag{31b}$$

At intermediate displacements, to determine the individual stress intensity factors the energy release rate must be decomposed into tension and bending parts. We propose the following mode decomposition

$$g_2 = \frac{\pi(1+\nu)k_2^2}{3(3+\nu)E} = \frac{D}{2}\int_{-L}^{L}(w'')^2\,\mathrm{d}y, \tag{32a}$$

$$G_{\mathrm{I}} = \frac{K_{\mathrm{I}}^2}{E} = \frac{N}{4}\int_{-L}^{L}(w')^2\,\mathrm{d}y, \tag{32b}$$

i.e., g_2 is associated with the strain energy of bending and G_{I} with the strain energy of tension. In normalised form, the results are

$$\hat{g}_2 = \frac{\alpha^4}{144(1-\nu^2)}\left[1+\frac{\cosh 2\alpha}{2}-\frac{3\sinh 2\alpha}{4\alpha}\right]^{-1}\left(\frac{\sinh 2\alpha}{2\alpha}-1\right), \tag{33a}$$

$$\hat{G}_{\mathrm{I}} = \frac{\alpha^4}{144(1-\nu^2)}. \tag{33b}$$

The stress intensity factors corresponding to this decomposition are

$$\frac{k_2L^{3/2}}{Eh^2} \cong \sqrt{\frac{3(3+\nu)}{\pi(1+\nu)}}\sqrt{\hat{g}_2}, \tag{34a}$$

$$\frac{K_{\mathrm{I}}L^{3/2}}{Eh^2} \cong \sqrt{\hat{G}_{\mathrm{I}}}. \tag{34b}$$

These mode decomposition results are compared with the FEM calculations in Figures 10 and 11. It is clear that (33) and (34) are not exact, but nevertheless provide an excellent approximation in the entire loading range. The asymptotic results are also plotted on Figure 10. An unexpected result is that extrapolations of the asymptotic results agree extremely well with the finite element results in the *entire loading range*. Therefore, they can replace the more complicated expressions given by (33a,b) and (34a,b). In other words, an excellent approximation for the stress intensity factors is

$$\frac{K_I L^{3/2}}{Eh^2} \cong \frac{\hat{w}_L^2}{2\sqrt{1-\nu^2}}, \qquad \frac{k_2 L^{3/2}}{Eh^2} \cong \frac{E\hat{w}_L}{L^{3/2}}\sqrt{\frac{3(3+\nu)}{4\pi(1-\nu^2)(1+\nu)}}. \tag{35}$$

3.4. Clamped circular plate with pressure loading (stress based solution)

The third example is that of a small crack of length $2a$ in a clamped circular plate of radius R under uniform pressure p. The approach here is simple. We first compute the membrane stress N and the bending moment M, *at the center of the uncracked plate*. The stress intensity factors are then computed by assuming that the crack is in an infinite plate which is subjected to a uniform far-field tension N and bending moment M. In such a situation (Sih et al. 1962)

$$K_I = N\sqrt{\pi a}/h, \qquad k_1 = 6M\sqrt{a}/h^2. \tag{36}$$

Clearly this approach is only valid if $a \ll R$. Otherwise the stress field in the region away from the crack will differ substantially from that of an uncracked plate.

Although there is a closed form solution for the uniformly loaded, uncracked clamped circular plate problem (Conway, 1957; Way, 1934), the results come out as complex expressions. To show very clearly the scalings of the mode couplings, a simple expression for N and M is desired. This is provided by the Foeppel interpolation approach (Conway, 1946; Foeppel and Foeppel, 1924) which assumes 'that the pressure is composed of two parts, p_1 and p_2, p_1 causing pure bending and p_2 pure stretching'. For pure bending,

$$\frac{w_0}{h} = \frac{3(1-\nu^2)p_1 R^4}{16Eh^4}, \tag{37a}$$

where w_0 is the deflection at the plate center. Hence the normalised P_1 is

$$P_1 = \frac{p_1 R^4}{Eh^4} = \frac{16 w_0}{3(1-\nu^2)h}. \tag{37b}$$

For pure stretching, we finding the stress by considering a membrane. For a uniformly loaded membrane, dimensional consideration shows that

$$\frac{w_0}{h} = \beta(\nu)\left(\frac{p_2 R^4}{Eh^4}\right)^{1/3}, \tag{38a}$$

where $\beta(\nu)$ is a dimensionless factor which must be determined by numerical or asymptotic analysis. An FEM analysis employing 10 axissymmetric quadratic shell elements (SAX2 in ABAQUS) was used to compute $\beta(\nu)$. Although these elements have bending and transverse

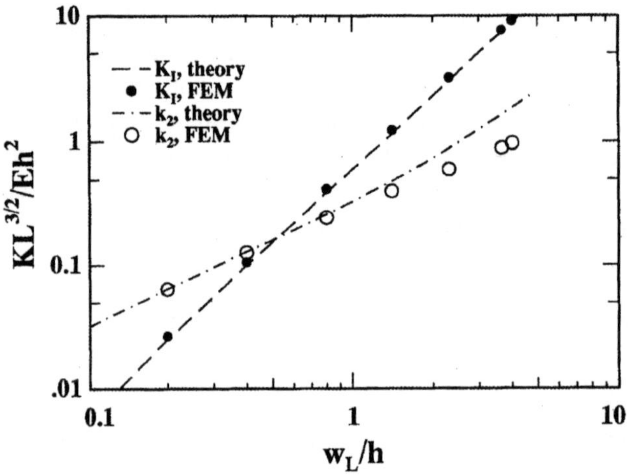

Figure 11. Stress Intensity factors versus out of plane shearing deflection for strip under shear loading.

shear stiffness, these were made negligible by setting $h/R = 10^{-4}$, and by examining the limiting case of large deflections where bending effects are negligible. By simply fitting the results, we find $\beta(\nu = 0.3) = 0.6407$, thus

$$P_2 = \frac{p_2 R^4}{Eh^4} = \left(\frac{w_0}{0.6407h}\right)^3, \quad \nu = 0.3 \quad \text{and} \tag{38b}$$

$$P = P_1 + P_2 = \frac{16w_0}{3(1-\nu^2)h} + \left(\frac{w_0}{0.6407h}\right)^3. \tag{39}$$

A comparison of the prediction of (39), which is based on Foeppel's interpolation method with the exact solution [24] shows excellent agreement. Thus, given the pressure, we can find w_0 by solving (39). The pressures P_1, P_2 can be computed using (37b) and (38b).

In the pure bending problem, the bending moment M at the plate center is

$$\frac{M}{Eh^2} = P_1 \left(\frac{h}{R}\right)^2 \frac{1+\nu}{16}. \tag{40}$$

For the membrane problem, using the known scaling that $N \propto P^{2/3}$, we write

$$\frac{NR^2}{Eh^3} = \gamma(\nu) P_2^{2/3}. \tag{41a}$$

Again by fitting to the axisymmetric FEM results it is found that

$$\frac{NR^2}{Eh^3} = 0.4164 P_2^{2/3}, \quad \nu = 0.3. \tag{41b}$$

The coefficients for the membrane deflection and stress in (38b) and (41b) are within a few percent of the values calculated by Hencky [27] using the finite difference method. The stress intensity factors can be estimated using (40), (41) and (36) and are found to be

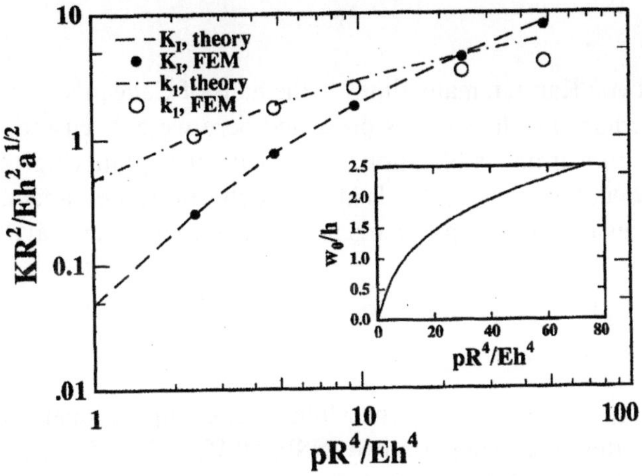

Figure 12. Stress Intensity factors for cracked, clamped circular plate under uniform pressure. The maximum deflection versus pressure for an *uncracked* plate is given in the inset for reference.

$$\frac{R^2 K_I}{Eh^2\sqrt{a}} = 0.4164\sqrt{\pi}P_2^{2/3}, \quad \nu = 0.3, \tag{42a}$$

$$\frac{R^2 k_1}{Eh^2\sqrt{a}} = \frac{6(1+\nu)P_1}{16}. \tag{42b}$$

The stress intensity factors are plotted versus pressure in Figure 12. Also shown, to help provide a physical interpretation of the pressure values, is the center point deflection of an uncracked plate. The FEM and analytical approximation results agree well up to a normalized pressure of 10. For $P > 10$, the theory predicts K_I well but underpredicts k_1. This is similar to the strip under shear loading, where the theory underpredicts k_2.

4. Discussion and conclusions

Our results show that the stress and deformation field near the crack tip is still comprised of a linear superposition of membrane behavior governed by the plane stress crack tip field and bending behavior governed by the Kirchhoff plate theory crack tip field. An implicit assumption is that the crack length must be substantially larger than the plate thickness and smaller than the in-plane dimensions so that these crack tip fields have a nonzero region of dominance.

The finite element calculations are based on a geometrically nonlinear shell model. The excellent agreements between the total energy release rate calculated using the finite element results and the von Karman theory show that the von Karman theory is a good analytical model for theoretical studies of plate fracture, as long as the deflections are on the order of no more than several plate thicknesses. As expected, our results show that mode coupling becomes important when the plate deflections are on the order of $h/2$ and there is no prestrain. The prestrain has a pronounced effect on the energy release rates. In our example, for a thin enough plate, the bending energy release rate is practically negligible even for a small amount of prestrain. The measure of prestrain is governed by the dimensionless parameter $\varepsilon/\delta^2 = L^2\varepsilon/h^2$ which is very sensitive to the plate thickness.

Epilogue

Surely Williams met von Karman many times in the halls of Caltech's Guggenheim Aeronautical Laboratories. Perhaps results such as presented here were discussed, perhaps they were too obvious to them to warrant publication. But for us, it is gratifying to know that ad-hoc assumptions made about the crack tip fields in large deflections are, in fact, correct, and hence that current approaches to practical problems in the fracture of plates have a sound theoretical basis.

Acknowledgements

The computations were carried out using the Multi-User Computational Facility of the Cornell Center for Materials Research, supported by NSF, DMR-9632275. The authors thank HKS Corporation for the use of ABAQUS under an academic license. The authors thank Professor H.D. Conway for bringing the Foeppel method to their attention.

References

Alwar, R.S. and Thiagarajan, S. (1989). Combined effect of shear and large deformation on stress intensity factors. *International Journal of Numerical Methods in Engineering* **28**, 1951–1964.

Boduroglu, H. and Erdogan, F. (1983). Internal and edge cracks in a plate of finite width under bending. *Journal of Applied Mechanics* **50**, 621–629.

Conway, H.D. (1946). Large deflection of circular and square plates. *Philosophical Magazine Series* 7 **37**, 757–767.

Conway, H.D. (1957). Notes on Way's large deflection solution for the uniformly loaded and clamped circular plate. *Journal of Applied Mechanics* **24**, 151–152.

Delale, F. and Erdogan, F. (1979). The effect of transverse shear in a crack plate under skew-symmetric loading. *Journal of Applied Mechanics* **46**, 618–624.

Foeppel, A. and Foeppl, L. (1924). *Drang und Zwang* **1**, 345.

Fung, Y.C. (1965). *Fundations of Solid Mechanics*, Prentice-Hall, Englewood Cliffs, New Jersey.

Hencky, H. (1915). *Z. Math. Physik* **63**, 311.

Hui, C.Y. and Zehnder, A.T. (1993). A theory for the fracture of thin plates subjected to bending and twisting moments. *International Journal of Fracture* **61**, 211–229.

Karman, von (1910). T. Festigkeitsprobleme in Maschinenbau. *Encyklopadie der mathematischen Wissenschaften* **IV**, Chap. 27.

Kirchhoff, G. (1850). Uber das gleichgewicht and die bewegung einer elastichen scheibe. *J. fuer Reine und Angewandte Mathematik* **40**, 51–88.

Knowles, J.K. and Wang, N.W. (1960). On the bending of an elastic plate containing a crack. *Journal of Mathematics and Physics* **39**, 223–236.

Murakami, Y. (1987). *Stress Intensity Factor Handbook* **2**, Pergamon Press, Elmford, New York.

Murthy, M.V.V., Raju, K.N. and Viswanath, S. (1981). On the bending stress distribution at the tip of a stationary crack from Reissner's theory. *International Journal of Fracture* **17**, 537–552.

Reissner, E. (1947). On the bending of elastic plates. *Quarterly of Applied Mathematics* **5**, 55–68.

Sih, G.C., Paris, P.C. and Erdogan, F. (1962). Crack tip stress intensity factors for plane extension and plate bending problems. *Journal of Applied Mechanics* **29**, 306–312

Simmonds, J.G. and Duva, J. (1981). Thickness effects are minor in the energy release rate integral for bent plates containing elliptical holes or cracks. *Journal of Applied Mechanics* **48**, 320–326.

Viz, M.J. (1996). *Fatigue Fracture of 2024-T3 Aluminum Plates under In-plane Symmetric and out of Plane Anti-symmetric Mixed-mode Deformations*, Ph.D thesis, Cornell University.

Viz, M.J. and Zehnder, A.T. (1994). *Fatigue Crack Growth in 2024-T3 Aluminum under Tensile and Transverse Shear Stresses*, in FAA/NASA International symposium on advanced structural integrity methods for airframe durability and damage tolerance, NASA CP-3274, C. Harris, eds, 891–910.

Viz, M.J., Potyondy, D.O., Zehnder, A.T., Rankin, C.C. and Riks, E. (1995a). Computation of membrane and bending stress intensity factors for thin cracked plates. *International Journal of Fracture* **72**, 21–38.

Viz, M.J., Zehnder, A.T. and Bamford, J.D. (1995b). Fatigue fracture of thin plates under tensile and tranverse shear stresses, in *Fracture Mechanics* **26**, ASTM STP 1256. Reuter et al, eds, American Society for Testing and Materials, Philadelphia, PA, 631–651.

Wang, N.M. (1970). Twisting of an elastic plate containing a crack. *International Journal of Fracture Mechanics* **6**, 367–378.

Way, S. (1934). Bending of circular plates with large deflections. *Transactions ASME* **56**, 627.

Williams, M.L. (1961). The bending stress distribution at the base of a stationary crack. *Journal of Applied Mechanics* **28**, 78–82.

Young, M.J. and Sun, C.T. (1993a). Crack plates subjected to out of plane tearing loads. *International Journal of Fracture* **60**, 1–18.

Young, M.J. and Sun, C.T. (1993b). On the strain energy release rate for crack plate subjected to out of plane bending moment. *International Journal of Fracture* **60**, 227–247.

Zehnder, A.T. and Hui, C.Y. (1994). Stress Intensity Factors for Plate Bending and Shearing problems. *Journal of Applied Mechanics* **61**, 719–722.

International Journal of Fracture

AUTHOR INDEX, VOL. 93 (1998)

MIX
Papier aus verantwortungsvollen Quellen
Paper from responsible sources
FSC® C105338

If you have any concerns about our products,
you can contact us on
ProductSafety@springernature.com

In case Publisher is established outside the EU,
the EU authorized representative is:
Springer Nature Customer Service Center GmbH
Europaplatz 3, 69115 Heidelberg, Germany

Printed by Libri Plureos GmbH
in Hamburg, Germany